New Advances in Carbon Nanomaterials

London, UK

1–3 September 2014

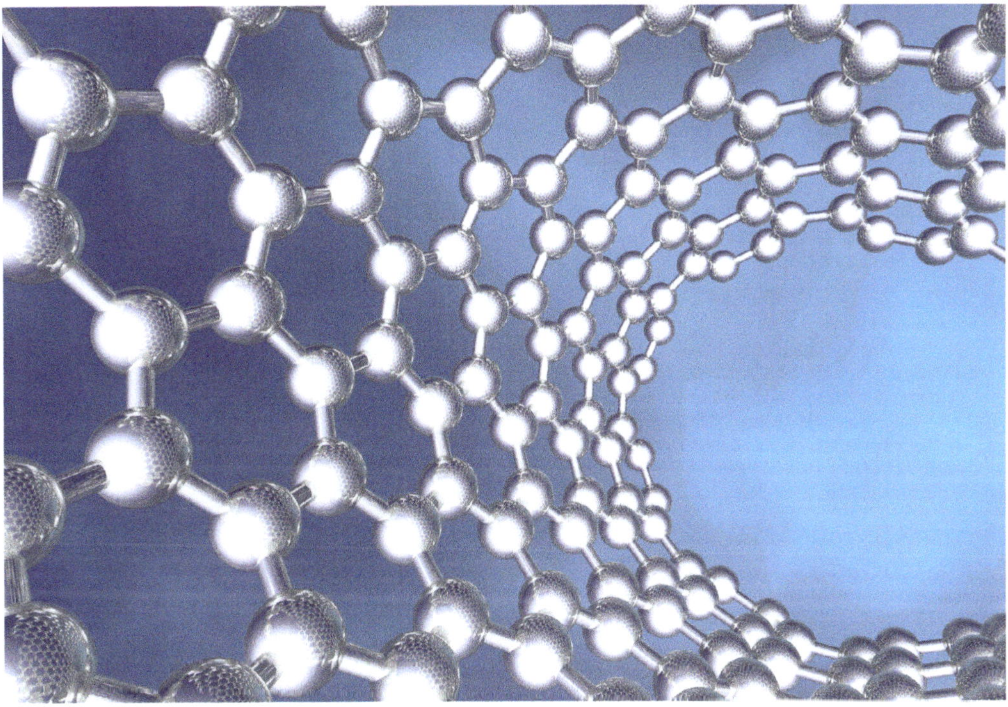

FARADAY DISCUSSIONS

Volume 173, 2014

ROYAL SOCIETY OF CHEMISTRY

The Faraday Division of the Royal Society of Chemistry, previously the Faraday Society, founded in 1903 to promote the study of sciences lying between Chemistry, Physics and Biology.

Faraday Discussions (Print ISSN 1359-6640, Electronic ISSN 1364-5498) is published 8 times a year by the Royal Society of Chemistry, Thomas Graham House, Science Park, Milton Road, Cambridge, UK CB4 0WF.

Volume 173 ISBN: 978-1-78262-044-0

2014 annual subscription price: print+electronic £827, US $1542; electronic only £785, US $1465. Customers in Canada will be subject to a surcharge to cover GST. Customers in the EU subscribing to the electronic version only will be charged VAT. All orders, with cheques made payable to the Royal Society of Chemistry, should be sent to RSC Order Department, Royal Society of Chemistry, Thomas Graham House, Science Park, Milton Road, Cambridge, CB4 0WF, UK.
Tel +44 (0) 1223 432398; E-mail www.orders@rsc.org

If you take an institutional subscription to any RSC journal you are entitled to free, site-wide web access to that journal. You can arrange access via Internet Protocol (IP) address at www.rsc.org/ip. Customers should make payments by cheque in sterling payable on a UK clearing bank or in US dollars payable on a US clearing bank.

Printed in the UK

Faraday Discussions documents a long-established series of Faraday Discussion meetings which provide a unique international forum for the exchange of views and newly acquired results in developing areas of physical chemistry,
biophysical chemistry and chemical physics.

New Advances in Carbon Nanomaterials

Faraday Discussions

www.rsc.org/faraday_d

A General Discussion on New Advances in Carbon Nanomaterials was held in London, UK on the 1st, 2nd and 3rd of September 2014.

CONTENTS

ISSN 1359-6640; ISBN 978-1-78262-044-0

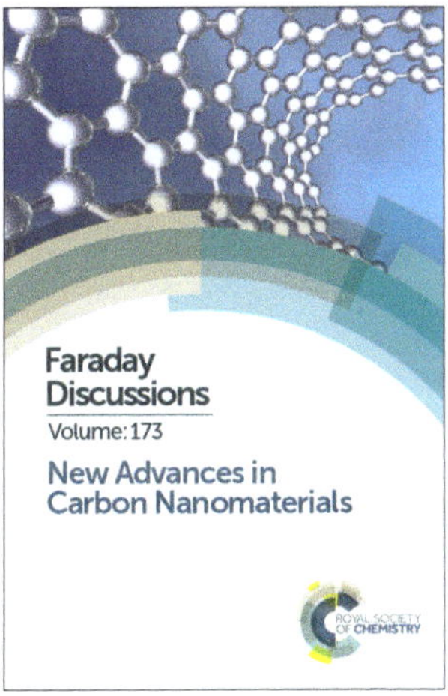

Cover
Front cover image courtesy of Shutterstock.

INTRODUCTORY LECTURE

PAPERS AND DISCUSSIONS

SPONSORS

Thomas Swan
Advanced Materials

Hiden Isochema
Advancing Sorption Analysis

PINE
RESEARCH INSTRUMENTATION

SUPPORTING ORGANISATIONS

The British Carbon Group

CONCLUDING REMARKS

ADDITIONAL INFORMATION

Faraday Discussions

PAPER

Spiers Memorial Lecture

Advances of carbon nanomaterials

Lulu Ma, Amelia H. C. Hart, Sehmus Ozden, Robert Vajtai and Pulickel M. Ajayan*

Received 1st November 2014, Accepted 6th November 2014
DOI: 10.1039/c4fd90039a

During the last two decades the exciting properties of carbon nanomaterials, *i.e.* fullerene, carbon nanotubes and graphene, have drawn pronounced attention. This brief review will discuss the recent advances in the science and applications of nanocarbons, mainly of nanotubes and graphene, and the opportunities and challenges that exist in future technologies based on these emerging materials. The review will discuss the growth and selection, chemical modifications, substitutional doping, and engineering three-dimensional structures, and demonstrations of possible applications. This is a perspective on how the advent of nanocarbons will enable the development of the next generation of carbon materials and technologies relevant for a broad range of applications.

1 Introduction

Carbon, an abundant element in nature, has long been known to exist in three forms: graphite, diamond, and amorphous carbon, in the form of soot and charcoal.[1] It is the discovery of fullerenes in 1985 that ignited a revolution for carbon materials.[2] Tremendous interest in this new material spawned further exploration, resulting in the synthesis of a series of larger carbon molecules, such as C_{70}, *etc.*[3] The subsequent discovery of carbon nanotubes (CNTs) attracted significantly pronounced attention to nano-scale science and technology and accelerated the development of materials science and nanotechnology.[1,4–10] The term nanocarbon emerged to specify these newly developed carbon materials whose structure and size are controlled at the nanometer scale.[11] Graphene, a one-atom-thick planar sheet of carbon, is the most recent member of the nanocarbon family.[12] As a building block for carbon nanotubes, fullerenes, and other carbon materials (see Fig. 1), the discovery of graphene has led to an explosion of interest in the central issues of nanoscience and nanotechnology in physics, chemistry, and materials science, producing a wide array of potential applications.[13–25] Due

Department of Materials Science and NanoEngineering, Rice University, Houston, TX 77005, USA. E-mail: ajayan@rice.edu; Fax: +1 7133485423; Tel: +1 7133485904

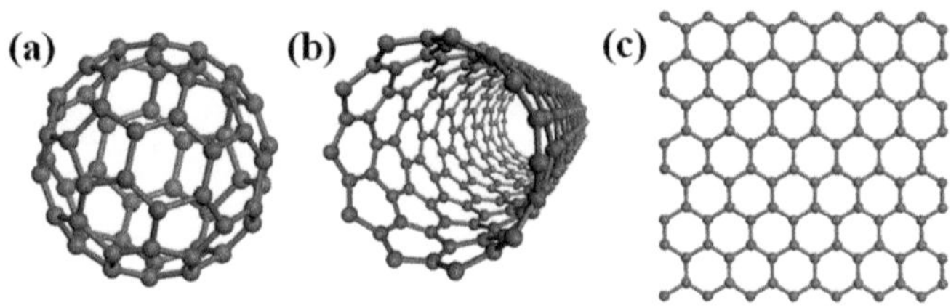

Fig. 1 Graphene is the building block for fullerenes and carbon nanotubes: (a) fullerene; (b) carbon nanotube; (c) graphene. (Courtesy, M. Liu).

to the large interest in nanocarbons over the last thirty years, the focus of this paper will be on the recent progress made in the areas of carbon nanotubes and graphene. Also to be discussed: advances in nanotube and graphene growth and selection, substitutional doping, chemical modification, three-dimensional (3D) structures, as well as their uses and potential applications.

2 Growth and selection

2.1 Carbon nanotubes

CNTs are cylindrically rolled graphene sheets with fullerene capped ends. These tubes are either single-walled carbon nanotubes (SWNTs), consisting of one rolled graphene sheet, or multi-walled carbon nanotubes (MWNTs), consisting of multiple concentric layers of varying diameter SWNTs. The first CNTs to be discovered were MWNTs found in carbon soot on graphite anodes made by arc discharging,[4,26] then years later SWNTs were synthesized.[5,27–29] SWNTs possess uniform diameters (0.8–2 nm), while MWNTs form in a large range of diameters, typically from 2 to 25 nm but can exceed 100 nm,[30,31] and all CNTs can range in lengths from less than 100 nm to several centimeters.

Though there are many methods to synthesize CNTs, [electric arc charge, laser ablation and chemical vapor deposition (CVD)] CVD is currently the most dominant method. CVD can be used with either liquid, solid, or gas carbon sources; combined either by using thin catalyst layers pre-deposited on substrates,[7,32–36] or through vapor-phase catalyst delivery;[37–42] and is practical for large-scale production of high-purity SWNTs,[7,43,44] double-walled CNTs (DWNTs)[45] and MWNTs.[46] DWNTs with heights of up to 2.2 mm were achieved with selectivity of 85% by precisely controlling the catalyst iron thickness.[45] The capacity for MWNT production worldwide in 2006 was estimated at approximately 200 tons per year. A variety of aligned CNT structures have been synthesized predominately on nonconducting substrates, posing limitations on applications where conductive substrates/contacts are required.[7,33,34,38,47] To circumvent this problem, aligned MWNTs can be directly grown on a metallic alloy, Inconel 600, using vapor-phase catalyst delivery (see Fig. 2).[48]

The growth mechanisms of CNTs have been explored since their discovery, and with each new discovery researchers are closer to enabling specific tailoring of CNT growth to very specialized applications. The latest news in CNT growth is the ability to control chiralities, or orientations of the lattice with respect to an arbitrary tube axis resulting from the way the graphene sheet is rolled during

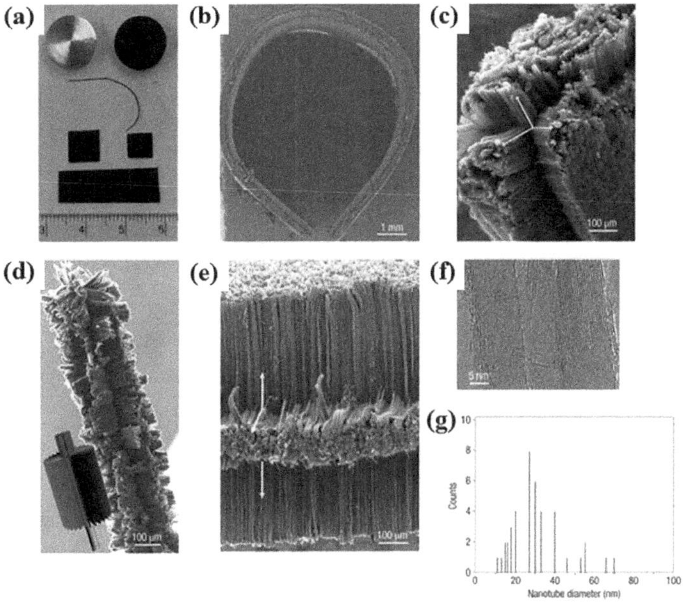

Fig. 2 Aligned MWNTs grown on different shapes of Inconel substrates. (a) Optical image showing various Inconel substrates used for CNT growth. Scale is in cm. (a and b) Scanning electron microscope (SEM) images of CNTs grown on a mechanically deformed Inconel sheet (b) and the corner of a CNT-grown Inconel sheet (c). (d) SEM image of CNTs grown on an Inconel wire (e) SEM image of one side of an Inconel sheet. (f) High resolution transmission electron microscope (HRTEM) image of a typical as-grown nanotube. (g) Typical diameter distribution of CNTs measured from the TEM data. Reproduced with permission from ref. 48. Copyright© 2006 Nature Publishing Group.

synthesis; by understanding the mechanisms of CNT growth, their chiralities can be more precisely controlled.

There are three possible CNT chiralities: zigzag, armchair, and chiral. These three different chiralities are defined by their chiral angle and wrapping integers (n, m), illustrated in Fig. 3, where $m \leq n$ at all times. By changing chiralities, the CNT's band gap and electrical conductivity changes between metallic and semiconducting. The bandgap of semiconducting SWNTs is inversely proportional to the tube diameter. MWNTs are typically metallic, due to wall numbers and the differing chiralities of each wall, and can carry currents of up to 109 A cm^{-2}.[49] SWNTs can exhibit metallic or semiconducting behaviors depending on the orientation of the graphene lattice with respect to the tube axis, indicating chirality.[4,5] Since their discovery, a large problem faced when producing CNTs is the inability to produce a specific chirality; recently, this problem has been solved using various types of catalyst and growth methods.[50,51]

With the CVD method, using a solid iron catalyst and gaseous carbon source, the mechanism of growth involves the deposited iron layer heating and separating to become iron nanoclusters. By diffusion and dissolution, the carbon atoms bond to the surface of the Fe nanocluster, and after having supersaturated the Fe nanocluster, the carbon atoms segregate and form carbonaceous structures from the catalyst surface. The controlled number of CNT walls can be achieved by varying the catalyst thickness.[52]

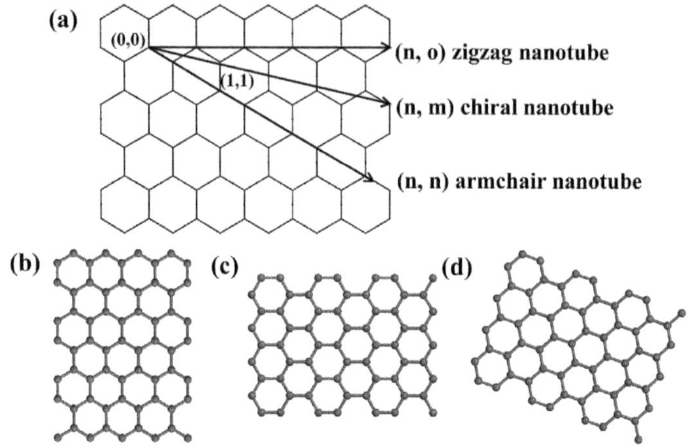

Fig. 3 (a) Schematic representation of how a graphene sheet is rolled to form three chiralities of nanotubes (b) zigzag, (c) armchair and (d) chiral nanotubes. (Courtesy, M. Liu).

Controlled synthesis has endowed SWNTs with narrow distributions of tube diameter and a large fraction of a predetermined tube type through composition and shape engineering of catalysts,[53,54] end-cap engineering[55] or cloning strategies,[56,57] and also tube sorting based on chromatography,[58–60] density-gradient centrifugation,[61–65] or electrophoresis.[64,52] Nanotube separation, based on the chromatography technique, for biological applications includes dispersing bundled SWNTs in water by sonication in the presence of single stranded DNA.[59] The density gradient ultracentrifugation technique is also widely used. During this process, surfactant-suspended metallic and semiconducting SWNTs, which have different mass densities, are dispersed in an aqueous suspension to travel through a mass density gradient *via* ultracentrifugation. The CNTs are driven by density differences to positions in the centrifuge tube where their buoyant density is equal to that of the surrounding fluid.[63,66,67]

The other, newly developed, method of easily obtaining a large amount of a required CNT chirality is through direct growth. Molecular precursors have been converted into ultrashort singly capped (6, 6) ‘armchair’ nanotube seeds using surface-catalyzed cyclodehydrogenation on a Pt (111) surface to produce single-chirality and essentially defect-free SWNTs with lengths up to a few hundred nanometers.[50] Similarly, (12, 6) chiral SWNTs were directly grown at >96% yield using a tungsten-based bimetallic catalyst system.[51] The high melting point of catalysts was anticipated to be crucial in chirality control of SWNTs.

Though these new developments have been a significant step forward on the path to be able to consistently control chirality during CNT synthesis, without the theory of the exact mechanisms of this growth this cannot be a practice that is very common. A few months ago, the exact formation mechanism of each type of chirality during CNT growth was found. It was postulated that during nucleation, as carbon nanotubes attach to a nascent CNT nucleus, adding new hexagonal and pentagonal rings, the chirality of a CNT becomes permanently “locked in” when the final 6th pentagon is added to the hemispherical cap. Each CNT “grows” by adding only hexagons, in a periodic fashion. It was also found that the catalyst and growth temperatures will yield different chiralities of CNTs: lower temperatures and solid catalysts yield near armchair nanotubes, whereas higher temperatures and liquid catalysts yield several types of chiralities but the abundance is nearly proportional to the chiral angle.[68]

Besides their electrical conductive properties, carbon nanotubes are known for a large variety of other outstanding properties such as: high specific surface area, 1314 m^2 g^{-1} for SWNTs;[10] ultrahigh elastic modulus of ~1 TPa and a tensile strength of 100 GPa for MWNTs.[69] SWNTs can possess an extraordinary thermal conductivity of ~3500 W m^{-1} K^{-1} at room temperature.[70]

In addition to organized CNT structures, a variety of tailored structures have been synthesized for special applications with enhanced properties such as: one-dimensional (1D) yarns, fibers, ropes, brushes, 2D sheets, buckypapers, and 3D foams and sponges. CNT brushes have been synthesized using silicon carbide fiber partially masked by a 15 nm Au layer and grown in a ferrocene–xylene CVD to produce ~60 μm aligned CNTs to form a radial brush.[43] For electrical contacts, CNT brushes were also created using MWNT forests mounted on flexible brass cantilevers. These brushes produced more steady and regular voltage and less noise than traditional carbon–copper brushes.[56] Spun fibers and ribbons of CNTs were synthesized directly by CVD using a liquid source of carbon and iron catalyst. In this method, the key requirement for continuous spinning is the rapid growth of high-purity nanotubes to form an aerogel in the furnace hot zone and the forcible withdrawal of the product from reaction by continuous wind up.[71] In 2007, CNT fiber was synthesized using the method of spun aerogel directly from CVD using mostly DWNTs and compared to the techniques of spinning from a lyotropic liquid crystalline suspension of nanotubes and spinning from MWNTs previously grown on a substrate as aligned carpets. It was found that the undensified direct spinning from aerogel is more mechanically advantageous.[72] However, in 2013, it was found that wet spinning ~5 μm "short" CNTs into high-performance multifunctional fibers produced a fiber possessing a higher modulus than previous wet spinning methods and comparable to solid state spun fibers while enabling an easier scale-up ability.[73] CNT yarns were created up to 30 cm in length by drawing out CNTs from super aligned 100 μm arrays.[50] The introduction of a twist during spinning is essential to create multi-ply, torque-stabilized yarns that have achieved yarn strengths of greater than 460 MPa.[51] In 1996, crystalline ropes of metallic carbon nanotubes were synthesized from a mat of randomly oriented SWNTs 10–20 nm in diameter.[74] CNT sheets have been synthesized using track patterns of exposed silica created in CNT films to grow longer CNT stripes.[41] In 2005, buckypaper, a paper-like material consisting of hexagonally packed bundles of coaxial DWNTs with very small diameter was created.[75] Monolayered and multilayered colloidal silica templates were used to fabricate porous 3D architectures of CNTs *via* CVD. Mechanically and thermally stable porous CNT membranes and three dimensional CNT foams were obtained after the removal of silica with hydrofluoric acid.[58] Macroscopic, monolithic CNT sponges were made using CVD where ferrocene dissolved in dicholorobenzene was injected at 860 °C for 4 hours, yielding CNTs self-assembled into porous interconnected 3D frameworks.[59] Densely packed SWNTs were fabricated for use in supercapacitor electrodes using the zipping effect of liquids to draw tubes together.[76] Ultralong SWNTs, over 18.5 cm long, were synthesized[77] for mass production and very recently, half-meter long CNTs were reported.[78] With the demand of well-defined architectures for integrated systems, predetermined orientations of CNT assemblies were synthesized, including: vertically aligned CNT pillars on selective areas, and flower-like patterns of multi-oriented, organized nanotube structures (see Fig. 4).[39]

2.2 Graphene

As the mother of CNTs, graphene has many similarities to CNTs, in structure and properties, including: a large measured surface area of 2630 $m^2\ g^{-1}$,[79] a Young's modulus of 1 TPa and intrinsic strength of 130 GPa,[80] a high thermal conductivity of above 3000 W m k^{-1},[22] and room-temperature electron mobility of 2.5×10^5 $cm^2\ V^{-1}\ S^{-1}$.[23] Graphene's superior properties have stimulated a great amount of research over the last decade, and currently there are several methods for producing graphene in terms of specific size, shape, and quality.

Single-layer graphene samples were first isolated by the mechanical exfoliation technique.[12] This approach can produce the highest quality of samples but yield a minute amount of graphene only useful for fundamental study. A number of alternative approaches to mechanical exfoliation have been explored, including liquid and thermal exfoliation, CVD growth and epitaxial growth.[12,81–96]

The liquid exfoliation method is based on immersing bulk graphite in solvents to yield single- and few-layer graphene sheets.[84,97] A similar method is to chemically modify graphite to produce water dispersible graphite oxide using the Hummers' method.[81] This is followed by thermal annealing or chemical reduction to produce reduced single- and few-layer graphene sheets,[83,98–100] and can be used to fabricate macroscopic structures.[101–107] This approach is cost effective and massively scalable, but the quality yielded is not appropriate for devices since a number of defects and impurities still remain in the graphene.

Graphene has also been synthesized from the high temperature reduction of silicon carbide by the epitaxial method,[82,87,108] which produces few-layer graphene sheets behaving like graphene. The epitaxial growth of graphene on transition metals has been also reported,[86] but the ability to achieve graphene domains with uniform thickness remains a challenge.

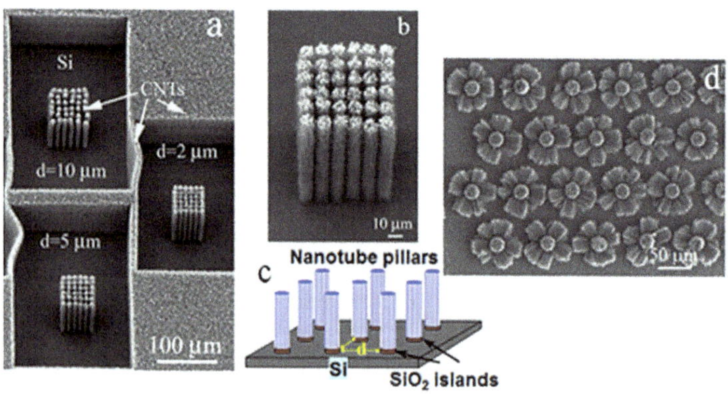

Fig. 4 (a) SEM image of three blocks of cylindrical pillars (~10 μm diameter) of vertically aligned CNT arrays. (b) CNTs were only grown on the silica islands. (c) Schematic of pillars shown in (b). (d) Growth in both vertical and horizontal directions from the top and the sides of silica, respectively. Figure adapted with permission from ref. 39. Copyright© 2002 Nature Publishing Group.

The CVD method can produce large-area uniform graphene films of single to few-layer graphene grown on different metal substrates using gas, solid and liquid precursors.[89,91,92,94,103] The production of tens of square meters of graphene film has been developed by an industrial-scale roll-to-roll CVD method, allowing for the possibilities of the commercial production of large-scale transparent electrodes. Besides 2D graphene, the CVD synthesis of 3D graphene with interconnected networks was reported recently.[109]

Since CVD-grown graphene is typically polycrystalline with many grain boundaries, which impede electron transport,[110–112] the greatest challenge is to be able to gain control of the early stages of nucleation and growth of graphene and to ultimately create large single crystalline graphene domains. The solution could be to create nucleation sites situated far enough from each other and to grow the single crystalline domains appropriately fast, so that no additional nucleation sites can emerge.[110]

Recently, it has emerged as an intense interest for materials scientists to grow large-size single crystalline graphene using CVD.[110–118] Electrochemical polishing and high-pressure annealing can be used to improve the quality of the copper foil and accordingly suppress graphene nucleation site density. Large single-crystal hexagonal single-layer graphene domains ($\sim$4.5 mm^2) were achieved (see Fig. 5),[118] and show a mobility of positive charge carriers of $\sim$11 000 cm^2 V^{-1} s^{-1} at room temperature. Also, a wafer-scale ($\sim$5 cm in diameter) wrinkle-free single crystalline single-layer graphene grown on silicon wafer using a hydrogen-terminated germanium buffer layer was reported.[119]

Other shapes of single crystalline graphene domains have also been grown such as hexagonal,[112–116,118,120] rectangular,[117,121] triangular,[122] and dendritic shapes.[113,123] Theoretical work explains that hexagonal graphene domains form under kinetic control conditions because the zigzag edges grow slowest on Cu substrates under vacuum.[124] Under etching, the kinetically controlled island shape is determined by the fastest-moving edges, producing peculiar dodecagonal crystals.[122] Lower-symmetry shapes of graphene islands, such as triangular or elongated, can also arise due to symmetry relations between graphene and the underlying metal surface.[125] Recent experimental and theoretical work discovered that oxygen on the copper surface was able to accelerate growth of large-size single crystalline graphene domains as well as shift the graphene domain shape from hexagonal to dendritic.[126]

Another challenge for the CVD growth of graphene is its transfer from the preferred metal substrates to a desired substrate,[103] which causes unexpected cracks, folds and ridges which decreases the graphene quality. Toward this end, a

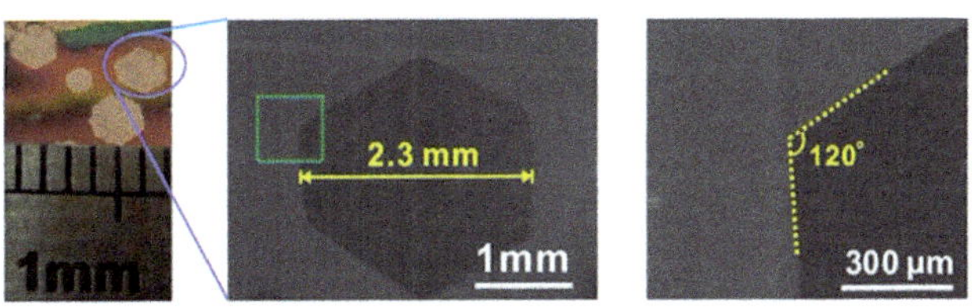

Fig. 5 Typical optical and SEM images of large-size ($\sim$2.3 mm) single crystalline graphene domains on Cu. Adapted with permission from ref. 118. Copyright 2013 American Chemical Society.

face-to-face transfer method for wafer-scale graphene films was reported, which could take advantage of nascent gas bubbles to form capillary bridges, so that graphene stays attached to the substrate while the copper is removed.[127]

Though considerable effort has been made to grow large-size graphene, there is growing interest in breaking the atomic structures within the graphitic materials to create graphene components with special shapes, sizes and edge structures.[128,129] For example, controlled nanocutting of graphene into pieces with specific edge configuration, using nickel nanoparticles as a knife which was driven by catalytic hydrogenation of graphene lattice, was reported.[130] Besides metal nanoparticles, nonmetal SiO_x nanoparticles were also used to tailor graphene into regular pieces with smooth edges.[131]

The graphene nanoribbon (GNR) is an ultra-narrow strip of graphene. It has been predicted that GNRs with zigzag edges are metallic, whereas armchair edges can make GNRs either metallic or semiconducting depending on their width. The bulk production of high-quality GNRs with the length of <20–30 μm and widths of 20–300 nm using the CVD method has been reported.[132] GNRs could be also produced by unzipping CNTs by solution-based oxidative processes and plasma etching,[133,134] which gives better control over the chemical functionalization and quality of the edges. A mechanical sonication of gas-phase oxidized MWNTs in organic solvent has also been reported to be able to make high-quality GNRs.[135] Recently, the unzipping of CNTs was reported using spherically shaped CNT pellets as bullets to shoot at an aluminum target at a velocity of 5.9 km s^{-1}. A large number of defects were generated in the nanotubes, as well as rapid atom evaporation, which led to the nanotubes unzipping along the axis.[136] In addition to the methods mentioned above, an exciting bottom-up method has been reported for producing high-quality GNRs of different geometries and widths using surface-assisted coupling of molecular precursors into linear polyphenylenes with subsequent cyclodehydrogenation.[137,138]

Graphene quantum dots (GQDs) with nanometer sizes have been fabricated from a variety of graphitic materials including graphite,[139] fullerene,[140] CNTs,[141] graphene,[142] carbon fibers,[143] and even coal, containing crystalline carbon within its structure,[144] based on acid treatment and chemical exfoliation. GQDs were produced by oxidation cutting of carbon fiber in the size range of 1–4 nm with zigzag edge structure dominance, showing the tailored photoluminescence through changing the size of the GQDs (see Fig. 6).[143]

Recently, graphene sheets were patterned into nanodisk and nanoring arrays using electron-beam lithography and oxygen plasma etching,[145] and gated tenability and hybridization of localized plasmons were observed in these nanostructured graphene pieces down to 3.7 μm light wavelengths.

Despite considerable efforts to explore the horizontal morphologies of graphene, there is increasing interest in the vertical features of graphene including stacking configuration and thickness, such as single- and bi-layer graphene.[91,95,146–151] Graphene is known to have a zero band gap, but bi- and tri-layer graphene have tunable band gaps depending on their interlayer stacking configurations, allowing for broad applications in electronic devices. For example, bilayer graphene sheets can be AB-(Bernal) or AA-stacked and trilayer can be mostly stacked in ABA and ABC configurations. The CVD growths of large-area uniform Bernal-stacked bi-, tri- and tetra-layer polycrystalline graphene films were produced through a precisely tuned total pressure and methane/hydrogen

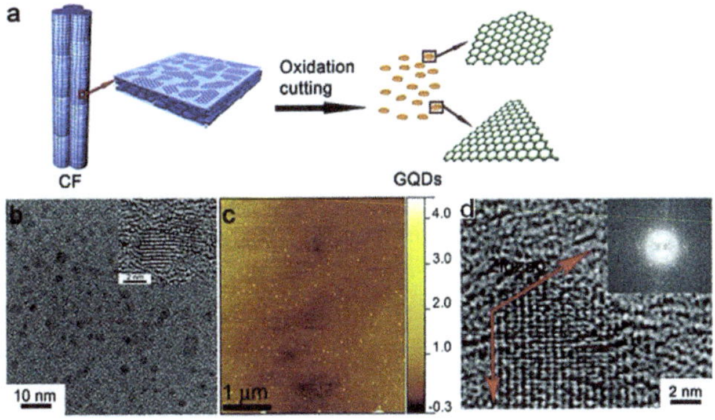

Fig. 6 (a) Schematic of oxidation cutting of carbon fiber into GQDs. (b) TEM images of GQDs, inset of (b) is the HRTEM of GQDs. (c) Atomic force microscopy (AFM) image of GQDs. (d) HRTEM image of the edge of a GQD, inset is the 2D fast Fourier transform of the edge in (d). Reproduced with permission from ref. 143. Copyright 2012 American Chemical Society.

gas ratio.[149] Recently, the synthesis of large hexagonal bi- and tri-layer graphene single crystals with 0° (Bernal stacked) or 30° interlayer rotations was reported.[152]

Additionally, it is critical to confirm the graphene thicknesses before any further study. Raman spectroscopy emerged as an unambiguous, high-throughput, nondestructive technique to determine the quality and thickness of graphene samples.[153] The major features, a D peak at ~1350 cm^{-1}, G peak at ~1580 cm^{-1}, and G′ peak at ~2700 cm^{-1} recorded in the Raman spectra, reflect the changes in the electronic structure and electron-phone interactions for single-, bi- and few-layer graphene, which are used to determine the number of graphene layers.

3 Chemical modification

When the intrinsic properties of CNTs and graphene alone do not meet the requirement of various applications, chemical modification creates a new interest area that involves altering the surface chemical properties and electronic properties of CNTs and graphene. The chemical modification of CNTs and graphene can be realized through both non-covalent and covalent modifications.

3.1 Carbon nanotubes

Although CNTs have outstanding properties, there are some challenges that bring limitations for their applications: they are not soluble, have limited dispersion in common solvents because of their surface energy, and their seamless surface does not provide good physical interactions with common chemical media. Because of these limitations, surface modifications of CNTs are desirable for their processability and specific applications. After their initial discovery, different surface modifications of CNTs were investigated and functionalization was found to be a unique method for these alterations. Functionalization has become one of the most important key routes to manipulate the properties of CNTs for desirable applications. For this reason, understanding the functionalization of CNTs is a crucial step for exploiting their application potential. Some effective methods for the surface functionalization of CNTs can be classified into two major categories: covalent and non-covalent functionalization (see Fig. 7).[154] In covalent functionalization, CNTs are functionalized by non-reversible attachment onto the surface of CNTs. Since CNTs are chemically inert materials, there are a few reactive chemicals, such as fluorine, which is able to link up a conjugated π–π system (see Fig. 7). One of the most common processes to functionalize CNTs is the creation of defects on the surface of CNTs *via* acid treatment, mainly generating surface attached carboxyl groups.[155,156] In this method, acid generated carboxyl groups are converted to acid chloride, and so can easily be functionalized with amine functional organic molecules.[157] Recently, CNTs were modified using the Suzuki cross-coupling reaction after carboxyl groups were converted to acyl chloride (see Fig. 8).[158]

Other methods of covalent functionalization of CNTs that have been reported include fluorination followed by substitution of alkyl groups,[159] diazonium functionalization,[160,161] and organic radicals.[162] In diazonium functionalization,[160] functional groups were directly connected to the side-wall of CNTs. The mechanism of this process includes the reduction of aryl diazonium yielding aryl radicals that covalently attach to the carbon surface. While the acid treatment method reduces the physicochemical and mechanical properties of CNTs, these properties are not as affected by the diazonium functionalization. CNTs can be functionalized by generating free radicals *via* decomposition of benzoyl peroxide in the presence of alkyl iodides.[162]

The second type of functionalization is non-covalent functionalization, which is an alternative way to modify CNTs without disturbing their straight tubular structures. This method relies on weak interactions such as: van der Waals,

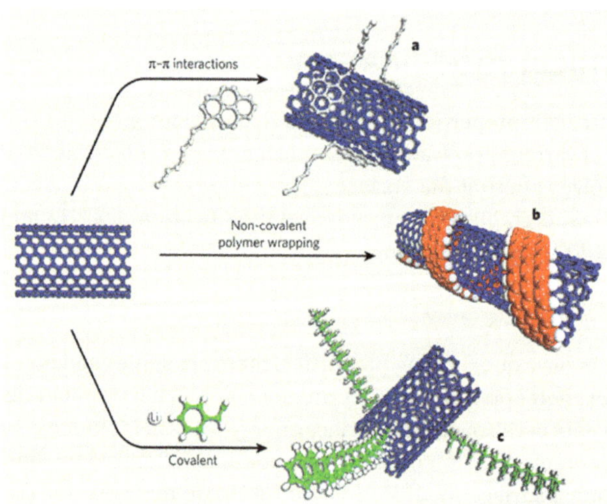

Fig. 7 Schematic of the three methods how CNTs can be made more soluble. (a) Aromatic molecules can be adsorbed to side-walls using certain non-covalent interactions. (b) Non-covalent interactions can be used to wrap polymers around nanotubes. (c) Chemical groups can be covalently attached to nanotubes. Reproduced with permission from ref. 154. Copyright© 2007 Nature Publishing Group.

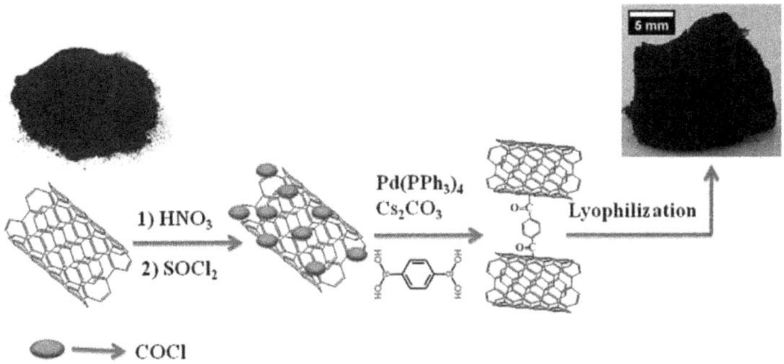

Fig. 8 Schematic of the 3D CNT solid macrostructure synthesis from CNT powder; using HNO_3 oxidation, Suzuki coupling, and freeze drying. Reproduced from ref. 158. Copyright© 2014 WILEY-VCH Verlag GmbH & Co. KGaA, Weinheim.

hydrophobic and π–π interactions. While covalent functionalization disturbs their straight tubular morphology, non-covalent functionalization does not perturb their structures. This type of CNT functionalization is an effective route for CNT dispersion in aqueous and non-aqueous solvents without changing their unique physical and mechanical properties.[163] Aromatic molecules such as conjugated polymers, pyrene and porphyrin can interact with the surface of CNTs by π–π interactions.[164,165] The most common application of non-covalent functionalization is the use of surfactant molecules to stabilize individual CNTs in aqueous media.[166] Because of their large surface area, they tend to aggregate into bundles by van der Waals interactions; this rebundling can be inhibited by using suitable surfactants *via* ultrasonication. CNTs can also be functionalized with various kinds of polymers to improve their properties, where polymers are wrapped around CNTs *via* physical interactions. Additionally, adsorption of surfactants, π–π interactions of small aromatic molecules and biomolecules such as DNA and peptides are other methods of non-covalent functionalization.[167,168]

Other types of CNT functionalization can be classified as endohedral functionalization.[169] In this method, the inner walls of CNTs are filled with small molecules and nanoparticles, such as fullerenes. For example, CNTs can be filled with paramagnetic iron oxide nanoparticles which showed excellent magnetic properties that can be applied in various fields, such as nanofluids.[170] Another type of CNT functionalization is asymmetrical end functionalization, which attaches two different functional groups to the opposing sides of the CNT.[171] CNTs with two differentiated ends create attractive building blocks for a variety of applications such as: rectifying diodes,[172] or asymmetrically functionalized CNT arrays used to harvest water from air.[171]

3.2 Graphene

Graphene oxide (GO) is known to be the derivative of graphene, which consists of oxygenated groups, with epoxy and hydroxyl functional groups present in the basal planes, carbonyl, and five- and six-membered-ring lactol functionalities attached to the edges (see Fig. 9).[173] The presence of these functionalities disrupts the π-conjugation network in graphene, thus making GO insulating, remarkably hydrophilic, and analogously chemically reactive. Owing to those chemical moieties, GO has been studied as the starting material for production of chemically functionalized graphene.[83,101,174–182]

Reduction of GO is expected to remove the functionalities and rehybridize the sp^3 carbon atoms to sp^2 carbon, making it aromatic and defect-free. However, despite a number of methods established to reduce GO, there are still many impurities and defects remaining.[83,98–100] Toward this end, a complete reduction and thermal annealing process has been developed to produce highly soluble and conductive graphene, with effective restoration of π-conjugation.[173] Only small amounts of impurities remain in the final product (less than 0.5 wt% of sulfur and nitrogen, which is less than about 3 wt% of other chemical reductions). GO is nonstoichiometric and inhomogeneous with a mixture of regions that are pristine and densely decorated, which is difficult to have precise chemical control of its functionalities. In this respect, a stepwise reduction of GO using a gas-based hydrazine was developed, to selectively improve the chemical homogeneity of GO.[183] The carbonyl groups are the easiest to remove and the tertiary alcohol takes the longest to be completely removed from the GO surface. This method contributes to a progressive tuning of the optical gap from 3.5 eV down to 1 eV.

Graphane is a stoichiometric derivative of graphene, which is a fully saturated hydrocarbon derived from graphene, with formula CH. Theoretical work first foresaw the presence of graphane with all of the carbon atoms in the sp^3 hybridization instead of sp^2 hybridization of pristine graphene, leading to a conversion to an insulator from metallic.[184] This theory was proven experimentally by exposing graphene to cold hydrogen plasma, which was generated from a remote discharged low-pressure hydrogen–argon mixture. The hydrogenation process is reversible, so that graphane can be converted back to graphene by annealing, without any changes in structure or properties.

Thanks to the strong carbon–fluorine bond, fluorinated graphene has been considered as a more stable graphene derivative candidate for graphane. Fluorographene, the fully fluorinated graphene with fluorine atom attached to each carbon has been synthesized by exposing graphene to atomic F formed by decomposition of xenon difuloride.[185] Instead of possible fluorination in plasma as used for hydrogenation of graphene,[184] this method circumvented the potential damage due to ion bombardment, The as-synthesized fluorographene is a high-quality insulator with an optical gap of 3 eV and inherits the good mechanical strength of graphene with a Young's modulus of 0.3 TPa and strains of 15%. A methodology of synthesizing highly fluorinated GO (FGO) by oxidizing the basal

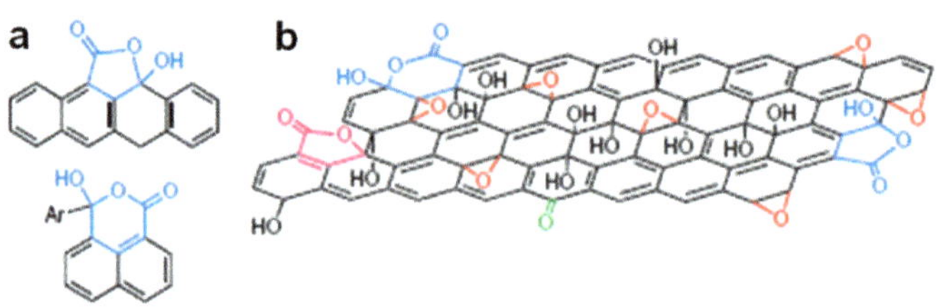

Fig. 9 (a and b) Structural models of GO sheets, taking into account the five- and six-membered lactol rings (a), ester of a tertiary alcohol, hydroxyl, epoxy and ketone functionalities. Figure reprinted with permission from ref. 173. Copyright© 2009 Nature Publishing Group.

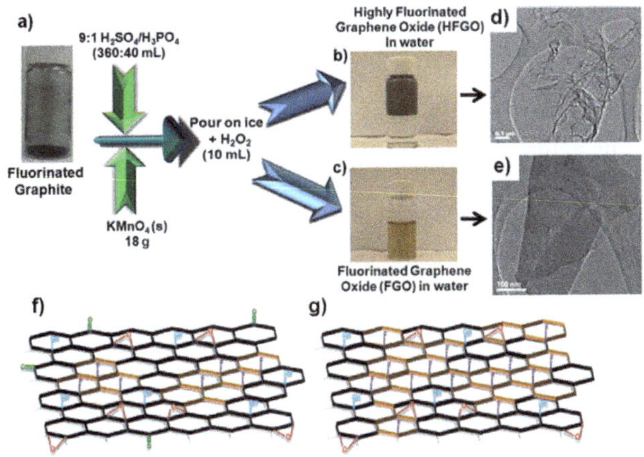

Fig. 10 (a) Schematic showing the reaction involved in the synthesis of both FGO and HFGO. (b and c) Photographs of HFGO (b) and FGO (c) in water. (d and e) TEM images of (d) HFGO and (e) FGO. (f and g) Structural models of FGO (f) and HFGO (g). Adapted with permission from ref. 186. Copyright© 2014 WILEY-VCH Verlag GmbH & Co. KGaA, Weinheim.

plane of fluorinated graphite, allowing for tunable hydrophobicity of GO, was also reported (see Fig. 10).[186] FGO was found to be able to serve as a magnetically responsive drug carrier (magnetic resonance imaging and photophilic contrast agent), under preclinical settings, as a type of photothermal therapy.[187]

4 Substitutional doping

Carbon atoms in CNTs and graphene can be substituted by heteroatoms, which are electron donors or acceptors, allowing them to be tailored to have electronic properties attractive for specific applications. Significant progress has been made in the synthesis of substitutional doping of CNTs and graphene.

4.1 Carbon nanotubes

Boron (B) and nitrogen (N) have been known to be able to substitute for carbon atoms in CNTs, which act as an electron acceptor and donor, respectively. Experimental and theoretical studies show that substituting carbon atoms with B and N render CNTs and graphene p-type and n-type semiconducting, respectively.[188–192] B and N atoms can be *in situ* doped into CNTs homogeneously during CNT synthesis. The CVD method has been widely used to produce doped CNTs using different B- and N-containing precursors.[190–195] Many attempts have been made to reach the maximum N incorporation. For example, N-doped MWNTs with a N content up to 20 atom% were synthesized based on an aerosol method.[196]

There are various types of N substitutions identified in CNTs, such as pyridinic N (N atoms in six-member ring bonded with two carbon atoms), pyrrolic N (N atoms in five member ring), graphitic N (N atoms within the graphene lattice bonded to three carbon atoms) and oxidized N species (N atoms bonded with oxygen atoms) (see Fig. 11).[197] Different types of N substitutions have different functionalities so that it is desirable to synthesize N-doped CNTs with controlled types of N for specific applications. With respect to this issue, metallic N-doped CNTs have been synthesized with "bamboo-like" structures by pyrolysis of ferrocene–melamine mixtures at 1050 °C. The as-synthesized N-doped CNTs contain a nitrogen content of 3–4 atom%.[191,198] Theoretical calculations indicate that the metallic behavior of CNTs originates from the pyridine-like N structures (see Fig. 12).

In addition to N-doped CNTs, B-doped DWNTs have been produced by the catalytic decomposition of a mixture of methane and diborane over a Fe–Mo/MgO catalyst at 900 °C.[190] The boron concentration is ~1 atom%, giving a high proportion of the large diameter DWNTs compared with undoped DWNTs synthesized by the similar method.

Doping of B and N into CNTs at the same time is also very interesting, in terms of their contrasting electronic properties. B and N co-doped carbon materials started with graphite.[199,200] Graphite with the composition $B_xC_yN_z$ was synthesized

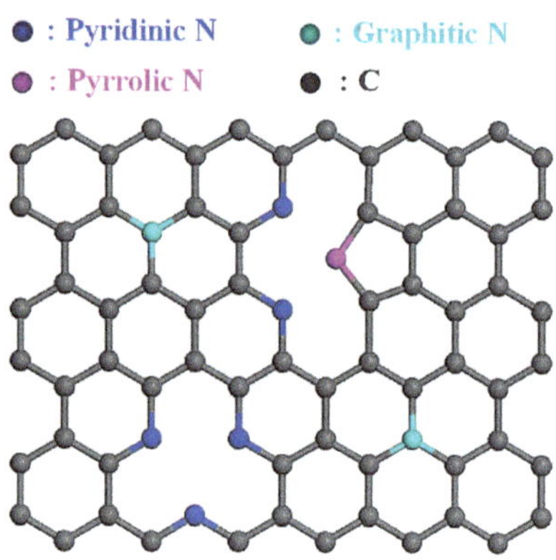

Fig. 11 Schematic diagram of the three types of doped N atoms in N doped graphene. (Courtesy, M. Liu).

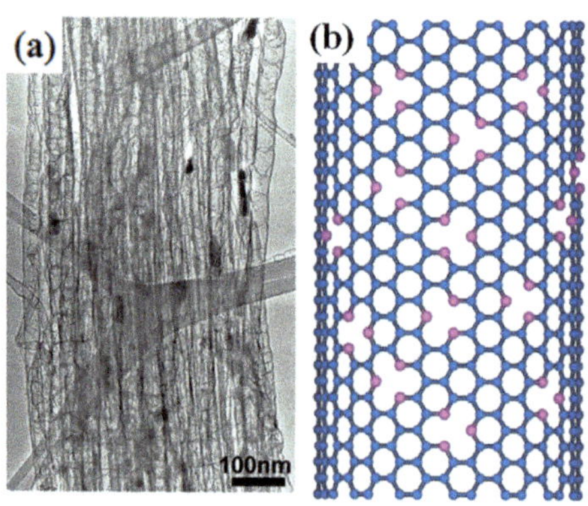

Fig. 12 (a) TEM image of a typical region showing densely packed aligned and hollow nanotubes with "bamboo-like" structures. (b) Molecular model of a N-doped nanotube containing pyridine-like N atoms. Adapted with permission from ref. 198. Copyright© 2002 Springer-Verlag.

from the interaction of boron trichloride, acetyloride, acetylene and ammonia at 400–700 °C. Each graphene layer in the structure is B and N co-doped. The material of approximate composition $B_{0.35}C_{0.30}N_{0.35}$ is a semiconductor. Subsequently, some efforts were made to study B and N co-doped CNTs theoretically and experimentally.[200–203] The synthesis of B and N co-doped CNTs with the composition of BC_2N were developed using an arc discharge method similar to those previously used for pure CNTs.[201,203] When the composition of B and N substitution in CNTs reaches 100 atom%, it will yield pure BN nanotubes.[204,205] BN nanotubes can be directly grown by the pyrolysis of ammonia and borane over cobalt powder at 1100 °C.[204] Multiwalled BN nanotubes were also synthesized *via* an oxidation treatment of multiwalled $B_xC_yN_z$ nanotubes.[205]

4.2 Graphene

Since graphene has zero-bandgap, various methods have been developed to modify and open its band structure in order to make it a potential material for high-performance electronic devices. In addition to the synthesis of bilayer graphene with different interlayer rotation, one of the most feasible methods to open up the band gap is doping, which could be used to control the electrical properties of graphene. N-doped graphene sheets have been synthesized *via* CVD using ammonia or acetonitrile as N precursors,[206–208] through thermal annealing of GO in ammonia,[209] or arc discharge.[210] B-doped graphene has been produced by arc discharge using graphite electrodes in the presence of hydrogen and diborane, or using boron-stuffed graphite electrodes.[210]

Besides heteroatoms B and N, atoms such as sulfur and phosphorus have been predicted to be able to dope graphene based on theoretical calculations.[211] The CVD growth of large-area sulfur-doped graphene on a copper substrate has been reported by using liquid organics (hexane in the presence of S) as the precursor.[212] Recently, a low-temperature method towards the synthesis of large-area heavily N and S co-doped graphene on copper foils was reported, which is based on a free radical reaction using polyhalogenated aromatic compounds.[213]

In addition to N-doped graphene sheets, there were many theoretical and experimental studies of N doped GQDs and GNRs.[209,214–217] N-doped GNRs have been synthesized by CVD using pyrazine as a N precursor[217] or through high power electrical annealing in ammonia.[218] N-doped GQDs have been prepared by an *in situ* electrochemical approach using N-containing tetrabutylammonium perchlorate in acetonitrile as the electrolyte. This approach could create N-doped GQDs continuously by cyclic voltammograms scanning over a potential window of ±3.0 V.[215]

Like the study of BN nanotubes, it is very important to mention that there is a growing interest in the synthesis of graphene-like materials, such as hexagonal boron nitride (h-BN), so-called "white graphene". h-BN has similar crystal structures with a lattice constant difference of only 2%. However, unlike graphene, the single-layer h-BN is a dielectric with a wide bandgap of ~5.9 eV. The CVD synthesis of h-BN films consisting of two to five layers, was reported.[219] As-synthesized h-BN shows a large optical energy bandgap of 5.5 eV and 2D elastic modulus in the range of 200–500 N m^{-1}.

Like the previously mentioned B and N co-doped nanotubes,[202,220] boron and nitrogen atoms can be co-doped into graphene lattices to form a variety of semiconducting layered structures with variable stoichiometry[199–201] as well as a tunable bandgap. Fig. 13 shows a schematic of pristine graphene, B- and N-doped graphene, B, N co-doped graphene and h-BN. It has been demonstrated that B–N and C–C bonds tend to segregate in the BNC systems.[221] Recently reported was a CVD growth of B and N co-doped atomic monolayers consisting of hybridized domains of h-BN and graphene (h-BNC) using a mixed methane and ammonia borane source.[222] The atomic ratio of B, C, and N can be tuned with the atomic percentage of C ranging from 10% to 100%. The domain size could reach ~42 nm at 50 atom% C, introducing a gap of 18 mV.

The size and shape of graphene and h-BN domains still need to be engineered more precisely for the fabrication of novel electronic devices. The fabrication of graphene and h-BN in-plane heterostructures with controlled domain sizes have been established by synthesizing graphene in lithographically patterned h-BN atomic layers (see Fig. 14).[223] It was shown that the as-fabricated domain size could be as small as 100 nm and the shapes of individual domains can be controlled precisely. Graphene and h-BN in-plane heterostructures were also produced using the inverse growth procedure by growing h-BN on photolithographically patterned graphene.[224]

Additionally, topological B and N substitutional co-doping of graphene was developed using a mixed gas of boric acid and ammonia as B, N precursors. The BN concentration can be tuned from 0 to 100% by controlling the growth time.[225] The first synthesis of B and N co-doped graphene quantum dot/sheet hybrid nanoplatelets was also recently demonstrated by hydrothermal self-assembly of GQD/graphene sheet hybrid nanoplatelets followed by B and N co-doping. The B, N co-doped hybrid material possesses advantages from both components, rendering it promising as a high-performance oxygen reduction electrocatalyst.[226] It is necessary to mention that there is an emergent interest in the fabrication of graphene and h-BN vertical heterostructures.[227–230]

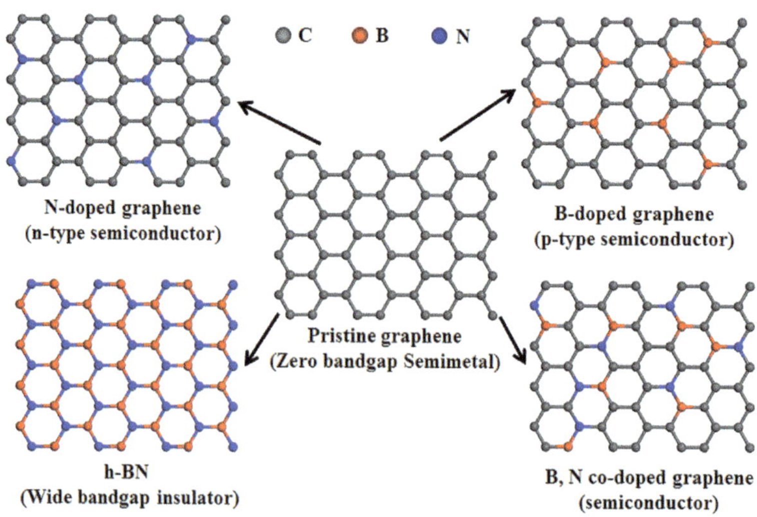

Fig. 13 Schematic of pristine graphene, B- and N-doped graphene, B, N co-doped graphene and h-BN. (Courtesy, M. Liu).

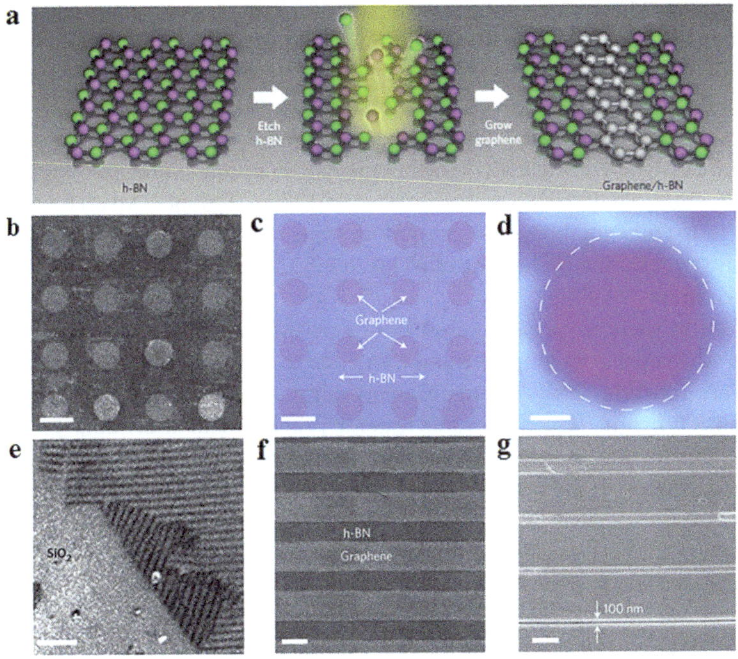

Fig. 14 (a) Schematic showing the fabrication procedure for in-plane graphene/h-BN heterostructures, (b–d) SEM image, optical image and Raman mapping of a graphene/h-BN array of circles, with graphene circles in an h-BN matrix. Scale bars, 50 μm (b and c) and 10 μm (d). (e and f) SEM images of graphene/h-BN stripes. Scale bars, 50 μm (e) and 10 μm (f). (g) SEM image of graphene/h-BN strip structure with (d) strip dimensions. The widths of each strip, from top to bottom, are 1 μm, 500 nm, 200 nm and 100 nm, respectively. Scale bar, 1 μm. Adapted with permission from ref. 223. Copyright© 2013 Nature Publishing Group.

Besides B and N doping, single iron and silicon atoms were discovered to substitute carbon atoms of graphene sheets, respectively.[231,232] They open up varieties of heteroatom substitutional doping although further theoretical and experimental efforts are needed to fully understand the substitution.

5 Engineered 3D nanocarbons

It is essential to integrate individual CNTs or graphene sheets into advanced 3D multifunctional architectures for practical macroscopic applications. Although a large number of CNT- or graphene-based macrostructures in different forms have been produced,[38,41,42,71–75,101–107,233–239] as mentioned in the section on growth and selection, there is still a challenge in realizing the full potential of the individual "nanoblock" components while constructing the macroscopic structures.[102] The covalent linkage between individuals is a recipe which can bring out the best in individuals for macroscopic structures.

Early experimental and theoretical studies showed that two CNTs with different atomic and electronic structures could be seamlessly fused together by introducing a pentagon and a heptagon into the hexagonal carbon lattice.[240–242] It has been found experimentally that dopant atoms such as N or S could induce morphology changes and covalent multi-junctions in CNTs such as different types of covalent Y-junctions.[243–245] The first demonstration of exploiting these heteroatom substitution effects on CNT morphology was in the creation of elastic 3D sponge-like B-doped MWNT macrostructures, which were grown *via* an aerosol-assisted catalytic CVD using triethylborane as the boron source (see Fig. 15).[246] Subsequently, Shan *et al.* reported 3D sponge-like N-doped MWNT architectures.[247] In this method, S was used to tune the diameters of nanotubes, accordingly resulting in the controlled mechanical and electrical properties of the architectures.

Recently, there is special interest in design and fabrication of hybrid macroscopic structures from graphene and CNTs due to their similar crystal structures and work functions.[248–255] Theoretical work suggests that a covalent bonded graphene/CNT hybrid structure would extend properties of individual graphene sheet or nanotubes to three dimensions,[249] and a number of attempts have been made to synthesize the hybrid structures based on CVD growth of CNTs on graphene sheets.[248,252,253,255] The direct growth of 3D CNT–graphene seamless hybrids was reported, where graphene, with an interconnected 3D network, was grown on porous nickel substrates (see Fig. 16).[255]

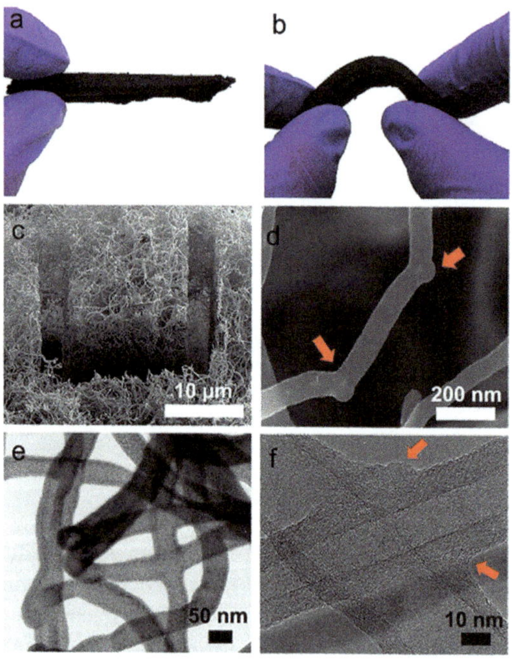

Fig. 15 Physical structure and morphology of B-doped MWNT 3D structure. (a) Photograph of 3D macroscopic B-doped MWNT sponge material. (b) Photograph demonstrating flexibility and mechanical stability. (c) SEM image after ion beam "slice and view" feature showing the interior porous structure of the entangled nanotube network. (d) Closer look at the "elbow" defects found in B-doped MWNT solids. (e) STEM image showing two, four-way covalent nanojunctions. (f) TEM image showing two overlapping B-doped MWNTs welded together assisted by boron doping. Reproduced with permission from ref. 246. Copyright© 2012 Nature Publishing Group.

Instead of direct covalent junctions between individual nanoblocks, 3D engineered architectures of CNTs and graphene have also been created *via* covalent crosslinking of individuals. A 3D scaffold produced by the covalent crosslinking of SWNTs using diazonium functionalization has been reported,[256] which shows twice the adsorption of hydrogen molecules compared with other macroporous carbon materials. In the work mentioned previously, the modified CNTs *via* Suzuki cross-coupling were covalently interconnected to form 3D macroscopic solids with increased surface area compared to pristine CNTs.[257]

As discussed above, CNTs and graphene can be chemically modified by incorporation with polymers or precursor resin matrices in order to make 3D reinforced composites since their discovery.[83,154,177,258–265] They act as matrix modifiers, providing multifunctional attributes to the matrices. Although the interfacial shear in nanotube-based composites is detrimental to stiffness and strength, it could produce high mechanical damping. In some previous work, the strong viscoelastic behavior with up to 1400% raise in loss factor (damping ratio) of the baseline epoxy was found in densely packed MWNT–epoxy composites. It indicates that nanotube fillers could be useful for improving structural damping without sacrificing mechanical properties and structural integrity.[261] Moreover, the self-stiffening behavior in vertically aligned CNT/poly(dimethylsiloxane) nanocomposites was observed.[266]

Graphene has higher surface-to-volume ratios than SWNTs due to the inaccessibility of the inner nanotube surface to polymer molecules, which allows for graphene to be potentially more promising for changing all matrix properties.[83,177] 3D graphene nanocomposites were explored by incorporating functionalized graphene sheets with various polymer hosts including poly(methyl methacrylate), poly(acrylic acid), polyacrylonitrile and polystyrene.[177,267] Also reported, was the production of 3D networked GO and FGO structures (noted as poly-GO and

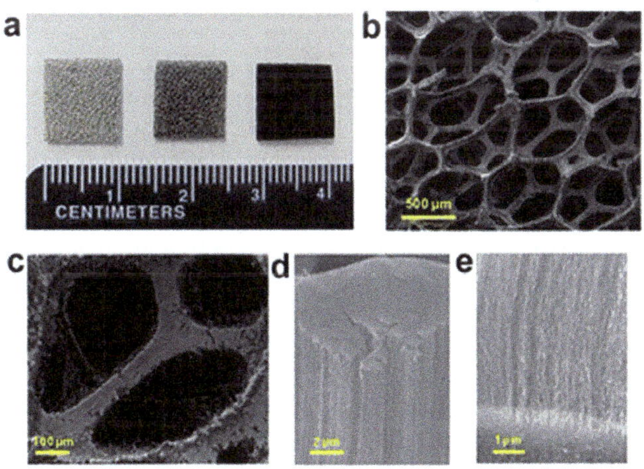

Fig. 16 Characterization of CNT–graphene hybrid materials grown on porous nickel substrates. (a) Photographs of porous nickel, graphene–porous nickel, and CNT–graphene–porous nickel (from left to right). (b–e) SEM images of the same samples. Side view of the CNT–catalyst interface (d) and CNT–graphene interface (e). For more details, see ref. 255. Reproduced with permission from ref. 255. Copyright 2013 American Chemical Society.

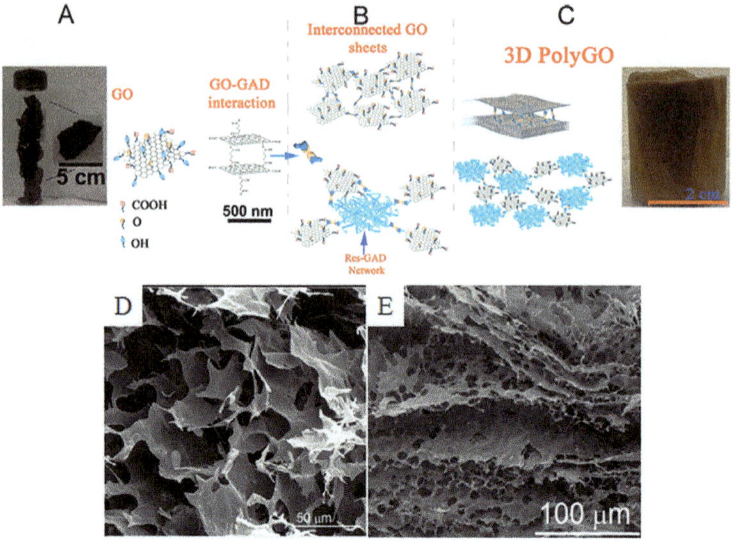

Fig. 17 (A–C) Schematic of the poly-GO synthesis process from initial GO. For a more detailed process see ref. 268. (D and E) SEM images of poly-GO (D) and poly-FGO (E). Reproduced with permission from ref. 268. Copyright 2013 American Chemical Society.

poly-FGO), which resulted from the polymerization of individual GO or FGO sheets with conventional polymer monomer (see Fig. 17).[268] In this approach, glutaraldehyde (GAD) was used to make crosslinked GO or FGO structures. The as-synthesized 3D networks with highly porous interconnected structures show high CO_2 gas adsorption capacity.

6 Applications

There are many applications of both CNTs and graphene, currently and for the future. Table 1 shows the most widely used currently proposed and somewhat developed applications.

6.1 Carbon nanotubes

In the last thirty years since their discovery, CNTs have been able to be investigated and developed for use in many applications that are industrially applicable. Their numerous exceptional properties make them ideal for replacements or improvements in many previously used technologies. The current problem, however, is their small dimensions and the necessary development of scale-up approaches for large industrial applications. To this end, CNTs can be found in many different composites as well as in the form of coatings and films, fibers, yarns, and sheets. As there are so many different applications of CNTs, many reviews have been specifically researched and published detailing their present and future commercial applications.[269]

In some industries, specifically in automotive and microelectronics industries, conductive CNT plastics are used for electrostatic-assisted painting of mirror housings, fuel lines, and filters that dissipate electrostatic charges, as well as

Table 1 Applications of CNTs and graphene

- Composites (structural, electrostatic discharge *etc.*)
- Coatings and paints
- Transparent conducting films
- Filtration/separation membranes
- Lithium ion batteries and supercapacitors
- Dye-sensitized solar cells and fuel cells (electrodes, catalyst support)
- Nanoelectronics and sensors (transistors, displays, touch screens)
- Biomedical applications (bio-sensors, drug-delivery systems)

electromagnetic interference (EMI)-shielding packages and wafer carriers.[270] CNT powders mixed with polymers or precursor resins are able to increase stiffness, strength, and toughness in load bearing applications.[271] In sporting goods industries, engineering nanoscale stick-slip amount CNTs and CNT–polymer contacts can increase material damping, which can be used to improve tennis racquets, baseball bats, and bicycle frames.[261] Wind turbine blades and hulls for maritime security boats can be made stronger and lighter using carbon fiber composites with CNT-enhanced resin.[271,272] Other composites include growing aligned CNT forests onto glass, SiC, alumina, or carbon fibers, for lightning-strike protection, deicing, and structural health monitory for aircraft.[262,271,273] Multilayered 3D composites were grown by using a high-temperature epoxy matrix to infiltrate 3D fabrics, which was produced through growing well-aligned MWNT forests perpendicular to 2D woven fabrics of SiC (see Fig. 18). The multilayered 3D composites provided indispensable interlaminar strength and toughness under various loading conditions.[262]

In some aerospace and automotive industrial structures, small amounts of CNTs added to metals have provided increased tensile strength and modulus.[274] When in the form of coatings and films, CNTs are able to be deposited and dispersed over large areas. In the marine industry, MWNT-containing paints reduce the attachment of algae and barnacles to ship hulls and are a possible alternative to environmentally hazardous biocide-containing paints.[275] When used with anticorrosion coatings for metals, CNTs can enhance coating stiffness and strength while providing an electric pathway for cathode protection. In electronics, the increasing scarcity of indium is creating a larger interest in flexible, cost-efficient CNT-based transparent conducting films as an alternative to indium tin oxide (ITO) for use in displays, touch screen devices, and photovoltaics.[276]

For thermal applications, laser patterned CNT films are transferred and mounted on the back side of the chips for chip thermal management.[277] CNTs can be used for gas ionization sensors, by using the sharp tips of CNTs to generate very high electric fields at low voltages, lower break down voltages and enabling compact, battery-powered, and safe operating sensors. Some applications of these CNT-based gas ionization sensors are environmental monitoring, sensing in chemical processing paths, and gas detection for counter-terrorism.[278] Carbon nanotube forests are also the blackest man-made material ever, with a diffused reflectance of 1×10^{-7} and a corresponding integrated total reflectance of 0.045%. Applications of this material include solar energy conversion and pyroelectric detectors.[279] Synthetic CNT-based gecko tape can be created by

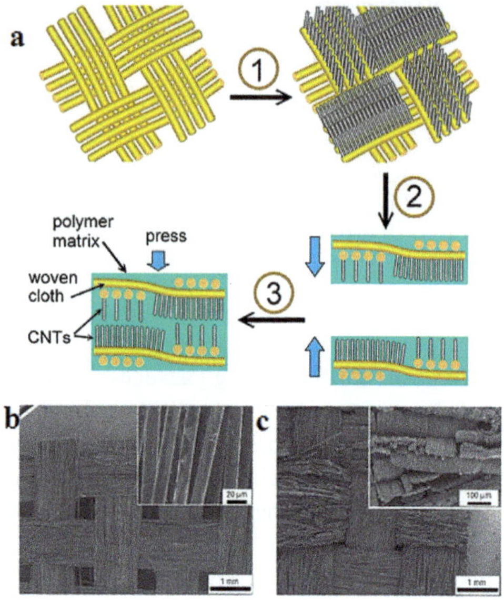

Fig. 18 (a) Schematic diagram of the hierarchical nanomanufacturing process of a 3D composite. (b) Plain-weave SiC fabric cloth. Inset: individual fibers of the woven cloth. (c) The cloth with CNTs grown perpendicularly on its surface. Inset: close-up. Adapted from ref. 262. Copyright© 2006 Nature Publishing Group.

transferring micropatterned CNT arrays onto flexible polymer tape on the basis of the hierarchical structure found on the foot of a gecko. This gecko tape can support shear stress of 36 N cm^{-2} and can be used as flexible adhesive tape able to hold infinitely long, under extreme conditions.[280] In 2013, the first CNT computer was designed and created using entirely CNT field-effect transistors. It runs an operating system that is capable of multitasking, and its instruction and data memories are implemented offchip following the convention of most computers today.[281]

6.2 Graphene

Graphene has numerous remarkable physical and chemical properties, which provide potential applications for the future. Like CNTs, graphene is also attractive for composite materials due to its excellent mechanical and chemical properties. They have been integrated into polymers to make light, stiff, and strong composites.[83,109,177] Graphene-enhanced composites were used to make bicycle race wheels and tennis rackets recently. The improved spoke-hole strength and lateral stiffness were reported in the bicycle race wheels made from graphene-based composites.

One of its most remarkable applications is electronic and optoelectronic devices. Graphene-based transistors have been considered and intensely developed as an alternative for silicon-based high-frequency transistors. Recently, a 67 nm channel-length graphene transistor with a cut-off frequency of 427 GHz has been reported, which approaches the best value of silicon-based transistor applications.[282] However, more effort is still needed to improve the maximum oscillation frequency, which is far from that of high-frequency silicon transistors (330 GHz).

Another important application of transistors is the logic switch. For such applications, the transistor works in the saturation region (On state) and cut-off region (Off state). This is a significant difference from the transistor oscillator. Due the high mobility of charge carriers in graphene, it can always provide a good logic "On" or "1" state. But because of the lack of bandgap, there is always current passing through the graphene, *i.e.* the graphene transistor cannot be turned off or at a good logic "0" state. To solve this issue, some attempts have been made to open a band gap including tailoring graphene into nanoribbons, modulation or doping with foreign atoms, or relying on the sole interaction with substrates.[283]

Since single-layer graphene is almost transparent with only 2.3% adsorption of incident white light, it has been widely studied as transparent conducting films for electronics, such as touch screen displays, organic light-emitting diodes as well as e-papers. Combining its high transmittance with high elasticity makes graphene also attractive for flexible electronics.[90]

The broadband absorption of graphene makes it a promising candidate for terahertz modulators and ultrafast photodetectors.[284–287] Photodetectors are based on transforming light signals into detectable electronic signals. There are several means to enhance the sensitivity of graphene photodetectors, such as by utilizing plasmonic nanostructures to enhance the local optical electric field. A photodetector was constructed by nanoscale antennas sandwiched between two graphene monolayers, which efficiently converts visible and near-infrared photons into electrons with strong enhancement (800%) of the photocurrent relative to the antenna-less graphene device.[285]

Graphene coating has been found to be wetting transparent, which does not significantly disrupt the intrinsic wetting behavior of substrate surfaces where van der Waals forces control the wetting. It means that without influencing the wettability of the substrate, other outstanding properties of graphene can be harnessed for the coated substrate. For example, its remarkable chemical inertness has made graphene a promising corrosion barrier against water and oxygen.[288–290] Retaining the wetting character of metals, graphene has been used as an ultrathin protective coating in protection of metals from degradation in marine or saline environments.[290]

With its 2D structures, excellent electric conductivity, high surface area as well as remarkable mechanical strength, graphene is indispensable for use in energy storage and conversion. The conventionally used electrode materials for lithium-ion batteries suffer from poor electrical conductivity and sluggish ion and electron transport kinetics or structural deformation, which have been circumvented by incorporating graphene with them.[291] A series of graphene-based lithium-ion batteries have been reported using graphene, doped graphene, and graphene–metal oxide electrodes.[208,292–297] Recently, progress was made in power density by using single-crystal vanadium dioxide ribbons-graphene ribbons as cathodes for lithium storage, which consume only 20 s in a full charge or discharge.[296]

Owing to the limited performance of CNT-based supercapacitors, graphene-based materials with a porous structure are currently intensely studied as an alternative carbon material for conventional supercapacitors. Graphene-based supercapacitors constructed with 3D structures of chemical modified graphene have demonstrated high capacitance and energy density as electrodes.[298–300] CNT and graphene hybrid materials with enhanced capacity compared with pure CNTs have been reported.[254,301–303] Besides, GO could also provide an interesting solid electrolyte for supercapacitors due to its insulating character and capability of trapping water.[304]

An intense interest has also been drawn in exploring graphene-based materials for solar cells because of their unique electrical and mechanical properties. Graphene and graphene hybrid materials have been utilized as counter electrodes for catalyzing the redox reaction in dye sensitized solar cells. Recently, graphene has also been used as an alternative material to widely used metal oxide window electrodes such as ITO due to graphene's distinguished mechanical properties in both bending and stretching tests.[305] Besides, the use of graphene sheets for fuel cells is continuously growing as a supporting material for platinum electro-catalysts or a nonmetal catalysts in place of platinum.[207,306–308]

There is special interest in investigating graphene for biomedical applications including cancer therapies, drug delivery, tissue engineering, biosensing and medical devices.[178,187,309,310] However, future research is very important in terms of the biodegradability or biopersistence as well as potential long-term toxicity concerns.[310,311] Graphite oxide has also found a niche in water purification. Graphite oxide functionalized with thiol groups *via* diazonium chemistry, has been used to make nanostructured graphite oxide-coated sand, which was shown capable of adsorbing 5-fold higher concentration of heavy metal and organic contaminants than pure sand.[312] Graphene and GO membranes have also been engineered for desired gas separation. GO membranes, with careful control of stacking structures, were able to provide high CO_2/N_2 selectivity in high relative humidity, allowing for great potential for industrial CO_2 separation processes.[313]

7 Conclusions

Since the discovery of fullerene in 1985, nanocarbon, specifically carbon nanotubes and graphene, has been an ever increasing research area that has helped, and continues to help, revolutionize so many industries. By exploring the growth, chemical modification, and doping of these exciting materials and using them in new configurations, more novel applications will become apparent. This is an important time to review what has been learned during the last two decades and, most important, to look into the future of carbon nanotubes and graphene.

Acknowledgements

This work has been supported by the office of Naval Research through the MURI program on graphene and by the U.S. Department of Defense: U.S. Air Force Office of Scientific Research (AFOSR) for the project "MURI: Synthesis and Characterization of 3-D Carbon Nanotube Solid Networks" through the award # FA9550-12-1-0035.

References

1 H. Marsh, E. A. Heintz and F. Rodríguez-Reinoso, *Introduction to Carbon Technologies*, Universidad de Alicante, 1997.

2 H. W. Kroto, J. R. Heath, S. C. O'Brien, R. F. Curl and R. E. Smalley, *Nature*, 1985, **318**, 162–163.
3 W. Kratschmer, L. D. Lamb, K. Fostiropoulos and D. R. Huffman, *Nature*, 1990, **347**, 354–358.
4 S. Iijima, *Nature*, 1991, **354**, 56–58.
5 S. Iijima and T. Ichihashi, *Nature*, 1993, **363**, 603–605.
6 F. Banhart and P. M. Ajayan, *Nature*, 1996, **382**, 433–435.
7 J. Kong, H. T. Soh, A. M. Cassell, C. F. Quate and H. Dai, *Nature*, 1998, **395**, 878–881.
8 P. M. Ajayan, *Chem. Rev.*, 1999, **99**, 1787–1800.
9 P. J. F. Harris, *Carbon Nanotubes and Related Structures: New Materials for the Twenty-first Century*, Cambridge University Press, 2001.
10 A. Peigney, C. Laurent, E. Flahaut, R. R. Bacsa and A. Rousset, *Carbon*, 2001, **39**, 507–514.
11 M. Inagaki and L. R. Radovic, *Carbon*, 2002, **40**, 2279–2282.
12 K. S. Novoselov, A. K. Geim, S. V. Morozov, D. Jiang, Y. Zhang, S. V. Dubonos, I. V. Grigorieva and A. A. Firsov, *Science*, 2004, **306**, 666–669.
13 A. K. Geim and K. S. Novoselov, *Nat. Mater.*, 2007, **6**, 183–191.
14 H. B. Heersche, P. Jarillo-Herrero, J. B. Oostinga, L. M. K. Vandersypen and A. F. Morpurgo, *Nature*, 2007, **446**, 56–59.
15 F. Schedin, A. K. Geim, S. V. Morozov, E. W. Hill, P. Blake, M. I. Katsnelson and K. S. Novoselov, *Nat. Mater.*, 2007, **6**, 652–655.
16 K. I. Bolotin, K. J. Sikes, J. Hone, H. L. Stormer and P. Kim, *Phys. Rev. Lett.*, 2008, **101**, 096802.
17 J. S. Bunch, S. S. Verbridge, J. S. Alden, A. M. van der Zande, J. M. Parpia, H. G. Craighead and P. L. McEuen, *Nano Lett.*, 2008, **8**, 2458–2462.
18 C. Lee, X. Wei, J. W. Kysar and J. Hone, *Science*, 2008, **321**, 385–388.
19 S. V. Morozov, K. S. Novoselov, M. I. Katsnelson, F. Schedin, D. C. Elias, J. A. Jaszczak and A. K. Geim, *Phys. Rev. Lett.*, 2008, **100**, 016602.
20 R. R. Nair, P. Blake, A. N. Grigorenko, K. S. Novoselov, T. J. Booth, T. Stauber, N. M. R. Peres and A. K. Geim, *Science*, 2008, **320**, 1308.
21 D. C. Elias, R. R. Nair, T. M. G. Mohiuddin, S. V. Morozov, P. Blake, M. P. Halsall, A. C. Ferrari, D. W. Boukhvalov, M. I. Katsnelson, A. K. Geim and K. S. Novoselov, *Science*, 2009, **323**, 610–613.
22 A. A. Balandin, *Nat. Mater.*, 2011, **10**, 569–581.
23 A. S. Mayorov, R. V. Gorbachev, S. V. Morozov, L. Britnell, R. Jalil, L. A. Ponomarenko, P. Blake, K. S. Novoselov, K. Watanabe, T. Taniguchi and A. K. Geim, *Nano Lett.*, 2011, **11**, 2396–2399.
24 K. S. Novoselov, V. I. Falko, L. Colombo, P. R. Gellert, M. G. Schwab and K. Kim, *Nature*, 2012, **490**, 192–200.
25 J. Rafiee, X. Mi, H. Gullapalli, A. V. Thomas, F. Yavari, Y. Shi, P. M. Ajayan and N. A. Koratkar, *Nat. Mater.*, 2012, **11**, 217–222.
26 T. W. Ebbesen and P. M. Ajayan, *Nature*, 1992, **358**, 220–222.
27 P. M. Ajayan, J. M. Lambert, P. Bernier, L. Barbedette, C. Colliex and J. M. Planeix, *Chem. Phys. Lett.*, 1993, **215**, 509–517.
28 D. S. Bethune, C. H. Klang, M. S. de Vries, G. Gorman, R. Savoy, J. Vazquez and R. Beyers, *Nature*, 1993, **363**, 605–607.
29 C. Journet, W. K. Maser, P. Bernier, A. Loiseau, M. L. de la Chapelle, S. Lefrant, P. Deniard, R. Lee and J. E. Fischer, *Nature*, 1997, **388**, 756–758.

30 R. G. Ding, G. Q. Lu, Z. F. Yan and M. A. Wilson, *J. Nanosci. Nanotechnol.*, 2001, **1**, 7–29.
31 Z. K. Tang, L. Zhang, N. Wang, X. X. Zhang, G. H. Wen, G. D. Li, J. N. Wang, C. T. Chan and P. Sheng, *Science*, 2001, **292**, 2462–2465.
32 W. Z. Li, S. S. Xie, L. X. Qian, B. H. Chang, B. S. Zou, W. Y. Zhou, R. A. Zhao and G. Wang, *Science*, 1996, **274**, 1701–1703.
33 Z. F. Ren, Z. P. Huang, J. W. Xu, J. H. Wang, P. Bush, M. P. Siegal and P. N. Provencio, *Science*, 1998, **282**, 1105–1107.
34 S. Fan, M. G. Chapline, N. R. Franklin, T. W. Tombler, A. M. Cassell and H. Dai, *Science*, 1999, **283**, 512–514.
35 B. Zheng, C. Lu, G. Gu, A. Makarovski, G. Finkelstein and J. Liu, *Nano Lett.*, 2002, **2**, 895–898.
36 K. Hata, D. N. Futaba, K. Mizuno, T. Namai, M. Yumura and S. Iijima, *Science*, 2004, **306**, 1362–1364.
37 M. Terrones, N. Grobert, J. Olivares, J. P. Zhang, H. Terrones, K. Kordatos, W. K. Hsu, J. P. Hare, P. D. Townsend, K. Prassides, A. K. Cheetham, H. W. Kroto and D. R. M. Walton, *Nature*, 1997, **388**, 52–55.
38 A. Cao, B. Wei, Y. Jung, R. Vajtai, P. M. Ajayan and G. Ramanath, *Appl. Phys. Lett.*, 2002, **81**, 1297–1299.
39 B. Q. Wei, R. Vajtai, Y. Jung, J. Ward, R. Zhang, G. Ramanath and P. M. Ajayan, *Nature*, 2002, **416**, 495–496.
40 Y. J. Jung, B. Wei, R. Vajtai, P. M. Ajayan, Y. Homma, K. Prabhakaran and T. Ogino, *Nano Lett.*, 2003, **3**, 561–564.
41 A. Cao, P. L. Dickrell, W. G. Sawyer, M. N. Ghasemi-Nejhad and P. M. Ajayan, *Science*, 2005, **310**, 1307–1310.
42 A. Cao, V. P. Veedu, X. Li, Z. Yao, M. N. Ghasemi-Nejhad and P. M. Ajayan, *Nat. Mater.*, 2005, **4**, 540–545.
43 H. Dai, A. G. Rinzler, P. Nikolaev, A. Thess, D. T. Colbert and R. E. Smalley, *Chem. Phys. Lett.*, 1996, **260**, 471–475.
44 D. N. Futaba, K. Hata, T. Yamada, T. Hiraoka, Y. Hayamizu, Y. Kakudate, O. Tanaike, H. Hatori, M. Yumura and S. Iijima, *Nat. Mater.*, 2006, **5**, 987–994.
45 T. Yamada, T. Namai, K. Hata, D. N. Futaba, K. Mizuno, J. Fan, M. Yudasaka, M. Yumura and S. Iijima, *Nat. Nanotechnol.*, 2006, **1**, 131–136.
46 M. Endo, T. Hayashi and Y.-A. Kim, *Pure Appl. Chem.*, 2006, **78**, 1703.
47 Z. J. Zhang, B. Q. Wei, G. Ramanath and P. M. Ajayan, *Appl. Phys. Lett.*, 2000, **77**, 3764–3766.
48 S. Talapatra, S. Kar, S. K. Pal, R. Vajtai, L. Ci, P. Victor, M. M. Shaijumon, S. Kaur, O. Nalamasu and P. M. Ajayan, *Nat. Nanotechnol.*, 2006, **1**, 112–116.
49 B. Q. Wei, R. Vajtai and P. M. Ajayan, *Appl. Phys. Lett.*, 2001, **79**, 1172–1174.
50 J. R. Sanchez-Valencia, T. Dienel, O. Groning, I. Shorubalko, A. Mueller, M. Jansen, K. Amsharov, P. Ruffieux and R. Fasel, *Nature*, 2014, **512**, 61–64.
51 F. Yang, X. Wang, D. Zhang, J. Yang, D. Luo, Z. Xu, J. Wei, J.-Q. Wang, Z. Xu, F. Peng, X. Li, R. Li, Y. Li, M. Li, X. Bai, F. Ding and Y. Li, *Nature*, 2014, **510**, 522–524.
52 S. Chakrabarti, H. Kume, L. Pan, T. Nagasaka and Y. Nakayama, *J. Phys. Chem. C*, 2007, **111**, 1929–1934.
53 W.-H. Chiang and R. Mohan Sankaran, *Nat. Mater.*, 2009, **8**, 882–886.

54 M. He, A. I. Chernov, P. V. Fedotov, E. D. Obraztsova, E. Rikkinen, Z. Zhu, J. Sainio, H. Jiang, A. G. Nasibulin, E. I. Kauppinen, M. Niemela and A. O. I. Krause, *Chem. Commun.*, 2011, **47**, 1219–1221.

55 J. Liu, A. G. Rinzler, H. Dai, J. H. Hafner, R. K. Bradley, P. J. Boul, A. Lu, T. Iverson, K. Shelimov, C. B. Huffman, F. Rodriguez-Macias, Y.-S. Shon, T. R. Lee, D. T. Colbert and R. E. Smalley, *Science*, 1998, **280**, 1253–1256.

56 R. E. Smalley, Y. Li, V. C. Moore, B. K. Price, R. Colorado, H. K. Schmidt, R. H. Hauge, A. R. Barron and J. M. Tour, *J. Am. Chem. Soc.*, 2006, **128**, 15824–15829.

57 Y. Yao, C. Feng, J. Zhang and Z. Liu, *Nano Lett.*, 2009, **9**, 1673–1677.

58 X. Tu, S. Manohar, A. Jagota and M. Zheng, *Nature*, 2009, **460**, 250–253.

59 M. Zheng, A. Jagota, E. D. Semke, B. A. Diner, R. S. McLean, S. R. Lustig, R. E. Richardson and N. G. Tassi, *Nat. Mater.*, 2003, **2**, 338–342.

60 H. Liu, D. Nishide, T. Tanaka and H. Kataura, *Nat. Commun.*, 2011, **2**, 309.

61 D. Jariwala, V. K. Sangwan, L. J. Lauhon, T. J. Marks and M. C. Hersam, *Chem. Soc. Rev.*, 2013, **42**, 2824–2860.

62 M. S. Arnold, S. I. Stupp and M. C. Hersam, *Nano Lett.*, 2005, **5**, 713–718.

63 A. A. Green and M. C. Hersam, *Adv. Mater.*, 2011, **23**, 2185–2190.

64 M. C. Hersam, *Nat. Nanotechnol.*, 2008, **3**, 387–394.

65 M. S. Arnold, A. A. Green, J. F. Hulvat, S. I. Stupp and M. C. Hersam, *Nat. Nanotechnol.*, 2006, **1**, 60–65.

66 E. H. Hároz, W. D. Rice, B. Y. Lu, S. Ghosh, R. H. Hauge, R. B. Weisman, S. K. Doorn and J. Kono, *ACS Nano*, 2010, **4**, 1955–1962.

67 S. Nanot, E. H. Hároz, J.-H. Kim, R. H. Hauge and J. Kono, *Adv. Mater.*, 2012, **24**, 4977–4994.

68 V. I. Artyukhov, E. S. Penev and B. I. Yakobson, *Nat. Commun.*, 2014, **5**, 4892.

69 B. Peng, M. Locascio, P. Zapol, S. Li, S. L. Mielke, G. C. Schatz and H. D. Espinosa, *Nat. Nanotechnol.*, 2008, **3**, 626–631.

70 E. Pop, D. Mann, Q. Wang, K. Goodson and H. Dai, *Nano Lett.*, 2005, **6**, 96–100.

71 Y.-L. Li, I. A. Kinloch and A. H. Windle, *Science*, 2004, **304**, 276–278.

72 K. Koziol, J. Vilatela, A. Moisala, M. Motta, P. Cunniff, M. Sennett and A. Windle, *Science*, 2007, **318**, 1892–1895.

73 N. Behabtu, C. C. Young, D. E. Tsentalovich, O. Kleinerman, X. Wang, A. W. K. Ma, E. A. Bengio, R. F. ter Waarbeek, J. J. de Jong, R. E. Hoogerwerf, S. B. Fairchild, J. B. Ferguson, B. Maruyama, J. Kono, Y. Talmon, Y. Cohen, M. J. Otto and M. Pasquali, *Science*, 2013, **339**, 182–186.

74 A. Thess, R. Lee, P. Nikolaev, H. Dai, P. Petit, J. Robert, C. Xu, Y. H. Lee, S. G. Kim, A. G. Rinzler, D. T. Colbert, G. E. Scuseria, D. Tománek, J. E. Fischer and R. E. Smalley, *Science*, 1996, **273**, 483–487.

75 M. Endo, H. Muramatsu, T. Hayashi, Y. A. Kim, M. Terrones and M. S. Dresselhaus, *Nature*, 2005, **433**, 476.

76 D. N. Futaba, K. Hata, T. Yamada, T. Hiraoka, Y. Hayamizu, Y. Kakudate, O. Tanaike, H. Hatori, M. Yumura and S. Iijima, *Nat. Mater.*, 2006, **5**, 987–994.

77 X. Wang, Q. Li, J. Xie, Z. Jin, J. Wang, Y. Li, K. Jiang and S. Fan, *Nano Lett.*, 2009, **9**, 3137–3141.

78 R. Zhang, Y. Zhang, Q. Zhang, H. Xie, W. Qian and F. Wei, *ACS Nano*, 2013, **7**, 6156–6161.

79 M. D. Stoller, S. Park, Y. Zhu, J. An and R. S. Ruoff, *Nano Lett.*, 2008, **8**, 3498–3502.

80 C. Lee, X. Wei, J. W. Kysar and J. Hone, *Science*, 2008, **321**, 385–388.

81 W. S. Hummers and R. E. Offeman, *J. Am. Chem. Soc.*, 1958, **80**, 1339.

82 C. Berger, Z. Song, X. Li, X. Wu, N. Brown, C. Naud, D. Mayou, T. Li, J. Hass, A. N. Marchenkov, E. H. Conrad, P. N. First and W. A. de Heer, *Science*, 2006, **312**, 1191–1196.

83 S. Stankovich, D. A. Dikin, G. H. B. Dommett, K. M. Kohlhaas, E. J. Zimney, E. A. Stach, R. D. Piner, S. T. Nguyen and R. S. Ruoff, *Nature*, 2006, **442**, 282–286.

84 Y. Hernandez, V. Nicolosi, M. Lotya, F. M. Blighe, Z. Sun, S. De, I. T. McGovern, B. Holland, M. Byrne, Y. K. Gun'Ko, J. J. Boland, P. Niraj, G. Duesberg, S. Krishnamurthy, R. Goodhue, J. Hutchison, V. Scardaci, A. C. Ferrari and J. N. Coleman, *Nat. Nanotechnol.*, 2008, **3**, 563–568.

85 A. Reina, X. Jia, J. Ho, D. Nezich, H. Son, V. Bulovic, M. S. Dresselhaus and J. Kong, *Nano Lett.*, 2008, **9**, 30–35.

86 P. W. Sutter, J.-I. Flege and E. A. Sutter, *Nat. Mater.*, 2008, **7**, 406–411.

87 K. V. Emtsev, A. Bostwick, K. Horn, J. Jobst, G. L. Kellogg, L. Ley, J. L. McChesney, T. Ohta, S. A. Reshanov, J. Rohrl, E. Rotenberg, A. K. Schmid, D. Waldmann, H. B. Weber and T. Seyller, *Nat. Mater.*, 2009, **8**, 203–207.

88 K. S. Kim, Y. Zhao, H. Jang, S. Y. Lee, J. M. Kim, K. S. Kim, J.-H. Ahn, P. Kim, J.-Y. Choi and B. H. Hong, *Nature*, 2009, **457**, 706–710.

89 X. Li, W. Cai, J. An, S. Kim, J. Nah, D. Yang, R. Piner, A. Velamakanni, I. Jung, E. Tutuc, S. K. Banerjee, L. Colombo and R. S. Ruoff, *Science*, 2009, **324**, 1312–1314.

90 S. Bae, H. Kim, Y. Lee, X. Xu, J.-S. Park, Y. Zheng, J. Balakrishnan, T. Lei, H. Ri Kim, Y. I. Song, Y.-J. Kim, K. S. Kim, B. Ozyilmaz, J.-H. Ahn, B. H. Hong and S. Iijima, *Nat. Nanotechnol.*, 2010, **5**, 574–578.

91 A. Srivastava, C. Galande, L. Ci, L. Song, C. Rai, D. Jariwala, K. F. Kelly and P. M. Ajayan, *Chem. Mater.*, 2010, **22**, 3457–3461.

92 Z. Sun, Z. Yan, J. Yao, E. Beitler, Y. Zhu and J. M. Tour, *Nature*, 2010, **468**, 549–552.

93 J. Zhao, S. Pei, W. Ren, L. Gao and H.-M. Cheng, *ACS Nano*, 2010, **4**, 5245–5252.

94 H. Gullapalli, A. L. Mohana Reddy, S. Kilpatrick, M. Dubey and P. M. Ajayan, *Small*, 2011, **7**, 1697–1700.

95 Z. Yan, Z. Peng, Z. Sun, J. Yao, Y. Zhu, Z. Liu, P. M. Ajayan and J. M. Tour, *ACS Nano*, 2011, **5**, 8187–8192.

96 T. Kobayashi, M. Bando, N. Kimura, K. Shimizu, K. Kadono, N. Umezu, K. Miyahara, S. Hayazaki, S. Nagai, Y. Mizuguchi, Y. Murakami and D. Hobara, *Appl. Phys. Lett.*, 2013, **102**, 023112.

97 P. Blake, P. D. Brimicombe, R. R. Nair, T. J. Booth, D. Jiang, F. Schedin, L. A. Ponomarenko, S. V. Morozov, H. F. Gleeson, E. W. Hill, A. K. Geim and K. S. Novoselov, *Nano Lett.*, 2008, **8**, 1704–1708.

98 I. Jung, D. A. Dikin, R. D. Piner and R. S. Ruoff, *Nano Lett.*, 2008, **8**, 4283–4287.

99 P. V. Kumar, N. M. Bardhan, S. Tongay, J. Wu, A. M. Belcher and J. C. Grossman, *Nat. Chem.*, 2014, **6**, 151–158.

100 H. C. Schniepp, J.-L. Li, M. J. McAllister, H. Sai, M. Herrera-Alonso, D. H. Adamson, R. K. Prud'homme, R. Car, D. A. Saville and I. A. Aksay, *J. Phys. Chem. B*, 2006, **110**, 8535–8539.
101 G. Eda, G. Fanchini and M. Chhowalla, *Nat. Nanotechnol.*, 2008, **3**, 270–274.
102 D. Li and R. B. Kaner, *Science*, 2008, **320**, 1170–1171.
103 X. Li, Y. Zhu, W. Cai, M. Borysiak, B. Han, D. Chen, R. D. Piner, L. Colombo and R. S. Ruoff, *Nano Lett.*, 2009, **9**, 4359–4363.
104 S. H. Lee, H. W. Kim, J. O. Hwang, W. J. Lee, J. Kwon, C. W. Bielawski, R. S. Ruoff and S. O. Kim, *Angew. Chem., Int. Ed.*, 2010, **49**, 10084–10088.
105 Y. Xu, K. Sheng, C. Li and G. Shi, *ACS Nano*, 2010, **4**, 4324–4330.
106 W. Gao, N. Singh, L. Song, Z. Liu, A. L. Reddy, L. Ci, R. Vajtai, Q. Zhang, B. Wei and P. M. Ajayan, *Nat. Nanotechnol.*, 2011, **6**, 496–500.
107 S. Yin, Z. Niu and X. Chen, *Small*, 2012, **8**, 2458–2463.
108 Y.-M. Lin, C. Dimitrakopoulos, K. A. Jenkins, D. B. Farmer, H.-Y. Chiu, A. Grill and P. Avouris, *Science*, 2010, **327**, 662.
109 Z. Chen, W. Ren, L. Gao, B. Liu, S. Pei and H. M. Cheng, *Nat. Mater.*, 2011, **10**, 424–428.
110 P. M. Ajayan and B. I. Yakobson, *Nat. Mater.*, 2011, **10**, 415–417.
111 P. Y. Huang, C. S. Ruiz-Vargas, A. M. van der Zande, W. S. Whitney, M. P. Levendorf, J. W. Kevek, S. Garg, J. S. Alden, C. J. Hustedt, Y. Zhu, J. Park, P. L. McEuen and D. A. Muller, *Nature*, 2011, **469**, 389–392.
112 Q. Yu, L. A. Jauregui, W. Wu, R. Colby, J. Tian, Z. Su, H. Cao, Z. Liu, D. Pandey, D. Wei, T. F. Chung, P. Peng, N. P. Guisinger, E. A. Stach, J. Bao, S.-S. Pei and Y. P. Chen, *Nat. Mater.*, 2011, **10**, 443–449.
113 X. Li, C. W. Magnuson, A. Venugopal, R. M. Tromp, J. B. Hannon, E. M. Vogel, L. Colombo and R. S. Ruoff, *J. Am. Chem. Soc.*, 2011, **133**, 2816–2819.
114 W. Wu, L. A. Jauregui, Z. Su, Z. Liu, J. Bao, Y. P. Chen and Q. Yu, *Adv. Mater.*, 2011, **23**, 4898–4903.
115 L. Gao, W. Ren, H. Xu, L. Jin, Z. Wang, T. Ma, L.-P. Ma, Z. Zhang, Q. Fu, L.-M. Peng, X. Bao and H.-M. Cheng, *Nat. Commun.*, 2012, **3**, 699.
116 D. Geng, B. Wu, Y. Guo, L. Huang, Y. Xue, J. Chen, G. Yu, L. Jiang, W. Hu and Y. Liu, *Proc. Natl. Acad. Sci. U. S. A.*, 2012, **109**, 7992–7996.
117 H. Wang, G. Wang, P. Bao, S. Yang, W. Zhu, X. Xie and W.-J. Zhang, *J. Am. Chem. Soc.*, 2012, **134**, 3627–3630.
118 Z. Yan, J. Lin, Z. Peng, Z. Sun, Y. Zhu, L. Li, C. Xiang, E. L. Samuel, C. Kittrell and J. M. Tour, *ACS Nano*, 2012, **6**, 9110–9117.
119 J.-H. Lee, E. K. Lee, W.-J. Joo, Y. Jang, B.-S. Kim, J. Y. Lim, S.-H. Choi, S. J. Ahn, J. R. Ahn, M.-H. Park, C.-W. Yang, B. L. Choi, S.-W. Hwang and D. Whang, *Science*, 2014, **344**, 286–289.
120 M. Terrones, *Proc. Natl. Acad. Sci. U. S. A.*, 2012, **109**, 7951–7952.
121 B. Zhang, W. H. Lee, R. Piner, I. Kholmanov, Y. Wu, H. Li, H. Ji and R. S. Ruoff, *ACS Nano*, 2012, **6**, 2471–2476.
122 T. Ma, W. Ren, X. Zhang, Z. Liu, Y. Gao, L.-C. Yin, X.-L. Ma, F. Ding and H.-M. Cheng, *Proc. Natl. Acad. Sci. U. S. A.*, 2013, **110**, 20386–20391.
123 Y. Zhang, L. Zhang, P. Kim, M. Ge, Z. Li and C. Zhou, *Nano Lett.*, 2012, **12**, 2810–2816.
124 V. I. Artyukhov, Y. Liu and B. I. Yakobson, *Proc. Natl. Acad. Sci. U. S. A.*, 2012, **109**, 15136–15140.

125 Y. H. Vasilii,I. Artyukhov, R. S. Ruoff, B. I. Yakobson, 2014DOI: arXiv:1405.5799v1.
126 Y. Hao, M. S. Bharathi, L. Wang, Y. Liu, H. Chen, S. Nie, X. Wang, H. Chou, C. Tan, B. Fallahazad, H. Ramanarayan, C. W. Magnuson, E. Tutuc, B. I. Yakobson, K. F. McCarty, Y.-W. Zhang, P. Kim, J. Hone, L. Colombo and R. S. Ruoff, *Science*, 2013, **342**, 720–723.
127 L. Gao, G.-X. Ni, Y. Liu, B. Liu, A. H. Castro Neto and K. P. Loh, *Nature*, 2014, **505**, 190–194.
128 P. M. Ajayan and B. I. Yakobson, *Nature*, 2006, **441**, 818–819.
129 J.-L. Li, K. N. Kudin, M. J. McAllister, R. K. Prud'homme, I. A. Aksay and R. Car, *Phys. Rev. Lett.*, 2006, **96**, 176101.
130 L. Ci, Z. Xu, L. Wang, W. Gao, F. Ding, K. Kelly, B. Yakobson and P. Ajayan, *Nano Res.*, 2008, **1**, 116–122.
131 L. Gao, W. Ren, B. Liu, Z.-S. Wu, C. Jiang and H.-M. Cheng, *J. Am. Chem. Soc.*, 2009, **131**, 13934–13936.
132 J. Campos-Delgado, J. M. Romo-Herrera, X. Jia, D. A. Cullen, H. Muramatsu, Y. A. Kim, T. Hayashi, Z. Ren, D. J. Smith, Y. Okuno, T. Ohba, H. Kanoh, K. Kaneko, M. Endo, H. Terrones, M. S. Dresselhaus and M. Terrones, *Nano Lett.*, 2008, **8**, 2773–2778.
133 L. Jiao, L. Zhang, X. Wang, G. Diankov and H. Dai, *Nature*, 2009, **458**, 877–880.
134 D. V. Kosynkin, A. L. Higginbotham, A. Sinitskii, J. R. Lomeda, A. Dimiev, B. K. Price and J. M. Tour, *Nature*, 2009, **458**, 872–876.
135 L. Jiao, X. Wang, G. Diankov, H. Wang and H. Dai, *Nat. Nanotechnol.*, 2010, **5**, 321–325.
136 S. Ozden, P. A. S. Autreto, C. S. Tiwary, S. Khatiwada, L. Machado, D. S. Galvao, R. Vajtai, E. V. Barrera and P. M. Ajayan, *Nano Lett.*, 2014, **14**, 4131–4137.
137 X. Yang, X. Dou, A. Rouhanipour, L. Zhi, H. J. Räder and K. Müllen, *J. Am. Chem. Soc.*, 2008, **130**, 4216–4217.
138 J. Cai, P. Ruffieux, R. Jaafar, M. Bieri, T. Braun, S. Blankenburg, M. Muoth, A. P. Seitsonen, M. Saleh, X. Feng, K. Mullen and R. Fasel, *Nature*, 2010, **466**, 470–473.
139 Y. Sun, S. Wang, C. Li, P. Luo, L. Tao, Y. Wei and G. Shi, *Phys. Chem. Chem. Phys.*, 2013, **15**, 9907–9913.
140 J. Lu, P. S. E. Yeo, C. K. Gan, P. Wu and K. P. Loh, *Nat. Nanotechnol.*, 2011, **6**, 247–252.
141 L. Lin and S. Zhang, *Chem. Commun.*, 2012, **48**, 10177–10179.
142 D. Pan, J. Zhang, Z. Li and M. Wu, *Adv. Mater.*, 2010, **22**, 734–738.
143 J. Peng, W. Gao, B. K. Gupta, Z. Liu, R. Romero-Aburto, L. Ge, L. Song, L. B. Alemany, X. Zhan, G. Gao, S. A. Vithayathil, B. A. Kaipparettu, A. A. Marti, T. Hayashi, J.-J. Zhu and P. M. Ajayan, *Nano Lett.*, 2012, **12**, 844–849.
144 R. Ye, C. Xiang, J. Lin, Z. Peng, K. Huang, Z. Yan, N. P. Cook, E. L. Samuel, C. C. Hwang, G. Ruan, G. Ceriotti, A. R. Raji, A. Martí and J. M. Tour, *Nat. Commun.*, 2013, **4**, 2943.
145 Z. Fang, S. Thongrattanasiri, A. Schlather, Z. Liu, L. Ma, Y. Wang, P. M. Ajayan, P. Nordlander, N. J. Halas and F. J. Garcia de Abajo, *ACS Nano*, 2013, **7**, 2388–2395.

146 Z.-S. Wu, W. Ren, L. Gao, B. Liu, C. Jiang and H.-M. Cheng, *Carbon*, 2009, **47**, 493–499.

147 A. Dimiev, D. V. Kosynkin, A. Sinitskii, A. Slesarev, Z. Sun and J. M. Tour, *Science*, 2011, **331**, 1168–1172.

148 Z. Peng, Z. Yan, Z. Sun and J. M. Tour, *ACS Nano*, 2011, **5**, 8241–8247.

149 Z. Sun, A.-R. O. Raji, Y. Zhu, C. Xiang, Z. Yan, C. Kittrell, E. L. G. Samuel and J. M. Tour, *ACS Nano*, 2012, **6**, 9790–9796.

150 H. Zhou, W. J. Yu, L. Liu, R. Cheng, Y. Chen, X. Huang, Y. Liu, Y. Wang, Y. Huang and X. Duan, *Nat. Commun.*, 2013, **4**, 2096.

151 W. Fang, A. L. Hsu, Y. Song, A. G. Birdwell, M. Amani, M. Dubey, M. S. Dresselhaus, T. Palacios and J. Kong, *ACS Nano*, 2014, **8**, 6491–6499.

152 Z. Yan, Y. Liu, L. Ju, Z. Peng, J. Lin, G. Wang, H. Zhou, C. Xiang, E. L. G. Samuel, C. Kittrell, V. I. Artyukhov, F. Wang, B. I. Yakobson and J. M. Tour, *Angew. Chem.*, 2014, **126**, 1591–1595.

153 A. C. Ferrari, J. C. Meyer, V. Scardaci, C. Casiraghi, M. Lazzeri, F. Mauri, S. Piscanec, D. Jiang, K. S. Novoselov, S. Roth and A. K. Geim, *Phys. Rev. Lett.*, 2006, **97**, 187401.

154 P. M. Ajayan and J. M. Tour, *Nature*, 2007, **447**, 1066–1068.

155 V. Datsyuk, M. Kalyva, K. Papagelis, J. Parthenios, D. Tasis, A. Siokou, I. Kallitsis and C. Galiotis, *Carbon*, 2008, **46**, 833–840.

156 H. Hu, B. Zhao, M. E. Itkis and R. C. Haddon, *J. Phys. Chem. B*, 2003, **107**, 13838–13842.

157 H. Hu, B. Zhao, M. A. Hamon, K. Kamaras, M. E. Itkis and R. C. Haddon, *J. Am. Chem. Soc.*, 2003, **125**, 14893–14900.

158 S. Ozden, T. N. Narayanan, C. S. Tiwary, P. Dong, A. H. C. Hart, R. Vajtai and P. M. Ajayan, *Small*, 2014, DOI: arXiv:1405.5799v1.

159 R. K. Saini, I. W. Chiang, H. Q. Peng, R. E. Smalley, W. E. Billups, R. H. Hauge and J. L. Margrave, *J. Am. Chem. Soc.*, 2003, **125**, 3617–3621.

160 J. L. Bahr, J. P. Yang, D. V. Kosynkin, M. J. Bronikowski, R. E. Smalley and J. M. Tour, *J. Am. Chem. Soc.*, 2001, **123**, 6536–6542.

161 C. A. Dyke and J. M. Tour, *Nano Lett.*, 2003, **3**, 1215–1218.

162 Y. M. Ying, R. K. Saini, F. Liang, A. K. Sadana and W. E. Billups, *Org. Lett.*, 2003, **5**, 1471–1473.

163 P. Bilalis, D. Katsigiannopoulos, A. Avgeropoulos and G. Sakellariou, *RSC Adv.*, 2014, **4**, 2911–2934.

164 C. Ehli, G. M. A. Rahman, N. Jux, D. Balbinot, D. M. Guldi, F. Paolucci, M. Marcaccio, D. Paolucci, M. Melle-Franco, F. Zerbetto, S. Campidelli and M. Prato, *J. Am. Chem. Soc.*, 2006, **128**, 11222–11231.

165 D. W. Steuerman, A. Star, R. Narizzano, H. Choi, R. S. Ries, C. Nicolini, J. F. Stoddart and J. R. Heath, *J. Phys. Chem. B*, 2002, **106**, 3124–3130.

166 V. C. Moore, M. S. Strano, E. H. Haroz, R. H. Hauge, R. E. Smalley, J. Schmidt and Y. Talmon, *Nano Lett.*, 2003, **3**, 1379–1382.

167 Y. L. Zhao and J. F. Stoddart, *Acc. Chem. Res.*, 2009, **42**, 1161–1171.

168 R. J. Chen, Y. G. Zhang, D. W. Wang and H. J. Dai, *J. Am. Chem. Soc.*, 2001, **123**, 3838–3839.

169 A. Hirsch, *Angew Chem Int Edit*, 2002, **41**, 1853–1859.

170 S. Kopyl, V. Bystrov, I. Bdikin, M. Maiorov and A. C. M. Sousa, *J. Mater. Chem. C*, 2013, **1**, 2860–2866.

171 S. Ozden, L. H. Ge, T. N. Narayanan, A. H. C. Hart, H. Yang, S. Sridhar, R. Vajtai and P. M. Ajayan, *ACS Appl. Mater. Interfaces*, 2014, **6**, 10608–10613.
172 K. M. Lee, L. C. Li and L. M. Dai, *J. Am. Chem. Soc.*, 2005, **127**, 4122–4123.
173 W. Gao, L. B. Alemany, L. Ci and P. M. Ajayan, *Nat. Chem.*, 2009, **1**, 403–408.
174 S. Gilje, S. Han, M. Wang, K. L. Wang and R. B. Kaner, *Nano Lett.*, 2007, **7**, 3394–3398.
175 C. Gómez-Navarro, R. T. Weitz, A. M. Bittner, M. Scolari, A. Mews, M. Burghard and K. Kern, *Nano Lett.*, 2007, **7**, 3499–3503.
176 H. A. Becerril, J. Mao, Z. Liu, R. M. Stoltenberg, Z. Bao and Y. Chen, *ACS Nano*, 2008, **2**, 463–470.
177 T. Ramanathan, A. A. Abdala, S. Stankovich, D. A. Dikin, M. Herrera Alonso, R. D. Piner, D. H. Adamson, H. C. Schniepp, X. Chen, R. S. Ruoff, S. T. Nguyen, I. A. Aksay, R. K. Prud'Homme and L. C. Brinson, *Nat. Nanotechnol.*, 2008, **3**, 327–331.
178 X. Sun, Z. Liu, K. Welsher, J. Robinson, A. Goodwin, S. Zaric and H. Dai, *Nano Res.*, 2008, **1**, 203–212.
179 H. Kang, A. Kulkarni, S. Stankovich, R. S. Ruoff and S. Baik, *Carbon*, 2009, **47**, 1520–1525.
180 G. Eda and M. Chhowalla, *Adv. Mater.*, 2010, **22**, 2392–2415.
181 S. Niyogi, E. Bekyarova, M. E. Itkis, H. Zhang, K. Shepperd, J. Hicks, M. Sprinkle, C. Berger, C. N. Lau, W. A. deHeer, E. H. Conrad and R. C. Haddon, *Nano Lett.*, 2010, **10**, 4061–4066.
182 D. Yu, Y. Yang, M. Durstock, J.-B. Baek and L. Dai, *ACS Nano*, 2010, **4**, 5633–5640.
183 A. Mathkar, D. Tozier, P. Cox, P. Ong, C. Galande, K. Balakrishnan, A. Leela Mohana Reddy and P. M. Ajayan, *J. Phys. Chem. Lett.*, 2012, **3**, 986–991.
184 J. O. Sofo, A. S. Chaudhari and G. D. Barber, *Phys. Rev. B: Condens. Matter Mater. Phys.*, 2007, **75**, 153401.
185 R. R. Nair, W. Ren, R. Jalil, I. Riaz, V. G. Kravets, L. Britnell, P. Blake, F. Schedin, A. S. Mayorov, S. Yuan, M. I. Katsnelson, H.-M. Cheng, W. Strupinski, L. G. Bulusheva, A. V. Okotrub, I. V. Grigorieva, A. N. Grigorenko, K. S. Novoselov and A. K. Geim, *Small*, 2010, **6**, 2877–2884.
186 A. Mathkar, T. N. Narayanan, L. B. Alemany, P. Cox, P. Nguyen, G. Gao, P. Chang, R. Romero-Aburto, S. A. Mani and P. M. Ajayan, *Part. Part. Syst. Charact.*, 2013, **30**, 266–272.
187 R. Romero-Aburto, T. N. Narayanan, Y. Nagaoka, T. Hasumura, T. M. Mitcham, T. Fukuda, P. J. Cox, R. R. Bouchard, T. Maekawa, D. S. Kumar, S. V. Torti, S. A. Mani and P. M. Ajayan, *Adv. Mater.*, 2013, **25**, 5632–5637.
188 R. Sen, B. C. Satishkumar, A. Govindaraj, K. R. Harikumar, M. K. Renganathan and C. N. R. Rao, *J. Mater. Chem.*, 1997, **7**, 2335–2337.
189 D. L. Carroll, P. Redlich, X. Blase, J. C. Charlier, S. Curran, P. M. Ajayan, S. Roth and M. Rühle, *Phys. Rev. Lett.*, 1998, **81**, 2332–2335.
190 B. C. Satishkumar, A. Govindaraj, K. R. Harikumar, J.-P. Zhang, A. K. Cheetham and C. N. R. Rao, *Chem. Phys. Lett.*, 1999, **300**, 473–477.
191 R. Czerw, M. Terrones, J. C. Charlier, X. Blase, B. Foley, R. Kamalakaran, N. Grobert, H. Terrones, D. Tekleab, P. M. Ajayan, W. Blau, M. Rühle and D. L. Carroll, *Nano Lett.*, 2001, **1**, 457–460.

192 F. Villalpando-Paez, A. Zamudio, A. L. Elias, H. Son, E. B. Barros, S. G. Chou, Y. A. Kim, H. Muramatsu, T. Hayashi, J. Kong, H. Terrones, G. Dresselhaus, M. Endo, M. Terrones and M. S. Dresselhaus, *Chem. Phys. Lett.*, 2006, **424**, 345–352.
193 S. van Dommele, A. Romero-Izquirdo, R. Brydson, K. P. de Jong and J. H. Bitter, *Carbon*, 2008, **46**, 138–148.
194 L. Yang, S. Jiang, Y. Zhao, L. Zhu, S. Chen, X. Wang, Q. Wu, J. Ma, Y. Ma and Z. Hu, *Angew. Chem.*, 2011, **123**, 7270–7273.
195 G.-L. Tian, M.-Q. Zhao, Q. Zhang, J.-Q. Huang and F. Wei, *Carbon*, 2012, **50**, 5323–5330.
196 M. Glerup, M. Castignolles, M. Holzinger, G. Hug, A. Loiseau and P. Bernier, *Chem. Commun.*, 2003, 2542–2543.
197 T. Sharifi, G. Hu, X. Jia and T. Wågberg, *ACS Nano*, 2012, **6**, 8904–8912.
198 M. Terrones, P. M. Ajayan, F. Banhart, X. Blase, D. L. Carroll, J. C. Charlier, R. Czerw, B. Foley, N. Grobert, R. Kamalakaran, P. Kohler-Redlich, M. Rühle, T. Seeger and H. Terrones, *Appl. Phys. A*, 2002, **74**, 355–361.
199 R. B. Kaner, J. Kouvetakis, C. E. Warble, M. L. Sattler and N. Bartlett, *Mater. Res. Bull.*, 1987, **22**, 399–404.
200 A. Y. Liu, R. M. Wentzcovitch and M. L. Cohen, *Phys. Rev. B: Condens. Matter Mater. Phys.*, 1989, **39**, 1760–1765.
201 Y. Miyamoto, A. Rubio, M. L. Cohen and S. G. Louie, *Phys. Rev. B: Condens. Matter Mater. Phys.*, 1994, **50**, 4976–4979.
202 Z. Weng-Sieh, K. Cherrey, N. G. Chopra, X. Blase, Y. Miyamoto, A. Rubio, M. L. Cohen, S. G. Louie, A. Zettl and R. Gronsky, *Phys. Rev. B: Condens. Matter Mater. Phys.*, 1995, **51**, 11229–11232.
203 P. Redlich, J. Loeffler, P. M. Ajayan, J. Bill, F. Aldinger and M. Rühle, *Chem. Phys. Lett.*, 1996, **260**, 465–470.
204 R. Sen, B. C. Satishkumar, A. Govindaraj, K. R. Harikumar, G. Raina, J.-P. Zhang, A. K. Cheetham and C. N. R. Rao, *Chem. Phys. Lett.*, 1998, **287**, 671–676.
205 W.-Q. Han, W. Mickelson, J. Cumings and A. Zettl, *Appl. Phys. Lett.*, 2002, **81**, 1110–1112.
206 D. Wei, Y. Liu, Y. Wang, H. Zhang, L. Huang and G. Yu, *Nano Lett.*, 2009, **9**, 1752–1758.
207 L. Qu, Y. Liu, J.-B. Baek and L. Dai, *ACS Nano*, 2010, **4**, 1321–1326.
208 A. L. M. Reddy, A. Srivastava, S. R. Gowda, H. Gullapalli, M. Dubey and P. M. Ajayan, *ACS Nano*, 2010, **4**, 6337–6342.
209 X. Wang, X. Li, L. Zhang, Y. Yoon, P. K. Weber, H. Wang, J. Guo and H. Dai, *Science*, 2009, **324**, 768–771.
210 L. S. Panchakarla, K. S. Subrahmanyam, S. K. Saha, A. Govindaraj, H. R. Krishnamurthy, U. V. Waghmare and C. N. R. Rao, *Adv. Mater.*, 2009, **21**, 4726–4730.
211 A. G. Garcia, S. E. Baltazar, A. H. R. Castro, J. F. P. Robles and A. Rubio, *J. Comput. Theor. Nanosci.*, 2008, **5**, 2221–2229.
212 H. Gao, Z. Liu, L. Song, W. Guo, W. Gao, L. Ci, A. Rao, W. Quan, R. Vajtai and P. M. Ajayan, *Nanotechnology*, 2012, **23**, 275605.
213 J. Zhang, J. J. Li, Z. L. Wang, X. N. Wang, W. Feng, W. Zheng, W. W. Cao and P. A. Hu, *Chem. Mater.*, 2014, **26**, 2460–2466.
214 B. Biel, X. Blase, F. Triozon and S. Roche, *Phys. Rev. Lett.*, 2009, **102**, 096803.

215 Y. Li, Y. Zhao, H. Cheng, Y. Hu, G. Shi, L. Dai and L. Qu, *J. Am. Chem. Soc.*, 2011, **134**, 15–18.

216 R. Lv and M. Terrones, *Mater. Lett.*, 2012, **78**, 209–218.

217 J. Ortiz-Medina, M. L. García-Betancourt, X. Jia, R. Martínez-Gordillo, M. A. Pelagio-Flores, D. Swanson, A. L. Elías, H. R. Gutiérrez, E. Gracia-Espino, V. Meunier, J. Owens, B. G. Sumpter, E. Cruz-Silva, F. J. Rodríguez-Macías, F. López-Urías, E. Muñoz-Sandoval, M. S. Dresselhaus, H. Terrones and M. Terrones, *Adv. Funct. Mater.*, 2013, **23**, 3755–3762.

218 X. Li, H. Wang, J. T. Robinson, H. Sanchez, G. Diankov and H. Dai, *J. Am. Chem. Soc.*, 2009, **131**, 15939–15944.

219 L. Song, L. Ci, H. Lu, P. B. Sorokin, C. Jin, J. Ni, A. G. Kvashnin, D. G. Kvashnin, J. Lou, B. I. Yakobson and P. M. Ajayan, *Nano Lett.*, 2010, **10**, 3209–3215.

220 D. Golberg, Y. Bando, P. Dorozhkin and Z.-C. Dong, *MRS Bull.*, 2004, **29**, 38–42.

221 M. Kawaguchi, T. Kawashima and T. Nakajima, *Chem. Mater.*, 1996, **8**, 1197–1201.

222 L. Ci, L. Song, C. Jin, D. Jariwala, D. Wu, Y. Li, A. Srivastava, Z. F. Wang, K. Storr, L. Balicas, F. Liu and P. M. Ajayan, *Nat. Mater.*, 2010, **9**, 430–435.

223 Z. Liu, L. Ma, G. Shi, W. Zhou, Y. Gong, S. Lei, X. Yang, J. Zhang, J. Yu, K. P. Hackenberg, A. Babakhani, J. C. Idrobo, R. Vajtai, J. Lou and P. M. Ajayan, *Nat. Nanotechnol.*, 2013, **8**, 119–124.

224 M. P. Levendorf, C. J. Kim, L. Brown, P. Y. Huang, R. W. Havener, D. A. Muller and J. Park, *Nature*, 2012, **488**, 627–632.

225 Y. J. Gong, G. Shi, Z. H. Zhang, W. Zhou, J. Jung, W. L. Gao, L. L. Ma, Y. Yang, S. B. Yang, G. You, R. Vajtai, Q. F. Xu, A. H. MacDonald, B. I. Yakobson, J. Lou, Z. Liu and P. M. Ajayan, *Nat. Commun.*, 2014, **5**, 3193.

226 H. Fei, R. Ye, G. Ye, Y. Gong, Z. Peng, X. Fan, E. L. Samuel, P. M. Ajayan and J. M. Tour, *ACS Nano*, 2014, **8**, 10837–10843.

227 C. R. Dean, A. F. Young, I. Meric, C. Lee, L. Wang, S. Sorgenfrei, K. Watanabe, T. Taniguchi, P. Kim, K. L. Shepard and J. Hone, *Nat. Nanotechnol.*, 2010, **5**, 722–726.

228 Z. Liu, L. Song, S. Zhao, J. Huang, L. Ma, J. Zhang, J. Lou and P. M. Ajayan, *Nano Lett.*, 2011, **11**, 2032–2037.

229 L. Britnell, R. V. Gorbachev, R. Jalil, B. D. Belle, F. Schedin, A. Mishchenko, T. Georgiou, M. I. Katsnelson, L. Eaves, S. V. Morozov, N. M. R. Peres, J. Leist, A. K. Geim, K. S. Novoselov and L. A. Ponomarenko, *Science*, 2012, **335**, 947–950.

230 G. Gao, W. Gao, E. Cannuccia, J. Taha-Tijerina, L. Balicas, A. Mathkar, T. N. Narayanan, Z. Liu, B. K. Gupta, J. Peng, Y. Yin, A. Rubio and P. M. Ajayan, *Nano Lett.*, 2012, **12**, 3518–3525.

231 W. Zhou, J. Lee, J. Nanda, S. T. Pantelides, S. J. Pennycook and J.-C. Idrobo, *Nat. Nanotechnol.*, 2012, **7**, 161–165.

232 J. Zhao, Q. Deng, A. Bachmatiuk, G. Sandeep, A. Popov, J. Eckert and M. H. Rümmeli, *Science*, 2014, **343**, 1228–1232.

233 K. Jiang, Q. Li and S. Fan, *Nature*, 2002, **419**, 801.

234 M. Zhang, K. R. Atkinson and R. H. Baughman, *Science*, 2004, **306**, 1358–1361.

235 S. Kaur, P. M. Ajayan and R. S. Kane, *J. Phys. Chem. B*, 2006, **110**, 21377–21380.

236 M. B. Bryning, D. E. Milkie, M. F. Islam, L. A. Hough, J. M. Kikkawa and A. G. Yodh, *Adv. Mater.*, 2007, **19**, 661–664.

237 G. Toth, J. Mäklin, N. Halonen, J. Palosaari, J. Juuti, H. Jantunen, K. Kordas, W. G. Sawyer, R. Vajtai and P. M. Ajayan, *Adv. Mater.*, 2009, **21**, 2054–2058.

238 X. Gui, J. Wei, K. Wang, A. Cao, H. Zhu, Y. Jia, Q. Shu and D. Wu, *Adv. Mater.*, 2010, **22**, 617–621.

239 J. Zou, J. Liu, A. S. Karakoti, A. Kumar, D. Joung, Q. Li, S. I. Khondaker, S. Seal and L. Zhai, *ACS Nano*, 2010, **4**, 7293–7302.

240 B. I. Dunlap, *Phys. Rev. B: Condens. Matter Mater. Phys.*, 1992, **46**, 1933–1936.

241 M. Terrones, H. Terrones, F. Banhart, J.-C. Charlier and P. M. Ajayan, *Science*, 2000, **288**, 1226–1229.

242 M. Terrones, F. Banhart, N. Grobert, J. C. Charlier, H. Terrones and P. M. Ajayan, *Phys. Rev. Lett.*, 2002, **89**, 075505.

243 B. G. Sumpter, V. Meunier, J. M. Romo-Herrera, E. Cruz-Silva, D. A. Cullen, H. Terrones, D. J. Smith and M. Terrones, *ACS Nano*, 2007, **1**, 369–375.

244 J. M. Romo-Herrera, B. G. Sumpter, D. A. Cullen, H. Terrones, E. Cruz-Silva, D. J. Smith, V. Meunier and M. Terrones, *Angew. Chem., Int. Ed.*, 2008, **47**, 2948–2953.

245 J. M. Romo-Herrera, D. A. Cullen, E. Cruz-Silva, D. Ramírez, B. G. Sumpter, V. Meunier, H. Terrones, D. J. Smith and M. Terrones, *Adv. Funct. Mater.*, 2009, **19**, 1193–1199.

246 D. P. Hashim, N. T. Narayanan, J. M. Romo-Herrera, D. A. Cullen, M. G. Hahm, P. Lezzi, J. R. Suttle, D. Kelkhoff, E. Muñoz-Sandoval, S. Ganguli, A. K. Roy, D. J. Smith, R. Vajtai, B. G. Sumpter, V. Meunier, H. Terrones, M. Terrones and P. M. Ajayan, *Sci. Rep.*, 2012, **2**, 363.

247 C. Shan, W. Zhao, X. L. Lu, D. J. O'Brien, Y. Li, Z. Cao, A. L. Elias, R. Cruz-Silva, M. Terrones, B. Wei and J. Suhr, *Nano Lett.*, 2013, **13**, 5514–5520.

248 R. K. Paul, M. Ghazinejad, M. Penchev, J. Lin, M. Ozkan and C. S. Ozkan, *Small*, 2010, **6**, 2309–2313.

249 V. Varshney, S. S. Patnaik, A. K. Roy, G. Froudakis and B. L. Farmer, *ACS Nano*, 2010, **4**, 1153–1161.

250 S. W. Hong, F. Du, W. Lan, S. Kim, H.-S. Kim and J. A. Rogers, *Adv. Mater.*, 2011, **23**, 3821–3826.

251 K. H. Kim, Y. Oh and M. F. Islam, *Nat. Nanotechnol.*, 2012, **7**, 562–566.

252 M.-Q. Zhao, X.-F. Liu, Q. Zhang, G.-L. Tian, J.-Q. Huang, W. Zhu and F. Wei, *ACS Nano*, 2012, **6**, 10759–10769.

253 Y. Zhu, L. Li, C. Zhang, G. Casillas, Z. Sun, Z. Yan, G. Ruan, Z. Peng, A. R. Raji, C. Kittrell, R. H. Hauge and J. M. Tour, *Nat. Commun.*, 2012, **3**, 1225.

254 W. Wang, S. Guo, M. Penchev, I. Ruiz, K. N. Bozhilov, D. Yan, M. Ozkan and C. S. Ozkan, *Nano Energy*, 2013, **2**, 294–303.

255 Z. Yan, L. Ma, Y. Zhu, I. Lahiri, M. G. Hahm, Z. Liu, S. Yang, C. Xiang, W. Lu, Z. Peng, Z. Sun, C. Kittrell, J. Lou, W. Choi, P. M. Ajayan and J. M. Tour, *ACS Nano*, 2013, **7**, 58–64.

256 A. D. Leonard, J. L. Hudson, H. Fan, R. Booker, L. J. Simpson, K. J. O'Neill, P. A. Parilla, M. J. Heben, M. Pasquali, C. Kittrell and J. M. Tour, *J. Am. Chem. Soc.*, 2008, **131**, 723–728.

257 P. M. Ajayan, P. Redlich and M. Rühle, *J. Microsc.*, 1997, **185**, 275–282.

258 L. S. Schadler, S. C. Giannaris and P. M. Ajayan, *Appl. Phys. Lett.*, 1998, **73**, 3842–3844.

259 P. M. Ajayan, L. S. Schadler, C. Giannaris and A. Rubio, *Adv. Mater.*, 2000, **12**, 750–753.
260 J. Suhr, N. Koratkar, P. Keblinski and P. Ajayan, *Nat. Mater.*, 2005, **4**, 134–137.
261 V. P. Veedu, A. Cao, X. Li, K. Ma, C. Soldano, S. Kar, P. M. Ajayan and M. N. Ghasemi-Nejhad, *Nat. Mater.*, 2006, **5**, 457–462.
262 W. Ma, L. Liu, Z. Zhang, R. Yang, G. Liu, T. Zhang, X. An, X. Yi, Y. Ren, Z. Niu, J. Li, H. Dong, W. Zhou, P. M. Ajayan and S. Xie, *Nano Lett.*, 2009, **9**, 2855–2861.
263 Y. Yang, W. Rigdon, X. Huang and X. Li, *Sci. Rep.*, 2013, **3**, 2086.
264 Y. Yang, Z. H. Xu, Z. Pan and X. Li, *Adv. Mater.*, 2012, **24**, 881–885.
265 B. J. Carey, P. K. Patra, L. Ci, G. G. Silva and P. M. Ajayan, *ACS Nano*, 2011, **5**, 2715–2722.
266 T. Ramanathan, S. Stankovich, D. A. Dikin, H. Liu, H. Shen, S. T. Nguyen and L. C. Brinson, *J. Polym. Sci., Part B: Polym. Phys.*, 2007, **45**, 2097–2112.
267 P. M. Sudeep, T. N. Narayanan, A. Ganesan, M. M. Shaijumon, H. Yang, S. Ozden, P. K. Patra, M. Pasquali, R. Vajtai, S. Ganguli, A. K. Roy, M. R. Anantharaman and P. M. Ajayan, *ACS Nano*, 2013, **7**, 7034–7040.
268 M. F. L. De Volder, S. H. Tawfick, R. H. Baughman and A. J. Hart, *Science*, 2013, **339**, 535–539.
269 W. Bauhofer and J. Z. Kovacs, *Compos. Sci. Technol.*, 2009, **69**, 1486–1498.
270 T.-W. Chou, L. Gao, E. T. Thostenson, Z. Zhang and J.-H. Byun, *Compos. Sci. Technol.*, 2010, **70**, 1–19.
271 J. N. Coleman, U. Khan, W. J. Blau and Y. K. Gun'ko, *Carbon*, 2006, **44**, 1624–1652.
272 E. J. Garcia, B. L. Wardle, A. John Hart and N. Yamamoto, *Compos. Sci. Technol.*, 2008, **68**, 2034–2041.
273 S. R. Bakshi and A. Agarwal, *Carbon*, 2011, **49**, 533–544.
274 A. Beigbeder, P. Degee, S. L. Conlan, R. J. Mutton, A. S. Clare, M. E. Pettitt, M. E. Callow, J. A. Callow and P. Dubois, *Biofouling*, 2008, **24**, 291–302.
275 Z. Wu, Z. Chen, X. Du, J. M. Logan, J. Sippel, M. Nikolou, K. Kamaras, J. R. Reynolds, D. B. Tanner, A. F. Hebard and A. G. Rinzler, *Science*, 2004, **305**, 1273–1276.
276 K. Kordás, G. Tóth, P. Moilanen, M. Kumpumäki, J. Vähäkangas, A. Uusimäki, R. Vajtai and P. M. Ajayan, *Appl. Phys. Lett.*, 2007, **90**, 123105.
277 A. Modi, N. Koratkar, E. Lass, B. Wei and P. M. Ajayan, *Nature*, 2003, **424**, 171–174.
278 Z.-P. Yang, L. Ci, J. A. Bur, S.-Y. Lin and P. M. Ajayan, *Nano Lett.*, 2008, **8**, 446–451.
279 L. Ge, S. Sethi, L. Ci, P. M. Ajayan and A. Dhinojwala, *Proc. Natl. Acad. Sci. U. S. A.*, 2007, **104**, 10792–10795.
280 M. M. Shulaker, G. Hills, N. Patil, H. Wei, H.-Y. Chen, H. S. P. Wong and S. Mitra, *Nature*, 2013, **501**, 526–530.
281 R. Cheng, J. Bai, L. Liao, H. Zhou, Y. Chen, L. Liu, Y.-C. Lin, S. Jiang, Y. Huang and X. Duan, *Proc. Natl. Acad. Sci. U. S. A.*, 2012, **109**, 11588–11592.
282 D. Jariwala, A. Srivastava and P. M. Ajayan, *J. Nanosci. Nanotechnol.*, 2011, **11**, 6621–6641.
283 C. H. Liu, Y. C. Chang, T. B. Norris and Z. Zhong, *Nat. Nanotechnol.*, 2014, **9**, 273–278.

284 Z. Fang, Z. Liu, Y. Wang, P. M. Ajayan, P. Nordlander and N. J. Halas, *Nano Lett.*, 2012, **12**, 3808–3813.
285 M. Liu, X. Yin, E. Ulin-Avila, B. Geng, T. Zentgraf, L. Ju, F. Wang and X. Zhang, *Nature*, 2011, **474**, 64–67.
286 F. Xia, T. Mueller, Y.-m. Lin, A. Valdes-Garcia and P. Avouris, *Nat. Nanotechnol.*, 2009, **4**, 839–843.
287 S. Chen, L. Brown, M. Levendorf, W. Cai, S.-Y. Ju, J. Edgeworth, X. Li, C. W. Magnuson, A. Velamakanni, R. D. Piner, J. Kang, J. Park and R. S. Ruoff, *ACS Nano*, 2011, **5**, 1321–1327.
288 D. Prasai, J. C. Tuberquia, R. R. Harl, G. K. Jennings and K. I. Bolotin, *ACS Nano*, 2012, **6**, 1102–1108.
289 R. K. Singh Raman, P. Chakraborty Banerjee, D. E. Lobo, H. Gullapalli, M. Sumandasa, A. Kumar, L. Choudhary, R. Tkacz, P. M. Ajayan and M. Majumder, *Carbon*, 2012, **50**, 4040–4045.
290 P. G. Bruce, B. Scrosati and J.-M. Tarascon, *Angew. Chem., Int. Ed.*, 2008, **47**, 2930–2946.
291 E. Yoo, J. Kim, E. Hosono, H. S. Zhou, T. Kudo and I. Honma, *Nano Lett.*, 2008, **8**, 2277–2282.
292 H. Wang, L.-F. Cui, Y. Yang, H. Sanchez Casalongue, J. T. Robinson, Y. Liang, Y. Cui and H. Dai, *J. Am. Chem. Soc.*, 2010, **132**, 13978–13980.
293 Y. Sun, Q. Wu and G. Shi, *Energy Environ. Sci.*, 2011, **4**, 1113–1132.
294 Y. J. Gong, S. B. Yang, Z. Liu, L. L. Ma, R. Vajtai and P. M. Ajayan, *Adv. Mater.*, 2013, **25**, 3979–3984.
295 S. Yang, Y. Gong, Z. Liu, L. Zhan, D. P. Hashim, L. Ma, R. Vajtai and P. M. Ajayan, *Nano Lett.*, 2013, **13**, 1596–1601.
296 Y. Gong, S. Yang, L. Zhan, L. Ma, R. Vajtai and P. M. Ajayan, *Adv. Funct. Mater.*, 2014, **24**, 125–130.
297 M. D. Stoller, S. Park, Y. Zhu, J. An and R. S. Ruoff, *Nano Lett.*, 2008, **8**, 3498–3502.
298 Y. Zhu, S. Murali, M. D. Stoller, K. J. Ganesh, W. Cai, P. J. Ferreira, A. Pirkle, R. M. Wallace, K. A. Cychosz, M. Thommes, D. Su, E. A. Stach and R. S. Ruoff, *Science*, 2011, **332**, 1537–1541.
299 M. F. El-Kady, V. Strong, S. Dubin and R. B. Kaner, *Science*, 2012, **335**, 1326–1330.
300 Z. Fan, J. Yan, L. Zhi, Q. Zhang, T. Wei, J. Feng, M. Zhang, W. Qian and F. Wei, *Adv. Mater.*, 2010, **22**, 3723–3728.
301 J. Lin, C. Zhang, Z. Yan, Y. Zhu, Z. Peng, R. H. Hauge, D. Natelson and J. M. Tour, *Nano Lett.*, 2012, **13**, 72–78.
302 J. Lin, C. Zhang, Z. Yan, Y. Zhu, Z. Peng, R. H. Hauge, D. Natelson and J. M. Tour, *Nano Lett.*, 2013, **13**, 72–78.
303 W. Gao, N. Singh, L. Song, Z. Liu, A. L. M. Reddy, L. Ci, R. Vajtai, Q. Zhang, B. Wei and P. M. Ajayan, *Nat. Nanotechnol.*, 2011, **6**, 496–500.
304 P. Dong, Y. Zhu, J. Zhang, C. Peng, Z. Yan, L. Li, Z. Peng, G. Ruan, W. Xiao, H. Lin, J. M. Tour and J. Lou, *J. Phys. Chem. C*, 2014, DOI: 10.1021/jp505735j.
305 Y. Li, W. Gao, L. Ci, C. Wang and P. M. Ajayan, *Carbon*, 2010, **48**, 1124–1130.
306 R. Lv, T. Cui, M.-S. Jun, Q. Zhang, A. Cao, D. S. Su, Z. Zhang, S.-H. Yoon, J. Miyawaki, I. Mochida and F. Kang, *Adv. Funct. Mater.*, 2011, **21**, 999–1006.
307 Y. Li, W. Zhou, H. Wang, L. Xie, Y. Liang, F. Wei, J. C. Idrobo, S. J. Pennycook and H. Dai, *Nat. Nanotechnol.*, 2012, **7**, 394–400.

308 W. Yang, K. R. Ratinac, S. P. Ringer, P. Thordarson, J. J. Gooding and F. Braet, *Angew. Chem., Int. Ed.*, 2010, **49**, 2114–2138.

309 L. Feng and Z. Liu, *Nanomedicine*, 2011, **6**, 317–324.

310 V. C. Sanchez, A. Jachak, R. H. Hurt and A. B. Kane, *Chem. Res. Toxicol.*, 2011, **25**, 15–34.

311 W. Gao, M. Majumder, L. B. Alemany, T. N. Narayanan, M. A. Ibarra, B. K. Pradhan and P. M. Ajayan, *ACS Appl. Mater. Interfaces*, 2011, **3**, 1821–1826.

312 H. W. Kim, H. W. Yoon, S.-M. Yoon, B. M. Yoo, B. K. Ahn, Y. H. Cho, H. J. Shin, H. Yang, U. Paik, S. Kwon, J.-Y. Choi and H. B. Park, *Science*, 2013, **342**, 91–95.

Faraday Discussions

PAPER

Spinning of carbon nanotube fibres using the floating catalyst high temperature route: purity issues and the critical role of sulphur†

T. S. Gspann,* F. R. Smail and A. H. Windle*

Received 11th April 2014, Accepted 20th May 2014
DOI: 10.1039/c4fd00066h

The CVD process for the spinning of carbon nanotube (CNT) fibres combines the nucleation, growth and aggregation of CNTs in the form of an aerogel with fibre spinning into a single process step. The optimisation of the process requires agility in multi-dimensional parameter space, so one tends to find parameter 'islands' where spinning is possible, while exploration tends to follow 'routes' through this space. Here, we follow two such routes, one of which drastically improves fibre purity, the other changes the nature of the nanotubes comprising the fibres from multiwall to single wall. In the first case there is only a modest enhancement of the mechanical properties, but in the second a very considerable improvement is seen. In terms of the conditions required to make fibres comprising predominately single wall CNTs, the key factor appears to be the rigorous control of the sulphur addition, in trace quantities, coupled with the availability of carbon atoms at the earliest stage after injection, typically in the range 400–500 °C. A model is presented for the role of sulphur in floating catalysts CNT synthesis.

Introduction

Direct-spun carbon nanotube (CNT) fibres show the values of tensile strength and stiffness[2,3] which are competitive with other high performance fibres, such as traditional carbon fibre and Kevlar®, while their yarn-like character endows them with a much higher knot efficiency,[4] meaning that they can be knotted or woven with little, if any loss of tensile strength. The central challenge, however, remains. We are unable to transfer a sensible fraction of the vastly superior mechanical performance of individual CNTs[5,6] into a macroscopic structural material, even when the nanotubes are aligned axially as in a fibre. The fibre structure itself is hierarchical comprising of aligned bundles of CNTs, where each bundle consists

Department of Materials Science and Metallurgy, University of Cambridge, UK. E-mail: ahw1@cam.ac.uk; tsg28@cam.ac.uk

† Electronic supplementary information (ESI) available. See DOI: 10.1039/c4fd00066h

of well aligned individual nanotubes.[1,7] The bundles are networked by a nanotube exchange. The other two major methods for CNT fibre production, such as liquid-crystal spinning[8,9] or spinning from forests[10,11] each generate very clean structures, as the CNTs are grown first and, at least in case of liquid-crystal spinning, can be cleaned very effectively prior to the spinning. However, these processes consist of at least two, if not more steps. Additionally, the CNTs can be shortened as a result of any cleaning or purification stages. The attraction of the direct CNT fibre spinning process, first published by Ya-Li Li *et al.*,[1] lies in the fact that it consists of one single process step, with extraction of an aligned aerogel of long nanotubes from a floating catalyst CVD system leading directly to the final yarn, which is wound up continuously. However, co-synthesised impurities and catalyst residues have been a problem for a long time. We present here methods for greatly reducing the impurity content of the fibre, and show how, through the critical control of the sulphur content, a fibre can be grown which comprises of only single wall nanotubes. In the latter case a model is presented for the role of sulphur in floating catalyst synthesis.

Experimental

Spinning under standard conditions: particulate defects

The observation that fibres were stronger if tested at short gauge lengths, coupled with the observation of particulate defects, led[3] to surmise that a property improvement would come simply from a purity improvement. In part, this approach is one motivation for this current study. Here, we refer to 'standard conditions' as those employed over much of our research on direct spun CNT fibres over the past five years, or more. The reactor conditions are set as described in Table 1.

Kumar[12] has summarised the different types of defects which can occur within the internal microstructure of CNT fibres. Of these, we focus on extraneous material largely in a particulate form, which can be classified as: iron catalyst residue which has not given rise to nanotubes, or particles consisting of carbonaceous material which appears to be a secondary growth of extremely highly deformed multiwalled CNTs which entangle together forming larger clusters several micrometres in diameter (Fig. 1a).

What we know about the direct spinning process

The direct spinning process has at its heart high temperature, floating catalyst, nanotube synthesis as a continuous flow (Fig. 2). A centrally important advantage

Table 1 Standard spinning conditions in a reactor tube with inner diameter 85 mm, reactor hot zone at 1290 °C

Flow rates				Overall atomic ratio	
H_2 [l min^{-1}]	Methane [ml min^{-1}]	Ferrocene [mg min^{-1}]	Thiophene [mg min^{-1}]	S : C	S : Fe
0.8	80	0.35	1.3	0.004	7.65

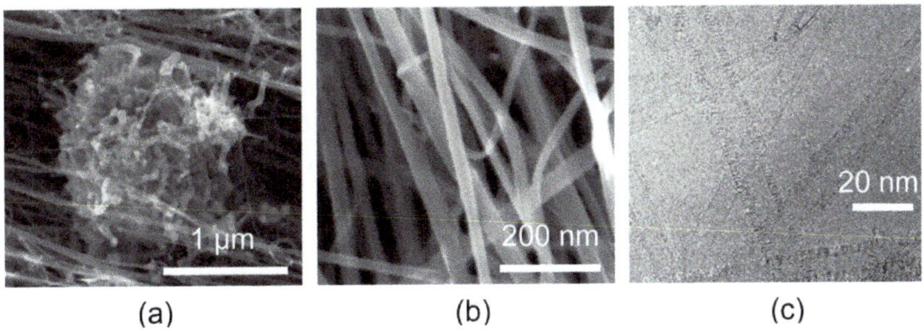

Fig. 1 (a) SEM of large impurity cluster composed of poorly formed MWCNTs and (b) well formed CNT bundles in a region remote from any particles. (c) TEM of nanotubes of the order of 10 nm in diameter, showing double and few walled examples.

of spinning at 1290 °C with added sulphur, rather than at 650–800 °C without sulphur, is that the nanotubes do not deposit on the reactor walls in the reaction zone. One reason, to be discussed below, is that the sulphur prevents the iron deposited on the reactor wall from nucleating and growing nanotubes. Another reason why this does not occur, certainly in the main hot zone of the reactor, is thermophoresis or the so-called 'Ludwig–Soret effect'; namely that within a mean free path of the hot reactor wall, gas molecules are leaving a little faster than they arrive, so any nano-scale particle within that distance of the wall will experience a repulsion away from the wall.

In the direct-spinning process, the CNTs grow tangentially on the catalyst particles, so the size of the catalyst particles is one of the most crucial factors controlling the type of CNTs grown. The size of the catalyst particles is mainly defined by two temperatures, namely the breakdown temperature of ferrocene and the temperature of the thermocatalytic breakdown of the carbon source on

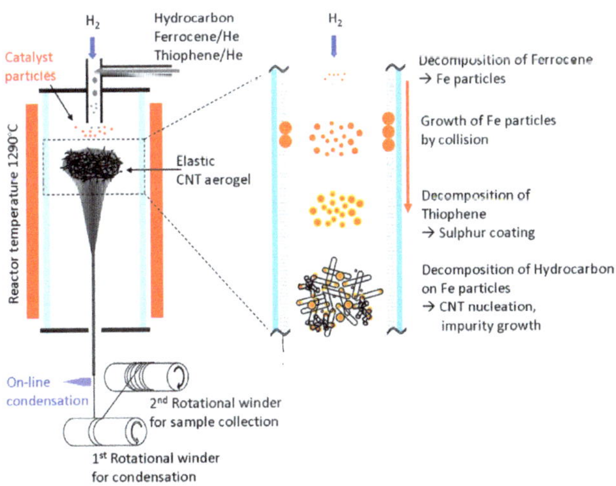

Fig. 2 Overview schematic of the CVD process for the direct spinning of CNT fibres and detailed schematic of the decomposition of the precursors leading to the formation of an elastic CNT aerogel.

the catalyst surface. Due to the temperature profile, the difference between these two temperatures corresponds to a certain vertical distance in the reactor tube and, for a given flow velocity profile, a particular journey time.

The modelled flow and heat transfer characteristics[13,14] indicated that there was very little temperature gradient across the reactor tube; and for the flow rates normally employed, the flow was laminar being an order of magnitude or more below the critical Reynolds number, so a Poiseuille velocity distribution was established within 1.5 tube diameters below the injector (little or no jetting). These papers also demonstrated that once the catalyst particles had grown by collision, to the point where their Brownian diffusion rate in the plane normal to the reactor axis became large compared with mean gas flow rate, the particles could be considered entrained in a particular region of the transverse velocity profile. Hence, those particles nearer the outside wall will be moving much slower, and thus have a longer residence time, and more opportunity for growth by collision before the temperature cross section is reached where nanotube growth commences. Fig. 3a is adapted from Conroy *et al.*[16] It emphasises that particle entrainment, and thus faster motion close to the reactor tube axis, becomes more significant with increasing particle size as the Brownian radial diffusion rates decrease. As the particles nearer to the walls travel more slowly, they will be bigger at any given cross section. The assumption also behind these calculations of iron particle coarsening is that particle contact definitely means coalescence, although it has been argued[15] that coalescence may be less rapid on account of sulphur on the particles.

There is another hydrodynamic effect which is possibly of significance. Segre and Silberberg[16,17] first showed experimentally that particles entrained within Poiseuille flow, after a certain distance from the entrance point into the reactor, would segregate into an annular region, close to the walls of the circular tube

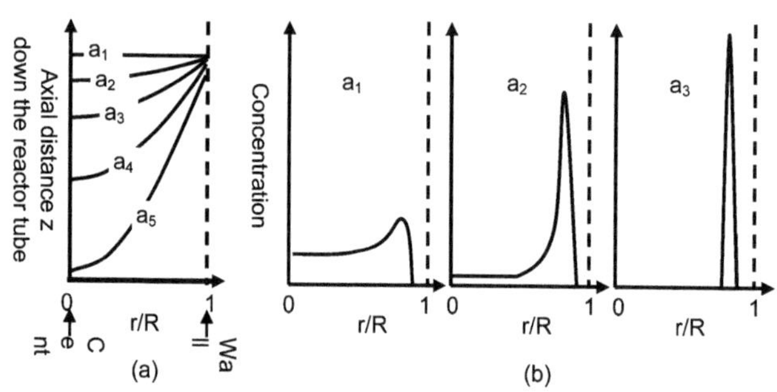

Fig. 3 (a) Distributions of catalyst particle diameters a_i in a parabolic flow profile ($a_1 < a_2 < a_n$) in a reactor tube with diameter $2R$ (adapted from ref. 13). For a given distance down the reactor, a cross section through the particle diameter contours gives the lateral particle size distribution with considerable larger sizes nearer to the reactor walls. The actual particle sizes, given in the Conroy paper are not included here, as the conditions for which they were calculated, especially a ferrocene decomposition temperature of 1000 °C, are not appropriate to the experimental conditions we explore here. (b) Derived concentration distribution of rigid particles of neutral density in a Poiseuille flow profile for $a_1 < a_2 < a_3$ (adapted from ref. 16). The bigger the particles the more distinct is the formation of an annular ring. However, these experiments were executed for particle sizes several orders higher than described here.

(typically at 0.63 times the radius). Their data are summarised in Fig. 3b. For smaller particles the segregation within the central half of the tube is not as pronounced, although the outer 10% of the radius remains essentially particle free. As their experiments were carried out with particles in the 0.1–1 mm range, larger by several orders of magnitude than our iron particles, we would not expect them to be relevant to the reaction described here. However, the concentration of nanotubes into an annular region is often observed in our reactor (we call the effect a 'sock' (SOM1)), so, with Conroy *et al.*, we are reluctant to dismiss the relevance of this segregation effect as established by fluid dynamicists.

Availability of components from precursors. The literature would suggest that ferrocene begins to thermally decompose to liberate iron at around 500 °C,[18,19] but as soon as Fe is available it catalyses the decomposition of further ferrocene (autocatalysis).[14,20] We have evidence of an iron deposit on the reactor tube at the point of reactant injection at 400 °C and thus believe that iron becomes available immediately on injection, whether through the action of this surface deposit or by the Brownian back diffusion of iron nuclei from hotter regions against the flow, is not yet clear.

There is extensive literature on the decomposition of thiophene catalysed by transition metals, although none specific to iron. In all the cases reported the metal surface will cause the breaking of the carbon–sulphur bonds at temperatures even as low as room temperature, while the remaining hydrocarbon fragments are broken down at temperatures below 400 °C.[21–23] The process is utilised to remove sulphur from natural hydrocarbons, with the preferred catalyst being molybdenum. We conclude that in the presence of iron, both as particles or present on the internal surfaces of the reactor, the thiophene will certainly be decomposed at the injection temperature of 400 °C. Hence, sulphur is available once iron has been liberated from the ferrocene. Furthermore, sulphur has long been understood as a growth accelerator[24,25] for carbon filaments on Fe particles, and we find it necessary to grow carbon nanotubes sufficiently quickly to form the coherent aerogel mandatory for the spinning process.

It should also be noted that the carbon component of the decomposition of both ferrocene and thiophene is not insignificant, and may be available for reconstruction on the surface of the growing iron particles. In this context we note a report that it is possible to synthesise carbon nanotubes from ferrocene alone,[18] albeit in an argon atmosphere rather than hydrogen, noting that for ferrocene 10 carbon atoms are supplied for each iron atom. However, while the early carbon from the remains of the thiophene and ferrocene molecule may enable nanotube nucleation, their growth to a length at which they will entangle, and thus form the aerogel capable of being spun into the fibre, will depend on the availability of carbon from the decomposition of the main carbon precursor.

The initial experiments described here depend on methane as the major source of carbon. Methane is known as the most stable hydrocarbon and thermocatalytic breakdown on the Fe particle surface will not even start below a temperature of 680 °C.[26–28] Toluene, which is used as the main carbon source in the final experiment described below, breaks down at a lower temperature, indeed it is routinely used as the carbon source in nanotubes synthesis at temperatures as low as 600 °C, where the metal catalyst is based on a substrate.[29]

Conversions efficiencies. The percentage of carbon converted into fibre is typically of the order of ~2% for the 'standard conditions' spinning. The lost carbon will consist of carbonaceous particles and nanotubes not incorporated into the fibre and removed from the reactor at the gas valve. There is also residual methane in the gas removed, showing that not all the hydrocarbon has reacted.[30] It should be added that the laboratory rig has not been optimised for efficiency of conversion. Perhaps more startling is the fraction of iron, added as ferrocene, which is actually used in growing nanotubes. It is comparatively straightforward to determine the fraction of the iron incorporated within the nanotubes, as iron, or its oxide, after its quantitative analysis by thermogravimetric analysis (TGA) in air. For standard conditions (Fig. 5a) there is some 10% by weight of iron (as opposed to 15% oxide recorded by the TGA) as catalyst residue within the fibre. For a ferrocene injection rate of 0.34 mg min^{-1} (0.10 mg min^{-1} Fe), we find that more than half of the iron injected as ferrocene can be accounted for in the fibre produced (SOM2). Making the reasonable assumption that decomposition of the ferrocene is complete after traversing the hot zone at 1290 °C, then the non-incorporated iron has either been removed by the gas valve in particulate form, or plated onto the inside of the reactor tube. Indeed, iron deposits are seen on the inside of the reactor tube starting level with the injector and then continuing until the temperature of the tube reaches ~1000 °C, where it becomes too sparse to detect. It is interesting to note that in the region of 700–800 °C there is a thin deposit of carbon on top of the iron. The deposit of iron on the inside of the tube underlines the fact that the reactor tube, while not being the substrate for nanotube growth as occurs in lower temperature nanotube synthesis where there is no sulphur in the feedstock, is nevertheless a factor in determining the exact reactor conditions.

However, with much of the iron being incorporated within the fibres, it is instructive to determine the fraction which is actually involved in the nucleation and growth of the long nanotubes comprising the fibre. If we assume that our nanotubes are at least 100 μm long – and we have made an estimate in the past of an upper limit of 1 mm, despite it being a rather difficult experiment to conduct,[3] further that they are on average triple wall and 10 nm in outside diameter,[7] then an estimate of the fractional weight of iron (Fe/(Fe + C)) to catalyse such a tube would be $\sim 5 \times 10^{-4}$. It thus appears that, for an iron content in the fibre of 10%, 99.5% of the iron by mass is not used in the actual generation of nanotubes, but is incorporated within the nanotube fibre as much larger particles often covered with carbon typically as very poorly formed short multiwall CNTs. However, any attempt to significantly reduce the proportion of iron injected causes the process, – in terms of producing a dense 'cloud' of nanotubes with sufficient mechanical integrity to be spun – , to stop working. It thus appears that the nucleation of a nanotube from an iron particle is a chancy affair, with a very low success rate, so it is necessary to have a huge excess of particles available.

An important aspect of the process can be described as 'mechanical fractionation', in that one could expect that only those nanotubes long enough to entangle to give a mechanically coherent aerogel will be capable of being wound out of the reactor and drawn into a fibre. Once the nanotubes qualify, by virtue of spatial density and length, to form an aerogel, they are removed from the reactor mechanically at a rate typically 20 times the mean velocity of the gas flow. However, the presence of a significant particulate content within the fibre, based largely on the excess and redundant iron present, means that this ideal of

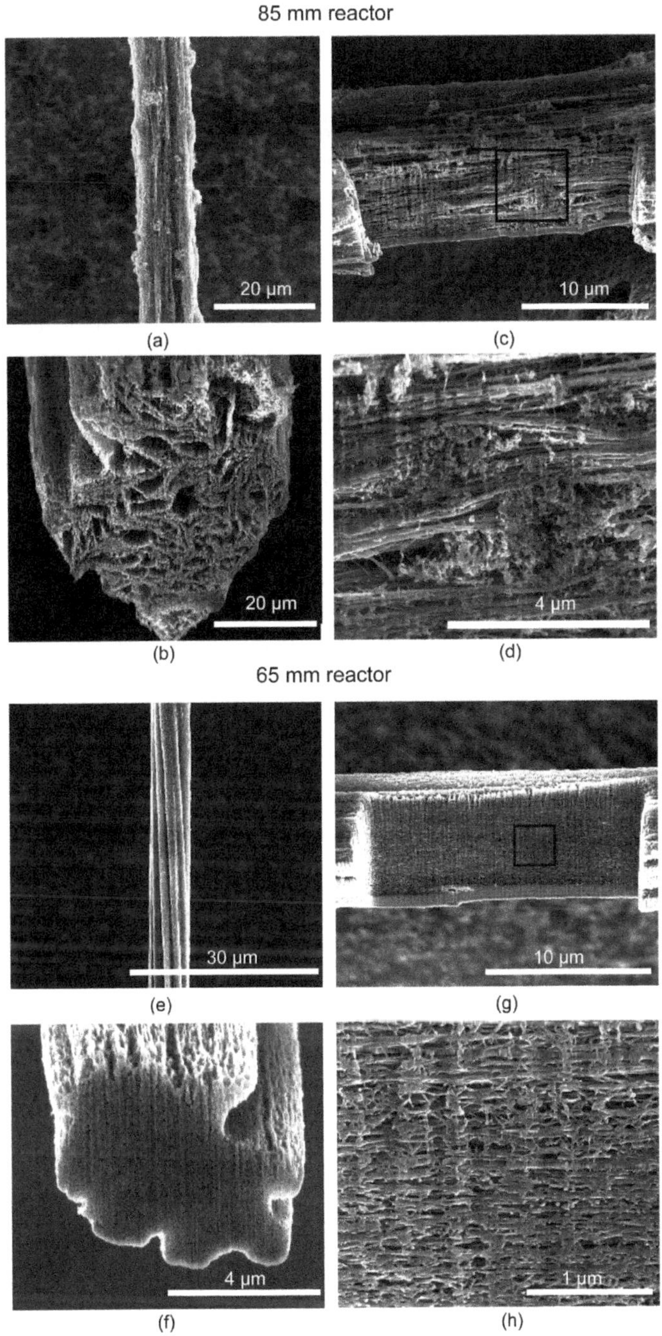

Fig. 4 SEM pictures of (a) the 'standard condition' impure fibre showing the big cluster impurities at the outside. FIB milled cross (b) and longitudinal (c and d) sections show the significant internal voids as the nanotubes 'flow' around the impurity particles. (e) The fibres spun with a steeper velocity profile resulting from the smaller diameter reactor tube show a much cleaner structure with no cluster impurities. The cross-(f) and longitudinal (g and h) sections show the highly improved density and uniformity of the cleaner fibre.

mechanical fractionation is not being achieved. A process development to improve the purity of the fibre will now be described.

Improvement in purity

We observed that if we increased the flow rate of the hydrogen gas, then the particle content of the fibres decreased. However, as these experiments approached the design limitation of the equipment, we achieved a faster flow through the alternative approach of reducing the reactor tube diameter from 85 to 65 mm, while maintaining the same volumetric gas flow rate, so in this case the ratio of hydrogen to reactants was maintained constant. This comparatively modest increase in the velocity through the reactor led to a dramatic reduction in the entrapped particles. This approach also had the additional advantage that, as the volumetric rate was maintained, we did not have to adjust the wind-up rate, which otherwise would have implications for nanotube orientation within the fibre.

SEM images of the standard conditions material already described above (Fig. 4a–d) show a high amount of co-synthesised impurity clusters which can be observed on the outside, while focussed ion beam (FIB) milled cross-sections and longitudinal sections show significant voids associated with embedded impurities. However, the samples spun in the smaller diameter reactor tube (Fig. 4e–h), *i.e.* with a steeper velocity profile, show a dramatically cleaner and denser structure, not disrupted by any big impurities.

Thermogravimetric analysis (TGA) supports the observation of enhanced cleanliness by particle fractionation. The standard conditions material (Fig. 5a), containing a high level of particle impurities, showed typically two combustion events which could be identified as an impurity burning event at ~560 °C and a clean bundles burning event approximately 100 °C higher (SOM3).

In contrast, the material from the smaller diameter reactor shows only one main combustion event around 650 °C (Fig. 5b) which is in the range of the 'clean bundles peak' of the impure fibre, although slightly shifted to lower

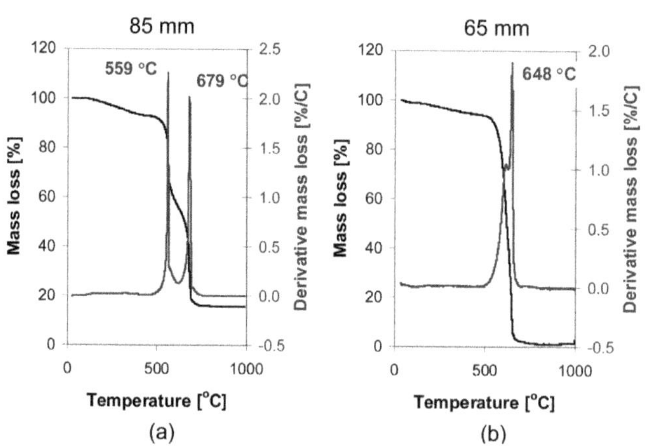

Fig. 5 Thermogravimetric analysis of (a) impure fibre material and (b) clean CNT fibre material, spun with a steeper velocity profile. In the impure material the carbon impurity clusters burn first, due to the highly defective structure and catalysed by intern.

temperatures. Additionally, the iron residue, for the same rate of iron input, is dramatically reduced to only 1.5% compared with >15% before. This order of magnitude decrease reflects the absence of the large Fe particles causing the impurity clusters.

In order to understand why the faster flow means greater purity, we need to consider how these clusters associated with larger iron particles come to be incorporated. We propose that it is the poorly formed multiwall CNTs which grow from their surfaces which provide the necessary opportunity for them to become entangled with the main aerogel and thus pulled from the reactor with the fibre and incorporated into it on condensation. We have observed these larger clusters forming chains on the surface of the fibre (Fig. 6).

For the faster flow in the 65 mm reactor tube the average residence time of the particles in the reaction zone is decreased by 70% (SOM4), which would appear to provide insufficient opportunity for slow moving, larger iron particles to grow their own nanotubes to a length sufficient for them to entangle both with their large neighbours and the main aerogel. In the 'clean' fibre we still observe (Fig. 6c, SOM1), albeit very occasionally, poorly formed nanotubes growing from larger particles, but the nanotubes there are much shorter, and do not appear to form clusters.

The critical role of sulphur

The process is dependent on the addition of sulphur, the role of which in high temperature floating catalyst CNT synthesis has been variously ascribed to:

(a) Acting as a promoter to enhance the addition of carbon atoms to the growing ends of the graphene tubes.[24]

(b) Acting as a surfactant to encourage tube nucleation and thus prevent carbon encapsulation of the catalyst particle. Sulphur is also understood as a

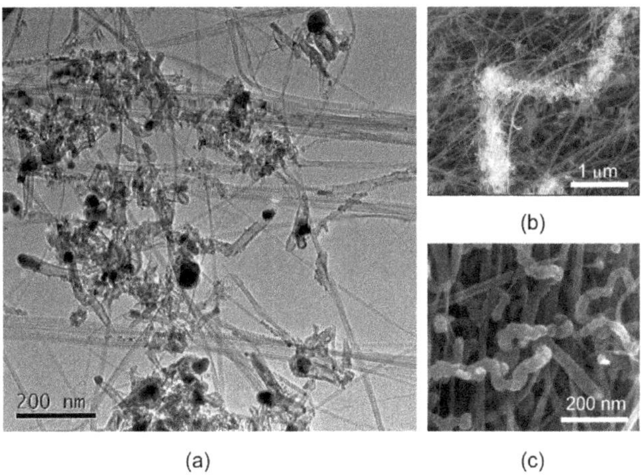

Fig. 6 (a) TEM of poorly formed MWCNTs grown from big Fe particles clustered together in standard material, (b) SEM of chained clusters in uncondensed standard material of several micrometres in length, (c) SEM of occasionally observed deformed but un-clustered tubular structures in clean fibre.

Table 2 Spinning conditions with low sulphur and carbon in a reactor tube with inner diameter of 65 mm

	Flow rates				Total atomic ratio			
	H_2 [l min^{-1}]	Carbon precursor	Ferrocene [mg min^{-1}]	Thiophene [mg min^{-1}]	S : C	S : Fe	Average linear density of fibre [tex]	Mass ratio of CNTs 'out' to total carbon 'in'
Standard chemistry, steeper velocity profile sample	0.8	Methane 80 ml min^{-1}	0.35	1.3	0.004	7.65	0.02	0.015
Early carbon (toluene), low sulphur sample	1.1	Toluene 1.37 mg min^{-1}	1.0	0.03	0.002	0.06	0.035	0.35

surfactant in affecting the growth of graphitic phase during the solidification of cast irons [ref. 25, SOM5].

(c) Limiting the rate at which the iron particles coarsen by collision.[15]

One major challenge in studying the process to make a CNT fibre is that one is operating in multi-dimensional parameter space, and changing any one parameter beyond a limited 'window' while keeping all other parameters constant usually moves out of the continuous spinning window. The switch from methane to toluene as the carbon source, with the same rate of input of carbon, prevented the process from working, with much carbonaceous material being formed, but no spinnable aerogel. We took the view that the greater availability of carbon from toluene as opposed to the much more stable methane, especially at lower temperatures, meant that there was an oversupply. Indeed, on reducing the nominal carbon input rate by a factor of 20, spinning was restored. Under these conditions it was found that the concentration of sulphur could be reduced by a factor as large as 42, while still maintaining stable spinning. The conditions are summarised in Table 2.

When starting the synthesis without any sulphur added, stable spinning could never be achieved. However, if spinning with thiophene ($S : C = 0.02$) and then terminating its injection, stable spinning continued for about 23 min, whereupon it ceased and could not be re-established again. We consider that as sulphur will be present on the inside of the reactor vessel, it takes some time before the equipment is complete purged of that component. We also noted a very considerable carbon build up in the reactor while the rig was operating at zero sulphur. We conclude that continuous spinning of CNT fibres using this process is not possible without the addition of sulphur.

Structural characterization of low-sulphur CNT fibres. The fibres were as clean as those generated using methane in the 65 mm reactor tubes. Fig. 7b and c show a surface view of the uncondensed material and a FIB section through a

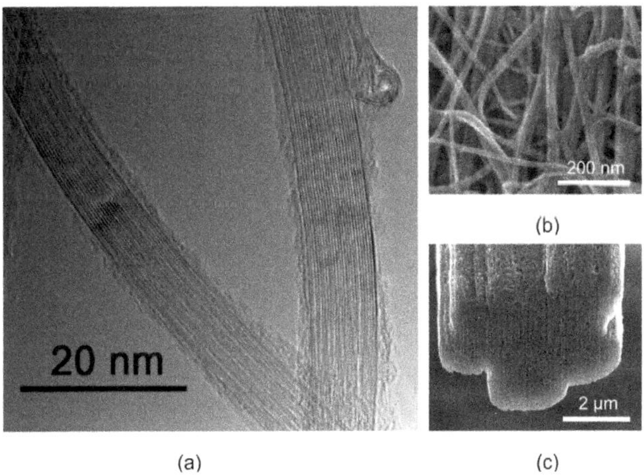

Fig. 7 (a) TEM image of SWCNT bundles, (b) SEM image of a toluene fibre with CNT bundle diameters of 8–27 nm, and (c) the cross section of a condensed fibre cut by focussed ion beam. The outside of the FIB milled fibre looks slightly molten as carbon re-deposits on the surface, whilst the cross section is cut by the ion beam.

condensed fibre. Along with ref. 31,32 we observed that as the amount of sulphur was reduced the mean number of walls in the nanotubes decreased, so that at the low sulphur limit, the nanotubes were predominately single wall.

Fig. 7a shows the single wall nanotubes resolved in the TEM, while Fig. 8a shows the Raman spectra which reveal strong radial breathing mode (RBM) peaks, which are a finger print of single wall nanotubes. Fig. 8b is the equivalent Raman spectrum of the standard conditions fibre discussed above. The D/G ratio, indicating the proportion of badly formed (or edge related) carbon, is much reduced for the low sulphur, single wall sample. The RBMs appear mainly at two wavenumbers, 161 and 252 cm^{-1}, indicating according to $\omega_{RBM} = (223.5\ cm^{-1}\ nm/d_{CNT}) + 12.5\ cm^{-1}$,[33] CNTs of 0.9 to 1.5 nm diameter, which is compatible with the measurements from the TEM. Our model for the critical role played by sulphur in the direct spinning of the CNT fibre draws also from the wider catalysis literature.

A model for the role of sulphur. Sulphur features largely in the literature associated with the catalysis of organic reactions by metal, particularly transition metal, particles. Sulphur has been known as a catalyst poison, where even trace amounts of the element in a hydrocarbon feedstock can drastically reduce the activity of a catalyst.[34,35] However, somewhat more recently it has become clear that distinctly beneficial effects can follow from the partial and well controlled poisoning of metals by sulphur.[25,36] This opens up the possibility that the control of sulphur to provide only partial coverage of the iron catalyst particles in the CNT floating catalyst process may provide distinct benefits.

It is now well established that the diameter of the carbon nanotube is closely related to the diameter of the catalyst particle generating it, at least in the smaller diameter range. In the case of a synthesis from methane, not only will the sulphur be available before carbon, but with eight sulphur atoms available per iron atom (standard conditions), for the case of a 1 nm iron particle where approximately two thirds of the iron atoms are surface atoms, the conditions would appear to be

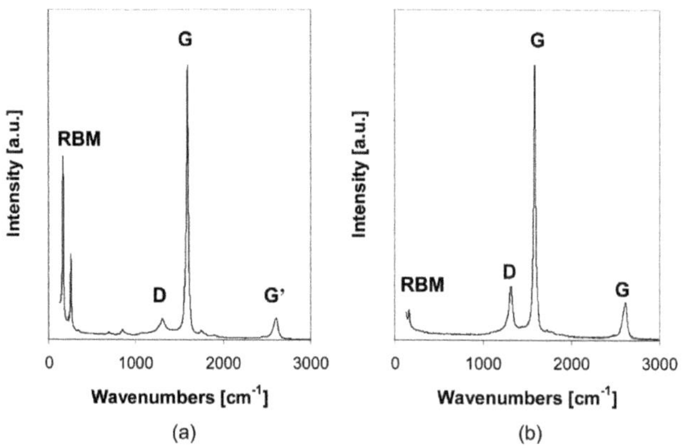

Fig. 8 (a) Raman spectra of toluene fibre showing the typical, very distinct RBM peaks, which correspond to CNTs with diameters of 0.9 and 1.5 nm and a very low D/G ratio of 0.05. (b) Raman spectra of standard material showing a D/G ratio of 0.19; infrequently, RBMs can be observed even in the standard material indicating the occasional SWCNTs, usually however there are no RBMs to be observed.

ripe for the total sulphur poisoning of the catalyst. Setting aside for the moment that fibres were nevertheless produced under these conditions, if we consider the other extreme, that of no sulphur, then we are unable to make nanotubes of sufficient quality and quantity to make a fibre. Carbon products are generated and some nanotubes too, however TEM examination shows a profusion of catalyst particles 'blinded' by a continuous carbon coating (Fig. 9).

The S : Fe ratio, 0.06, at which we obtained a single wall CNT fibre, with a high carbon conversion efficiency, would mean the partial covering (~12%) for iron particles, 1 nm in diameter. These conditions would avoid poisoning and yet mean that sulphur is available in its surfactant role for the early available carbon whether from thiophene, ferrocene or toluene. However, the fact that we were not successful in spinning with low sulphur using methane instead of toluene would imply that it is the combination of critical sulphur delivery and a comparatively early carbon availability which is key to the generation of nanotube nucleation before the iron particles have grown by collision. The process is illustrated in Fig. 10a.

A previous report from this laboratory demonstrated the spinning of single wall CNT fibres using carbon disulphide as the sulphur source.[15] Here, the amount of sulphur added was larger than in this case. However, we note that the catalysed thermolysis of carbon disulphide would liberate available carbon at the same time as sulphur, and thus the key condition of having iron particles partially coated with sulphur may still be met. The single wall CNTs in the fibres generated in this experiment were also predominately 'arm chair' and thus metallic in nature. The single wall material generated here, although showing two or perhaps three predominant RBM peaks in the Raman (Fig. 8a), are not singularly of one chirality. The reason for 'armchair' dominance in the previous case is the subject of on-going studies.

We now need to return to the question as to why, if the recently nucleated iron particles are completed coated and 'blinded' by sulphur, which will occur when the sulphur is available in significant quantities before carbon, as for 'standard conditions', we are successful in growing nanotubes at all, given that they are now of significantly larger diameter with only occasional single wall examples detected.

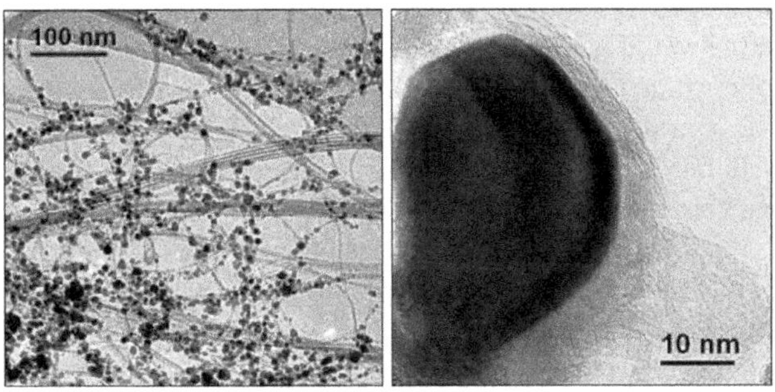

Fig. 9 TEM micrographs of the carbonaceous material generated in the reactor under conditions of zero sulphur where it was impossible to spin fibres. Although there are some nanotubes, the material is dominated by catalyst particles completely encased in graphene layers (courtesy Prof. I. A. Kinloch).

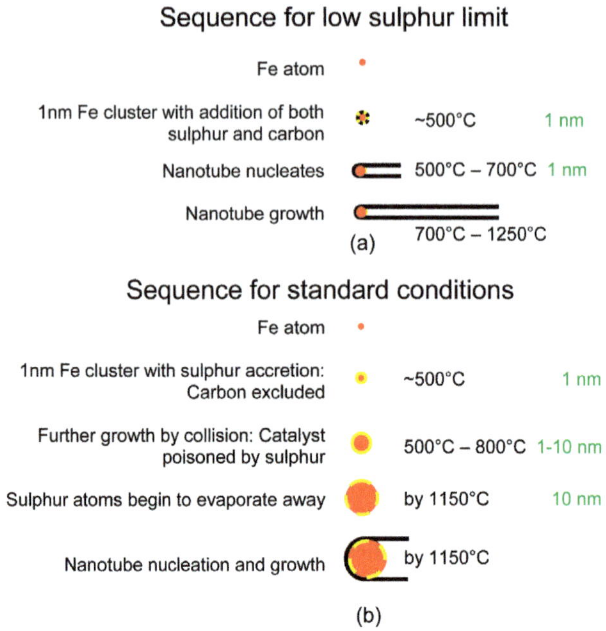

Fig. 10 (a) For low sulphur conditions, the 1 nm Fe clusters (estimated particle diameters in green) experience a balanced arrival of both sulphur and carbon. The nanotubes nucleate, with sulphur preventing carbon encapsulation. Further iron growth by collision is thought to be prevented by the nanotube nucleus. (b) For standard conditions, from the point of availability of sulphur, the sulphur accretes on the 1 nm Fe cluster. Further particle growth by collision can occur despite the sulphur on the surface, the catalytic property is however hemmed by sulphur poisoning. At temperatures above 1000 °C, the sulphur atoms begin to evaporate away, hence, carbon atoms have access to Fe and nucleate a CNT.

An indication of the answer to this question comes from a consideration of the thermodynamic stability of iron sulphide (FeS). The free energy 'Ellingham' diagram (SOM5) can be read to show that for standard conditions, where the maximum possible value of H_2S/H_2 is 4×10^{-4}, hydrogen would extract sulphur from FeS above 1150 °C to form H_2S. This temperature depends on the ratio of sulphur to hydrogen injected into the reactor. For higher sulphur/hydrogen ratios the temperature would increase and *vice versa*. However, as the sulphur is surface adsorbed rather than incorporated within the crystal structure, these temperatures must be regarded as a first estimate.

The possibility of sulphur being lost from a catalyst particle at higher temperatures than that at which it was deposited was also pointed out in a paper from before the CNT era.[25] We thus propose that the larger diameter, multiwall nanotubes which comprise the fibres spun under high sulphur regimes, occur because the growing iron particles, initially poisoned by sulphur, become active, after further growth by collision, when the sulphur begins to be lost to H_2S as the reactants move into the higher temperature regions of the reactor. At this point the active particles are in the presence of available hydrogen, the two conditions for nanotube growth.

The issue of whether sulphur will lower, or even prevent, the coarsening of the iron particles by collision was discussed in ref. 15. While the current observations suggest that coarsening is not prevented completely, it is very possible that sulphur coated particles may not coarsen as rapidly as uncoated ones, simply on the basis that the primary sulphur bonds are satisfied by their contact with iron, so that a sulphur coated particle may be less 'sticky' in bonding terms.

The model for the growth of multiwall nanotubes from the floating catalyst process, involving first complete poisoning by sulphur followed by its partial loss in the presence of available carbon, is shown in Fig. 10b.

Mechanical properties

So far this study has related processing conditions to both purity and the structure of the carbon nanotubes comprising the fibres. However, one major potential application of the CNT fibre will stem from its unique spectrum of mechanical properties. The experiments described above, based on enhanced control of the process, have provided samples which provide new data on the influence of both purity of fibre and diameter of nanotubes on mechanical strength.

Stress strain curves comparing standard conditions fibre (impure), fibre based on similar nanotube structure but with greatly reduced particulate impurities,

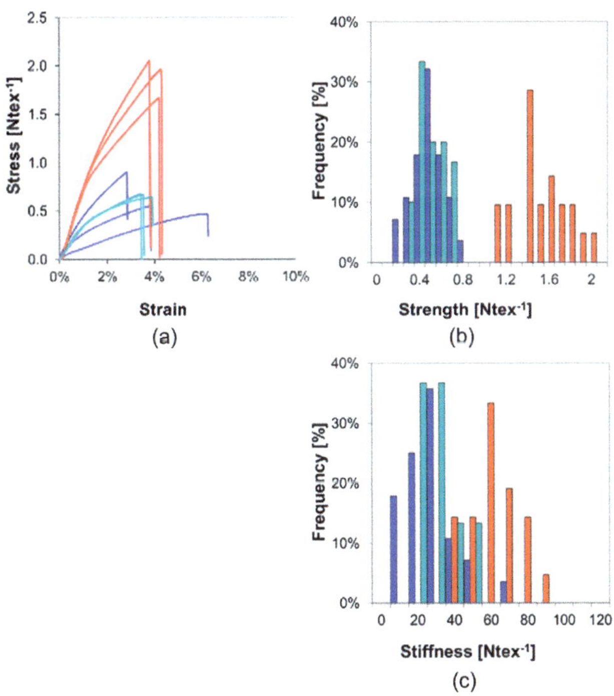

Fig. 11 (a) Selected typical stress–strain curves at 20 mm gauge length are shown for standard conditions (sample 1, dark blue), standard chemistry, but cleaner fibre (sample 2, turquoise), and SWCNT fibre (sample 3, red). (b) and (c) show the average strength and stiffness of 25 tests each. Low sulphur–toluene fibre compared to the standard fibre showed strength and stiffness increased by a factor of 2.8 and 2.5, respectively.

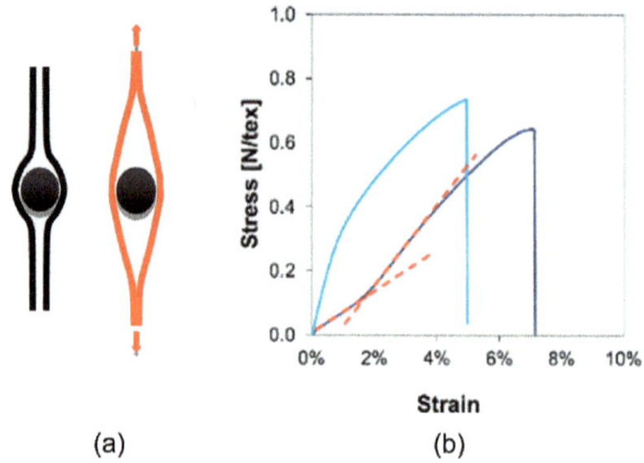

Fig. 12 (a) In a tensile test of fibres containing impurities the stretching of the bundles bent by the impurities can cause an initial region with a less steep slope (dark blue in (b)). Clean fibres (turquoise) are already optimally condensed and aligned, therefore leading to a higher slope of the tensile curve.

and fibre also with low particulate impurities but comprising almost entirely of single wall nanotubes are compared in Fig. 11.

Fig. 11b and c compare the strength and stiffness of the three fibre classes, in each case the values being the averages of 25 stress strain curves. The tests were carried out on samples of gauge length approximately 2000 times the fibre diameter, using a Favimat dedicated fibre tester, in which the linear density of each tensile specimen was measured by detecting the natural vibration frequency.

Comparing first the standard conditions fibre and the much purer derivative (dark blue and turquoise in Fig. 11), it is interesting that the much cleaner fibre is no stronger than the one full of particles. The reasonable expectation, based on virtually all other high strength fibres, that the elimination of second phase (particulate) impurities would increase the strength does not apply in this case. The fact that the aligned yet flexible nanotube bundles ‘flow’ around impurity particles, rather as wood grain does around knots, gives a rationale as to why the particles do not initiate fracture at lower stresses. Interestingly though, there is a marked increase in the initial stiffness on purification. The reason for this is illustrated in Fig. 12. The condensation process produces a structure which wraps around the particle, but is capable of extension on stressing thus reducing the initial elastic modulus. The standard deviation in the values of stiffness was decreased from 46% for to 14% by the elimination of particulate impurities.

The samples produced under conditions of sulphur control which produce single wall nanotubes show markedly different mechanical properties, as apparent from Fig. 11 (red curves). Although the microstructure of the fibre, as observed by SEM, is virtually indistinguishable from the purified version of the standard fibre, the strength is nearly trebled with values approaching 2 Ntex^{-1}. Furthermore, the stiffness is increased by approximately a factor of two. While this result is encouraging in terms of the applicability of the fibre, we do not yet definitely understand the origin of the increased properties. They do not seem to be related to the level of particulate impurities, so we must assume that they are associated with the more efficient sharing of stress in bundles of single wall CNTs compared with the case of larger diameter multiwall tubes, some of which are collapsed.

Conclusions

One of the key aspects of the continuous process to spin carbon nanotube fibres directly from the nanotube synthesis zone is the fact that the nanotubes, as they are synthesised, only qualify for removal as a fibre once they are long enough to entangle and create an aerogel. This 'mechanical fractionation' of the synthesised material should, ideally, ensure that the fibres are entirely of long nanotubes. In general, this condition has not been met. The origin of impurities based on larger catalyst particles growing poorly formed multiwall nanotubes has been identified, firstly with the slow moving component of the Poiseuille flow near to the reactor walls, and then with the engagement of the 'hairy clusters' with the main aerogel (and indeed sometimes with each other), and their resultant removal as unwanted components of the fibre. An increase in gas stream mean velocity for identical injection rates and proportions, has been achieved through the device of reducing reactor tube diameter. It has been demonstrated that the poorly formed clusters are no longer incorporated in the fibre, presumably because their occasionally occurring defective nanotube overgrowths are not sufficiently developed to cause this problem. The reduction in particle count is dramatic, while the amount of residual iron as determined by TGA and the proportion of graphitic material which is defective in terms of a high D/G ratio on the Raman spectrum is also marked. Interestingly though, the very considerable increase in perfection does not lead to a concomitant improvement in fibre strength, although both the stiffness and the variability in this parameter from sample to sample is definitely increased.

It has proved possible to spin fibres consisting almost entirely of single wall CNTs limiting sulphur addition to a critically controlled low level, and also supplying carbon as a precursor (in this case toluene) which has a lower pyrolysis temperature than methane. Again the fibres do not contain extraneous particulate clusters, and the residual iron content is low. We determined also that it is impossible to spin fibre in the complete absence of sulphur, although it is important to completely purge the reactor from residual sulphur otherwise one might be initially misled. The fact that nanotubes could be nucleated from iron particles before they had grown to beyond 1 nm diameter by collision is explained in terms of the precepts of the catalyst science, where partial coating of metal catalyst by sulphur enhances activity while complete coverage effectively passivates the particle, as indeed does complete encapsulation by a graphene shell (at least in the case of floating catalysts). This model, however, needs to be applicable to standard conditions fibres. Here we propose a model based on the premise that where sulphur is added in larger amounts the metal particles are initially fully passivated by sulphur, while continuing to grow by collision (albeit possibly at reduced rate). However, as the reactants move into the higher temperature zone of the reactor, the sulphur is predicted to be lost as H_2S, exposing at a critical juncture partially sulphur covered iron particles to available carbon, again achieving the conditions for nanotube nucleation and growth. The mechanical properties of the single wall CNT fibres are greatly improved compared with the multiwall 'standard' material, with the mean strength of the samples increased by nearly a factor of 3. The exact reasons for this marked and extremely useful enhancement are still being explored.

Acknowledgements

The authors wish to thank the *U. S. Office of Naval Research Global*, and *Innovative Material Synergies and Composite Processing Strategies* for funding.

Notes and references

1 Y.-L. Li, I. A. Kinloch and A. H. Windle, *Science*, 2004, **304**, 276.
2 M. Motta, Y.-L. Li, I. Kinloch and A. Windle, *Nano Lett.*, 2005, **5**, 1529.
3 K. Koziol, J. Vilatela, A. Moisala, M. Motta, P. Cunniff, M. Sennett and A. Windle, *Science*, 2007, **318**, 1892.
4 J. J. Vilatela and A. H. Windle, *Adv. Mater.*, 2010, **22**, 4959.
5 B. G. Demczyk, Y. M. Wang, J. Cumings, M. Hetman, W. Han, A. Zettl and R. O. Ritchie, *Mater. Sci. Eng.*, 2002, **A334**, 173.
6 M.-F. Yu, O. Lourie, M. J. Dyer, K. Moloni, T. F. Kelly and R. S. Ruoff, *Science*, 2000, **287**, 637.
7 M. Motta, A. Moisala, I. A. Kinloch and A. H. Windle, *Adv. Mater.*, 2007, **19**, 3721.
8 A. B. Dalton, S. Collins, J. Razal, E. Munoz, V. H. Ebron, B. G. Kim, J. N. Coleman, J. P. Ferraris and R. H. Baughman, *J. Mater. Chem.*, 2004, **14**, 1.
9 N. Behabtu, C. C. Young, D. E. Tsentalovich, O. Kleinerman, X. Wang, A. W. K. Ma, E. A. Bengio, R. F. Waarbeek, J. J. de Jong, R. E. Hoogerwerf, S. B. Fairchild, J. B. Ferguson, B. Maruyama, J. Kono, Y. Talmon, Y. Cohen, M. J. Otto and M. Pasquali, *Science*, 2013, **339**, 182.
10 M. Zhang, K. R. Atkinson and R. H. Baughman, *Science*, 2004, **306**, 1358.
11 M. D. Lima, S. Fang, X. Lepró, C. Lewis, R. Ovalle-Robles, J. Carretero-González, E. Castillo-Martínez, M. E. Kozlov, J. Oh, N. Rawat, C. S. Haines, M. H. Haque, V. Aare, S. Stoughton, A. A. Zakhidov and R. H. Baughman, *Science*, 2011, **331**, 51.
12 H. G. Chae and S. Kumar, *Science*, 2008, **319**, 908.
13 D. Conroy, A. Moisala, S. Cardoso, A. Windle and J. Davidson, *Chem. Eng. Sci.*, 2010, **65**, 2965.
14 K. Kuwana and K. Saito, *Carbon*, 2005, **43**, 2088.
15 R. M. Sundaram, K. K. Koziol and A. H. Windle, *Adv. Mater.*, 2011, **23**, 5064.
16 (*a*) G. Segre and A. Silberberg, *Nature*, 1961, **189**, 209; (*b*) G. Segre and A. Silberberg, *J. Fluid Mech.*, 1962, **14**, 136.
17 R. V. Repetti and E. F. Leonard, *Nature*, 1964, **203**, 1346.
18 A. Barreiro, S. Hampel, M. H. Rulmmeli, C. Kramberger, A. Grüneis, K. Biedermann, A. Leonhardt, T. Gemming, B. Bulchner, A. Bachtold and T. Pichler, *J. Phys. Chem. B*, 2006, **110**, 20973.
19 K. Elhin, *Dissertation "Synthesis of carbon-coverediron nanoparticles by photolysis of ferrocene"*, University Uppsala, 2002.
20 L. M. Dyagileva, V. P. Mar'in, E. I. Tsyganova and G. A. Razuvaev, *J. Organomet. Chem.*, 1979, **175**, 63.
21 S. G. Gagarin, Y. A. Teterin and Y. V. Plekhanov, *Theor. Exp. Chem.*, 1989, **25**, 335.
22 F. Zaera, E. B. Kollin and J. L. Gland, *Surf. Sci.*, 1987, **184**, 75.
23 M. Zonnevylle, R. Hoffmann and S. Harris, *Surf. Sci.*, 1988, **199**, 320.

24 G. G. Tibbetts, C. A. Bernardo, D. W. Gorkiewicz and R. L. Alig, *Carbon*, 1994, **32**, 569.
25 M. S. Kim, N. M. Rodriguez and R. T. K. Baker, *J. Catal.*, 1993, **143**, 449.
26 M. A. Ermakova, D. Y. Ermakov, A. L. Chuvilin and G. G. Kuvshinov, *J. Catal.*, 2001, **201**, 183.
27 T. V. Reshetenko, L. B. Avdeeva, Z. R. Ismagilov, V. A. Ushakov, A. L. Chuvilin and Y. T. Pavlukhin, *Chem. Sustainable Dev.*, 2003, **11**, 239.
28 A. Konieczny, K. Mondal, T. Wiltowski and P. Dydo, *Int. J. Hydrogen Energy*, 2008, **33**, 264.
29 X. Ma, A. A. Lall, N. Aronhime and M. R. Zachariah, *Int. J. Hydrogen Energy*, 2010, **35**, 7476.
30 M. Motta and A. Moisala, 2008, unpublished report.
31 W. Ren, F. Li, S. Bai and H.-M. Cheng, *J. Nanosci. Nanotechnol.*, 2006, **6**, 1339.
32 V. Reguero, B. Alemán, B. Mas and J. J. Vilatela, *Chem. Mater.*, 2014, **26**, 3550.
33 S. M. Bachilo, M. S. Strano, C. Kittrell, R. H. Hauge, R. E. Smalley and R. B. Weisman, *Science*, 2002, **298**, 2361.
34 R. A. Dalla Betta, A. G. Piken and M. J. Shelef, *J. Catal.*, 1975, **40**, 173.
35 J. Oudar, *Catal. Rev. Sci. Eng.*, 1980, **22**, 171.
36 C. H. Bartholomew, G. D. Weatherbee and G. A. Jarvi, *J. Catal.*, 1979, **60**, 257.

Faraday Discussions

PAPER

Boundary layer chemical vapour synthesis of self-organised ferromagnetically filled radial-carbon-nanotube structures

Filippo S. Boi,[ab] Rory M. Wilson,[c] Gavin Mountjoy,[d] Muhammad Ibrar[a] and Mark Baxendale*[a]

Received 17th April 2014, Accepted 13th May 2014
DOI: 10.1039/c4fd00071d

Boundary layer chemical vapour synthesis is a new technique that exploits random fluctuations in the viscous boundary layer between a laminar flow of pyrolysed metallocene vapour and a rough substrate to yield ferromagnetically filled radial-carbon-nanotube structures departing from a core agglomeration of spherical nanocrystals individually encapsulated by graphitic shells. The fluctuations create the thermodynamic conditions for the formation of the central agglomeration in the vapour which subsequently defines the spherically symmetric diffusion gradient that initiates the radial growth. The radial growth is driven by the supply of vapour feedstock by local diffusion gradients created by endothermic graphitic-carbon formation at the vapour-facing tips of the individual nanotubes and is halted by contact with the isothermal substrate. The radial structures are the dominant product and the reaction conditions are self-sustaining. Ferrocene pyrolysis yields three common components in the nanowire encapsulated by multiwall carbon nanotubes, Fe_3C, α-Fe, and γ-Fe. Magnetic tuning in this system can be achieved through the magnetocrystalline and shape anisotropies of the encapsulated nanowire. Here we demonstrate proof that alloying of the encapsulated nanowire is an additional approach to tuning of the magnetic properties of these structures by synthesis of radial-carbon-nanotube structures with γ-FeNi encapsulated nanowires.

1 Introduction

Ferromagnetic transition-metal-based nanowires encapsulated by carbon nanotubes have a long history of investigation owing to a plethora of potential applications.[1–9] The carbon nanotube serves to chemically passivate and minimise

[a]School of Physics and Astronomy, Queen Mary University of London, Mile End Road, London E1 4NS, UK. E-mail: m.baxendale@qmul.ac.uk
[b]School of Physical Science and Technology, Sichuan University, 29 Wangjiang Road, Chengdu, 610061China
[c]School of Materials Science and Engineering, Queen Mary University of London, Mile End Road, London E1 4NS, UK
[d]School of Physical Sciences, University of Kent, Canterbury CT2 7NH, UK

mechanical degradation of the central nanowire. The magnetic functionality of these structures can be tuned through the shape and magnetocrystalline anisotropies of the nanowire, which in turn can be controlled by process parameters.[10] These structures self-organise on smooth, inert substrates exposed to a steady-state vapour flow comprising the products of metallocene ($M(C_5H_5)_2$, where M is a transition metal) pyrolysis at high temperatures in a conventional CVD reactor. The details of the formation mechanism are controversial; descriptions usually evoke the same vapour-, liquid-, and solid-phase chemistry and catalysis at transition-metal surface models that underpin production methods for unfilled carbon nanotubes by catalytic CVD.[11–14] The elemental iron from the products of ferrocene pyrolysis ($Fe + H_2 + CH_4 + C_5H_6 + \ldots$) forms a particle on the inert substrate which then reacts with the local hydrocarbon environment. The carbon-to-metal ratio in this local environment is carbon-rich relative to those used in single-wall nanotube production and therefore the outcome is growth of a multiwall carbon nanotube (MWCNT). The subsequent growth of the MWCNT and the filling of its central capillary by elemental iron or an iron-carbide is driven by the supply of species from the vapour either to the base of the structure or to the open tip, or both. This is an entirely self-organised production process. It is important to note that the nucleation of the MWCNT growth is heterogeneous, *i.e.* occurs at the iron-particle/substrate interface. The final reaction product is a closely-packed array of individual MWCNT-encapsulated single-crystal nanowires oriented perpendicular to the substrate. The most commonly observed encapsulated crystal structures are Fe_3C, α-Fe, and γ-Fe.[1–9] The surprising observation of γ-Fe, which is a high-temperature, high-pressure phase in bulk, has been attributed to frustration of a relaxation to α-Fe due to confinement by the elastic MWCNT.

The landmark publication by Boi *et al.* demonstrated vapour-phase, homogeneous nucleation of equivalent structures through use of intentionally roughened substrates to produce powders of spherically symmetric, radial filled-carbon-nanotube structures departing from a central particle,[15] Fig. 1 and 2.

Detailed analysis of the central particle revealed it to be an agglomeration of smaller Fe_3C nanocrystals encapsulated by spherical graphitic-carbon shells so it is more correctly referred to as the core-agglomeration. The primary significance of this work was that the role of the roughened substrate was to intentionally introduce fluctuations in the flowing vapour and in so doing create thermal pathways within the reactor capable of a reproducible yield of a single, dominant reaction product that nucleates and grows entirely in the vapour.

An early vapour-composition/product-morphology mapping in transition-metal/carbon systems by Elliott *et al.* identified the local vapour-phase carbon-to-metal ratio in a given thermal trajectory as the significant factor which determines the product composition/morphology.[16] Since the thermal trajectories in the boundary layer chemical vapour synthesis are defined by fluctuation-induced nucleation and the carbon-to-metal ratio by the selection of the decomposed molecule, we propose the method as a route to the controlled production of transition-metal/carbon structures with unusual composition and morphologies.

Ferromagnetically filled MWCNTs have a wide range of potential applications owing to the ability to tune the magnetic properties through the control of composition, magnetocrystalline- and shape-anisotropy of the encapsulated single-crystal nanowires: magnetic storage media and quantum disk fabrications

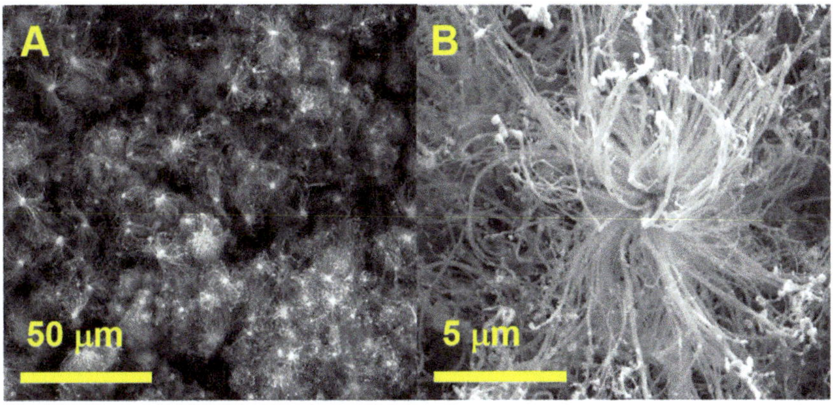

Fig. 1 Scanning electron micrographs, (A) an ensemble of radial structures comprising iron-based nanowires encapsulated by multiwall carbon nanotubes departing from a central particle deposited on a roughened quartz substrate under an Ar/ferrocene vapour flow in a conventional horizontal CVD reactor at 990 °C, and (B) detail of an individual structure.

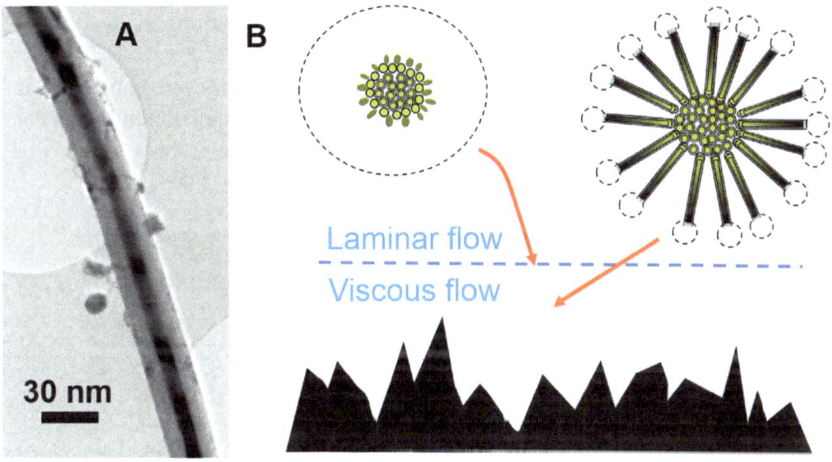

Fig. 2 (A) Transmission electron micrograph of a typical individual radial multiwall carbon nanotube, the nanowire is typically continuous for several micrometres. (B) A schematic diagram depicting the principles of boundary layer chemical vapour synthesis; the pyrolysed metallocene flow is horizontal over the roughened substrate (black); spherically encapsulated nanocrystals rapidly nucleate and agglomerate at the boundary between laminar and viscous flow (dotted black lines represent the symmetry of diffusion gradients); peripheral nanocrystals are elongated by the supply of feedstock species from the incoming vapour, these nucleate the growth of the radial nanotubes; the growth of the radial structure is driven by the diffusion gradient at the nanotube tips.

(owing to the high coercivity displayed by single magnetic domain ferromagnetic systems); drug delivery and nanoscale containers for biomedical therapy, monitoring and diagnosis (these exploit the high mechanical strength and hysteretic dissipation in presence of a time varying magnetic field); magnetic force microscopy probes; nanocomposite magnetic filler particles; buckypaper; microwave absorption materials, and nanoscale inductors.[17–30] In addition some applications exploit the torque in a constant magnetic field, or the oscillatory motion in a time-varying magnetic field.[23,25]

The principle reasons for considering radial structures, rather than the counterpart produced by conventional CVD, for these applications are: (i) the method of synthesis is conducive to bulk production, (ii) the nanowires are continuous on the micrometre scale (at least one order of magnitude greater than can be achieved by conventional methods), and (iii) the morphology facilitates ease of dispersion in polymeric matrices.

The filling of nanotubes with Fe–Ni or Co–Fe alloys has been recently shown to increase the degree to which these systems can be magnetically tuned, opening new ways of controlling the Curie temperature, coercivity, saturation magnetization, and facilitating the fabrication of nanotube-based spin-valve prototypes.[31–38] Much of this interest has focused on γ-FeNi alloys. Sufficient control of the Fe content could allow tuning the Curie temperature to the ablation temperature for magnetic hyperthermia cancer therapy (~45 °C).[39,40] The central difficulty prohibiting the translation of prototypes to industrial production is the lack of achievable compositional and dimensional control with existing CVD methods. Ultimately, this poor control is a manifestation of the heterogeneous nucleation (metal/substrate interaction) and natural fluctuations in steady-state diffusion gradients.

Here we demonstrate the principle of production of alloyed nanowires in radial structures by boundary layer chemical vapour synthesis. Radial γ-FeNi alloy-filled-MWCNTs were synthesised in vapour produced by the thermal

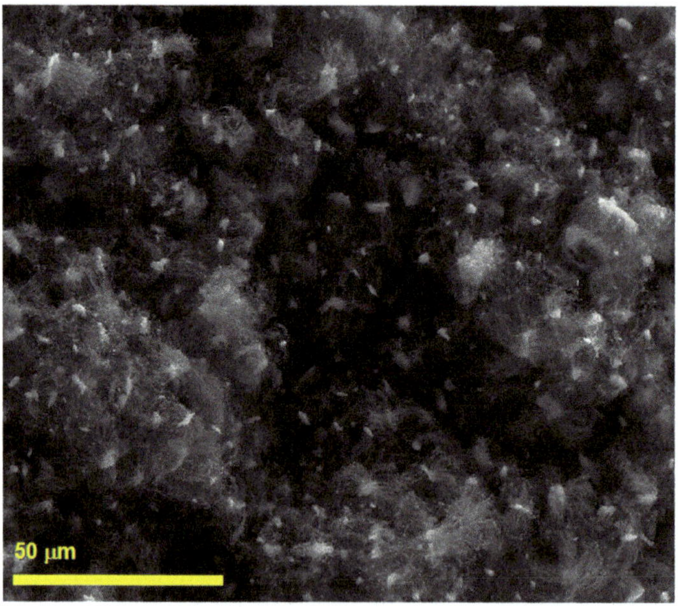

Fig. 3 Scanning electron micrograph of an ensemble of the radial structures deposited on a roughened quartz substrate under an Ar/ferrocene–nickelocene vapour flow in a conventional horizontal CVD reactor at 990 °C.

decomposition of ferrocene and nickelocene. Through X-ray diffraction, scanning electron microscopy, and transmission electron microscopy we obtain the composition and the morphology of these structures.

2 Experimental procedure

2.1 Synthesis

A mixed powder of 30 mg of ferrocene and 70 mg of nickelocene was sublimated using a coil preheater (maximum temperature ~180 °C) in a 12.5 ccm argon flow upstream of a horizontal CVD reactor comprising a quartz tube (length: 2 m, outer diameter: 22 mm, inner diameter: 19 mm, wall thickness: 1.5 mm) positioned inside a one-zone electrical furnace with the isothermal reaction zone at 990 °C. The roughened substrates (100 μm peak-to-valley fractal roughness) were placed in the isothermal reaction zone. The duration of the reaction was 4.5 minutes. The reactor was cooled to room temperature at the natural rate of the furnace.

2.2 Characterisation

Scanning electron microscopy (SEM) was performed using an FEI Inspect F microscope. X-ray diffraction (XRD) analyses were performed with both a Siemens D5000 and Xpert-Pro diffractometers (both with Cu K_α source). The samples were prepared by evaporation of the alcohol in which the structures were previously

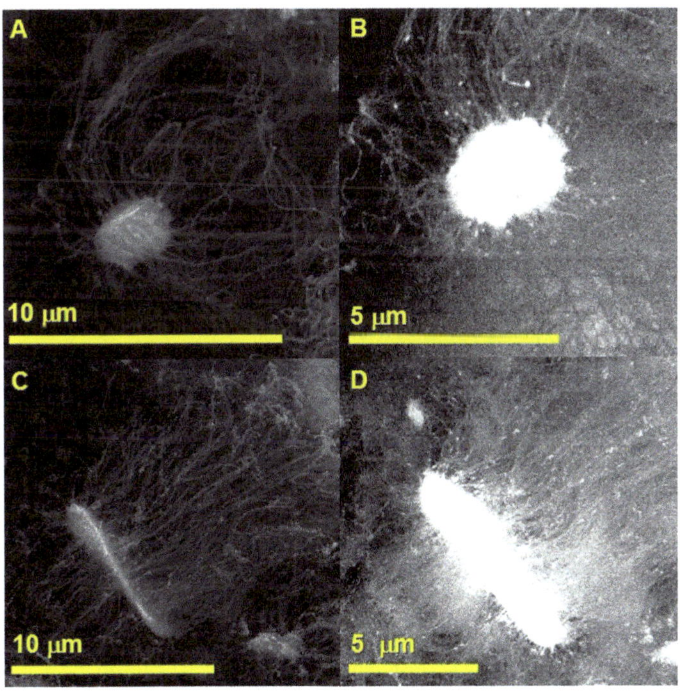

Fig. 4 (A) and (C) scanning electron micrographs of individual structures. (B) and (D) backscattered electron micrographs in individual structures. (C) and (D) are an examples of elongated core-agglomerations, the multiwall carbon nanotubes depart from the flattened faces.

dispersed. The diffractograms were analysed using the ICSD chemical database cards for the commonly observed products of the ferrocene and nickelocene pyrolysis. The Rietveld refinement method, which uses the least-squares approach to match a theoretical line profile to the diffractogram, was used to identify and estimate the relative abundances of the encapsulated phases from the area enclosed by the diffraction peaks. The oxides Fe_3O_4 and NiO result from spontaneous oxidation of non-encapsulated iron or nickel when the sample is handled in air (confirmed by electron microscopy). Transmission electron microscopy (TEM) was performed by using a 200 kV Jeol Jem 2010 with a LaB_6 filament. The TEM samples were prepared by transferring the structures to carbon-coated copper grids. The magnetisation measurements were performed at 5 K with a MPMS-7 magnetometer (Quantum Design) on powder extracted from the reactor.

3 Results and discussion

Scanning electron micrographs of the substrates extracted from the reactor revealed thick deposits comprising ensembles of individual radial structures, Fig. 3 and 4. In some cases the central core-agglomeration is elongated and flattened, relative to those observed in ferrocene-only syntheses, and the MWCNTs grow mainly from the flattened faces, Fig. 4C and D.

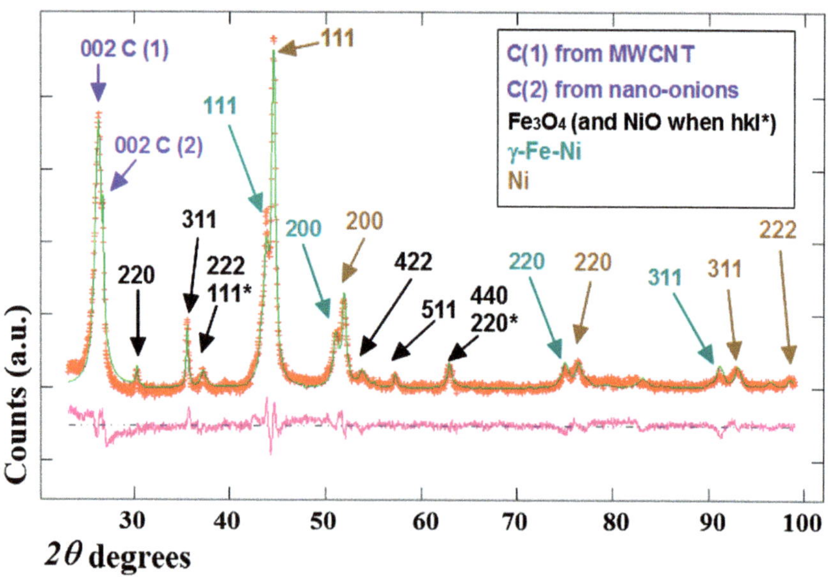

Fig. 5 XRD diffractogram (red) and Rietveld refinement (green) from the powder extracted from the reactor. The violet line corresponds to the difference between the diffractogram data and the refinement. The diffraction peaks indicated by the arrows are labelled with the indices of the corresponding crystallographic reflection from the element or compound indicted in the key. C1 and C2 refers to the two species of graphitic carbon observed on the shell of the encapsulated nanocrystals (here referred to as 'nano-onions') and multiwall carbon nanotubes.[15]

XRD analyses of the deposit revealed the presence of γ-FeNi and elemental Ni; Fig. 5. The relative abundance of the phases obtained from this data was 7.4% γ-FeNi (space group $Fm\bar{3}m$), 16.2% Ni (space group $Fm\bar{3}m$), 67.4% C(1) (space group $P6_3/mmc$), 3.4% Fe_3O_4 ($Fd\bar{3}mz$), 3.0% NiO (space group $Fm\bar{3}m$) and 2.6% C(2) (space group $P6_3/mmc$). The presence of oxides is the result of the oxidation of traces of elemental metal on the exterior of the structures due to post-synthesis handling in air.

No evidence for presence of Fe_3C and Ni_3C phases was found. This is surprising since Fe_3C and Ni_3C are considered essential intermediates for

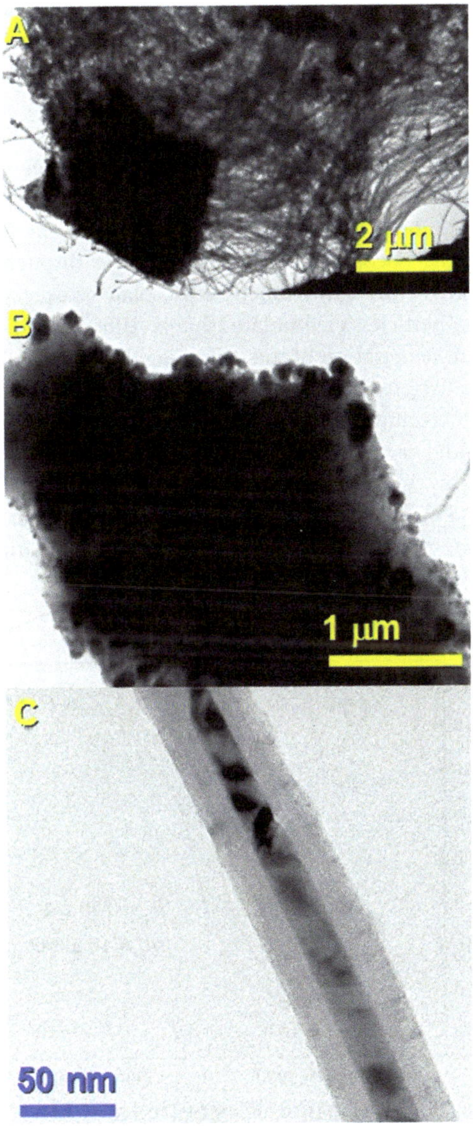

Fig. 6 Transmission electron micrographs, (A) an individual radial structure, (B) a detail of a core-agglomeration, (C) detail of a typical MWCNT-encapsulated nanowire.

MWCNT growth.[41–50] Previous reports of this synthesis using conventional solid- and liquid-source CVD do not comment on the absence of Ni_3C and Fe_3C in the final product nor attempt to describe the growth mechanism.[31–38] The absence of Fe_3C and Ni_3C diffraction peaks in Fig. 5 suggests that there is direct formation of FeNi alloys without the intermediate metal carbide formation, or MWCNT growth is driven by the carbon supply from an unstable carbide that decomposes at high rate.

TEM micrographs revealed the core-agglomeration to be composed of approximately spherical nanoparticles, while the MWCNTs emerge from elongated particles on the periphery of the core, Fig. 5. Individual spherical nanoparticles were found to have diameters in the range 500–1000 nm. Energy dispersive X-ray analysis of a low number of spherical particles confirmed the presence of either Ni or FeNi central nanocrystals. In view of the central-agglomeration formation process described above, this is likely to be a consequence of local vapour composition variation in the fluctuating eddies. Similar analysis of the encapsulated nanowires revealed local regions of either Ni or FeNi; measurement of the spatial extent of these regions requires further work. We speculate that the observed elongation and flattening of the central agglomeration (Fig. 4C and D and 6A) is due either (i) elongation of the fluctuating eddies due to lowering of the viscosity of the boundary layer through multiple metallocene vapour usage, or (ii) a micrometre-scale platelet exists at the centre on which the spherical particles accumulate (this is difficult to quickly confirm by TEM since the beam penetration of the central-agglomeration is quite low).

Fig. 7 shows the hysteresis loop for a powder sample of the radial structures; the saturation magnetization is 18.5 emu g^{-1} and the coercivity is 500 Oe at 5 K. The latter value is higher than those previously measured for counterpart structures produced by steady-state CVD using dichlorobeneze (90 Oe), trichlorobenzene (415 Oe), and benzene (214 Oe) precursors.[33,37] Similar comparisons of the saturation magnetisation value are difficult since the

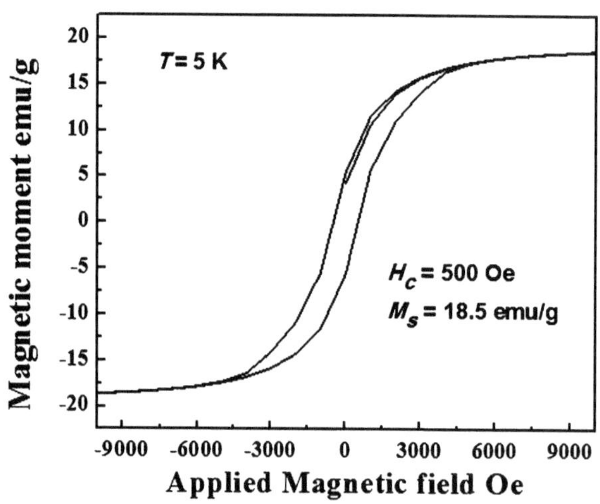

Fig. 7 Ferromagnetic hysteresis loop for a powder sample of material extracted from the reactor indicating the coercive field H_c, and saturation magnetisation M_s.

ferromagnetic oxide and diamagnetic MWCNT contributions are rarely considered. The corrected saturation magnetization value of the ferromagnetic fraction (30%) of this sample (*i.e.* the γ-FeNi, Ni, Fe_3O_4, NiO components) was estimated to be 61.7 emu g^{-1} by subtracting the diamagnetic contribution of MWCNTs (−0.7 emu g^{-1} at 50 kOe) from the measured value.[51]

4 Conclusion

Boundary layer chemical vapour synthesis using ferrocene and nickelocene sources was successfully employed to synthesise radial structures comprising ferromagnetic alloy nanowires encapsulated by multiwall carbon nanotubes departing from a central agglomeration. This result is proof that alloying of the encapsulated nanowire is an additional approach to tuning of magnetic properties of these structures.

Acknowledgements

We acknowledge the financial support from the South-East Physics Network. We also acknowledge Geoff Gannaway and Ken Scott for the help in the preparation of some parts of the reactor, Richard Thorogate for the help in the magnetic measurements and George Nevill for his help in the preparation of the quartz substrates.

References

1 R. Sen, A. Govindaraj and C. N. R. Rao, *Chem. Phys. Lett.*, 1997, **267**, 276–280.

2 C. N. R. Rao, R. Sen, B. C. Satishkumar and A. Govindaraj, *Chem. Commun.*, 1998, 1525–1526.

3 J. F. Marco, J. R. Gancedo, A. Hernando, P. Crespo, C. Prados, J. M. González, N. Grobert, M. Terrones, D. R. M. Walton and H. W. Kroto, *Hyperfine Interact.*, 2002, **139–140**, 535–542.

4 U. Weissker, S. Hampel, A. Leonhardt and B. Büchner, *Materials*, 2010, **3**, 4387–4427.

5 C. Prados, P. Crespo, J. M. González, A. Hernando, J. F. Marco, R. Gancedo, N. Grobert, M. Terrones, D. R. M. Walton and H. W. Kroto, *Phys. Rev. B: Condens. Matter*, 2002, **65**, 113405–113409.

6 D. Golberg, M. Mitome, Ch. Müller, C. Tang, A. Leonhardt and Y. Bando, *Acta Mater.*, 2006, **54**, 2567–2576.

7 C. Müller, D. Golberg, A. Leonhardt, S. Hampel and B. Büchner, *Phys. Status Solidi A*, 2006, **203**, 1064–1068.

8 U. Weissker, M. Löffler, F. Wolny, M. U. Lutz, N. Scheerbaum, R. Klingeler, T. Gemming, T. Mühl, A. Leonhardt and B. Büchner, *J. Appl. Phys.*, 2009, **106**, 054909–054914.

9 M. U. Lutz, U. Weissker, F. Wolny, C. Müller, M. Löffler, T. Mühl, A. Leonhardt, B. Büchner and R. Klingeler, *J. Phys.: Conf. Ser.*, 2010, **200**, 072062.

10 N. Grobert, W. K. Hsu, Y. Q. Zhu, J. P. Hare, H. W. Kroto, D. R. M. Walton, M. Terrones, H. Terrones, P. Redlich, M. Ruhle, R. Escudero and F. Morales, *Appl. Phys. Lett.*, 1999, **75**, 3363–3365.

11 R. S. Wagner and W. C. Ellis, *Appl. Phys. Lett.*, 1964, **4**, 89–90.

12 A. R. Harutyunyan, O. A. Kuznetsov, C. J. Brooks, E. Mora and G. Chen, *ACS Nano*, 2009, **3**, 379–385.
13 A. Moisala, A. G. Nasibulin and E. I. Kauppinen, *J. Phys.: Condens. Matter*, 2003, **15**, S3011–S3035.
14 M. Kumar and Y. Ando, *J. Nanosci. Nanotechnol.*, 2010, **10**, 3739–3758.
15 F. S. Boi, G. Mountjoy and M. Baxendale, *Carbon*, 2013, **64**, 516–526.
16 B. R. Elliott, J. J. Host, V. P. Dravida, M. H. Teng and J.-H. Hwang, *J. Mater. Res.*, 1997, **12**, 3328–3344.
17 S. Philippi, U. Weißker, T. Muhl, A. Leonhardt and B. Buchner, *J. Appl. Phys.*, 2011, **110**, 084319.
18 M. U. Lutz, K. Lipert, Y. Krupskaya, S. Bahr, A. Wolter and A. El-Gendy, *Carbon Nanotubes for Biomedical Applications*, Springer (Berlin Heidelberg), 2011, pp. 97–124.
19 A. Taylor, K. Lipert, K. Kramer, S. Hampel, S. Fussel and A. Meye, *J. Nanosci. Nanotechnol.*, 2009, **9**, 709–716.
20 M. A. A. Elhissi, W. Ahmed, I. U. Hassan, V. R. Dhanak and A. D'Emanuele, *J. Drug Delivery*, 2012, **2012**, 837327.
21 B. Sitharaman and L. J. Wilson, *Int. J. Nanomed.*, 2006, **1**, 291–295.
22 F. Wolny, T. Muhl, U. Weissker, K. Lipert, J. Schumann and A. Leonhardt, *Nanotechnology*, 2010, **21**, 435501.
23 S. Hudziak, A. Darfeuille, R. Zhang, T. Peijs, G. Mountjoy, G. Bertoni and M. Baxendale, *Nanotechnology*, 2010, **21**, 125505.
24 H. Lin, H. Zhu, H. Guo and L. Yu, *Mater. Lett.*, 2007, **61**, 3547–3550.
25 S. Noriaki, N. Masakazu and K. Takeyuki, *Carbon*, 2007, **45**, 78–82.
26 P. C. P. Watts, W. K. Hsu, V. Kotzeva and G. Z. Chen, *Chem. Phys. Lett.*, 2002, **366**, 42–50.
27 P. C. P. Watts, W. K. Hsu, D. P. Randall, V. Kotzeva and G. Z. Chen, *Chem. Mater.*, 2002, **14**, 4505–4508.
28 P. C. P. Watts and W. K. Hsu, *Appl. Phys. A: Mater. Sci. Process.*, 2004, **78**, 79–83.
29 H. Terrones, F. López-Urías, E. Munoz-Sandoval, J. A. Rodríguez-Manzo, A. Zamudio and A. L. Elías, *Solid State Sci.*, 2006, **8**, 303–320.
30 R. Lv, S. Tsuge, X. Gui, K. Takai, F. Kang and T. Enoki, *Carbon*, 2009, **47**, 1141–1145.
31 N. Grobert, M. Mayne, M. Terrones, J. Sloan, R. E. Dunin-Borkowski and R. Kamalakaran, *Chem. Commun.*, 2001, 471–472.
32 R. Lv, F. Kang, W. Wang, J. Wei, J. Gu, K. Wang and D. Wu, *Carbon*, 2007, **45**, 1433–1438.
33 R. Lv, F. Kang, D. Cai, C. Wang, J. Gu, K. Wang and D. Wu, *J. Phys. Chem. Solids*, 2008, **69**, 1213–1217.
34 J. Wang, Z. Dong, J. Huang, J. Li, X. Jin, J. Niu, J. Sun, J. Jin and J. Ma, *Appl. Surf. Sci.*, 2013, **270**, 128–132.
35 R. Kozhuharova, M. Ritschel, D. Elefant, A. Graff, I. Monch, T. Muhl, C. M. Schneider and A. Leonhardt, *J. Magn. Magn. Mater.*, 2005, **290–291**, 250–253.
36 H. Aurich, A. Baumgartner, F. Freitag, A. Eichler, J. Trbovic and C. Schonenberger, *Appl. Phys. Lett.*, 2010, **97**, 153116.
37 R. Lv, A. Cao, F. Kang, W. Wang, J. Wei, J. Gu, K. Wang and D. Wu, *J. Phys. Chem. C*, 2007, **111**, 11475–11479.

38 N. Grobert, M. Mayne, M. Terrones, J. Sloan, R. E. Dunin-Borkowski, R. Kamalakaran and T. Seeger, *AIP Conf. Proc.*, 2001, **591**, 287.
39 K. L. McNerny, Y. Kim, D. E. Laughlin and M. E. McHenry, *J. Appl. Phys.*, 2010, **107**, 09A312.
40 K. J. Miller, M. Sofman, K. McNerny and M. E. McHenry, *J. Appl. Phys.*, 2010, **107**, 09A305.
41 M. A. Ermakova, D. Y. Ermakov, A. L. Chuvilin and G. G. Kuvshinov, *J. Catal.*, 2001, **201**, 183–197.
42 C. Muller, D. Golberg, A. Leonhardt, S. Hampel and B. Buchner, *Phys. Status Solidi A*, 2006, **203**, 1064–1068.
43 S. Groudeva-Zotova, R. Kozhuharova, D. Elefant, T. Muhl, C. M. Schneider and I. Monch, *J. Magn. Magn. Mater.*, 2006, **306**, 40–50.
44 S. A. Moshkalev and C. Verissimo, *J. Appl. Phys.*, 2007, **102**, 044303.
45 O. A. Louchev and H. Kanda, *J. Chem. Phys.*, 2004, **121**, 446–456.
46 W. Ren, F. Li and H. M. Cheng, *J. Phys. Chem. B*, 2006, **110**, 16941–16946.
47 J. Y. Raty, F. Gygi and G. Galli, *Phys. Rev. Lett.*, 2005, **95**, 096103.
48 S. Helveg, C. Lopez-Cartes, J. Sehested, P. L. Hansen, B. S. Clausen, J. R. Rostrup-Nielsen, F. Abild-Pedersen and J. K. Nørskov, *Nature*, 2004, **427**, 426–429.
49 A. K. Schaper, H. Hou, A. Greiner and F. Phillipp, *J. Catal.*, 2004, **222**, 250–254.
50 A. Schneider, *Corros. Sci.*, 2002, **44**, 2353–2365.
51 V. Likodimos, S. Glenis, N. Guskos and C. L. Lin, *Phys. Rev. B: Condens. Matter Mater. Phys.*, 2003, **68**, 045417.

Faraday Discussions

PAPER

Effect of post-annealing on the plasma etching of graphene-coated-copper†

L. S. Hui,[a] E. Whiteway,[b] M. Hilke[b] and A. Turak*[a]

Received 24th May 2014, Accepted 13th June 2014
DOI: 10.1039/c4fd00118d

High temperature deposition of graphene on Cu by chemical vapor deposition can be used to produce high quality films. However, these films tend to have a non-equilibrium structure, with relatively low graphene adhesion. In this study, samples of graphene grown on copper foils by high temperature CVD were post-deposition annealed at temperatures well below the critical temperature of Cu. Resistance to etching under plasma was examined to assess the mechanical robustness of the graphene on the Cu surface, analyzed using optical and Raman microscopies. We found a correlation between the post-annealing time and etching time for the complete removal of graphene from Cu. Etching rates, minimum etch times, and surface appearance were observed to vary depending on the etching plasma (air, oxygen or nitrogen). Oxygen plasmas were found to be the least aggressive, emphasizing the improved adhesion with post-annealing treatments. Our results imply that the etching of graphene on Cu, and hence the adhesion of graphene, can be controlled by proper annealing and choice of plasma gas.

Introduction

Due to its chemical inertness, excellent mechanical strength, and high electrical conductivity graphene has been a widely studied material since its discovery in 2004.[1–3] Potential applications include its use as electrodes,[4–6] charge carrier transport layers,[7–9] lubricants,[10] impermeable barriers,[11,12] and in nano-electromechanical systems (NEMS).[13]

Though much research has focused on understanding the electronic and thermal properties of graphene,[1–3] the mechanical properties have been less widely explored. However, recent work suggests that the mechanical and friction properties of graphene are just as impressive as the electronic properties.[1,14,15] Despite recent excellent work, there still exists some controversy on the strength

[a]Department of Engineering Physics, McMaster University, Hamilton, Ontario, L8S 4L7, Canada. E-mail: turaka@mcmaster.ca

[b]Department of Physics, McGill University, Montréal, H3A 2T8, Canada

† Electronic supplementary information (ESI) available. See DOI: 10.1039/c4fd00118d

of adhesion of chemical vapor deposited (CVD) graphene, especially on Cu surfaces.[16–18]

CVD is an excellent low cost method of producing large area graphene with controlled thickness and high quality.[19,20] Though the film quality is high, CVD graphene grown on Cu foil at high deposition temperatures (>1000 °C) tends to have a non-equilibrium structure,[21] characterized by wrinkles and folds.[16,21–23] Deposition at high temperatures leads to rapid contraction of graphene upon cooling[21,23,24] due to the mismatch in the thermal expansion coefficients between Cu and graphene.[25,26] The Chen group[27,28] has recently used high temperature post-deposition annealing to recover the equilibrium structure of graphene on Cu.[22] This equilibrium structure has been found to have much higher adhesion energies[17] than those found on wrinkled samples.[16]

This study examined the impact of post-deposition annealing (post-annealing) on the graphene–Cu system when exposed to etching plasmas. Post-deposition annealing, even at modest temperatures (<200 °C), appears to be key in manipulating the adhesion strength of graphene on Cu, reflected by the increased etching times required to completely remove CVD graphene from the Cu surface. Plasma etching, therefore, may be an easy and effective method of examining the impact of post-deposition treatments on graphene systems.

Experimental

Graphene was synthesized by chemical vapour deposition (CVD) on commercially available 25 μm copper foils (Alfa Aesar). The foils were first chemically treated with acetic acid and annealed for 4 hours at 1078 °C under a flow of 8 sccm hydrogen gas in order to prepare the copper surface. The temperature was maintained at 1078 °C during a 4 min CVD growth phase where a gas flow of 1.2 sccm methane was introduced. A long annealing time and short growth period were chosen in order to limit the size and density of the graphene domains. To observe the graphene flakes, oxidation of the Cu surface is necessary to provide contrast between the graphene coated and uncoated portions of the Cu surface. Samples were annealed in batches on a hot plate (Pace Heat Wave) in air at 180 °C with different annealing times of $t = 2, 4, 9, 15, 30$, and 60 min. The groups of samples with different annealing times were then separated into batches for air (unfiltered laboratory air), nitrogen (Alphagaz, 99.999%) and oxygen (Alphagaz, 99.999%) plasma etching. The plasma etching was performed in a Harrick Plasma PDC-001 plasma cleaner system and a Thierry Plasma Zepto Low Pressure System with a 13.56 MHz R.F. generator at 29.6 W. The plasma chamber was held at a base pressure of 70 mTorr prior to target gas filling. During etching, a gas flow rate of 30 sccm was maintained. The samples were progressively etched from 2–260 min with time intervals of a minimum of 1 min and were observed under an optical microscope (Leica DMR microscope) to detect the first disappearance of contrast. For the case of N_2 plasma etching, a second post-annealing step of the etched samples was needed to show the contrast of Cu–graphene after treatment. The time intervals from the etching were accumulated to determine the minimum time required to completely remove the graphene from the Cu surface, designated as the critical etching time. Raman spectra were collected using a Renishaw inVia spectrometer at 514 nm laser excitation, with a 20× objective and 1200 l per mm grating. The laser power was 20 mW to prevent damage to the graphene. All of the spectra were background corrected to remove the Cu photoluminescence excitation at 514 nm (ref. 29) (see the ESI†).

Results and discussion

Though exposure to plasma is often used to modify the electronic properties of graphene layers,[30–34] under sufficiently high doses bombardment with energetic particles can also completely etch away the graphene layer. The etching time for the complete removal of any residue is generally correlated to the strength of the adhesion of the coating on the surface.[35] Therefore, plasma interactions are a potential method for probing the mechanical strength of graphene–copper interfaces. As our interest concerns the robustness of the graphene against plasma etching, partially covered substrates were desired. The CVD process was tuned to produce square graphene flakes on the Cu foil with ~30–100 μm widths.

Impact of etching times

Fig. 1 shows the background subtracted Raman spectra for the as-grown graphene, and the resultant spectra after etching in O_2 for various times. The pristine sample shows both the characteristic features of graphene, the G and 2D peaks at 1587 cm^{-1} and 2705 cm^{-1} respectively. The G peak corresponds to the first order degenerate phonon energy, E_{2g} mode at the Γ point of graphene;[36] the sharp 2D peak is a result of the second order phonon intervalley scattering, from the π–π^* transition.[36–38] The absence of a peak between ~1280 and 1360 cm^{-1} suggests that the graphene islands are mostly defect free.[36,39] The $I_G/I_{2D(\text{peak height})}$ ratio of 0.4 indicates that the graphene flakes on the Cu surface are mostly few or even single layered.[36,40] The 2D peak can be fit by a single Lorentzian function (not shown), further supporting the idea that the CVD grown graphene films are mostly single layer, defect free patches on the Cu surface.[36] Some islands were also observed

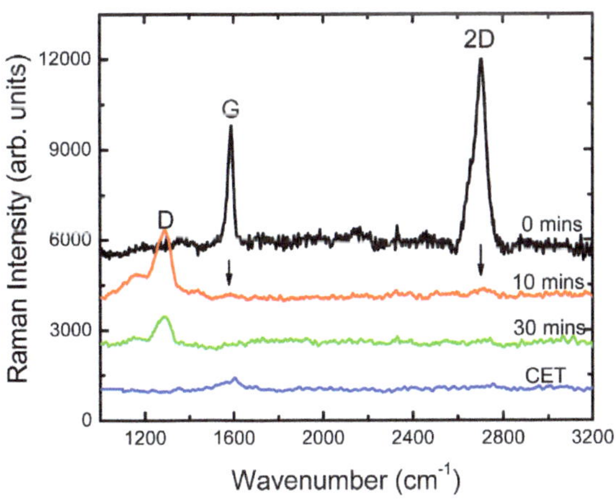

Fig. 1 Raman spectra for CVD graphene on Cu foil at etching times of 0, 10, 20 and 201 (CET) min, showing the evolution of the etching process for O_2 plasmas. The critical etching time is defined as the complete removal of graphene, as determined by optical methods.

with slightly higher I_G/I_{2D} ratios and a shoulder on the 2D peak, suggesting that some flakes are likely multilayered.[36]

After a short etch in O_2 plasma, significant sp^3 defects were introduced with a strong D band appearing at 1290 cm^{-1}.[36,41] Though much reduced, both the G and 2D bands are still barely visible as shown by the arrows in Fig. 1. There may also be a merging of the G and D′ broad bands at 1590 cm^{-1}, which is attributed to the C–H sp^3 hybridization defect and the overtone of the D mode activated after plasma treatment.[22,27] These changes suggest that the plasma initially introduces defects into the graphene rather than removing it from the Cu, as is often seen in lower dose plasma processing used to tune the electronic configuration of graphene flakes.[30] Another explanation for the appearance of the strong D band is the formation of graphene oxide, though that is generally accompanied by an equally large G peak, which was not observed here.[42] Further etching eliminates the G, D′ and 2D signals and reduces the scattering intensity of the D band, indicating that graphene is being removed from the surface. Finally, after the critical etching time (CET), there is no more evidence of graphene in the Raman spectrum. This fully etched condition of the graphene was confirmed by optical microscopy.

As received, the graphene on Cu foil is without any contrast in the forward scattering condition. Cu, however, is prone to oxidation in air even at low temperatures. Graphene has been widely reported as an excellent passivation layer,[11,12] protecting the surface from chemical oxidation as the densely packed benzene-ring structures make graphene impermeable to most gases, including helium.[43,44] Mild oxidation, from short post-annealing treatments in air or even exposure to oxidizing plasmas, can therefore be used to visualize the graphene islands (see Fig. 2(a)). The contrast results from the different light scattering intensities and refractive indices of cuprous oxide and the unoxidized Cu protected by the graphene. The oxidation contrast, therefore, can be used as a method of tracking the progress of the etching. Fig. 2 shows the effects of the

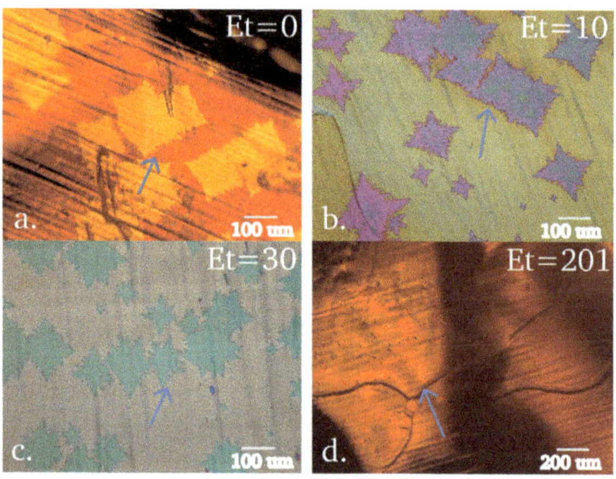

Fig. 2 Optical micrographs of CVD graphene on Cu foils. (a) As-grown sample, annealed to show the contrast between bare and graphene covered regions. (b–d) The same sample after etching in O_2 plasma for 10, 30 and 201 min. Etching time is denoted as "Et" in the figures. The blue arrows indicate the contrast showing the graphene islands. The final panel was used to determine the critical etching time (CET).

etching procedure on the graphene islands, supporting the findings suggested by the Raman data. After short etching times, the graphene is still visible, and then becomes progressively less visible as the etching progresses. Finally, the critical etching time (CET) is determined as the point where almost no contrast is visible after etching, suggesting complete removal of the graphene layer (Fig. 2(d)).

After the CET, there is still a Raman feature visible at the G/D′ location of ~1580 cm^{-1} (Fig. 1). However, the eigenvector of the G mode involves the in-plane bond-stretching motion of pairs of sp^2 C atoms, and does not require the presence of six-fold rings.[45] Without the accompanying 2D peak, which needs six-fold rings to be excited,[45] this cannot be taken as a feature of graphene, and is most likely due to C=C residues on the surface. As can be seen in Fig. 3, after the CET as determined from optical observations, Raman measurements show no indication of graphene on the surface for N_2, O_2 and unfiltered laboratory air etching plasmas. This suggests that the optical method is adequate to determine the CET for complete removal of graphene from the Cu surface, and the CET can be considered to be roughly proportional to the strength of the interaction between graphene and the Cu surface. Optical determination of the CET is straightforward for O_2 and air plasmas, as exposure to the plasma also oxidized the Cu surface; however, identification of the CET for N_2 plasma required a post-etching anneal of the etched samples to show the contrast of the Cu–graphene after treatment. It is likely therefore that the visual method leads to an over estimation of the real CET for N_2 plasmas. This was confirmed with Raman measurements where graphene patches were visible optically, but there was no graphene signature (see next sections).

Effect of post-deposition annealing

CVD graphene grown on Cu foil at high deposition temperatures, such as those in this study, tends to have a non-equilibrium structure.[21] Egberts *et al.*[16] recently showed that the work of adhesion for graphene on Cu is low for such systems. The Chen group[27,28] have been able to use high temperature post-deposition annealing

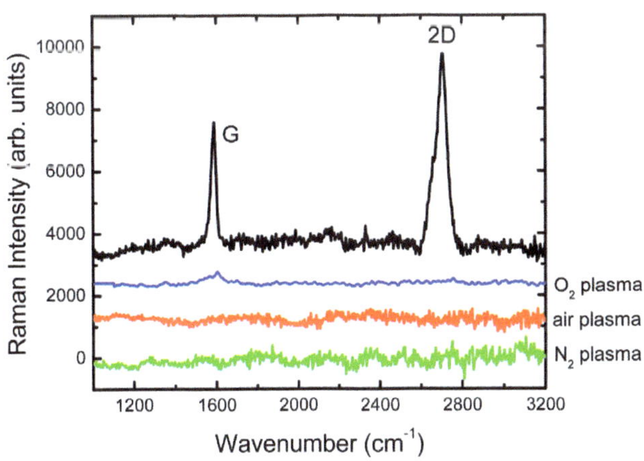

Fig. 3 Raman spectra of as-grown CVD graphene on Cu and after etching with different gases.

to recover the crystallinity of the Cu surface, approaching the equilibrium structure of graphene on Cu.[22] High quality equilibrium films have been reported to have very high adhesion strengths.[17] This suggests that post-deposition annealing might increase the interaction strength between graphene and Cu.

To examine this effect, we performed low temperature (180 °C) post-deposition annealing on our graphene–Cu samples for various annealing times. As described above, short annealing times can make the graphene islands visible. Under further annealing (Fig. 4(b)) the contrast increases, likely due to the enhanced oxidation of unprotected Cu in the presence of graphene, recently described by

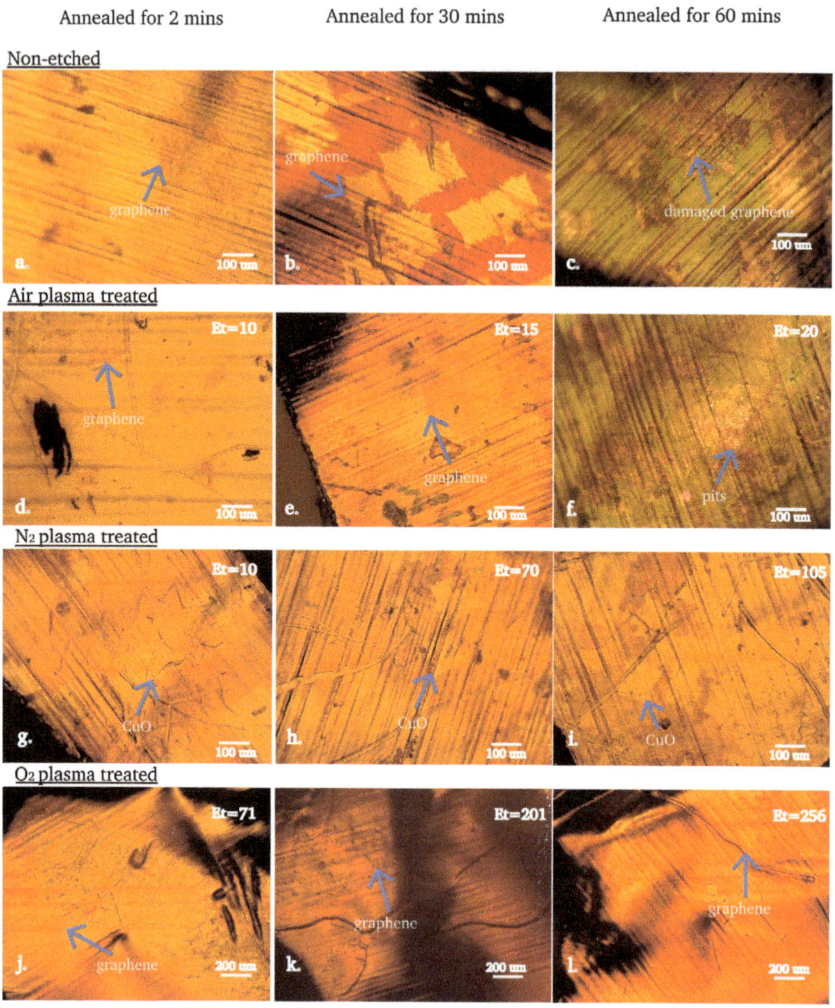

Fig. 4 (a–l) Optical micrographs of CVD graphene on Cu foil annealed and etched for various times. (a, d, g and j) Samples post-annealed for 2 min; (b, e, h and k) for 30 min and (c, f, i and l) for 60 min. (a–c) The samples have not been plasma treated; (d–f) treated with air plasma, (g–i) N_2 plasma treated and (j–l) O_2 plasma treated with the etching time denoted as "et" in the figures. The blue arrows indicate the contrast showing the graphene islands.

Zhou *et al.*[46] Eventually, after annealing for 60 min, in some areas of the sample the graphene island edges are no longer clear, and the islands appear damaged. This is likely to be a result of graphene exfoliation from the surface, due to a mismatch in the thermal expansion coefficients of graphene ($(-8.0 \pm 0.7) \times 10^{-6}$ K^{-1}) and the Cu crystals (14×10^{-6} K^{-1}).[25,26] With longer annealing times, more material is allowed to expand and the rigid graphene structure is not fully bonded to the surface, exposing some of the Cu underneath the surface, resulting in a more "messy" oxidation of the Cu foil.[47]

Fig. 5 shows the Raman spectra for the post-deposition annealed films. In all cases the G and 2D bands are both visible, without any D band development. This suggests that there was neither significant defect nor graphene oxide formation.[42,48] Post-deposition annealing in fact seems to have very little impact on the graphene islands, even when optically significant damage appears to have occurred. The only notable change is a slight blue shift of the 2D band observed to occur systematically with annealing time. Though this is characteristic of multi-layered graphene,[49] no peak broadening or change in the I_G/I_{2D} ratio (which varies between 0.36 and 0.46 depending on the graphene flakes) were observed. The observed features can be well fitted with a single Lorentzian peak shape with a similar full width at half maximum (FWHM), pointing to defect free, monolayer graphene in all cases. Another possible explanation for the observed shift is the change of strain at the interface. Wang *et al.*[50] saw shifts in the 2D band resulting from the lattice mismatch between SiC and the carbon layer/buffer layer, which has a graphene-like honeycomb lattice that is covalently bonded to the SiC substrate.[50] Yu *et al.*[37] also observed strain induced shifts in the 2D band during temperature tuning of growth that they attributed to strain buildup at the interface. A change in the mechanical interaction between the graphene and Cu during post-annealing appears to relieve strain at the interface.

Upon exposure to plasma however, significant differences were observed with differing post-annealing times. Optically, as can be seen in Fig. 4, systematically longer etching times were necessary to fully remove the graphene from the

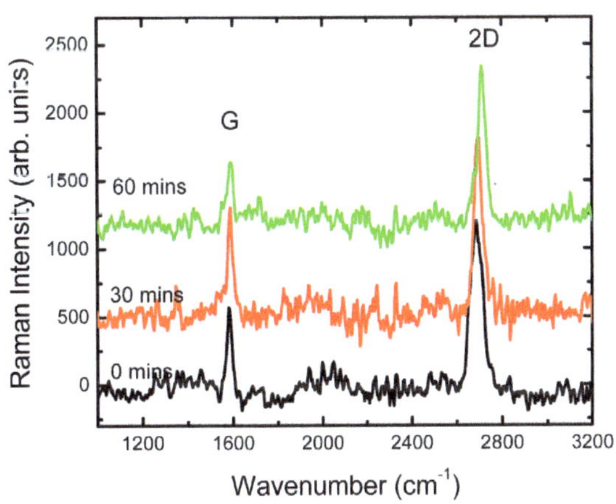

Fig. 5 The Raman spectra of CVD graphene on Cu samples with different post-annealing times at 180 °C (0, 30 and 60 min). The spectra were normalized to the 2D peak intensity at ~2700 cm^{-1} for comparison.

substrate with post-annealing, for all of the etching gases. The observed CET values for all the process gases are summarized in Fig. 6. In all cases, an increased annealing time protects the system against etching. The visually different etching times as a function of annealing, as shown in Fig. 4(d)–(l) support the premise that annealing can be used to control the removal of graphene, by making the structure more robust. It takes much longer in all cases to see the complete removal of graphene from the surface. Fig. 7 shows the Raman spectra for the optically determined CET for O_2 plasma with various post-annealing times. Though there is no evidence of graphene or even C residue on the as-grown samples, even short post-deposition annealing changed the resulting spectra. With only 2 min of annealing, the defect or graphene oxide D band observed previously with 10 min plasma exposure was still visible after 75 min plasma exposure (see the ESI† for non-normalized CET values). Further annealing showed an increasing signal from the G/D′ band, which may be due to C=C residues that are not removed even with extremely long etch times (250 min etching for 60 min post-annealing time). The extremely long etching times and the inability of the plasma to remove carbon residues indicates a strong interaction between the Cu and the graphene.

Effect of plasma gas

Though Fig. 6 shows that the trend of increasing CET with annealing holds for all of the process gases, there are significant differences in the measured CET values for the different etch gases. These effects are visible even with normalization to remove sample variation. Using the CET estimations, air is the most aggressive etchant, with O_2 and N_2 plasmas having decreasing aggressiveness. The effect of post-annealing in protecting the graphene from etch removal is more enhanced the less aggressive the etchant. However the optical measurements are slightly misleading. Raman measurements for the same etching time, shown in Fig. 8,

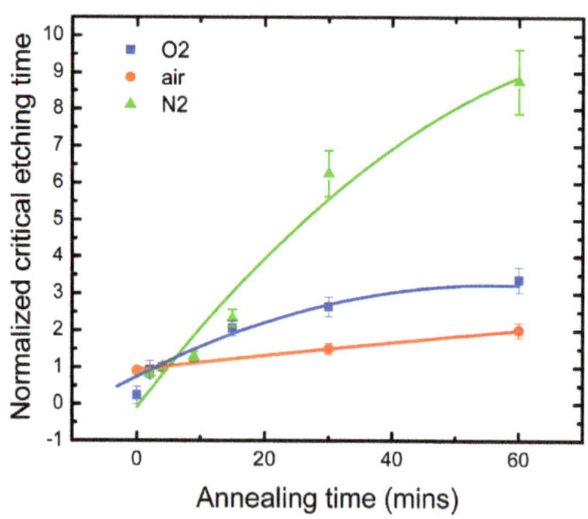

Fig. 6 Normalized CET values for graphene with different post-annealing treatment for various plasmas. The lines are the least squares fit on the data. The CET values have been normalized using the value of annealing time at 4 min, where contrast is visible, to eliminate possible sample discrepancies and emphasize the common trend for all of the process gases.

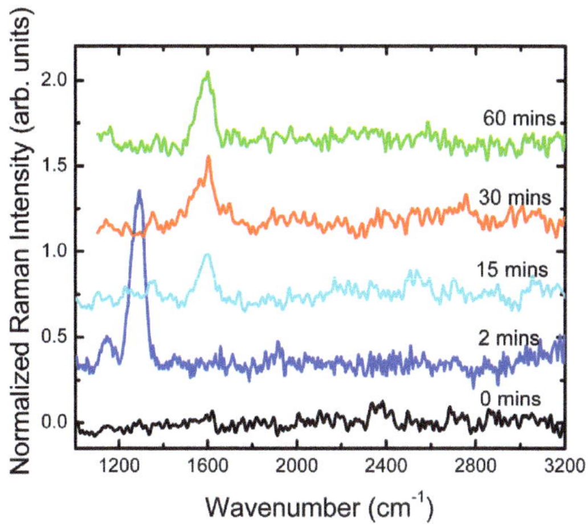

Fig. 7 Raman spectra for CET on CVD graphene on Cu foils with O_2 plasma for various post-annealing times (0–60 min) at 180 °C.

support the order of the O_2 and air plasmas, but completely contradict the behavior with N_2 etching.

Fig. 8 confirms that the different process gases have different impacts on the graphene for the same etching time. In this case, N_2 plasmas appear to be the most aggressive. No Raman signal is visible from graphene nor are there any C–O or C=O bands after 10 min etching, suggesting a total destruction of graphene and carbon residues. N_2 plasma etching is a physical sputtering process, that has previously been used for nitrogen doping of graphene[51] and is thought to lead to unsaturated dangling carbon bonds at the edges of the graphene.[51] The dangling carbon bonds can react with oxygen atoms once exposed to air. However, in this study, no evidence of graphene or graphene oxide was observed, only complete sputter removal of the graphene.

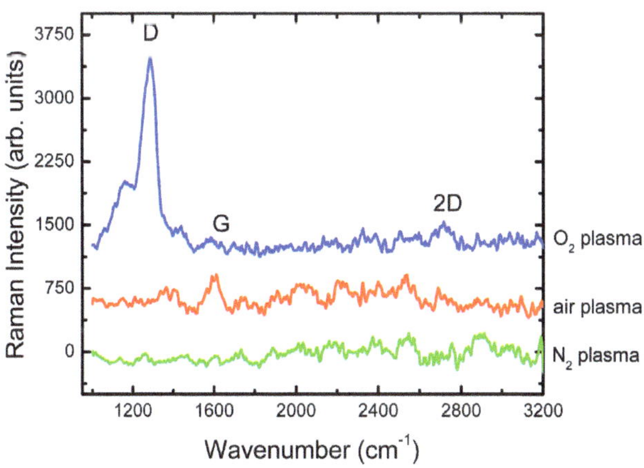

Fig. 8 Raman spectra for CVD graphene on Cu foils with the same post-annealing (30 min) and etching times (10 min) with different gases.

It appears that, while the optical detection for determining the CET works well for air and oxygen plasmas, values of the CET as determined optically are highly overestimated for the N_2 plasma. Optically Fig. 4(h) shows that the contrast initially assigned to graphene is still visible for a sample that has been etched for 75 min, under the same annealing conditions as the 10 min etching in Fig. 8. As nitrogen plasma cannot oxidize Cu during the etching, this contrast is more likely to result from the varying degrees of Cu oxidation rather than a signature of graphene on the surface. The first annealing step oxidizes the regions without graphene. After interaction with the plasma the graphene is removed, leaving behind regions with different levels of oxidation. During low temperature oxidation, metastable Cu_3O_2 forms first followed by CuO, the thermodynamically stable phase of cuprous oxide.[52] Possibly the regions covered with graphene form the metastable oxide during the second annealing step (*i.e.* after plasma etching to expose the graphene), while the previously oxidized regions continue to oxidize further. Eventually appearing to be "fully-etched", without any contrast, actually indicates a uniform oxidation or the complete removal of the initially formed oxide by the vigorous physical sputtering action of the plasma. The trend in the observed CET values for N_2 plasma may suggest that it takes longer for the cuprous oxide to be sputtered away at longer annealing times.

The Raman data indicate that air plasma is also very aggressive, leaving only the C=C residue/G band at 1590 cm^{-1}. Air plasma etching is a very vigorous process, with the CET reached quickly with visible pitting observed on the Cu–graphene surface (see Fig. 4(f)). This pitting is likely to be due to the effect of hydrogen plasma in air, as Diankov *et al.*[53] observed the high selectivity of hydrogen gas on graphene. Pits are not present on the samples etched by N_2 (Fig. 4(i)) or O_2 plasmas (Fig. 4(l)) for the same annealing time and even for much longer etching times. It is likely that there is true etching occurring with the air plasma, due to the highly reactive O^- and H^- species, as well as the sputtering observed with nitrogen.

O_2 plasmas were the least aggressive, showing the defect structure, D band, and with some G and 2D peaks barely visible. As described previously, there could also be a broad D′ band, merging with the G band. The presence of both D′ and D peaks suggests that significant doping and defects are introduced, but likely with little removal of graphene. The gas with the slowest etching rate is thus oxygen, where only chemical etching takes place with O^- radical species.

The rates observed for the various plasmas can be attributed to the mechanisms of plasma etching for the different gases. Different plasma gases can have varied interaction mechanisms and diverse ions present in the plasma.[53–55] Plasma is ignited by the large voltage applied across the electrodes outside the plasma chamber, which causes the free electrons to accelerate rapidly towards the anode. When the electrons collide inelastically with gas molecules, an electron will be removed from the molecules and thus the gas will then become dissociated or ionized. The removed electron continues to be accelerated to the anode and collides with other gas molecules, resulting in a cascading effect. The gas discharge causes excitation and relaxation of gas molecules that release energy in the form of photons, thus causing the glow inside the chamber.

The process of etching is generally considered to be the result of radicals actively breaking chemical bonds and reacting with the substrate surface.[56] Ions, however, can also cause energetic bombardment or ejection of atoms from the surface, a process referred to as sputtering. Oxygen will form highly reactive radicals by molecule dissociation, resulting in etching of the graphene, while nitrogen can be ionized and form ions,[57–59] thus removing graphene primarily through sputtering. Under exposure to air plasma, graphene is expected to be both etched and sputtered by oxygen, hydrogen, nitrogen, carbon and other atoms of trace gases present in air.[55] As there is high a percentage of nitrogen in air (78%), one might expect the air plasma etching process to be dominated by nitrogen. However, pure N_2 plasma showed a much faster etching compared to air and oxygen plasmas.

Possible mechanisms for increased adhesion strength with annealing

The time required for the complete removal of any residue by etching can be broadly correlated to the strength of the adhesion of the coating on the surface.[35] The observed increase in CET values with annealing for Cu–graphene systems therefore suggests a significant improvement in the interaction between graphene and Cu and an improvement in the mechanical properties of the interface. There is some controversy in the literature regarding the value of the adhesion strength between graphene and Cu,[16–18] and on the nature of the interaction leading to this value. There are two main mechanisms that might be used to describe the observed behavior with plasma etching of the graphene–Cu interface: the process of Cu oxidation itself, or the reconstruction of the Cu surface with annealing.

The CET as determined for N_2 etching seems to be due to the removal of Cu oxides rather than the graphene layer. Yet the trend of increasing resistance to etch removal with annealing is observed. As Cu is highly susceptible to oxidation, even at low temperatures, it may be that the oxidation process itself leads to the increased strength of the interaction between Cu and graphene.

Fundamentally, the oxidation of Cu can be described by the Cabrera–Mott theory.[60] Cu ions and electrons migrate to form the oxide layer on the surface, leaving cationic Cu sites behind.[61] The strong electric field that results promotes further diffusion of Cu ions to continue oxidation.[61]

In the composite Cu–graphene system, prolonged annealing may provide more kinetic energy for Cu ion migration to areas where graphene is absent (Fig. 9). As the vacant Cu cation sites draw Cu ions to the oxide–metal interface, they may pile up under the graphene flake. This could result in a stronger coupling of C-π orbitals and Cu-d orbital hybridization of the graphene–Cu interface. This bunching of the Cu under the graphene layers as a result of the oxidation electric field could explain the blue shift observed in the 2D band with annealing (see Fig. 5), attributed to relief of interfacial strain.[37,50]

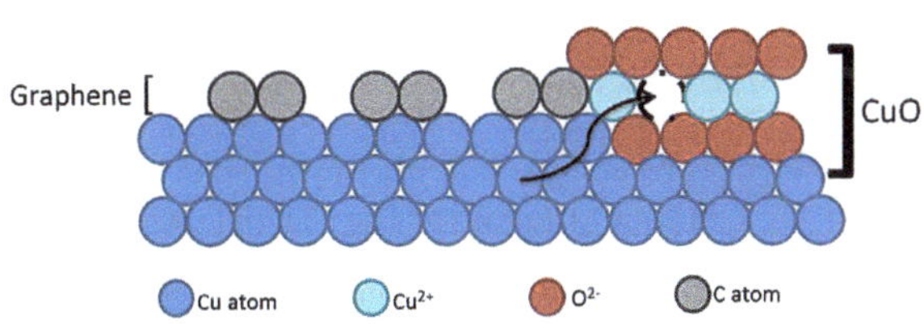

Fig. 9 Schematic of graphene–Cu interface oxidation.

Another possible mechanism influencing the etch resistance of the post-annealed system is the surface reconstruction of the Cu due to annealing. Rapid contraction of the graphene upon cooling is observed with high temperature (>1000 °C) graphene deposition[21,23,24] due to the mismatch in the thermal expansion coefficients.[25,26] This is one of the possible mechanisms for the formation of wrinkles widely observed in graphene–Cu systems.[16,21–23] Cho *et al.* and Tian *et al.*[27,28] have shown that high temperature post-annealing is able to partially heal the non-equilibrium structure that results from high temperature growth. Zhang *et al.*[21] showed that the existence of steps in Cu under the graphene lowers the adhesion. With annealing induced reconstruction recovering the steps within a Cu grain[27] the graphene wrinkles can relax, recovering the Moiré reconstruction pattern.[22] In the equilibrium structure graphene is unwrinkled, and flat graphene flakes are likely to have higher adhesion to the surface compared to wrinkled ones.[17,62] The observed blue shift, attributed to relief of interfacial strain,[50] could also result in this case, where the lattice mismatch is changed by the reconstruction of the Cu surface. Though we used lower temperatures for our post-annealing study, it is possible that a similar, though incomplete, reconstruction occurred in our samples. The increase in interaction strength, as suggested by the robustness to etching, with annealing time therefore could be correlated to the amount of reconstruction.

Further work using friction measurements to clarify these mechanisms is currently under way.

Conclusions

The graphene grown on Cu foil by high temperature CVD was annealed after deposition, and exposed to various plasmas. The post-annealing treatment alters the mechanical strength of graphene adhering on the Cu foil, as shown by the increased resistance to plasma etching. Raman spectroscopy confirms that the graphene is damaged after annealing and plasma treatments, but typically not oxidized during these treatments. The samples were etched and sputtered by air, oxygen, and nitrogen plasmas, yielding different etch rates, appearance and uniformity of the graphene. Nitrogen plasma has a much faster etch rate than air and oxygen due to the mechanism of physical sputtering. For any given plasma however, with post-annealing before plasma exposure, the graphene is harder to remove due to the bonding between the graphene and Cu substrate. Our results imply that etching of graphene on Cu can be controlled by proper annealing and the right choice of plasma gas. Plasma etching, with a suitable process gas, appears to be an effective method of estimating the impact of post-processing effects on the mechanical properties of CVD graphene.

Acknowledgements

The authors acknowledge financial support from 384889-2010 CREAT. The authors are grateful to J. Moran-Mirabal for the Harrick Plasma system, P. Jonasson for the annealing system, S. Kornic for the Raman instrument and CEDT for the microscope support at McMaster University. P. Egberts of U. Calgary is gratefully acknowledged for the insightful discussions.

Notes and references

1 C. Lee, X. Wei, J. W. Kysar and J. Hone, *Science*, 2008, **321**, 385–388.
2 Y. S. Dedkov, M. Fonin, U. Rüdiger and C. Laubschat, *Appl. Phys. Lett.*, 2008, **93**, 022509.
3 K. S. Novoselov, A. K. Geim, S. V. Morozov, D. Jiang, Y. Zhang, S. V. Dubonos, I. V. Grigorieva and A. A. Firsov, *Science*, 2004, **306**, 666–669.
4 H. Park, R. M. Howden, M. C. Barr, V. Bulović, K. Gleason and J. Kong, *ACS Nano*, 2012, **6**, 6370–6377.
5 Y. Wang, S. W. Tong, X. F. Xu, B. Ozyilmaz and K. P. Loh, *Adv. Mater.*, 2011, **23**, 1514–1518.
6 Y. J. Jeong, J. Jang, S. Nam, K. Kim, L. H. Kim, S. Park, T. K. An and C. E. Park, *ACS Appl. Mater. Interfaces*, 2014, **6**, 6816–6824.
7 S. Zhong, J. Q. Zhong, H. Y. Mao, R. Wang, Y. Wang, D. C. Qi, K. P. Loh, A. T. S. Wee, Z. K. Chen and W. Chen, *ACS Appl. Mater. Interfaces*, 2012, **4**, 3134–3140.
8 M. M. Stylianakis, E. Stratakis, E. Koudoumas, E. Kymakis and S. H. Anastasiadis, *ACS Appl. Mater. Interfaces*, 2012, **4**, 4864–4870.
9 Q.-H. Wu, G. Hong, T. W. Ng and S. T. Lee, *Appl. Phys. Lett.*, 2012, **100**, 161603.
10 D. Berman, A. Erdemir and A. V. Sumant, *Mater. Today*, 2014, **17**, 31–42.
11 S. Chen, L. Brown, M. Levendorf, W. Cai, S.-Y. Ju, J. Edgeworth, X. Li, C. W. Magnuson, A. Velamakanni, R. D. Piner, J. Kang, J. Park and R. S. Ruoff, *ACS Nano*, 2011, **5**, 1321–1327.
12 E. Sutter, P. Albrecht, F. E. Camino and P. Sutter, *Carbon*, 2010, **48**, 4414–4420.
13 J. S. Bunch, A. M. van der Zande, S. S. Verbridge, I. W. Frank, D. M. Tanenbaum, J. M. Parpia, H. G. Craighead and P. L. McEuen, *Science*, 2007, **315**, 490–493.
14 T. Filleter, J. McChesney, A. Bostwick, E. Rotenberg, K. Emtsev, T. Seyller, K. Horn and R. Bennewitz, *Phys. Rev. Lett.*, 2009, **102**, 086102.
15 Y. J. Shin, R. Stromberg, R. Nay, H. Huang, A. T. S. Wee, H. Yang and C. S. Bhatia, *Carbon*, 2011, **49**, 4070–4073.
16 P. Egberts, G. H. Han, X. Z. Liu, A. T. C. Johnson and R. W. Carpick, *ACS Nano*, 2014, **8**, 5010–5021.
17 T. Yoon, W. C. Shin, T. Y. Kim, J. H. Mun, T.-S. Kim and B. J. Cho, *Nano Lett.*, 2012, **12**, 1448–1452.
18 X.-Z. Liu, Q. Li, P. Egberts and R. W. Carpick, *Adv.Mater. Interfaces*, 2014, DOI: 10.1002/admi.201300053.
19 X. Wang, H. You, F. Liu, M. Li, L. Wan, S. Li, Q. Li, Y. Xu, R. Tian, Z. Yu, D. Xiang and J. Cheng, *Chem. Vap. Deposition*, 2009, **15**, 53–56.
20 W. Liu, H. Li, C. Xu, Y. Khatami and K. Banerjee, *Carbon*, 2011, **49**, 4122–4130.
21 Y. Zhang, T. Gao, Y. Gao, S. Xie, Q. Ji, K. Yan, H. Peng and Z. Liu, *ACS Nano*, 2011, **5**, 4014–4022.
22 L. Gao, J. R. Guest and N. P. Guisinger, *Nano Lett.*, 2010, **10**, 3512–3516.
23 W. Zhu, T. Low, V. Perebeinos, A. A. Bol, Y. Zhu, H. Yan, J. Tersoff and P. Avouris, *Nano Lett.*, 2012, **12**, 3431–3436.
24 L. Zhao, K. T. Rim, H. Zhou, R. He, T. F. Heinz, A. Pinczuk, G. W. Flynn and A. N. Pasupathy, *Solid State Commun.*, 2011, **151**, 509–513.
25 D. Yoon, Y. Son and H. Cheong, *Nano Lett.*, 2011, **11**, 3227–3231.

26 F. C. Nix and D. MacNair, *Phys. Rev.*, 1941, **60**, 597–605.
27 J. Cho, L. Gao, J. Tian, H. Cao, W. Wu, Q. Yu, E. N. Yitamben, B. Fisher, J. R. Guest, Y. P. Chen and N. P. Guisinger, *ACS Nano*, 2011, **5**, 3607–3613.
28 J. Tian, H. Cao, W. Wu, Q. Yu, N. P. Guisinger and Y. P. Chen, *Nano Lett.*, 2012, **12**, 3893–3899.
29 S. D. Costa, A. Righi, C. Fantini, Y. Hao, C. Magnuson, L. Colombo, R. S. Ruoff and M. A. Pimenta, *Solid State Commun.*, 2012, **152**, 1317–1320.
30 J. Bai, X. Zhong, S. Jiang, Y. Huang and X. Duan, *Nat. Nanotechnol.*, 2010, **5**, 190–194.
31 Y. Wang, Y. Shao, D. W. Matson, J. Li and Y. Lin, *ACS Nano*, 2010, **4**, 1790–1798.
32 M. Wojtaszek, N. Tombros, A. Caretta, P. H. M. van Loosdrecht and B. J. van Wees, *J. Appl. Phys.*, 2011, **110**, 063715.
33 D. G. Matei, N.-E. Weber, S. Kurasch, S. Wundrack, M. Woszczyna, M. Grothe, T. Weimann, F. Ahlers, R. Stosch, U. Kaiser and A. Turchanin, *Adv. Mater.*, 2013, **25**, 4146–4151.
34 A. Zandiatashbar, G.-H. Lee, S. J. An, S. Lee, N. Mathew, M. Terrones, T. Hayashi, C. R. Picu, J. Hone and N. Koratkar, *Nat. Commun.*, 2014, **5**, 3186.
35 G. N. Taylor and T. M. Wolf, *Polym. Eng. Sci.*, 1980, **20**, 1087–1092.
36 A. C. Ferrari, J. C. Meyer, V. Scardaci, C. Casiraghi, M. Lazzeri, F. Mauri, S. Piscanec, D. Jiang, K. S. Novoselov, S. Roth and A. K. Geim, *Phys. Rev. Lett.*, 2006, **97**, 187401.
37 V. Yu, E. Whiteway, J. Maassen and M. Hilke, *Phys. Rev. B: Condens. Matter Mater. Phys.*, 2011, **84**, 205407.
38 J. D. Jones, W. D. Hoffmann, A. V. Jesseph, C. J. Morris, G. F. Verbeck and J. M. Perez, *Appl. Phys. Lett.*, 2010, **97**, 233104.
39 I. Pocsik, M. Hundhausen, M. Koos and L. Ley, *J. Non-Cryst. Solids*, 1998, 1083–1086.
40 Z. Sun, Z. Yan, J. Yao, E. Beitler, Y. Zhu and J. M. Tour, *Nature*, 2010, **468**, 549–552.
41 F. Tuinstra and J. L. Koenig, *J. Chem. Phys.*, 2003, **53**, 1126–1130.
42 O. Akhavan, *Carbon*, 2010, **48**, 509–519.
43 J. S. Bunch, S. S. Verbridge, J. S. Alden, A. M. Van Der Zande, J. M. Parpia, H. G. Craighead and P. L. Mceuen, *Nano Lett.*, 2008, **8**, 2458–2462.
44 J. Iijima, J.-W. Lim, S.-H. Hong, S. Suzuki, K. Mimura and M. Isshiki, *Appl. Surf. Sci.*, 2006, **253**, 2825–2829.
45 A. Ferrari and J. Robertson, *Phys. Rev. B: Condens. Matter Mater. Phys.*, 2000, **61**, 14095–14107.
46 F. Zhou, Z. Li, G. J. Shenoy, L. Li and H. Liu, *ACS Nano*, 2013, **7**, 6939–6947.
47 A. Siokou, F. Ravani, S. Karakalos, O. Frank, M. Kalbac and C. Galiotis, *Appl. Surf. Sci.*, 2011, **257**, 9785–9790.
48 K. N. Kudin, B. Ozbas, H. C. Schniepp, R. K. Prud'homme, I. A. Aksay and R. Car, *Nano Lett.*, 2008, **8**, 36–41.
49 L. M. Malard, M. A. Pimenta, G. Dresselhaus and M. S. Dresselhaus, *Phys. Rep.*, 2009, **473**, 51–87.
50 Y. Y. Wang, Z. H. Ni, T. Yu, Z. X. Shen, H. M. Wang, Y. H. Wu, W. Chen and A. T. S. Wee, *J. Phys. Chem. C*, 2008, **112**, 10637–10640.
51 Y. Shao, S. Zhang, M. H. Engelhard, G. Li, G. Shao, Y. Wang, J. Liu, I. A. Aksay and Y. Lin, *J. Mater. Chem.*, 2010, **20**, 7491–7496.

52 M. Lenglet, K. Kartouni, J. Machefert, J. M. Claude, P. Steinmetz, E. Beauprez, J. Heinrich and N. Celati, *Mater. Res. Bull.*, 1995, **30**, 393–403.

53 G. Diankov, M. Neumann and D. Goldhaber-Gordon, *ACS Nano*, 2013, **7**, 1324–1332.

54 V. M. Donnelly and A. Kornblit, *J. Vac. Sci. Technol., A*, 2013, **31**, 050825.

55 J. Herron and D. Green, *Plasma Chem. Plasma Process.*, 2001, **21**, 459–481.

56 C. Cardinaud, M. Peignon, P. Tessier and L. Materiaux, *Appl. Surf. Sci.*, 2000, **164**, 72–83.

57 D. Manos, D. Flamm, *Plasma Etching: An Introduction*, 1989.

58 C. Riccardi, R. Barni and M. Fontanesi, *J. Appl. Phys.*, 2001, **90**, 3735.

59 A. Hojabri, N. Haghighian, K. Yasserian and M. Ghoranneviss, *IOP Conf. Ser.: Mater. Sci. Eng.*, 2010, **12**, 012004.

60 N. Cabrera and N. Mott, *Rep. Prog. Phys.*, 1949.

61 T. N. Rhodin, *J. Am. Chem. Soc.*, 1950, **72**, 5102–5106.

62 Z. Ye, C. Tang, Y. Dong and A. Martini, *J. Appl. Phys.*, 2012, **112**, 116102.

Faraday Discussions

PAPER

Synthesis of face-centred cubic Cs_3C_{60} in THF†

Alec G. McLennan,[a] Alexey Y. Ganin,[ac] Yasuhiro Takabayashi,[b] Ross H. Colman,[b] Ruth H. Zadik,[b] Matthew J. Rosseinsky*[a] and Kosmas Prassides*[bd]

Received 30th April 2014, Accepted 11th July 2014

DOI: 10.1039/c4fd00085d

A solution chemistry synthetic route yields Cs_3C_{60} with a face-centred cubic structure. The described method uses well-established Schlenk techniques and THF as a solvent. The controlled addition of an organo-metallic salt reducing agent prevents the formation of C_{60}^{4-} salts. The final product can be precipitated from the solution using hexane as an anti-solvent.

Introduction

Solution chemistry methods have been an important source of novel alkali-metal fullerides and proven especially advantageous for the synthesis of ternary phases of sodium intercalated compounds where better control over stoichiometry and homogeneity of the product is required.[1-6] The viability of these methods was heralded by the successful synthesis of the long-sought after highly-expanded Cs_3C_{60} fulleride which surprisingly, depending on the synthesis conditions, adopts two polymorphic structures which become superconducting under pressure.[7-10] While reaction of three equivalents of Cs with C_{60} in liquid methylamine produced *ca.* 78% of a phase with body-centred cubic arrangement of C_{60}^{3-} anions (A15-structured), replacement of the solvent by liquid ammonia yielded up to 86% of the face-centred cubic (FCC) polymorph (Fig. 1).[7,8]

Regardless of the solvent used both as-prepared products contained Cs_1C_{60} and Cs_4C_{60} body-centred orthorhombic (BCO) phases as impurities. *In situ* and *ex situ* PXRD measurements on repeatedly annealed samples confirmed that the

[a]*Department of Chemistry, University of Liverpool, Crown Street, Liverpool L69 7ZD, UK. E-mail: m.j.rosseinsky@liverpool.ac.uk; Fax: +44 (0) 151 794 3588; Tel: +44 (0) 151 794 3499*

[b]*Department of Chemistry, Durham University, South Road, Durham DH1 3LE, UK. E-mail: k.prassides@durham.ac.uk*

[c]*School of Chemistry, University of Glasgow, Glasgow G12 8QQ, UK*

[d]*WPI Research Center, Advanced Institute for Materials Research, Tohoku University, Sendai 980-8577, Japan*

† Electronic supplementary information (ESI) available: PXRD patterns of $Mn(Cp^*)_2$ and $Cs[Mn(Cp^*)_2]$ products; tables with atomic parameters; Raman spectra; and high pressure SQUID measurements. See DOI: 10.1039/c4fd00085d

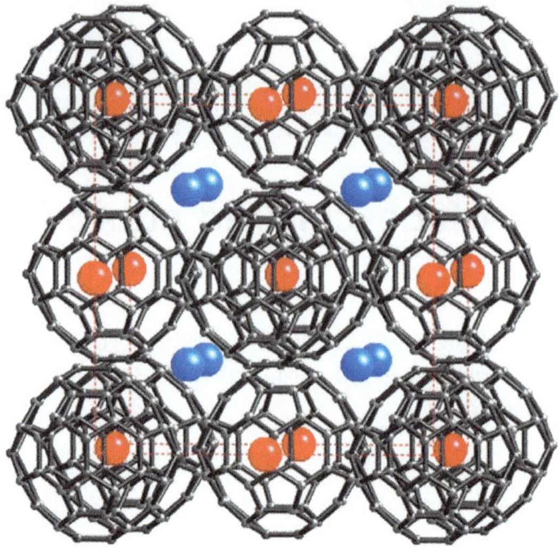

Fig. 1 Face-centred cubic structure (space group $Fm\bar{3}m$) of Cs_3C_{60} represented using only one ($Fm\bar{3}$) of the two disordered C_{60} orientations. The Cs cations are situated in both the tetrahedral (blue, mean Cs–C distance 3.45 Å) and octahedral (red, mean Cs–C distance 3.90 Å) interstitial sites.

amount of Cs_4C_{60} could be reduced upon annealing.[7,10] This was in contrast to the earlier work proposing disproportionation of Cs_3C_{60} after annealing at higher temperatures.[11] Therefore, it is possible that Cs_4C_{60} originates from the synthesis in liquid amine due to the large reduction potential of metallic caesium. Cyclic voltammetry measurements indicate that reduction potentials in the range of $E_{1/2}{}^{0/1-} = -1.5, -2.15$ and -3.02 V *vs.* SCE are required to achieve $C_{60}{}^{-3}$, $C_{60}{}^{-4}$ and $C_{60}{}^{-6}$ oxidation states respectively and caesium ($E_{1/2}{}^{+/0} = -2.69$ V *vs.* SCE in THF) is a potent reducing agent to reduce C_{60} to the $C_{60}{}^{-4}$ charge state completely in solution.[12–15] As a result, precise control of composition within the stable Cs_xC_{60} phases may be challenging. Broderick *et al.* proposed an elegant solution synthesis of K_3C_{60} by using K[Mn(Cp*)$_2$] as a reducing agent.[16] With a reduction potential of $E_{1/2}{}^{0/1-} = -2.17$ *vs.* SCE, K[Mn(Cp*)$_2$] was a milder reagent than potassium metal ($E_{1/2}{}^{+/0} = -2.7$ V *vs.* SCE) and produced a phase-pure superconducting K_3C_{60} fulleride.[16,17] It is reasonable to anticipate that a similar approach if applied to the Cs–C_{60} system could reduce or completely eliminate the competing Cs_4C_{60} phase. In this work we have demonstrated that control over stoichiometry by using Cs[Mn(Cp*)$_2$] as the reducing agent can produce high quality samples of the Cs_3C_{60} FCC polymorph.

Results and discussion

Tetrahydrofuran (THF) was used as an alternative to liquid ammonia (NH_3) and methylamine (CH_3NH_2) as it offered the opportunity to perform the reaction at ambient conditions without the use of cryogenics and permitted the manipulation of the reaction media by standard Schlenk techniques facilitating administration of the reducing agent in a controlled fashion.

The initial step of preparing Cs_3C_{60} required synthesis of the $Cs[Mn(Cp^*)_2]$ reducing agent. To the best of our knowledge, there have been no previous literature reports on the synthesis of $Cs[Mn(Cp^*)_2]$. Therefore, a synthetic protocol for $Cs[Mn(Cp^*)_2]$ was developed based on a modification of the method outlined for the synthesis of $Na[Mn(Cp^*)_2]$ (Scheme 1 and Experimental section).[17]

Compared to the sodium naphthalide its caesium analogue is less soluble in THF (solubility evaluated in this work 0.35 mol L^{-1}) but a fourfold increase in solvent volume ensured that the initial reaction between Cs and naphthalene produced a fully dissolved product. Adding the solution to the $Mn(Cp^*)_2$ gave a 92.8% yield of $Cs[Mn(Cp^*)_2]$ similar to those quoted before for $Na[Mn(Cp^*)_2]$.[17] The composition of the final bright orange highly air-sensitive product corresponded to $Cs_{1.0(1)}[Mn(Cp^*)_2]$, according to the microanalysis results. Powder X-ray diffraction studies showed that the product was free from $Mn(Cp^*)_2$ impurities (Fig. S1†).

The synthesis of the Cs_3C_{60} material required slow addition over 90 minutes of the $Cs[Mn(Cp^*)_2]$ reducing agent to a C_{60} suspension in THF (Scheme 2).

The slow addition of the reducing agent in a controlled manner was key to the success of the synthetic procedure (Experimental section). This was done to avoid the risk that rapid addition of the reducing agent would reduce C_{60} beyond an oxidation state of −3 to Cs_4C_{60}. The antisolvent precipitation of the $Cs_3C_{60}\cdot x$THF solvated product by hexane was the final and rather sensitive step in which the yield and crystallinity of the product could be affected. In this step, hexane played two roles; firstly it precipitated the solvated product and secondly it dissolved the $Mn(Cp^*)_2$ starting material. The as-made material was very sensitive to solvent loss, quickly losing its red colouring and crystallinity while becoming black. Annealing of the as-made material at 350 °C was required in order to obtain high quality solvent-free crystalline material (Sample 1) suitable for synchrotron X-ray powder diffraction studies.

Synthesis at twice this scale was also undertaken (Experimental section). The reaction itself proceeds in exactly the same fashion as its predecessor, first leading to a red solution and then on precipitation to a red-brown powder, which quickly turns black upon loss of the solvent. The final annealing step at 350 °C was required for the isolation of a high quality material (Sample 2).

The annealed products were characterised by synchrotron X-ray powder diffraction with the diffraction profiles shown in Fig. 2. The synchrotron data were refined by Rietveld methods against the $Fm\bar{3}m$ structure model. The comparison of the refined parameters with those reported before[7] for samples synthesized *via* ammonia and amine chemistry is summarized in Table 1. Both products showed reduction in the unit cell volumes by 0.11 and 0.28% for 1 and 2 respectively when compared with those reported in the literature. Although the

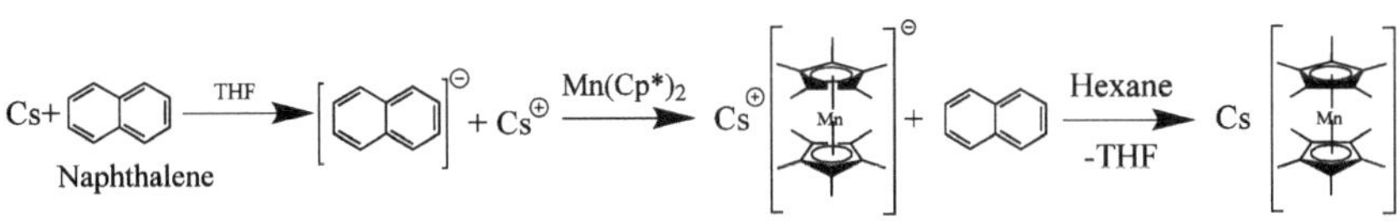

Scheme 1 Reaction scheme for the synthesis of the $Cs[Mn(Cp^*)]_2$ reducing agent.

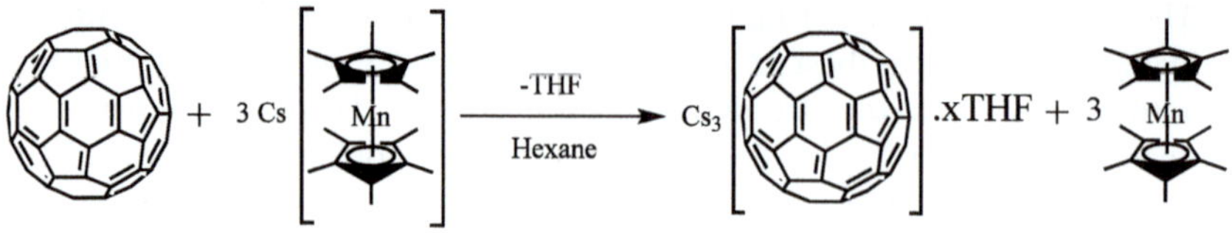

Scheme 2 Schematic representation of the reaction between Cs[Mn(Cp*)$_2$] and C_{60} to afford solvated Cs_3C_{60}.

diffraction patterns appear to be free from any Cs_4C_{60} phase peaks, they contain reflections which can be associated with the presence of a different impurity phase. This phase was accounted for in the course of the refinements by the pattern decomposition Pawley method using a monoclinic unit cell (space group *C*2/*m*) based on the structure model proposed for the solvated potassium analogue, $K_3C_{60}\cdot 14THF$.[18]

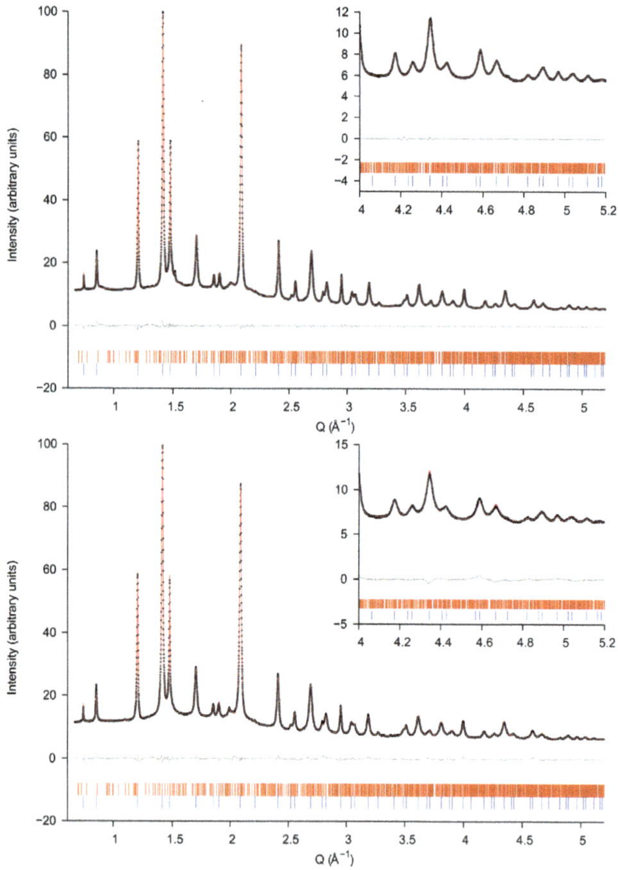

Fig. 2 Observed (black dots) and calculated (red line) synchrotron X-ray ($\lambda = 0.82716$ Å) powder diffraction patterns of the Cs_3C_{60} samples 1 (top) and 2 (bottom) at room temperature. The grey solid line shows the difference profile and the ticks mark the reflection positions of the FCC (blue) and solvated (orange) Cs_3C_{60} phases. Refined parameters are given in Table 1.

Table 1 Composition, phase fractions, unit cell parameters and volume per C_{60} for the FCC (space group $Fm\bar{3}m$) Cs_3C_{60}. The unit cell parameters of the monoclinic (space group $C2/m$) solvated impurity phase were: Sample 1: $a = 19.306(2)$, $b = 18.450(2)$, $c = 14.163(2)$ Å, $\beta = 110.53(1)°$; Sample 2: $a = 19.293(2)$, $b = 17.484(2)$, $c = 14.244(2)$ Å, $\beta = 110.69(1)°$. The refined atomic parameters are summarized in Tables S1 and S2†

Parameter	Sample 1	Sample 2	Ref. 7
Scale (mg)	100	200	278
λ (Å)	0.82716	0.82716	0.40004
BCO (wt%)	0	0	6.7(2)
FCC (wt%)	100	100	85.88(2)
A15 (wt%)	0	0	3.31(5)
CsC_{60} (wt%)	0	0	4.10(6)
a, Å	14.7560(2)	14.7479(2)	14.76151(8)
FCC V per C_{60} (Å^3)	803.24(3)	801.92(3)	804.14(1)
FCC XRD composition	$Cs_{2.942(2)}C_{60}$	$Cs_{2.947(1)}C_{60}$	$Cs_{2.901(6)}C_{60}$

The refined composition of the majority FCC phase is close to that of the ideal Cs_3C_{60} structure as is also evident from Raman spectroscopy (Table 2 and Fig. S2†) which showed a single peak in the 1400 cm^{-1} < $\bar{\nu}$ < 1600 cm^{-1} region, assigned to the $A_g(2)$ mode of C_{60} consistent with −3 oxidation state. The $A_g(2)$ pentagonal pinch mode has been found particularly sensitive to fulleride charge state, showing empirically an ~6 cm^{-1} shift with each additional integer C_{60} charge.[19] Unlike the literature reference material, which was prepared by amine chemistry and showed splitting of the $A_g(2)$ peak due to the presence of the impurity CsC_{60} and Cs_4C_{60} phases, the products of THF chemistry display only single peaks in their Raman spectra. However, a significant amount of peak broadening present could indicate some disorder within the system.

Temperature-dependent magnetisation measurements under pressure revealed that both Sample 1 and 2 become superconducting upon pressurisation (Fig. S3†). The transition into a superconducting state is initially observed at T_c = 29 K under a pressure of 0.26 GPa which are comparable to the values observed for the products prepared by amine/ammonia chemistry routes (Fig. 3). T_c increases to the maxima of 33.5 and 33.0 K for Sample 1 and 2 respectively. Although the materials show typical superconducting "dome"-type behaviour due to the tuning of the interfulleride separation by the pressure, both samples display lower superconducting temperatures compared with the literature analogues. The

Table 2 Summary of the results of Lorentzian fits to the Raman data. ω_0 represents the peak position, Γ the peak width and A the peak area

ID Parameter	Ganin *et al.*[7]		Sample 1	Sample 2
Peak	1	2	1	1
ω_0 (cm^{-1})	1446.70(2)	1456.34(4)	1448.15(3)	1446.25(2)
Γ (cm^{-1})	6.89(8)	13.9(8)	14.8(1)	12.37(8)
A (cm^{-1} counts)	10.3(1)	3.4(2)	23.6(2)	18.5(1)

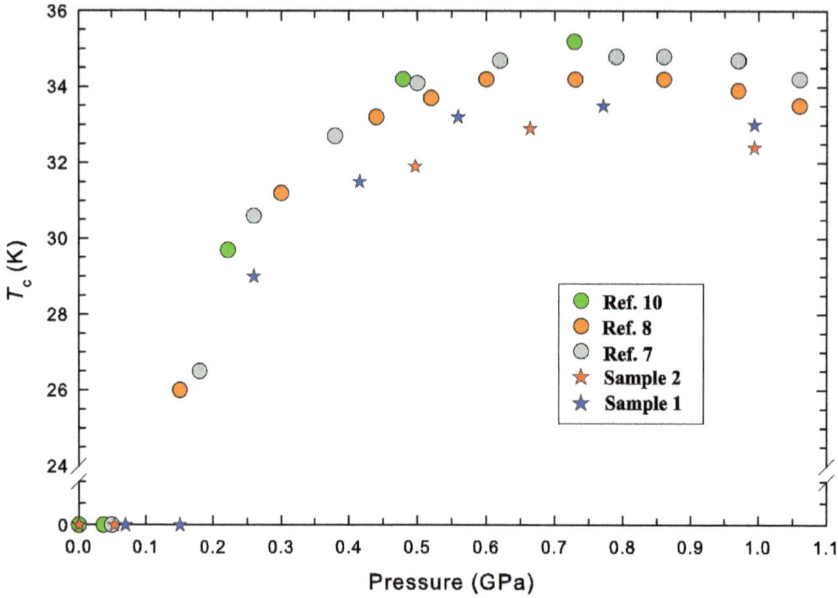

Fig. 3 Superconducting transition temperatures plotted as a function of the applied pressure for Cs_3C_{60} samples 1 (blue stars) and 2 (red stars). The circles represent the results from the literature .[7,8,10] The data were collected on increasing pressure.

disorder (Raman linewidth) and reduced unit cell dimensions could be possible explanations for the inferior superconducting properties of the materials prepared using Cs[Mn(Cp*)$_2$] as a reducing agent in THF. Further optimisation of the proposed synthetic procedure is required to produce Cs_3C_{60} materials with improved superconducting properties, but the present work demonstrates that control of the redox behaviour of the reducing agents reduces contamination by other C_{60} charge states.

Conclusions

The major importance of this work is the solution-based synthesis of the high T molecular superconductor Cs_3C_{60} with the readily available and handleable solvent THF, enabled by the use of the organometallic reducing agent. As well as opening up new routes to single crystal growth, the use of alternative counterions may permit new alkaline earth and lanthanide derivatives of fullerenes to be synthesized. The approach may also be applicable to the chemistry of reducible hydrocarbons such as the acenes[20] which display superconductivity but are challenging to prepare in single-phase form.

Experimental

Synthesis of the reducing agent Cs[Mn(Cp*)$_2$]

Caesium metal (548.0 mg, 4.12 mmol, Sigma-Aldrich, 99.5%) and naphthalene (530.0 mg, 4.14 mmol, Alfa Aesar, 99.6%) were reacted together for 1 hour at 30 °C in THF (120 ml, Fisher, GLC grade, 99.8+%) purified by distillation from sodium

(Sigma Aldrich, 99.5%)-benzophenone ketyl (Sigma-Aldrich, ReagentPlus, 99%). This solution was then transferred by cannula to a vessel containing $Mn(Cp^*)_2$ (additionally purified by sublimation, 1.34 g, 4.12 mmol, Sigma Aldrich, 97%) under constant stirring and left for 15 minutes to react. THF was then evaporated off into a liquid nitrogen trap and the resulting orange solid suspended with hexane (40 ml). The suspension was then separated using a frit and the precipitate washed with three lots of hexane (20 ml) before being dried under dynamic vacuum until a pressure of at least 1×10^{-1} mbar was achieved. The dried material was ground in a glove box giving a yield of 92.8%. The composition of the reducing agent was ascertained by microanalysis (CHN) by sealing tin cans with the sample inside in a He glove box. Theory (C: 52.42; H: 6.60; N: 0)/experiment (C: 52.21; H: 6.54; N: 0).

Synthesis of Cs_3C_{60} from the $Cs[Mn(Cp^*)_2]$ reducing agent and C_{60}

The optimised synthesis of the Cs_3C_{60} material was carried out by making two suspensions of $Cs[Mn(Cp^*)_2]$ (137.5 mg, 0.3 mmol) and C_{60} (72.0 mg, 0.1 mmol) in 160 ml and 20 ml of THF respectively. The $Cs[Mn(Cp^*)_2]$ suspension was then added drop wise to the C_{60} through a cannula over 90 minutes. The use of a cannula allowed the slow addition of reducing agent to the C_{60} by a pressure differential. The reaction mixture quickly turned from orange to the dark red colour indicative of the $C_{60}{}^{-3}$ anion. The mixture was then stirred for 30 minutes and the volume subsequently reduced to 20 ml. The $Cs_3C_{60}\cdot x$THF material was precipitated from the solution by the slow addition of dry hexane (20 ml). Initially the black crystalline product became visible leaving a pale yellow solution remaining. The slow addition of the remaining hexane produced large particles of $Cs_3C_{60}\cdot x$THF leaving a clear yellow solution. Once precipitated, the product was filtered through a frit, washed three times with hexane (10 ml) and dried by passing argon over the material. The product consisting 109.2 mg of the red brown material, quickly losing its appearance by turning into a black solid, was ground up in the glove box. In order to improve the crystallinity of the as made material and remove the solvent, 100 mg was annealed under dynamic vacuum (1×10^{-5} mbar) for 15 hours at 350 °C (heating protocol: RT $\rightarrow$ 5° min^{-1}, 350 °C, 15 h $\rightarrow$ 5° min^{-1} RT). After the annealing protocol, 79.5 mg of a black product was recovered and stored in the glovebox.

Scaled-up synthesis of Cs_3C_{60} from the $Cs[Mn(Cp^*)_2]$ reducing agent and C_{60}

The reaction protocol was carried out as described above by scaling up the amount of reagents and solvents. The amount of the as made product was 171.5 mg. Annealing of 160 mg of the as made sample yielded 153.1 mg of black powder which was stored in the glove box.

Synchrotron X-ray diffraction

High-resolution synchrotron powder X-ray diffraction data were collected at ambient temperature using the high-resolution powder diffractometers on beamline I11 at Diamond, UK ($\lambda = 0.82716$ Å).

TOPAS Academic V5 software was used for refinement of the collected PXRD patterns where the input files were written in JEdit.[21]

The Rietveld refinement of the Cs_3C_{60} phase was carried out based on the structural model used by Ganin *et al.* in the space group $Fm\bar{3}m$.[8] In order to minimize the effect of parameter correlations on the accuracy of refinements they were carried out in consecutive steps as described by McCusker *et al.*[22] The background was fitted using a Chebyshev polynomial function of no more than 13 terms for data obtained using synchrotron radiation. The peak shapes of the PXRD patterns of the materials presented in this work were fitted using standard or combinations of standard analytical peak shapes. Diffraction patterns of fullerides, like the Cs_3C_{60} materials reported here, have complex anisotropic profile broadening, which are dealt with using an anisotropic peak shape which was developed on this type of material.[23]

A Pawley refinement was used for the impurity phase as the simplest method of determining a suitable peak shape function, so that the peak profile could be determined in the absence of information about atomic parameters. The crystal structure of $K_3C_{60}\cdot(THF)_{14}$ and space group $C2/m$ was used for the fit. The background, zero offset, axial divergence and cell parameters were refined along with a Thompson–Cox–Hastings (TCHZ) function.[24]

Raman spectroscopy

Raman spectra of samples 1 and 2 were collected at ambient temperature using a 514.5 nm laser with a JY LabRam-HR Spectrometer operated in back-scattered geometry over a sample area of 500 μm^2. The typical acquisition times were 20×6 s. Calibration was performed by referencing the spectrometer to the 520.07 cm^{-1} silicon line. C_{60} and Cs_6C_{60} samples were used as standards ($A_g(2)$ peaks at 1468 and 1430 cm^{-1}, respectively). The resulting Raman mode shift per one electron transferred to C_{60} was found to be $\sim$6.33 cm^{-1}.

Magnetic measurements

High-pressure magnetic measurements were undertaken using an easyLab Technologies piston-cylinder high pressure cell (Mcell10) with an upper limit of hydrostatic pressure of 10.6 kbar; high-purity Sn was used as an *in situ* manometer with Daphne oil as pressure transmitting medium, and magnetization data were collected on $\sim$20 mg samples loaded in an argon-filled glovebox under ZFC and FC protocols at 20 Oe.

Acknowledgements

K.P. and M.J.R. thank the financial support by EPSRC (EP/G037132 and EP/G037949) and by the European Union FP7-NMP-2011-EU-Japan project LEMSUPER under contract no. 283214. We also thank STFC for access to the synchrotron X-ray facilities at Diamond (where we thank C. Tang for his assistance on beamline I11). K.P. is a Royal Society Wolfson Research Merit Award holder and M.J.R. is a Royal Society Research Professor.

References

1 O. Zhou, R. M. Fleming, D. W. Murphy, M. J. Rosseinsky, A. P. Ramirez, R. B. Vandover and R. C. Haddon, *Nature*, 1993, **362**, 433.

2 S. Margadonna, E. Aslanis and K. Prassides, *J. Am. Chem. Soc.*, 2002, **124**, 10146.

3 Y. Iwasa, H. Shimoda, Y. Miyamoto, T. Mitani, Y. Maniwa, O. Zhou and T. T. M. Palstra, *J. Phys. Chem. Solids*, 1997, **58**, 1697.

4 H. Shimoda, Y. Iwasa, Y. Miyamoto, Y. Maniwa and T. Mitani, *Phys. Rev. B: Condens. Matter Mater. Phys.*, 1996, **54**, 15653.

5 K. Prassides, *Curr. Opin. Solid State Mater. Sci.*, 1997, **2**, 433.

6 M. J. Rosseinsky, *Chem. Mat.*, 1998, **10**, 2665.

7 A. Y. Ganin, Y. Takabayashi, P. Jeglic, D. Arcon, A. Potocnik, P. J. Baker, Y. Ohishi, M. T. McDonald, M. D. Tzirakis, A. McLennan, G. R. Darling, M. Takata, M. J. Rosseinsky and K. Prassides, *Nature*, 2010, **466**, 221.

8 A. Y. Ganin, Y. Takabayashi, Y. Z. Khimyak, S. Margadonna, A. Tamai, M. J. Rosseinsky and K. Prassides, *Nat. Mater.*, 2008, **7**, 367.

9 Y. Takabayashi, A. Y. Ganin, P. Jeglic, D. Arcon, T. Takano, Y. Iwasa, Y. Ohishi, M. Takata, N. Takeshita, K. Prassides and M. J. Rosseinsky, *Science*, 2009, **323**, 1585.

10 M. T. McDonald, PhD thesis, Durham University, 2010.

11 T. T. M. Palstra, O. Zhou, Y. Iwasa, P. E. Sulewski, R. M. Fleming and B. R. Zegarski, *Solid State Commun.*, 1995, **93**, 327.

12 L. Echegoyen and L. E. Echegoyen, *Acc. Chem. Res.*, 1998, **31**, 593.

13 D. Dubois, G. Moninot, W. Kutner, M. T. Jones and K. M. Kadish, *J. Phys. Chem.*, 1992, **96**, 7137.

14 D. L. Reger, S. R. Goode and D. W. Ball, *Chemistry: Principles and Practice*, Brooks/Cole, 2009.

15 N. G. Connelly and W. E. Geiger, *Chem. Rev.*, 1996, **96**, 877–910.

16 X. H. Liu, W. C. Wan, S. M. Owens and W. E. Broderick, *J. Am. Chem. Soc.*, 1994, **116**, 5489.

17 J. L. Robbins, N. M. Edelstein, S. R. Cooper and J. C. Smart, *J. Am. Chem. Soc.*, 1979, **101**, 3853.

18 C. Janiak, S. Muhle, H. Hemling and K. Kohler, *Polyhedron*, 1996, **15**, 1559.

19 M. G. Mitch and J. S. Lannin, *Phys. Rev. B: Condens. Matter Mater. Phys.*, 1995, **51**, 6784.

20 R. Mitsuhashi, Y. Suzuki, Y. Yamanari, H. Mitamura, T. Kambe, N. Ikeda, H. Okamoto, A. Fujiwara, M. Yamaji, N. Kawasaki, Y. Maniwa and Y. Kubozono, *Nature*, 2010, **464**, 76.

21 Bruker AXS, Karlsruhe, 2003.

22 L. B. McCusker, R. B. Von Dreele, D. E. Cox, D. Louer and P. Scardi, *J. Appl. Crystallogr.*, 1999, **32**, 36–50.

23 P. W. Stephens, *J. Appl. Crystallogr.*, 1999, **32**, 281–289.

24 R. A. Young, *The Rietveld method*, Oxford University Press, Oxford, 1993.

Faraday Discussions

PAPER

Bulk properties of solution-synthesized chevron-like graphene nanoribbons

Timothy H. Vo,[a] Mikhail Shekhirev,[a] Alexey Lipatov,[a] Rafal A. Korlacki[b] and Alexander Sinitskii*[ac]

Received 13th June 2014, Accepted 3rd July 2014
DOI: 10.1039/c4fd00131a

Graphene nanoribbons (GNRs) have received a great deal of attention due to their promise for electronic and optoelectronic applications. Several recent studies have focused on the synthesis of GNRs by the bottom-up approaches that could yield very narrow GNRs with atomically precise edges. One type of GNRs that has received a considerable attention is the chevron-like GNR with a very distinct periodic structure. Surface-assisted and solution-based synthetic approaches for the chevron-like GNRs have been developed, but their electronic properties have not been reported yet. In this work, we synthesized chevron-like GNRs in bulk by a solution-based method, characterized them by a number of spectroscopic techniques and measured their bulk conductivity. We demonstrate that solution-synthesized chevron-like GNRs are electrically conductive in bulk, which makes them a potentially promising material for applications in organic electronics and photovoltaics.

1 Introduction

Graphene nanoribbons (GNRs), few-nm-wide strips of graphene, possess interesting electronic and optical properties.[1–6] While GNRs have long been considered theoretically,[1,2] the active experimental research in this area started less than a decade ago. Several recent studies have focused on the synthesis of GNRs by the bottom-up approaches that rely on building GNRs from smaller molecular species.[7–20] The resulting GNRs are very narrow (typically with widths, $w < 2$ nm) and have atomically precise edges, which is very important for potential applications.[3–6,21,22] One type of synthetic GNRs that has received considerable theoretical[23–25] and experimental attention[9,16,19,20,26] is the chevron-like GNR that has a very distinct periodic structure (Fig. 1a). Chevron-like GNRs were first synthesized by Cai *et al.* by the surface-assisted approach that involved coupling molecular precursors into linear polyphenylenes on Au(111) single crystals followed by

[a]*Department of Chemistry, University of Nebraska – Lincoln, Lincoln, NE 68588, USA. E-mail: sinitskii@unl.edu*
[b]*Department of Electrical Engineering, University of Nebraska – Lincoln, Lincoln, NE 68588, USA*
[c]*Nebraska Center for Materials and Nanoscience, University of Nebraska – Lincoln, Lincoln, NE 68588, USA*

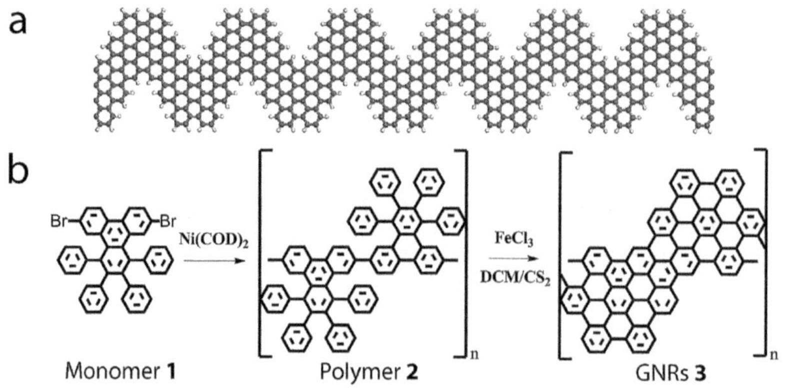

Fig. 1 Synthesis of chevron-like GNRs. (a) Schematic of a chevron-like GNR. (b) Scheme of the GNR synthesis; see text for details.

cyclodehydrogenation.[9] Later the solution-based synthetic approach for bulk quantities of chevron-like GNRs that relies on Yamamoto coupling of molecular precursors followed by cyclodehydrogenation *via* a Scholl reaction has also been demonstrated.[19,20]

Considering the potential use of GNRs in electronics[27–29] it is very important to understand their electronic properties. Despite some recent attempts to measure the electronic properties of GNRs synthesized by the bottom-up approaches[30,31] the electrical characterization of GNRs remains challenging in general and no conductivity measurements have been performed on chevron-like GNRs in particular. In this work, we synthesized chevron-like GNRs in bulk by a solution-based method,[19] confirmed their quality by scanning tunnelling microscopy (STM) and a number of spectroscopic techniques, and measured their bulk conductivity.

2 Experimental

All starting materials and solvents were purchased and used as received without any purification. Sodium hydroxide (97%), nitromethane (98%), iron(III) chloride (anhydrous, 98%), carbon disulphide and gold foil (0.025 mm, 99.985%) were purchased from Alfa Aesar. Methanol and mesitylene were purchased from Sigma-Aldrich. Acetone and dichloromethane were purchased from Fisher Scientific, and concentrated hydrochloric acid was purchased from Macron. Four-inch heavily p-doped silicon wafers with a 300 nm-thick layer of SiO_2 were purchased from Silicon Quest International.

Fig. 1b shows the scheme of the solution synthesis of chevron-like GNRs. We used the synthetic procedure that we disclosed in our previous work,[19] although with some modifications. We first synthesized the precursor monomer **1** and then converted it to the polymer **2** using bis(1,5-cyclooctadiene)nickel(0) (Ni(COD)$_2$) as described in our previous work.[19] We previously performed the cyclodehydrogenation of the polymer **2** with $FeCl_3$ to form GNRs using dichloromethane (DCM) as a solvent.[19] In the present work we preformed this reaction in a DCM–CS_2 mixture (1 : 1), since this mixture was shown to be an effective solvent for cyclodehydrogenation of other polyphenylenes *via* the Scholl reaction.[32] More specifically, 50 mg of polymer **2** was added to 80 ml of DCM–CS_2 mixture (1 : 1) and sonicated until well-dispersed. The reaction mixture was degassed with nitrogen gas. Iron(III) chloride (1.666 g) was dissolved in nitromethane (5 ml) and added to the reaction mixture. The nitrogen gas was bubbled through the reaction mixture for 24 h. After that another portion of DCM–CS_2 (1 : 1) (60 ml) was added to the reaction mixture and the reaction was continued for another 7 h. The reaction mixture was then filtered and washed with concentrated hydrochloric acid, acetone, water, methanol, a methanol–sodium hydroxide mixture, methanol, water, and finally acetone. The resulting black solid was dried at 65 °C, yielding 49 mg of a black powder.

A Raman spectrum of GNRs deposited on Si/SiO_2 substrate was recorded using a Thermo Scientific DXR Raman microscope with a 532 nm laser.

UV-vis-NIR measurements of GNRs were performed using a Jasco V-670 spectrophotometer. GNR powder (2 mg) was added to 7 ml of mesitylene and sonicated at 80–90 °C for 90 s. The mixture was then refluxed for 10 min and sonicated at 80–90 °C for 30 s. The reflux and sonication steps were repeated one more time. Finally, the GNR dispersion in mesitylene was heated to reflux and transferred to a cuvette for UV-vis-NIR measurements while still hot.

X-ray photoelectron spectroscopy (XPS) was performed using a PHI Quantera SXM scanning X-ray microprobe. To prepare XPS samples several droplets of a GNR suspension in DCM were deposited on a gold foil and dried in air.

GNRs were deposited on a Au(111) substrate and imaged by scanning tunnelling microscopy (STM). To prepare a sample for STM measurements ~2 mg of GNRs powder was mixed with ~5 ml of mesitylene in a 100 ml round-bottom flask. The mixture was sonicated for 1 min and then heated to reflux while agitated for 5 min and then sonicated for another 30 s. The reflux-sonication procedure was repeated several times to the total of ~2.5 min of sonication time. Using a glass pipette, two drops of the GNR suspension in mesitylene were deposited while still hot ($T > 100$ °C) on the Au(111) substrate. The STM imaging was performed using a commercial Pt–Ir tip and a home-built ultrahigh vacuum (UHV) variable temperature STM system at the Center for Nanophase Materials Science (CNMS) at Oak Ridge National Laboratory (ORNL); the base pressure was lower than 1×10^{-10} mbar. The GNR samples were degassed in UHV for several hours before performing the STM measurements.

Fourier transform infrared (FTIR) spectroscopy of GNRs was preformed using a Nicolet Avatar 360 FTIR instrument. The measurements were performed on a small amount of a GNR powder that was placed directly on the spectrometer sample stage.

The electrical measurements of GNRs were performed using an Agilent 4155C Semiconductor Parameter Analyzer that was connected to a computer through 82357B USB/GPIB Interface and controlled using National Instruments LabView software. GNRs were measured in the form of pressed pellets (diameter, $d = 7$ mm, thickness, $t \sim 0.1$ mm) that were prepared using a hand press (International Crystal Laboratories). Prior to some of the electrical measurements the GNR pellets were annealed at 200 °C in a tube furnace (Lindberg/Blue M) in H_2/Ar (5 sccm of H_2, 5 sccm of Ar, total pressure ~500 mTorr) for 2 h.

Theoretical calculations were performed using a plane-wave density functional theory code Quantum ESPRESSO,[33] with the combination of the exchange–correlation functional of Perdew, Burke, and Ernzerhof (PBE)[34] together with norm-conserving pseudopotentials developed by the Rappe Group.[35] The smallest repetition unit of the studied nanoribbon was placed in an orthorhombic cell, propagating along the *X*-direction and maintaining separation of at least 20 bohr (approx. 10.6 Å) on the vacuum sides (*Y* and *Z*). A large cutoff of 100 Ry for the electronic wavefunction and a tight convergence threshold of 1×10^{-12} for self-consistency were used throughout the calculations. The phonon frequencies and Raman activities were computed at the Γ point of the Brillouin zone using density-functional perturbation-theory for phonons[36] and second-order response for Raman activities,[37] for a structure previously relaxed to the level of reasonably small forces ($<6.0 \times 10^{-4}$ Ry per bohr).

3 Results and discussion

In order to confirm the high quality of GNRs that were prepared using the modified synthetic procedure, they were characterized by a number of analytical techniques. First of all, we used Raman spectroscopy, a technique known to be very sensitive to the disorder in carbon materials. Fig. 2a shows a comparison of theoretical and experimental Raman spectra of chevron-like GNRs. According to the results of the theoretical simulations, the Raman spectrum of chevron-like GNRs should exhibit not only the D and G bands (the most intense peaks in the spectrum around 1300 cm^{-1} and 1600 cm^{-1}, respectively) that are standard for graphene-based materials,[38] but also a series of smaller bands that are characteristic for this particular nanoribbon; these smaller bands are shown by arrows in the inset in Fig. 2a. These smaller bands include the fine structure of three peaks at the low-energy side of the G band, as well as the series of peaks around the D band. These features could be observed in the experimental spectrum of the chevron-like GNRs, which indicates their high structural quality.

We also used FTIR spectroscopy to confirm the efficiency of the oxidative cyclodehydrogenation of the polymer **2** to form GNRs **3** (see the last step in the reaction scheme shown in Fig. 1b) and the structural quality of the chevron-like GNRs. Fig. 2b demonstrates representative spectral regions for the material before and after cyclodehydrogenation, as well as the theoretical spectrum of the GNRs **3**. The left panel in Fig. 2b shows the spectral range from 3100 to 2850 cm^{-1} where the FTIR spectrum of the polymer **2** contains three characteristic peaks at 3077, 3053 and 3024 cm^{-1} that originate from the aromatic C–H stretching.[14,39] These peaks are not observed in the FTIR spectrum of GNRs **3**, which is in agreement with the theoretical spectrum, according to which the chevron-like GNRs should exhibit no peaks in this spectral range. Similar conclusions could be drawn from the analysis of the FTIR spectra of the polymer **2** and GNRs **3** in the spectral range from 900 to 650 cm^{-1}, see the right panel in Fig. 2b. The theoretical spectrum of the chevron-like GNRs is very different from the experimental FTIR spectrum of the polymer **2**, but, as expected, is in a good agreement with the experimental FTIR spectrum of GNRs **3**. In the polymer **2** spectrum the most intense peak at 697 cm^{-1}, which is typical for monosubstituted benzene rings,[14,39] nearly disappears after the oxidative cyclodehydrogenation of the polymer **2** to form GNRs **3** (Fig. 2b), and the remaining small peak actually matches well the similarly small peak in the theoretical spectrum of the chevron-like GNRs.

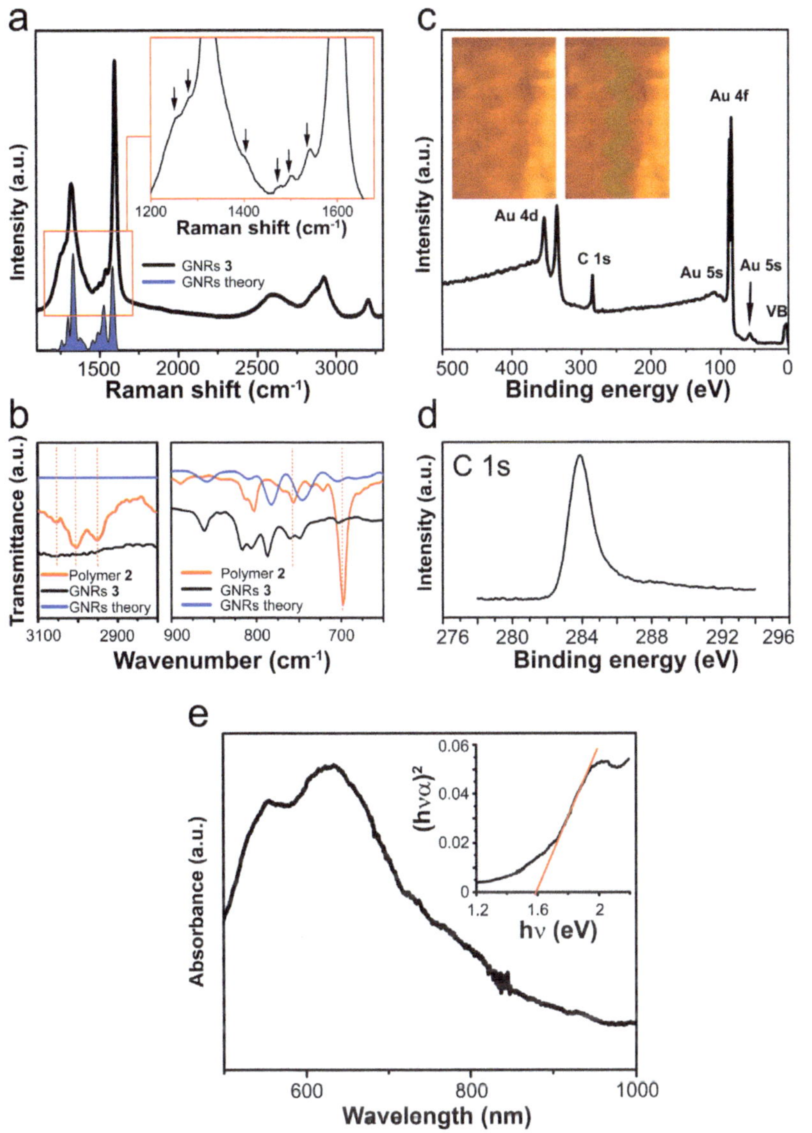

Fig. 2 Spectroscopic and microscopic characterization of GNRs. (a) Raman spectrum of GNRs. The black arrows in the inset show small peaks around D and G bands. (b) Comparison of the experimental FTIR spectra of the polymer **2**, GNRs **3**, and theoretical FTIR spectrum of GNRs **3**. (c and d) XPS spectra of GNRs deposited on a gold substrate: (c) survey spectrum, (d) C 1s spectrum. The inset in (c) shows 2.8 × 4.3 nm^2 STM image of a GNR deposited on Au(111) substrate; the left panel shows the original image, and the right panel shows the same image where the GNR is highlighted for the sake of clarity. (e) UV-vis-NIR absorbance spectrum of GNRs suspended in mesitylene. The inset shows the analysis of the energy-dependent absorbance of GNRs suspended in mesitylene, demonstrating an optical bandgap of ~1.6 eV.

Overall, the FTIR spectroscopy results confirm the efficient transformation of the polymer **2** to GNRs **3** (Fig. 1b), and the structural quality of the resulting nanoribbons.

To further demonstrate the high quality of these GNRs we visualized them by STM. The inset in Fig. 2c shows STM image of a GNR that was deposited on Au(111) substrate; the left panel shows the original image, while in the right panel the GNR is highlighted in green for the sake of clarity. The structure of the GNR observed in this STM image is in perfect agreement with the atomic structure shown in Fig. 1a.

The chemical purity of the GNRs was confirmed using XPS spectroscopy. Fig. 2c shows the XPS survey spectrum of GNRs deposited on a gold substrate. No peaks other than the C 1s peak and the peaks associated with the gold substrate could be observed in this spectrum, suggesting that possible impurities could be present in the sample only in amounts below the XPS detection limit. Fig. 2d shows the XPS C 1s spectrum of GNRs, demonstrating only a single sharp component at 284.5 eV corresponding to the sp^2 carbon atoms, as expected for a graphene-based material.

Fig. 2e shows a UV-vis-NIR spectrum of the chevron-like GNRs dispersed in mesitylene. The GNRs have strong absorption in UV and visible regions, which then decreases toward the NIR region and becomes negligible at $\lambda \sim 950$ nm ($E \sim$ 1.3 eV), The linearization of the energy-dependent absorbance of the GNRs (see the inset in Fig. 2e) gives an optical band gap of ~1.6 eV, which is very close to the theoretical band gap value of 1.57 eV that was previously calculated using the density functional theory (DFT) method.[9,19]

The electrical measurements were performed on pressed GNR pellets using the home-built setup shown in Fig. 3a. A GNR pellet was placed between two aluminium electrodes surrounded by a dielectric cylinder ("inner shell"). The dielectric outer shell was gently pressed to the bottom electrode using two screws to ensure good contact between the electrodes and GNR pellet. The electrodes were connected to a parameter analyzer for electrical measurements. Fig. 3b shows a typical GNR pellet that is 7 mm in diameter and has a thickness of ~0.1 mm. Interestingly, although the GNR pellet has a black colour (Fig. 3b, left panel), it also has a characteristic metallic luster (Fig. 3b, right panel), thus somewhat resembling a graphite flake.

Two-terminal voltage–current (I–V) curves were measured in the voltage range from −5 to +5 V. Fig. 3c shows that the as-prepared GNR pellet is fairly conductive, showing a current of the order of a few microamperes in the given voltage range. The conductivity could be improved if the GNR pellet is annealed at 200 °C in H_2/Ar, as shown in Fig. 3c. The fact that the chevron-like GNRs are electrically conductive is an important result, considering the promise of bottom-up synthesized GNRs for organic electronics.[40,41]

It should also be noted that Fig. 3c illustrates the bulk conductivity of ribbons rather than represents the intrinsic electronic properties of an individual chevron-like GNR. An appropriate analogy can be made with carbon nanotubes (CNTs) that, similarly to GNRs, are also one-dimensional graphene-based materials. It is well-known that while CNTs have remarkable electronic properties, the electrical conductivities of CNTs fibers and composites are much lower than those of individual CNTs by at least one order of magnitude even for the rigorously optimized samples.[42] The electrical conductivity of a macroscopic CNT sample is affected by various structural imperfections that include poor intertube junctions, insufficient packing density, poor alignment of the CNTs, *etc.* In analogy with CNT fibers and composites, a bulk GNR sample (Fig. 3b) should have similar

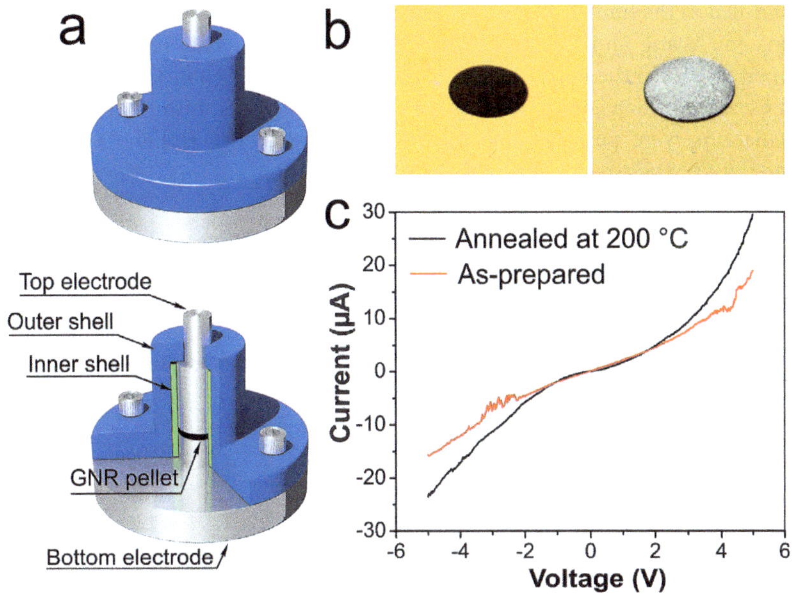

Fig. 3 Electrical measurements of GNRs. (a) Scheme of experimental setup for electrical measurements of GNRs. Top and bottom panels show the side view and the cross-section view of the setup, respectively. (b) Optical photographs of the as-prepared GNR pellet at different angles of observation. (c) Current–voltage (*I–V*) plots for the as-prepared and annealed GNR pellets.

structural imperfections, such as inefficient ribbon-to-ribbon junctions, low packing density of GNRs, and their poor alignment. Therefore, electrical measurements of individual chevron-like GNRs, as well as additional studies of different arrangements of GNRs, such as thin films and fibers, will be important to reveal the potential of this material for future applications.

4 Conclusions

In summary, we synthesized chevron-like GNRs in bulk by a solution-based method, visualized them by STM and characterized them by a number of spectroscopic techniques. We demonstrate that the as-synthesized chevron-like GNRs are electrically conductive in bulk, and the bulk conductivity of GNRs could be further improved by annealing in H_2/Ar. Electrical measurements of individual nanoribbons are still in order to better understand their electronic properties. However, the fact that the chevron-like GNRs are electrically conductive makes them a promising object for the future studies of the bulk applications of bottom-up synthesized GNRs in areas, such as GNR-based electronics and photovoltaics.

Acknowledgements

This work was supported by the Nebraska Center for Energy Sciences Research (#12-00-13), the Nebraska Research Initiative and the NSF through Nebraska MRSEC (DMR-0820521) and EPSCoR (EPS-1004094). STM imaging of GNRs was performed in the framework of the CNMS ORNL user project (CNMS2013-349); we thank Jun Wang and Minghu Pan for STM assistance. This research was performed in part in the Central Facilities of the Nebraska Center for Materials and Nanoscience, which is supported by the Nebraska Research Initiative. Theoretical calculations were performed using the resources of the Holland Computing Center at the University of Nebraska-Lincoln.

References

1 M. Fujita, K. Wakabayashi, K. Nakada and K. Kusakabe, *J. Phys. Soc. Jpn.*, 1996, **65**, 1920.
2 K. Nakada, M. Fujita, G. Dresselhaus and M. S. Dresselhaus, *Phys. Rev. B: Condens. Matter Mater. Phys.*, 1996, **54**, 17954.
3 V. Barone, O. Hod and G. E. Scuseria, *Nano Lett.*, 2006, **6**, 2748.
4 L. Yang, C.-H. Park, Y.-W. Son, M. L. Cohen and S. G. Louie, *Phys. Rev. Lett.*, 2007, **99**, 186801.
5 L. Yang, M. L. Cohen and S. G. Louie, *Nano Lett.*, 2007, **7**, 3112.
6 D. Prezzi, D. Varsano, A. Ruini, A. Marini and E. Molinari, *Phys. Rev. B: Condens. Matter Mater. Phys.*, 2008, **77**, 041404.
7 X. Y. Yang, X. Dou, A. Rouhanipour, L. J. Zhi, H. J. Rader and K. Mullen, *J. Am. Chem. Soc.*, 2008, **130**, 4216.
8 Y. Fogel, L. Zhi, A. Rouhanipour, D. Andrienko, H. J. Räder and K. Müllen, *Macromolecules*, 2009, **42**, 6878.
9 J. M. Cai, P. Ruffieux, R. Jaafar, M. Bieri, T. Braun, S. Blankenburg, M. Muoth, A. P. Seitsonen, M. Saleh, X. L. Feng, K. Mullen and R. Fasel, *Nature*, 2010, **466**, 470.
10 L. Dossel, L. Gherghel, X. L. Feng and K. Mullen, *Angew. Chem., Int. Ed.*, 2011, **50**, 2540.
11 A. Chuvilin, E. Bichoutskaia, M. C. Gimenez-Lopez, T. W. Chamberlain, G. A. Rance, N. Kuganathan, J. Biskupek, U. Kaiser and A. N. Khlobystov, *Nat. Mater.*, 2011, **10**, 687.
12 T. W. Chamberlain, J. Biskupek, G. A. Rance, A. Chuvilin, T. J. Alexander, E. Bichoutskaia, U. Kaiser and A. N. Khlobystov, *ACS Nano*, 2012, **6**, 3943.
13 S. Blankenburg, J. M. Cai, P. Ruffieux, R. Jaafar, D. Passerone, X. L. Feng, K. Mullen, R. Fasel and C. A. Pignedoli, *ACS Nano*, 2012, **6**, 2020.
14 M. G. Schwab, A. Narita, Y. Hernandez, T. Balandina, K. S. Mali, S. De Feyter, X. L. Feng and K. Mullen, *J. Am. Chem. Soc.*, 2012, **134**, 18169.
15 Y.-C. Chen, D. G. de Oteyza, Z. Pedramrazi, C. Chen, F. R. Fischer and M. F. Crommie, *ACS Nano*, 2013, **7**, 6123.
16 C. Bronner, S. Stremlau, M. Gille, F. Braußе, A. Haase, S. Hecht and P. Tegeder, *Angew. Chem., Int. Ed.*, 2013, **52**, 4422.
17 K. T. Kim, J. W. Lee and W. H. Jo, *Macromol. Chem. Phys.*, 2013, **214**, 2768.
18 A. Narita, X. Feng, Y. Hernandez, S. A. Jensen, M. Bonn, H. Yang, I. A. Verzhbitskiy, C. Casiraghi, M. R. Hansen, A. H. R. Koch, G. Fytas, O. Ivasenko, B. Li, K. S. Mali, T. Balandina, S. Mahesh, S. De Feyter and K. Mullen, *Nat. Chem.*, 2014, **6**, 126.
19 T. H. Vo, M. Shekhirev, D. A. Kunkel, M. D. Morton, E. Berglund, L. M. Kong, P. M. Wilson, P. A. Dowben, A. Enders and A. Sinitskii, *Nat. Commun.*, 2014, **5**, 3189.

20 T. H. Vo, M. Shekhirev, D. A. Kunkel, F. Orange, M. J. F. Guinel, A. Enders and A. Sinitskii, *Chem. Commun.*, 2014, **50**, 4172.
21 D. A. Areshkin, D. Gunlycke and C. T. White, *Nano Lett.*, 2006, **7**, 204.
22 E. R. Mucciolo, A. H. Castro Neto and C. H. Lewenkopf, *Phys. Rev. B: Condens. Matter Mater. Phys.*, 2009, **79**, 075407.
23 E. Costa Girão, L. Liang, E. Cruz-Silva, A. G. S. Filho and V. Meunier, *Phys. Rev. Lett.*, 2011, **107**, 135501.
24 S. D. Wang and J. L. Wang, *J. Phys. Chem. C*, 2012, **116**, 10193.
25 L. Liang and V. Meunier, *Phys. Rev. B: Condens. Matter Mater. Phys.*, 2012, **86**, 195404.
26 S. Linden, D. Zhong, A. Timmer, N. Aghdassi, J. H. Franke, H. Zhang, X. Feng, K. Müllen, H. Fuchs, L. Chi and H. Zacharias, *Phys. Rev. Lett.*, 2012, **108**, 216801.
27 X. R. Wang, Y. J. Ouyang, X. L. Li, H. L. Wang, J. Guo and H. J. Dai, *Phys. Rev. Lett.*, 2008, **100**, 206803.
28 A. Sinitskii, A. Dimiev, D. V. Kosynkin and J. M. Tour, *ACS Nano*, 2010, **4**, 5405.
29 J. Baringhaus, M. Ruan, F. Edler, A. Tejeda, M. Sicot, A. Taleb-Ibrahimi, A. P. Li, Z. G. Jiang, E. H. Conrad, C. Berger, C. Tegenkamp and W. A. de Heer, *Nature*, 2014, **506**, 349.
30 P. B. Bennett, Z. Pedramrazi, A. Madani, Y.-C. Chen, D. G. de Oteyza, C. Chen, F. R. Fischer, M. F. Crommie and J. Bokor, *Appl. Phys. Lett.*, 2013, **103**, 253114.
31 A. N. Abbas, G. Liu, A. Narita, M. Orosco, X. Feng, K. Müllen and C. Zhou, *J. Am. Chem. Soc.*, 2014, **136**, 7555.
32 C. D. Simpson, G. Mattersteig, K. Martin, L. Gherghel, R. E. Bauer, H. J. Rader and K. Mullen, *J. Am. Chem. Soc.*, 2004, **126**, 3139.
33 P. Giannozzi, S. Baroni, N. Bonini, M. Calandra, R. Car, C. Cavazzoni, D. Ceresoli, G. L. Chiarotti, M. Cococcioni, I. Dabo, A. Dal Corso, S. de Gironcoli, S. Fabris, G. Fratesi, R. Gebauer, U. Gerstmann, C. Gougoussis, A. Kokalj, M. Lazzeri, L. Martin-Samos, N. Marzari, F. Mauri, R. Mazzarello, S. Paolini, A. Pasquarello, L. Paulatto, C. Sbraccia, S. Scandolo, G. Sclauzero, A. P. Seitsonen, A. Smogunov, P. Umari and R. M. Wentzcovitch, *J. Phys. Condens. Matter.*, 2009, **21**, 395502.
34 J. P. Perdew, K. Burke and M. Ernzerhof, *Phys. Rev. Lett.*, 1996, **77**, 3865.
35 http://www.sas.upenn.edu/rappegroup/research/pseudo-potential-gga.html.
36 S. Baroni, S. de Gironcoli, A. Dal Corso and P. Giannozzi, *Rev. Mod. Phys.*, 2001, **73**, 515.
37 M. Lazzeri and F. Mauri, *Phys. Rev. Lett.*, 2003, **90**, 036401.
38 A. C. Ferrari and D. M. Basko, *Nat. Nanotechnol.*, 2013, **8**, 235.
39 A. Centrone, L. Brambilla, T. Renouard, L. Gherghel, C. Mathis, K. Müllen and G. Zerbi, *Carbon*, 2005, **43**, 1593.
40 S. A. Jensen, R. Ulbricht, A. Narita, X. Feng, K. Müllen, T. Hertel, D. Turchinovich and M. Bonn, *Nano Lett.*, 2013, **13**, 5925.
41 M. El Gemayel, A. Narita, L. F. Dossel, R. S. Sundaram, A. Kiersnowski, W. Pisula, M. R. Hansen, A. C. Ferrari, E. Orgiu, X. Feng, K. Mullen and P. Samori, *Nanoscale*, 2014, **6**, 6301.
42 N. Behabtu, C. C. Young, D. E. Tsentalovich, O. Kleinerman, X. Wang, A. W. K. Ma, E. A. Bengio, R. F. ter Waarbeek, J. J. de Jong, R. E. Hoogerwerf, S. B. Fairchild, J. B. Ferguson, B. Maruyama, J. Kono, Y. Talmon, Y. Cohen, M. J. Otto and M. Pasquali, *Science*, 2013, **339**, 182.

Faraday Discussions

DISCUSSIONS

Synthesis in gas and liquid phase: general discussion

Philip Davies, Pagona Papakonstantinou, Nazario Martin, Irena Kratochvílová, Chris Ewels, Milo Shaffer, Toshiaki Enoki, Malcolm Heggie, Thurid Gspann, Ayse Turak, Oana Andreea Bârsan, David Zitoun, Pulickel Ajayan, Lyndsey Mooring, Varsha Khare, Alexander Zöpfl, Andreas Hirsch, Mark Baxendale, Pedro Costa, Vladimir Falko, Juan Casado, Lok Shu Hui, Alan Windle, Alexander Sinitskii, Matthew Rosseinksy and Thanh Nguyen

DOI: 10.1039/c4fd90042a

Varsha Khare opened the discussion of the paper by Pulickel M. Ajayan: The environmental applications of graphene based materials in terms of heavy metal detoxification and particularly the aspect of recyclability, different micro-structures of graphene and doping using solution chemistry leads to better absorption efficiency and recyclability.

Philip Davies said: Could you comment on the present state of knowledge of the interface between carbon nanomaterials and other components.

Pulickel Ajayan responded: This is still a big challenge in my opinion. There has been quite a few single nanotube studies on pull-out from polymer matrices and several of these studies have shown that the interface between nanotubes and polymers can be made strong *via* chemical modification of the nanotube surface (*via* covalent or non-covalent functionalization). Relatively limited number of studies have focused on interfaces between nanotubes and metal matrices or nanotubes and ceramics. The electrical contacts between nanotubes and metals have been studied in detail, some leading to good ohmic contacts. The characterization of such interfaces *via* direct microscopy or spectroscopy has not been very successful.

Thanh Nguyen asked: Could you please comment on the potential biomedical applications of these carbon nanomaterials?

Pulickel Ajayan answered: Carbon nanotubes could be directly used in the form of scaffolds in applications such as tissue engineering or indirectly as sensors or probes in the detection of biomolecules. Several ideas have been conceived like drug delivery, electrodes, membranes *etc.* but one has to be careful about the long term effects, toxicity, biocompatibility *etc.* before these applications are implemented. I think this area has huge potential and will be one of the important applications of nanomaterials (including nanotubes) in the future.

Pagona Papakonstantinou addressed Pulickel Ajayan, Milo Shaffer and Ian Kinloch: The direct growth of carbon nanotubes on carbon fibers offers new avenues for the reinforcement of laminated polymer composites. What are the current bottlenecks of this technology?

Milo Shaffer answered: The combination of carbon nanotubes with traditional structural fibres to create hierarchical composites, is a very promising and practical approach to improving state of the art materials. In these systems, the objective is to address the critical matrix-dominated failures that normally limit the performance of structural fibre composites (see for example our review by Qian *et al.*[1]). There are a range of strategies for introducing the nanotubes. The specific example of direct growth of the CNTs on the carbon fibres offers potential advantages in processing and in the location/orientation of the CNTs. One central challenge is that growing CNTs on carbon fibre tends to damage the underlying carbon fibre; fundamentally, a good catalyst for CNT growth tends to dissolve carbon. Very recently, strategies have begun to emerge to tackle this problem; we ourselves have been able to grow grafted CNTs directly on carbon fibre in a continuous process, to produce carbon fibres that retain their original tensile properties. The resulting CNTs have a relatively small diameter and are short, as desired to obtain a high volume packing fraction of primary fibres in the final composite; it's worth noting that most CNTs grown on fibres are too long from this perspective. We plan to publish our process shortly. Intrinsically it is scalable and compatible with existing carbon fibre technology, and therefore could be practically implemented, in principle.

1 Qian *et al.*, *J. Mater. Chem.*, 2010, **20**, 4751–4762.

Pulickel Ajayan answered: I think the biggest challenge is the interface between the carbon fibers and the grown nanotubes These are typically weak van der Waals forces and hence delamination of the nanotubes during processing limits the effectiveness of reinforcement. The control of length, density and size of nanotubes and the infiltration of polymers into nanotube forests, all determine the efficacy of this approach.

Philip Davies opened the discussion of the paper by Alan Windle: Could the role of the sulfur be to control the rate of growth of the iron nanoparticles in the early stages of the reactor?

Alan Windle responded: As we have described, it appears that sulphur has a profound influence on the growth of carbon nanotubes from nano-scale iron particles. Furthermore, the key to obtaining single walled nanotubes of good length and with a satisfactory yield, is to enable carbon and sulphur to be available while the iron particles are still very small, ~1 nm. Too much sulphur appears to prevent carbon interacting with the iron, while too much carbon will simply encapsulate the particle. We are suggesting that too much sulphur will initially prevent CNT formation, but as it 'boils off' in the hotter part of the tube reactor, available carbon will begin to grow nanotubes, but now from iron particles which will have had greater opportunity to grow by collision. The question posed is important. One might expect that a coating of sulphur would make iron less sticky so that collisions would not lead to merging of the particles with such a high probability. However, under conditions where the surface iron and sulphur are highly mobile, I would not expect sulphur to preclude particle growth by collision. However, to put numbers on this statement is a future challenge.

Pulickel Ajayan asked: What is the diameter (or diameter distribution) of the nanotubes you obtain by CVD? Does sulfur make the average diameter larger?

Alan Windle replied: The diameter, and in a related way, the number of walls does seem to depend on the sulphur content. While it is impossible to spin without sulphur (and it is vital that the equipment is completely purged from sulphur to establish this fact) small controlled amounts of sulphur do lead to single walled CNTs in the diameter range 1-1.5 nm. With increasing sulphur there appears to be a jump to collapsed double walled nanotubes some 10 nm in diameter, futher sulphur leads to more traditional multiwall tubes of a wide range of diameters. The central point of our paper is that the controlled amounts of sulphur available, along with carbon, as the first nano-sized particles of iron form, appear to be key to the achievement of wholly single-walled nanotubes. Excess sulphur blinds the iron particles, which while continuing to grow by collision, will not become active to grow nanotubes until the sulphur boils off as the reactants heat towards the maximum temperature (~1300 °C) of the continuous flow reactor.

Pulickel Ajayan asked: Are SWNTs the best material for spinning fibers and ensuring mechanical properties of the fiber? People have spun fibers with double walled nanotubes and gotten good results, especially because of the flattening observed for the large double walled tubes, increasing inter-tube adhesion.

Alan Windle responded: Most of our studies so far have concentrated on collapsed double walled CNTs of comparatively large diameter.[1] Indeed, we have pointed out[2] that the collapsed state would maximise the contact area between the tubes, to the benefit of strength where this was limited by the tubes pulling out from each other under tension. We now find that fibres made wholly from single walled CNTs of diameter of the order of 1 nm show yet better strength. What we do not yet know whether it is because the tubes are single walled or whether they are simply longer, although measuring the length of single waled nanotubes where the axial ratio may be of the order of 106 is a particularly difficult task. Alternatively, there is evidence of a very thin carbonaceous deposit on the outside of the single walled nanotube bundles which is more uniform than in the case of multiwalled tubes. Perhaps this coating has adhesive properties.

1 M. Motta, A. Moisala, I. A. Kinloch and A. H. Windle, *Advanced Materials*, 2007, **48**, 16, 4824–4836.
2 J. J. Vilatela, J. A. Elliott and A. H. Windle, *ACS Nano*, 2011, **5**, 1921–1927.

David Zitoun said: Could you comment on the difference between the CNT fibres produced with your single pot method and those produced with post-CNT-synthesis coagulation spinning?

Alan Windle replied: Coagulation spinning, developed by the Bordeaux group of Philip Poulin, was probably the first successful method for the production of continuous carbon nanotube fibre. However, the essential difference in terms of the product is that in the case of coagulation spinning the resultant fibre is a polymer/CNT composite material. In terms of the process, as you point out, it is a two stage process requiring first the synthesis of the nanotubes as a powder, and their suspension in a liquid before the actual spinning. The original polymer used was polyvinyl alcohol, and property improvements have been reported as a result of post spinning treatments. We all await the commercialization of the process. It appears that a company, Canoe, has this in hand as a part of a wider program to spin fibres enabled with nanocomposites, so we must watch this space.

Chris Ewels asked: Is your sulphur model transferable to other growth impurities/heteroatoms, such as oxygen?

Alan Windle responded: I have been attracted to the chemical similarities between oxygen and sulphur, spurred on perhaps by research into cast irons, where the influence of sulphur and oxygen on the growth form of graphite on cooling the metal from the melt, is assessed as a 'sulphur/oxygen equivalent'. Our experience is that if we supply modest oxygen, perhaps by using ethanol as our main source of carbon, we still need sulphur for the synthesis to work, although the actual synthesis process worked very smoothly with alcohol feedstocks in general (except of course methanol, where all the carbon tended to elope with the oxygen). However, it is probably worth pointing to a major difference between sulphur and oxygen in the context of the process. Namely that carbon's affinity for oxygen is very considerable, with carbon monoxide being one of the most stable of molecules, at least in terms of bond strength, whereas carbon disulphide has, in comparative free energy terms, only marginal stability. I am sure we can learn more by ranging more widely across the periodic table, especially with respect to the choice of catalyst metal, although we have to confess we need to keep one eye on the commercial scaleability of the process, and ferrocene is cheap and widely available.

Thurid Gspann addressed Chris Ewels and Alan Windle regarding question 11: We're talking about oxygen and sulphur analogies and probably also have to consider nitrogen in this context. These analogies are known *e.g.* from metallurgy regarding cast irons. However, in our process regarding floating nano-catalyst particles, there appear to be marked differences in their second role. We have not been able to spin without sulphur, even with oxygen in the process. In the case of sulphur, we typically grow either single walled tubes with small diameters or, if the system is overloaded with sulphur, we grow multi walled CNTs, both uncollapsed.

On the other hand, the double-walled collapsed tubes were mainly observed in alcohol spun CNTs, thus we can assume the oxygen acts as an etchant that leads to the destruction of or prohibition of formation of small diameter tubes, hence favouring thin walled big diameter CNTs which can collapse under atmospheric pressure.[1,2]

1 M. Motta, A. Moisala, I. A. Kinloch, and A. H. Windle, *Adv. Mater.*, 2007, **19**, 3721–3726.
2 J. A. Elliott, J. K. W. Sandler, A. H. Windle, R. J. Young, and M. S. P. Shaffer, *Phys. Rev. Lett.*, 2004, **92**, 9, 095501-1.

Ayse Turak said: Since the particulates decorate the outside of the CNTs, it was unclear how the wood knot analogy you used to explain that they do not affect the strength would apply. Your analogy would suggest that the defect should be inside the fiber. Could you please clarify?

Alan Windle responded: Your question raises an important matter of detail regarding the continuous spinning process that we have developed at Cambridge. The aligned aerogel as it emerges from the reactor is dilute, with a density aporoximately 1% of that of the final fibre. Condensation is achieved by spraying the aligned aeorgel with atomised acetone, where surface tension effects achieve condensation. However, the process by which it achieves its final density, is not exactly affine, and is a combination of both shrinkage and folding. It is the folding component of this process which takes particles, originally on the outside of the aerogel, into the inner regions of the final fibre.

Thurid Gspann regarding question 13 addressed to Prof. A. Windle: We are reaching the very interesting point, where we possibly have to change our view of what is considered to be the ideal material for macroscopically usable, mechanically strong materials. One reason for the presented study (paper 2) was that impurities were considered to be the main reason that CNT fibres did not reach the high strength values expected. However, with the achievement of pure fibres, yet without changing the fibre comprising CNTs themselves, the expected strength increase did not appear; although the consistency of properties was improved significantly. On the other hand it becomes clearer, that constructing fibres out of pure CNTs with a finite length and pure carbon-carbon contacts means intending to construct a fibre out of a lubricant. Hence, a certain amount of defects increasing the roughness and friction between the CNTs, also counting covalent cross-links which would show in Raman in the so called D(defect)-peak, or as well a certain amount of entanglement without disrupting the bundle alignment too much, can help revealing the CNT properties in macroscopically useful objects.

Milo Shaffer enquired: At several points, you discuss the issue of trapping of particles within the fibres, both when considering the incorporation of catalyst agglomerates and the possibility of length fractionation. Is there any effect of 'sieving' through the growing networked 'sock' of nanotubes, such that material outside the sock is rejected in the exhaust and material inside the sock is collected in the fibre? In your model, the catalyst agglomerates form near the furnace wall and can be rejected, if they do not become sufficiently 'hairy' to entangle with the 'sock'. Shorter nanotubes might grow in the faster flowing central region of the furnace, but the opportunity for fractionation might be more limited. Is it possible to estimate the 'mass balance' of the process, namely what proportions of the feedstock ends up in the final fibre, in rejected solids/nanoparticles, or gaseous waste?

Alan Windle replied: I cannot answer this question in full, as the developed process which all but eliminates the particles in the fibre is comparatively new and we have not carried out the full mass balance yet. It is perhaps worth making a note about the 'sock' here. The carbon nanotube aerogel, as it moves down the reactor appears to form a cloud, concentric with the reactor which is darker at its edges than in the centre, hence the term 'sock'. However, when running under conditions which lead to minimal particle encapsulation in the eventual fibre after condensation, then we do not see a sock but simply a 'sausage' of aerogel along the reactor axis. At this stage we speculate that the dark perifery of the 'sock' was in fact caused by the additional particulate materials from the slower moving gas stream near the walls which, because they were 'hairy' with badly formed CNTs became attached to the aerogel proper. The mechanical fractionation, ideally, would mean that only CNTs of greater than the entanglement length would appear in the fibre. We, of course, cannot rule out that much shorter nanotubes would still become incorporated in the fibre. But, very detailed examination by high resulution TEM does not show any population of shorter nanotubes either within the bundles or between them, indeed nanotube ends are only seen on very rare occasions. It may be that the process simply does not make shorter tubes within the aerogel itself.

Milo Shaffer asked: Do you yet have a full mass balance for the process, to determine the fate of all of the injected carbon feedstock, including both collected fibre and the waste gas stream? For example, it would be interesting to know if shorter CNTs were escaping with the waste gas stream rather than becoming entangled in the 'aerogel', or if other nanoparticles or complex aromatics were forming.

Alan Windle commented: I think this point was partly addressed in my response to your question 15. There is no doubt that the waste gas drawn out through the gas valve contains particulate matter including nanotubes. It is difficult to determine the relative amounts of nanotubes and approximately equiaxial particles present, as the filters will always discriminate in favour of nanotubes, but that is not to say that it shouldn't be attempted.

Nazario Martin asked: Could the methodology developed for the synthesis of the CNTs be applicable for preparing chiral CNTs? What parameters should be changed? Do you think that a new scenario would be needed? I mean, for instance, the use of nanorings, *etc.*

Alan Windle answered: This is an important issue. Viewed from outside the chemistry community, one cannot but wonder that if classical synthetic chemistry can make almost any polymer molecule that can be imagined, then why cannot they create single walled carbon nanotubes, in a process potentially scaleable to industrial levels. Of course it is not as easy as that, but there has been progress using a variety of synthetic schemes, some involving ring assembly. Perhaps the most recent advance was reported in Nature last month by Sanchez-Valencia *et al.*,[1] where a particular aromatic arrangement laid on a Pt catalyst, was made to spontaneously reassemble into the end cap combined with a very short starting length of an armchair, single walled nanotube. The tube was then lengthened from this seed using catalytically decomposed hydrocarbon gas. I plan to say more about this different route to nanotubes at the end of the meeting (Paper 25).

1 J. R. Sanchez-Valencia *et al.*, *Nature*, 2014, **512**, 61–64.

Pagona Papakonstantinou said: Could graphene platelets be synthesized using the floating catalyst method?

Alan Windle responded: Wow! Thank you for making us think out side the box. One tends to want graphene flat, and covering a decent area, perhaps a few square microns or more. If one could generate graphene in the gas phase not coated onto a catalyst, and one is thinking in terms of single layered material, then I do not see how one could stop it rolling up, or joining up with other sheets perhaps in an unfortunate, disorganised way. If one can coat facets of the catalyst particles with graphene, particles which are small enough to be entrained in a gas flow, yet large enough to produce graphene of an acceptable area, then it might be possible. Perhaps another route would be to use larger particles and form a fluidised bed through which the reactants are passed. Of course the graphene would still have to be subsequently removed from the particles and transferred.

Pedro Costa opened the discussion of the paper by Mark Baxendale: How are the different iron structures (alpha and gamma) arranged structurally along the nanotubes? Is there a preferential presence of one of the phases at the tips?

Mark Baxendale responded: The gamma phase is most frequently observed at the tips but it is also observed along the length but much less frequently.

Pedro Costa asked: Are the arms in the sea-urchin like structures fully filled? What would be their relative filling yield across the sample?

Mark Baxendale replied: It is difficult to give figures in answer to both questions because the sea-urchin 'arms' are not easily accessible with the TEM. The best statement that we can make is that continuous filling on the micrometre scale is frequently observed whereas a surface-grown structure prepared under similar vapour conditions is dominated by 10–100 nm scale encapsulated nanowires with 100–1000 nm gaps between them.

Thurid Gspann asked: You mentioned that substrate grown MWCNTs are inconsistently filled with ferromagnetic material, while the MWCNTs grown on spherical particles grown by boundary layer chemical vapour synthesis are very consistently filled. Do you see any differences in the ferromagnetic properties depending on the consistency of the filling, and is there a minimum size the nanowires have to reach to show any ferromagnetic properties at all?

Mark Baxendale answered: The saturation magnetisation is relatively unaffected by the continuity of the MWCNT-encapsulated ferromagnetic nanowire.

However, owing to the influence of shape anisotropy, the coercive field is enhanced by better continuity. If the nanowire dimensions are less than that of a single ferromagnetic domain, *circa* 10 nm, we would expect to observe a superparamagnetic response.

Pulickel Ajayan remarked: In the Fe filled nanotubes produced by CVD reported before, people have observed the gamma Fe phase and found them to be ferromagnetic due to the presence of carbon impurities and lattice distortion? How do you tell about the ferromagnetic phases in your sample?

Mark Baxendale responded: The evidence for antiferromagnetic components is largely from the observation of exchange bias features, that is, an offset hysteresis loop and pinned moments in the ferromagnetic components at low temperature, as one would expect from antiferromagnetic and ferromagnetic phases in contact. The former is difficult to confirm experimentally since magnetometers have a small remnant field which comes from the components in the measurement system. The latter can be determined by comparing a weighted sum of the saturation magnetisations for each component, calculated using the bulk values and the relative abundances of the phases (antiferromagnetic components have a small or no net magnetisation), with the measured value. Typically, the measured values are a factor of two to three times less than the calculated value, indicating significant pinning. If, within the nanowire, there are significant ferromagnetic gamma-Fe components, the measured value would be greater than the calculated value. Our magnetisation measurements are conducted on powder extracted from the reactor so these comments apply to the net magnetisation, it does not exclude a small ferromagnetic component from gamma-Fe. There are several reports of exchange bias phenomena in iron-filled nanotube systems, the debate is whether the junctions are arranged sequentially along the length of the nanowire or coaxially.

Toshiaki Enoki asked: Magnetic particles are included in the sample. Some are ferromagnetic. I am expecting that the application of external magnetic field will affect the orientation of magnetization and subsequently the orientation of nanoparticles. So the growth progress is modified by applying the magnetic field. My question is what happens in the growth process under the application of the magnetic field?

Mark Baxendale answered: There is always an unintentional magnetic field in a reactor heated by resistive heating elements which carry a large current. For our reactor, we calculate that this field is too weak to orient magnetic domains at the formation temperature of the structures (900 °C) so thermal randomization of the moments will erase any preferred orientation due this magnetic field. We have not tried to intentionally introduce a strong magnetic field to influence the growth process. We anticipate that such an axial field would keep the particles in the vapor for longer and, therefore, increase the length of the radial nanotube 'arms'. But there are other ways to achieve this, for example by introducing trace quantities of sulfur to accelerate the growth and selecting precursor molecules with higher metal-carbon ratios (this ratio in ferrocene is 1 : 10).

Thanh Nguyen said: Could you please comment on the potential of these iron-based nanowires encapsulated by multiwalled carbon nanotubes for treating cancer by magnetic hyperthermia?

Mark Baxendale responded: Surface-grown equivalent structures are receiving a lot of attention in the literature for this application. The advantage of the MWCNT-encapsulation is that the nanowire is passivated and the surface is relatively straightforwardly functionalised with desirable molecular groupings (antibodies, drugs, *etc.*). For surface-grown structures, the localised heating in an applied r.f. magnetic field comes from the rotation of the structure in response to the field but immobilisation can allow the hysteretic dissipation to be usable. The structures also exhibit MRI contrast agent functionality. The dimensions of the radial structures that we have so far produced are *circa* 10 micrometres; scaling to the 1 micrometre is required for this application. The radial morphology has the advantage of minimising the torque in the r.f. field and having a component of the magnetic easy axis in the direction of the external field for all orientations (structures disperse randomly in a tumour), this means that the hysteretic component of heating is likely to be better than an equivalent rod-like structure.

Milo Shaffer queried: The driving force for nanotube growth has been much discussed, especially in the earlier carbon filament literature (see for example Baker's review in Carbon 1989)[1]. Whilst both temperature and concentration gradients have been mooted, the insensitivity to the exothermic/endothermic nature of the carbon source decomposition tends to favour concentration arguments. Do you have evidence yet, in your experiments to favour temperature over concentration effects, or indeed steric interactions thought to influence CNT array growth? Have you been able to prove whether tip or base growth dominates?

1 R. T. K. Baker, *Carbon*, 1989, **27**, 315-323.

Mark Baxendale answered: The radial structures nucleate and grow in both concentration and thermal gradients in the reactor but only when the flow rate is low (15 ccm). The relevance of the low flow rate is that local thermal gradients will not be damped; the nucleation imposes a spherical diffusion gradient as the central particle consumes vapour feedstock and feedstock to the tip growth fronts is directed by spherical diffusion gradients created by the endothermal graphite carbon formation and exothermal hydrocarbon decomposition at the tips. We exclude base growth - by which I mean the growth of the nanotube 'arms' is fed by vapour supply to the central particle - because the outer diameter of the radial structures does not depend on residence time, that is, the duration of exposure to the vapour. This observation is consistent with the tip growth model because the growth is 'switched off' by contact with the isothermal surface, that is, the thermal gradients at the tips are damped by this contact. The degree of order seen in the radial structures, particularly the continuity of the central nanowire, is better than that observed with surface-grown structures in the same reactor conditions, so we can conclude that the absence of steric interactions improves order; in other words, the perturbation of the vapour feedstock flow to tips created by a near neighbour creates the discontinuities in the central nanowire.

Pulickel Ajayan commented: In terms of growing graphene ribbons or corrugated graphitic filaments grown by CVD, perhaps these could be done by vapor phase growth without using metal catalysts; we have previously used non-catalytic methods, such as the arc discharge, to produce platelets and filaments along with tubes. The metal catalysts seem to produce tubular forms in general.

Juan Casado responded: I fully agree with this statement.

Pulickel Ajayan commented: The difficulty in measuring mechanical properties, and the role played by defects such as edges and grain boundaries *etc.*, is that it leads to a scatter in the measured values of the mechanical properties

Pulickel Ajayan commented: The nomenclature issue is a tough one since there is so much variability in structure and this is closely related to physical properties like mechanical strength.

Pulickel Ajayan asked: Since there are so many structural and dimensional variations in the nanotubes grown by various techniques, it may be a difficult task to categorize them (in assigning nomenclature) by structure or dimensions. Wouldn't it be more interesting to categorize them by conductivity or strength, similar to what has been done for carbon fibers?

Varsha Khare asked: What are the possibilities of the formation of metallobenzene after the synthesis or after dissociation of ferrocene *i.e.* after complete synthesis? Was this verified by XPS.

Alan Windle replied: In working to understand the important factors which control our synthesis of CNTs from floating catalysts, it has become very clear that the decomposition of the reactant precursors is complicated. With respect to ferrocene, it is clear from the literature (as referenced in our paper) that autocatalysis is an important feature, with the pre-existence of iron particles or surface deposits totally changing the kinetics of the decompositions. In lay terms one might see that iron present as the metal will steal the iron atom away from a ferrocene molecule, whereas if the molecule is simply left to its own devices, there is evidence that the carbon rings start to decompose with the iron atom *in situ*. The adsorption of benzene on the iron catalyst surface, is an important point as it could play a part in CNT assembly at the surface. I would be eager to see any experimental evidence for this, bearing in mind that modelling is also likely to provide insights in the near future. Overall, the studies reported in our paper underline that the iron particles, once they start to form, are not only catalysts for the growth of nanotubes, but also play a very important role in catalysing not only the further breakdown of ferrocene, but also the breakdown of thiophene and the plain hydrocarbon feedstock too. This fact has been articulated by chemists since the inception of CVD routes to make nanotubes, but it has also been overlooked from time to time.

Varsha Khare commented: I agree with the idea of edge-enhancement by the coexistence of sulfur and oxygen as demonstrated in our work on sulfur doped graphene.

Philip Davies said: Sulfur chemisorption on metal surfaces is very complicated (see for example ref. 1 and 2). The introduction of another adsorbate (carbon in this case) will add a whole new level of complexity (see ref. 3 for example) and so its not surprising that the chemistry involved in your system is poorly understood.

1 J. Oudar, *Bulletin De La Societe Francaise Mineralogie Et De Cristallographie*, 1971, **94**, 225.
2 A. F. Carley, P. R. Davies, R. V. Jones, *et al.*, *Surface Science*, 2000, **447**, 39-50.
3 A. F. Carley, P. R. Davies, R. V. Jones, *et al.*, *Chemical Communications*, 2000, 185-186.

Alan Windle responded: Thank you for this input. An understanding of the complexities of the role of sulphur as a surface promotor/surfactant will be necessary before we can have full scientific confidence in the floating catalyst synthesis of carbon nanotubes. However, we appear to have a further complicating factor in that, compared with, for example the system described in the questioner's reference 3, especially at small sizes, as the metal particles will not be crystalline, although there could be ordered faceting of their surfaces. Hence, the level of atomic disorder which will have to be faced before any exact explanation can emerge, is much higher.

Malcolm Heggie remarked: You mentioned you had been asked if your aerogel technique could be applied to make graphene. According to our DFT calculations sulfur should be capable of breaking open folded graphene.[1] Thus the wide nanotubes you made in early work - the ones which collapsed like a fireman's hose and have been termed 'dog bone' might be capable of being opened up by exposure to sulfur, forming sulfur terminated graphene ribbons.

1 J.-J. Adjizian, C. D. Latham, S. Öberg, P. R. Briddon, M. I. Heggie, *DFT study of the chemistry of sulfur in graphite, including interactions with defects, edges and folds*, *Carbon*, 2013, **62** 256–262.

Alan Windle answered: Thank you for drawing attention to this study. Modelling carried out by James Elliott some time back on 'dog-bone' collapsed nanotubes, indicated that the bonds, where the bending was sharpest had distinct sp3 character. So one might expect these to be the most vulnerable to chemical opening and stabilisation, along the lines you predict. However, the fact remains that when we spin carbon nanotube fibres in the presence of sulphur, the nanotubes we make are not unzipped, even in the case of 'dog bone' nanotubes.

Pulickel Ajayan said: Has anyone looked into how sulfur dopes substitutionally into the graphene lattice?

Lyndsey Mooring opened the discussion of the paper by L. S. Hui: A question to the carbon nanomaterials research community was raised. Discussions are covering the next generation of carbon nanotmaterials and preparation, along with mixed views on what are desirable attributes. Is there a sense among the community of when this new generation of materials will be available? Whether that be chirally pure or defect controlled materials, for example.

Milo Shaffer answered: It seems to be quite typical that new materials often take 20–30 years to develop from initial observation to practical products. The majority of the commercial nanotube production currently available in bulk in some way draws on earlier work on carbon filaments in the 1970s and 80s. We should therefore be entering an interesting time for the commercialisation of newer nanocarbons and associated applications, drawing on the explosion of scientific interest from 1991. For example, the price of single walled carbon nanotubes, clearly identified only in 1993, has fallen by around 2–3 orders of magnitude just this year. The continuing development of separated metallic and especially semiconducting SWNTs is also encouraging, especially since semiconducting SWNTs so dramatically outperform other conjugated polymers, particularly in terms of mobility. In an analogous fashion, many current commercialised 'graphene' materials draw on graphite oxide or other exfoliated graphite approaches that pre-date the wave of interest that has emerged since 2004. Practical materials exploiting the understanding of more perfect single layers, established in the last decade, may still take some time to emerge.

Pulickel Ajayan commented: Small start-ups are making specialty products using nanotubes or graphene and these products could become widely available in the next 5–10 years.

Mark Baxendale said: Convincing manufacturers that a synthesis can be scaled to large volume production is one thing but answering the question of why should the manufacturer do so if there is no immediate market for the product is another. We need an application, often called the 'killer application', for which there is an existing large market and the nanocarbon-based technology performs at least one order of magnitude better than the alternative technologies. Considerable benefits will follow from large volume production, such as quality control and economies of scale, which will open up possibilities for nanocarbon technologies that are unrelated to the 'killer'.

Pulickel Ajayan opened the discussion of the paper by A. Turak: When a graphene/copper system is annealed, do you observe any evidence for carbon diffusion into the copper substrate?

Ayse Turak responded: We have not done a systematic study of carbon diffusion into the copper substrate. However, a recent study by Leong *et al.*,[1] has shown that significant C diffusion can be seen for moderate temperature annealing (300°C) of Ni-graphene contacts. In their case, they saw significant growth of the D band with annealing. In our tracking of the impact of low temperature annealing, we did not see such an effect in the Raman spectra. Additionally, they did not see the shift in the 2D band that we observed for neither Ni-graphene (where there was determined to be C diffusion) nor Au-graphene (where there was no C diffusion). We believe that though C diffusion may be a potential explanation for the increased resistance to etching with annealing, it is likely not occurring in the case of Cu. We are planning to perform a follow-up experiment using Ni-graphene to further clarify this mechanism.

1 Leong *et al.*, *Nano Letters*, 2014, **14**, 3840–3847.

Milo Shaffer remarked: Both mechanisms which you propose imply a very high degree of copper mobility; are the diffusion rates reasonable at the temperatures you have studied? Is etching at the flake boundaries a possibility?

Ayse Turak responded: Though we have not measured the diffusion directly, Cu is known to be highly mobile at room temperature, especially in the presence of carbon based molecules (see for example ref. 1) or with oxygen (see for example ref. 2). STM measurements especially have shown the high mobility of Cu atoms at room temperature.[2] We are confident that the diffusion rates of Cu would be reasonable under the moderate annealing that we performed to allow either of the two proposed mechanisms. The strain relief mechanism in particular does not require much movement of Cu atoms. The observed shift in the 2D band of the Raman spectrum corresponds to only about 0.167% strain relief (~ +10cm^{-1}), using the values reported for $\delta\omega_{2D}/\delta\varepsilon$ of ~−66 cm^{-1}/% strain for deposition induced tensile strain[3] and ~−64cm^{-1}/% for applied tensile strain.[4] Additionally, strain is known to enhance diffusion (see for example Spjut *et al.*)[5], so a strained Cu surface from high temperature CVD growth would have more mobile surface atoms than the equilibrium surfaces measured with STM. These suggest that the strain relief mechanism might be the correct one for the enhanced coupling with low temperature annealing. With regards to the etching at flake boundaries, we would also expect that etching would begin at the edges of the graphene flakes. However, our visual observations do not support that, as we do not see the graphene flakes getting smaller with etching time. Further study is necessary to understand the impact of the plasma on the graphene itself.

1 Rosei *et al.*, *Science*, 2002, **296**, 328–331.
2 Frerichs and Liberman, *Phys. Rev.*, 1961, **121**, 991; Chapter 17 *Scanning Probe Microscopy and Spectroscopy* in *Characterization of Solid Materials and Heterogeneous Catalysts*, Wiley, 2012.
3 Yu *et al.*, *Phys. Rev. B*, 2011, **84**, 205407.
4 Mohiuddin *et al.*, *Phys Rev B*, 2009, **79**, 205433.
5 H. Spjut and D. A. Faux, *Surf. Sci.*, 1994, **306**, 233–239.

Alexander Zöpfl communicated: Full etching was confirmed *via* optical microscopy. In the case of oxygen- or air-etching, we could also confirm this with Raman spectroscopy or mapping, since the resulting CuO has specific peaks around 500 cm^{-1}. Would it be possible to monitor the etching process itself by comparing the intensity ratios of graphene and CuO peaks?

Lok Shu Hui communicated in reply: Thank you for your question, it is an excellent suggestion. In our Raman measurements, however, we did not observe significant signal peaks around 520, 640 or 280cm^{-1}, which have all been attributed to Cu oxides.[1] This is true even for scans near the graphene edge or on smaller flakes where a signal is expected both from the graphene and the unprotected Cu surface. One possible explanation could be the formation of metastable cuprous oxide, as suggested by Lengelet *et al*,[2] which has a peak between 150 and 250 cm^{-1}, below our detection range. Though we have not been successful in tracking the Cu oxidation using the Raman so far, it might be possible to lengthen the scan time and region to collect more of the copper oxide scattering to enhance the signal. Further studies with XPS are also planned to examine the evolution of Cu oxides.

1 Schennach and Gupper, *Copper oxidation studied by* in situ *Raman spectroscopy*, *MRS Proc.*, 2003, **766**, E3.2.
2 M. Lenglet, K. Kartouni, J. Machefert, J. M. Claude, P. Steinmetz, E. Beauprez, J. Heinrich and N. Celati, *Mater. Res. Bull.*, 1995, **30**, 393–403.

Thurid Gspann asked: Regarding the post-annealing CET (Fig. 4), assuming the proposed model that by annealing a CuO layer is built up around the graphene edges (Fig. 9), and further assuming that oxygen etches preferentially from the graphene edges or might creep under the graphene, should then the annealing not very effectively protect the graphene etching by oxygen and air? While the CET of N_2on the other hand, N_2 sputtering being a high kinetic energy process anyway and not necessarily preferential to edges, would not be expected to show such a massive increase by annealing?

Ayse Turak replied: You raise an excellent point with regards to the effectiveness of annealing in oxygen and air if the plasma attacks from the graphene edges. Though we too would expect that the highly reactive edges would interact with the plasma first, our visual observation with air sputtering, where pits were seen at many locations on the graphene flake, suggests that the plasma attacks the whole graphene flake equally. We don't really see a creeping removal of graphene from the edges with any plasma exposure (*i.e.* the graphene flakes do not get smaller and smaller with etching). Rather the whole flake seems to become progressively less visible. More work needs to be done to fully understand the impact that the plasma is having on the graphene, especially with regards to the plasma intensity. We speculate that increasing the plasma dose should shift the CET, though the general trend might remain. N_2 also is a special case, and one where the visual observation of the CET is very misleading. It appears in the case of N_2 what the CET is really tracking is the removal of different copper oxides. This gives us some clues regarding the possible mechanism for the improvement of the mechanical coupling - potentially non-equilibrium forms of copper oxide are being formed during the low temperature annealing, as suggested by Lengelet *et al.*[1] These could be protective of the graphene edges and lead to improved coupling.

1 M. Lenglet, K. Kartouni, J. Machefert, J. M. Claude, P. Steinmetz, E. Beauprez, J. Heinrich and N. Celati, *Mater. Res. Bull.*, 1995, **30**, 393–403.

Philip Davies remarked: Even at room temperature, I would expect a lot of movement from the copper atoms at the surface in the presence of oxygen. Is the bonding of the graphene to the surface strong enough to protect it from oxidation?

Ayse Turak replied: The graphene-Cu bond is expected to be very weak. Yet, graphene is widely reported as a passivation layer,[1-4] which is also supported by the visual observation in our work of contrast in regions protected by graphene *versus* those unprotected. However, Zhou *et al.*[5] did show that over the long term, graphene seems to enhance the oxidation of Cu, so it may be that the mobility of Cu atoms is a factor in such behaviour.

1 S. Chen, L. Brown, M. Levendorf, W. Cai, S.-Y. Ju, J. Edgeworth, X. Li, C. W. Magnuson, A. Velamakanni, R. D. Piner, J. Kang, J. Park and R. S. Ruoff, *ACS Nano*, 2011, **5**, 1321–1327.
2 E. Sutter, P. Albrecht, F. E. Camino and P. Sutter, *Carbon*, 2010, **48**, 4414–4420.
3 J. Iijima, J.-W. Lim, S.-H. Hong, S. Suzuki, K. Mimura and M. Isshiki, *Appl. Surf. Sci.*, 2006, **253**, 2825–2829.
4 A. Ferrari and J. Robertson, *Phys. Rev. B: Condens. Matter Mater. Phys.*, 2000, **61**, 14095–14107.
5 Zhou *et al.*, *ACS Nano*, 2013, 7, 6939.

Toshiaki Enoki said: The interaction between graphene and Cu substrate is weak, while other substrates such as Ni and Pd have stronger interactions with graphene. My question is if the results you obtained for the Cu substrate are modified when we employ Ni or Pd as the substrate?

Ayse Turak replied: Our interest in the mechanical coupling was due to the controversy that exists on the strength of the coupling between Cu and graphene. However, other metal systems have also shown a change in properties with low temperature annealing. The metal-graphene contacts most commonly annealed at temperatures similar to ours are usually made from exfoliated graphene, with the improvement in properties often attributed to the removal of resists needed to facilitate the transfer of graphene onto metal contacts. A recent study by Leong *et al.*,[1] has suggested that annealing with Ni leads to significant C diffusion, which could be a potential mechanism for improvement of the mechanical coupling. Though we did not see the increase in the D band with annealing that they observed, a follow-up experiment with Ni is planned to further clarify the potential mechanisms. One caveat of using Ni or Pd is that we can no longer apply our visual observation method to confirm the etching of graphene. Cu conveniently oxidizes with a change in the refractive index at low temperatures and under the plasma. Neither Ni nor Pd has a similar behaviour. However, tracking the graphene removal with Raman spectroscopy should be relatively easy in those cases.

1 Leong *et al.*, *Nano Letters*, 2014, **14**, 3840–3847.

Andreas Hirsch opened the discussion of the paper by Matthew Rosseinsky: How do you get rid of the metallocene?

Matthew Rosseinksy responded: It is removed in the final antisolvent precipitation step as it is soluble in the hexane used.

Irena Kratochvilova queried: How do you control/understand the electronic structure of a superconducting material? Please, describe in more detail the superconductivity process in this particular case.

Matthew Rosseinksy replied: This is a very broad question, as the mechanism of superconductivity is well-understood in only a very few materials. In our case, the proximity of magnetically ordered states to the superconductors makes it likely that the electronic correlations are important in the superconducting pairing, although this does not mean that conventional electron-phonon coupling is not also involved. There are recent reviews (*e.g.* from Tossatti, Capone and colleagues in Trieste)[1] on this interesting and controversial topic.

1 M. Capone, M. Fabrizio, C. Castellani and E. Tosatti, *Reviews of Modern Physics*, 2009, **81**, 943–958.

Chris Ewels remarked: If a Jahn–Teller distortion is splitting the triplet, to reduce this effect would it be possible to "rigidify" the fullerene somehow, for example by using endofullerenes?

Matthew Rosseinksy remarked: This is a very interesting idea, and the recent advances by the Shinohara[1] group in making larger quantities of endohedral fullerenes will enable investigating it, as the challenge to this point has been having enough endohedral material available to systematically identify the chemistry.

1 K. Akiyama, T. Hamano, Y. Nakanishi, E. Takeuchi, S. Noda, Z. Wang, S. Kubuk and H. Shinohara, *J. Am. Chem. Soc.*, 2012, **134**, 9762–9767.

Malcolm Heggie said: The HOMO of C60 is fivefold degenerate and therefore gives a higher density of states at the Fermi level, with possibly higher T_c - why is there no emphasis on hole (super-)conductivity?

Matthew Rosseinksy replied: It is chemically challenging to make cationic compounds of C60, and the anions that are chemically compatible with these highly oxidising cations tend to be large, and place the cations too far apart, producing insulating behaviour.

Malcolm Heggie asked: For hole conductivity, you say the problem is the large anions - surely these push molecules further apart, reducing the band width. This increases the density of states at the Fermi level and hence T_c, surely?

Matthew Rosseinksy commented : If the material remains metallic and the superconducting pairing is BCS, that is true. The problem is that if the overlap becomes less than the interelectron repulsion, the electrons become localised in the Mott–Hubbard insulating state and the metallic behaviour, and thus the superconductivity is lost. If we could control the spacing between the cations better, we might access the exciting regime just beyond the Mott–Hubbard transition. We are currently limited by the chemistry.

Thurid Gspann asked: Do you think there could be another solvent that would give you better material properties?

Matthew Rosseinksy responded: It's entirely possible, the investigation would need the commitment of a lot of resources though.

Milo Shaffer asked: In your synthesis, you used the specific 3 : 1 stoichiometry as well as controlling the redox potential of your reagent. Can you say whether either of these measures is necessary or sufficient?

Matthew Rosseinksy replied: This is a very interesting point for future investigation – as your question implies, if we are redox-limited then an excess of the metallocene anion would not matter.

Chris Ewels enquired: A question was raised concerning potential standardisation of carbon nanoform/allotrope nomenclature. We recently tried to address this issue through a categorisation of various carbon nanoobjects based on topological transformations of graphene required to produce them. These are 'thought experiment' transformations such as introducing wedge disclinations or dislocations, cutting, rolling, *etc.* Qualifiers could then be added to these base-form definitions to describe local/periodic structural variation (such as waviness).[1] The same paper also proposes a chemical structure nomenclature for functionalisation, following literature standard practise for endohedral filling of species B with species A described as A@B, adding species B with species A covalently attached to its surface as A//B. This convention has been developed further by M. Monthioux in his book "Carbon Meta-Nanotubes" to include X:CNTs to indicate carbon nanotubes (CNTs) doped by species X, X-CNTs to indicate CNTs functionalised by species X, X/CNTs to indicate CNTs decorated/coated by X, X@CNTs to indicated CNTs filled with X, and X*CNTs, X*BNNTs to indicate heterogenous nanotubes. Other (more experimental) nomenclature attempts include those by Inagaki and notably Gogotski who focused on growth-based naming.[2] IUPAC proposed in 1991 a list of 110 individual definitions for carbon materials, but with no attempt to coherently link these into a structured nomenclature. There are now ISO and IUPAC committees looking at this issue.

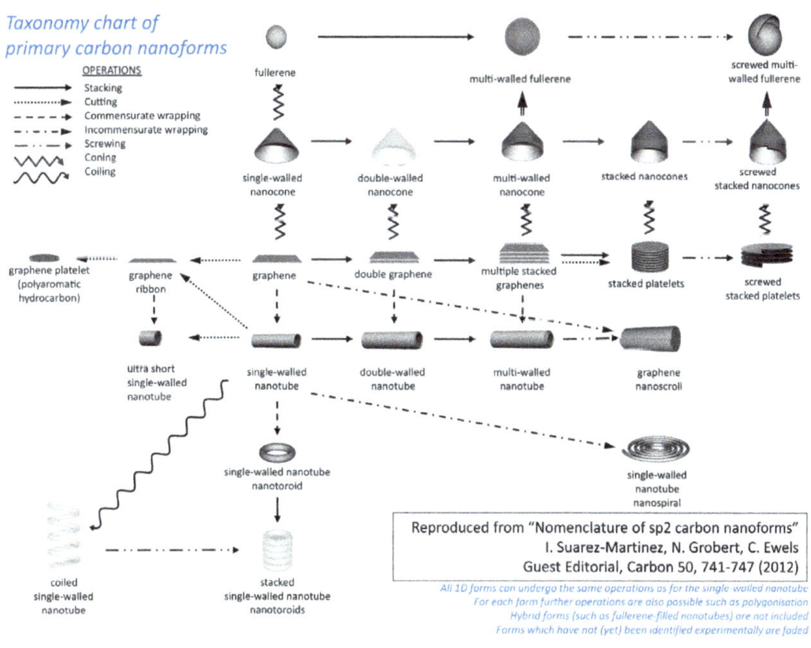

Fig. 1 Classification of the different "primary carbon nanoforms".

1 C. Ewels *et al.*, *Carbon*, 2012, **50**, 741.
2 Gogotski *et al.*, *Mat. Res. Innov.*, 2003, **7**, 192.

Vladimir Falko responded: Standardisation is an important issue for graphene applications. To mention, the European Graphene Flagship has a work package dedicated specifically to standardisation of graphene for various applications, including development of characterisation protocols suitable for industry and mass production.

Juan Casado commented: Fully in agreement with the needed uniformization of the nomenclature.

Milo Shaffer replied: There was some discussion about the relevance of topology to carbon materials classification. Whilst not applied to nanomaterials, it may be interesting to note that graph theory has been applied to the determination and classification of all possible periodic carbon allotropes within certain bonding criteria (see for example Strong *et al.*).[1]

1 Strong *et al.*, *Phys. Rev. B*, 2004, **70**, 045101.

Pedro Costa addressed Pulickel Ajayan and Chris Ewels: Are we any closer to having a taxonomy chart for carbon nanostructures? This is a recurrent point in the field with journal editorials and considerable discussion going on. Before even establishing metrology standards we should have a consensus of what we are standardising (and how we name it). With the advent of industrial applications of nanotubes and graphene, may a forum to establish a roadmap like the ITRS be considered?

Chris Ewels replied: In case it's useful, here is the diagram (Fig. 1) showing the classification of the different "primary carbon nanoforms", reproduced from ref. 1 in the paper but at a higher resolution.

1 C. Ewels *et al.*, *Carbon*, 2012, **50**, 741.

Pulickel Ajayan replied: As I have repeatedly mentioned during discussions, it may be better to create an Ashby like chart where the different types of nanotubes are plotted in a stiffness-electrical conductivity (or other properties) map. Rather than separating in terms of stucture/dimension/morphology, this type of property based classification (as used for carbon fibers for example) might be more interesting and useful. Of course the roadmap for nanotube based technologies is still premature in my opinion.

Pulickel Ajayan commented: Regarding nomenclature again, is it possible to consider factors other than topology for classification of nanotubes?

Chris Ewels replied: Please see my earlier response.

Malcolm Heggie remarked: I think this is a brave attempt. Where is the theory of topology that applies here, combining different operations? Has this been worked out or does this presage such topological understanding?

Chris Ewels replied: A better term might be morphology; we were using topology somewhat loosely to indicate underlying geometrical forms structurally distinct from one another, to which other structural variations, such as irregular local deformation, waviness, *etc.*, could then be applied. I'm sure higher level topological theory could be usefully applied to these pictures, but with my grounding in materials science I would suggest that the key driver should always be the utility of the classification system for practical materials identification and classification.

Pulickel Ajayan asked: Providing specific history of synthesis or manufacturing is as important as nomenclature - labs need to know what they are buying from vendors. Are there any MSDS (information) sheets for nanotubes, similar to what is available for chemical products?

Nazario Martin enquired: Do you think that the many shown carbon nanoforms should be considered as different carbon allotropes or do they belong to the allotropes of fullerenes? This should be clarified by the IUPAC.

Chris Ewels responded: There is a paper, "The Origin of the Term Allotrope",[1] which traces the origins of the concept of allotropism and its changing meaning over time. This states that Berzelius first used the term allotrope in 1841 to describe the difference between two forms of the same element. Later, in 1877, Otto Lehmann, suggested generalisation to describe any variation of a given substance (element or compound), "traceable to variations in the substance's intermolecular organisation". IUPAC have not followed that and stay with the definition, "Allotropes are different structural modifications of an element". It seems then that allotrope is appropriate if there are distinct structural differences, and if the materials in question are elemental (following IUPAC). So topologically "ideal" forms, such as single- or multi-walled nanotubes, fullerenes, or nanocones could be (should be?) described as different carbon allotropes. I'm not so sure it remains within this definition if these are real materials with *e.g.* varying impurity content, and structural variation within a given form (such as length, diameter, degree of crystallinity and purity in SWNTs). There is sometimes discussion of the difference between nano-*objects* and nano-*materials*, with the focus of the second on the collective properties of multiple similar nanoobjects. In a similar vein, and following the spirit of Otto Lehmann's suggestion, one solution might be to stick with "nanoobject" when the individual object structure and behaviour are under study without reference to it necessarily as a representative of a family of alike nanoobjects - in which case allotrope could be used. (Just to muddy the waters still further, Jensen observes that the meaning of allotrope is closer to "other behaviour" while allomorph means "other form". So maybe nanoobjects and carbon nanoscale allotropes should all actually be allomorphs! Following your question, taking allotrope to indicate 'other behaviour' other nanoobjects would not be allotropes of fullerenes once their behaviour was different.)

1 W. B. Jensen, *J. Chem. Educ.*, 2006, **83**(6), 838.

Andreas Hirsch asked: Some carbon nanoforms are not real allotropes of carbon since they contain heteroelements at the boundary.

Nazario Martin commented: I basically agree with the comment. However, in some way, it is the same occurring issue as with other well-known allotropes such as diamond and graphite.

Pulickel Ajayan said: Once again, regarding classification of nanotubes, it might be interesting to put together an Ashby plot (say, conductivity *vs.* stiffness/strength) where different types of nanotubes (based on properties) can plotted in a two-dimensional map.

Philip Davies opened the discussion of the paper by Alexander Sinitskii† by communicating: Do you have any evidence for the functionalisation of the graphene ribbon edges; does the IR show any OH stretches in the 3300–3600 cm^{-1} region for example or does the XPS show any oxygen containing species in the 531–534 eV region?

Alexander Sinitskii communicated in reply: There is no evidence for OH functionalization, as there are no peaks in the 3300–3600 cm^{-1} region in the IR spectra. We do not see any signs of C=O bonds either. In the XPS spectra we do see a tiny O1s peak in the 531–534 eV region, but based on the shape of the C1s peak it is unlikely that this oxygen represents C–O or C=O bonds. Based on FTIR and XPS data this tiny oxygen peak likely corresponds to molecular oxygen that was trapped in a porous GNR powder.

Oana Andreea Bârsan communicated: Based on your I/V curve presented in Fig. 3c, did you calculate a sheet/bulk resistance value for your sample or a conductivity value based on the dimensions of the sample? How do these compare to similarly prepared graphene or graphene nanoribbons samples in the literature?

Alexander Sinitskii communicated in reply: Estimates that are based on the data provided in the paper result in a very high resistance of a GNR pellet (~50 kOhm m), which is higher than in many other graphene materials. Considering the nanoscopic size of GNRs and their intrinsic semiconductor properties, as well as the assembly problems that we discuss in the manuscript, such high resistance is not surprising.

Oana Andreea Bârsan communicated: On page 6, line 45 you mention that "the electrical conductivities of CNTs fibers and composites are much lower than those of individual CNTs by at least one order of magnitude." While composite materials require the addition of another component (often a polymer) to the carbon nanomaterial which indeed affects their conductivity, many studies have shown that CNTs fibers, bucky paper, and graphene films have a lower resistance with increased thickness of the films,[1-3] values that are further lowered when applying

† The author was not available to present at the meeting, but questions relating to this paper were submitted afterwards.

even moderate pressures on the films. Could you elaborate on why you would you expect your 0.1 mm thick, compressed pellets of pure graphene nanoribbons to have a higher resistance than a single layer?

1 F. Bonaccorso, Z. Sun, T. Hasan and A. C. Ferrari, *Graphene photonics and optoelectronics*, *Nature Photonics*, 2011, **4**, 611–622.
2 J. Liu, M. Notarianni , G. W. Vincent, T. Tiong, H. Wang and N. Motta, *Electrochemically Exfoliated Graphene for Electrode Films: Effect of Graphene Flake Thickness on the Sheet Resistance and Capacitive Properties*, *Langmuir*, 2013, **29**, 13307–13314.
3 B. Marinho, M. Ghislandi, E. Tkalya , C. E. Koning, G. de With, *Electrical conductivity of compacts of graphene, multi-wall carbon nanotubes, carbon black, and graphite powder*, *Powder Technology*, 2012, **221**, 351–358.

Alexander Sinitskii communicated in reply: You probably confuse conductivity/resistivity (intrinsic properties of a material) with conductance/resistance (that do depend on thickness, *etc.*).

Alexander Zöpfl communicated: After heating to 200 °C you see an improvement of conductivity of your GNR pellet. Do you expect any change in the structure or is this maybe due to impurities which are evaporated during this step? Since you are only measuring the bulk conductivity, the intrinsic properties of a single nanoribbon would be interesting. How do you perform/plan such an experiment in terms of fixing and contacting a single nanoribbon?

Alexander Sinitskii communicated in reply: As demonstrated in the papers on the surface-assisted bottom-up synthesis of GNRs, these ribbons are stable at 200 °C, so there should be no change in their structure upon annealing. It is more likely that the conductivity of a pellet improves because of the removal of residual solvent molecules. We are highly interested in performing electrical measurements on a single nanoribbon, but such experiments would be very challenging due to the small size of a GNR.

Faraday Discussions

PAPER

Moiré pattern as a magnifying glass for strain and dislocations in van der Waals heterostructures

Diana A. Cosma, John R. Wallbank, Vadim Cheianov and Vladimir I. Fal'ko*

Received 2nd July 2014, Accepted 9th July 2014
DOI: 10.1039/c4fd00146j

We consider the role of deformations in graphene heterostructures with hexagonal crystals (including strain, wrinkles and dislocations) on the geometrical properties of moiré patterns characteristic for a pair of two incommensurate misaligned isostructural crystals. By employing a phenomenological theory to describe generic moiré perturbations in van der Waals heterostructures of graphene and hexagonal crystals we investigate the electronic properties of such heterostructures.

Two overlaying isostructural two-dimensional crystals with slightly different lattice constants, or two equivalent crystals deposited on each other with a slight misalignment, produce a long-range quasi-periodic structure[1,2] known as the moiré pattern. A rich library of moiré patterns has been found in graphene van der Waals (vdW) heterostructures on hexagonal catalysts,[3–7] silicon carbide,[8] hexagonal boron nitride (hBN),[9,10] or deposited with a small misalignment angle on graphite.[11] For graphene electrons a moiré pattern translates into a superlattice with a period magnified from the atomic scale of the lattice constant a by the large factor $M = [\theta^2 + \delta^2]^{-1/2}$ where $\delta \ll 1$ is the lattice mismatch and $\theta \ll 1$ is the misalignment angle. We show that a similar factor magnifies the appearance of a dislocation in crystals composing a vdW heterostructure, in the form of a dislocation in the moiré superlattice, and also that a small uniaxial strain $w \sim 1/M$ transforms a hexagonal moiré pattern into a quasi-one-dimensional incommensurate superlattice.

The upper panels of Fig. 1 offer a geometrical illustration of how a small uniaxial strain, $w \ll 1$ (either in graphene, or in the underlying material) induces qualitative changes in the moiré superlattice in a graphene–hBN heterostructure. The moiré superlattice of the unstrained heterostructure (left hand side panel) transforms from hexagonal to oblique in the right hand side panel, taking a quasi-1D form at $w \sim \delta$ (central panel). The moiré lattice vectors ($m = 0, 1,\ldots 5$),

Physics Department, Lancaster University, LA1 4YB, UK. E-mail: v.falko@lancaster.ac.uk

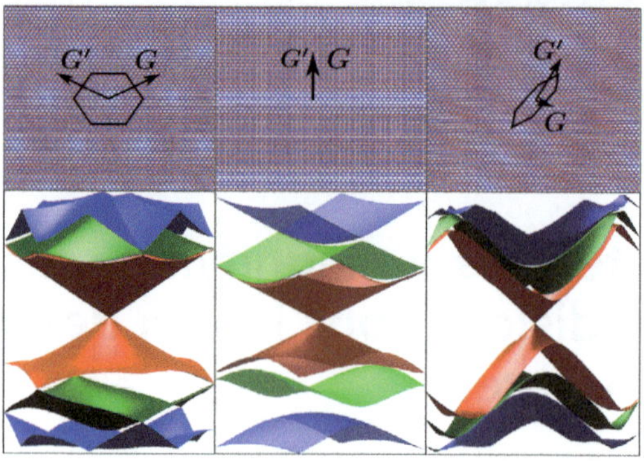

Fig. 1 Real space moiré pattern, with the superlattice Brillouin zones (top), and corresponding band structures (bottom). Geometrical parameters mimic $\delta = 1.8\%$ with $(w, \theta) = (0\%, 0°)$, $(1.8\%, 0°)$ and $(2\%, 0.92°)$ (left to right) but are exaggerated for clarity. Band structures are calculated using eqn (2) with $U_{i=0,1,3} = (v|\boldsymbol{G}'|/15)\{1/2, -1, -\sqrt{3}/2\}$.

$$\boldsymbol{A}_m = \hat{M}\boldsymbol{a}_m, \hat{M} = \frac{1}{\delta'^2 + \theta^2 - w'^2}\begin{pmatrix} \delta' + \left(l_y{}^2 - l_x{}^2\right)w' & \theta - 2l_x l_y w' \\ -\theta - 2l_x l_y w' & \delta' - \left(l_y{}^2 - l_x{}^2\right)w' \end{pmatrix}, \tag{1}$$

are magnified from graphene's lattice vectors $\boldsymbol{a}_m$ by the matrix $\hat{M}$, where, $\boldsymbol{l} = (l_x, l_y)$ is the principle axis of the strain tensor, $\delta' = \delta + w(1-\sigma)/2$, $w' = w(1+\sigma)/2$, and σ is the Poisson ratio. Also, these expressions are quoted for strain applied to the substrate; for strain applied to the graphene layer we substitute $w \to -w$ and use $\sigma = 0.165$.[12] The set of six vectors $\boldsymbol{a}_m$ is obtained by a $m\pi/3$ rotation of $\boldsymbol{a}_0 = (a, 0)$.

The perturbation produced by the moiré superlattice results in the formation of minibands for Dirac electrons in graphene.[13–20] The latter can be found by numerical diagonalisation of the moiré superlattice Hamiltonian,

$$\hat{H} = v\boldsymbol{p}\cdot\boldsymbol{\sigma} + \sum_m \left[U_0 + (-1)^m\left(iU_3\sigma_3 + U_1\frac{\boldsymbol{a}_m}{a}\cdot\boldsymbol{\sigma}\right)\right]e^{i\boldsymbol{G}_m\cdot\boldsymbol{r}}e^{i\boldsymbol{G}_m\cdot\hat{M}\boldsymbol{u}_\mathrm{d}(\boldsymbol{r})}, \tag{2}$$

where Pauli matrices $\sigma_{i=1,2,3}$ act on the electron amplitudes on graphene's A and B sublattices, and parameters U_0, U_3 and U_1 take into account a smooth potential, A–B sublattice asymmetry and the modulation of A–B hopping in graphene produced by the underlying material.[19] This moiré potential contains six harmonics characterised by Bragg vectors $\boldsymbol{G}_m = \frac{4\pi(\delta'^2 + \theta^2 - w'^2)}{\sqrt{3}a^2}\boldsymbol{l}_z \times \boldsymbol{A}_m$ dependent on the misalignment angle and lattice mismatch, with an additional phase factor taking into account lattice deformations,† $\boldsymbol{u}_\mathrm{d}(\boldsymbol{r})$, magnified by the matrix $\hat{M}$.

The anisotropy of the moiré superlattice in strained graphene–hBN heterostructures modifies the characteristic miniband spectrum for graphene electrons, as illustrated in the lower panels of Fig. 1. For the unstrained heterostructure (left) the spectrum exhibits a secondary Dirac point[13–15,17–19] in the valence band. Strain,

† The (Eulerian) displacement field $\boldsymbol{u}_\mathrm{d}(\boldsymbol{r})$ describes the displacements due to defects in the substrate. For defects in the graphene layer $\boldsymbol{u}_\mathrm{d}(\boldsymbol{r}) \to -\boldsymbol{u}_\mathrm{d}(\boldsymbol{r})$.

as low as $w \sim 1$–3% (right), is sufficient for the secondary Dirac point to become obscured by overlapping spectral branches. Most interestingly, a relatively weak strain,

$$w_c = \frac{(1-\sigma)\delta - \text{sign}(\delta)\sqrt{(1+\sigma)^2\delta^2 + 4\sigma\theta^2}}{2\sigma}, \tag{3}$$

transforms a hexagonal moiré superlatice into a quasi-1D lattice shown in the middle panel of Fig. 1. We find that such critical strain is independent of the orientation of the principal axes of the strain tensor and that, at $w = w_c$, all reciprocal lattice vectors are aligned along the same direction at an angle $\arctan\left(\frac{-\theta}{\delta + w_c}\right)$ from $\boldsymbol{l}$. In general such a quasi-1D superlattice is characterised by two incommensurate lattice periods and a pair of two shortest Bragg vectors, $\boldsymbol{G}$ and $\boldsymbol{G}'$. When the ratio $|\boldsymbol{G}|/|\boldsymbol{G}'| = p/q$ is rational, the electron spectrum is described in terms of plane waves with wavevector $\boldsymbol{k} = (k_x, k_y)$ where $|k_y| < |\boldsymbol{G}'|/q$ is counted along $\boldsymbol{G}$, and $|k_x|$ is unbound and counted perpendicular to $\boldsymbol{G}$. The diagonalisation of the Hamiltonian (eqn (2)) in the basis of Wanier wavefunctions corresponding to such wavenumbers returns a fractal spectrum[21,22] of "bands" and "gaps", for every value of k_x, as exemplified in Fig. 2. For a crystallographically

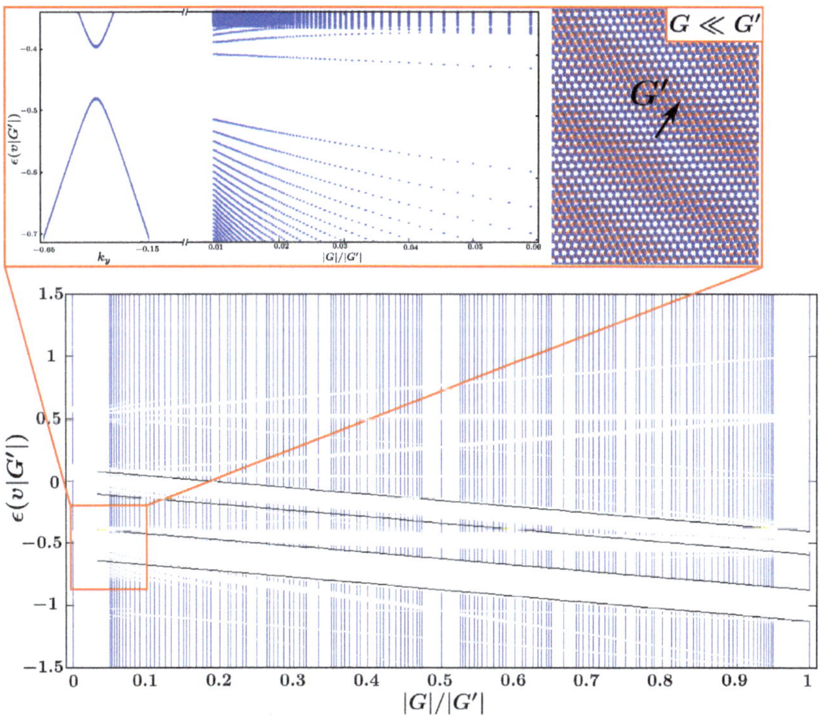

Fig. 2 Main panel: the spectral support of $\hat{H}$ at $k_x = 0$ calculated for a quasi-1D moiré superlatice with rational $|G|/|G'| = p/q$ and the same perturbation parameters as in Fig. 1. Black lines show the top and bottom of bands, calculated analytically for the two largest gaps. Top: full dispersion at $k_x = |G|/|G'| = 0$, and the spectral support for $|G|/|G'| \leq 0.06$ (left). The moiré superlattice for $|G|/|G'| = 0.01$ is shown on the right.

aligned graphene–hBN heterostructure ($\theta = 0$), strain $w = \delta = 1.8\%$ exactly compensates the lattice mismatch. This results in the aligned reciprocal lattice vectors, with $|\boldsymbol{G}|/|\boldsymbol{G}'| = 1$, as shown in Fig. 1 (centre). For a finite misalignment angle θ and $\boldsymbol{l} = \boldsymbol{a}_0/|\boldsymbol{a}_0|$, the moiré pattern can be viewed as two superimposed incommensurate superlattices with an arbitrary ratio of periods, corresponding to

$$\frac{|\boldsymbol{G}|}{|\boldsymbol{G}'|} = \sqrt{\frac{f - 1.14|\theta|}{f + 1.14|\theta|}},$$

where for the graphene–hBN pair $f \approx -0.0595 + 3.17\sqrt{4.4 \times 10^{-3} + 0.66\theta^2}$. Hence, in general, the incommensurate structures possible in strained moiré superlattices have periods with $|\boldsymbol{G}|/|\boldsymbol{G}'|$ spanning across a considerable range of $0 < |\boldsymbol{G}|/|\boldsymbol{G}'| < 1$.

Fractal spectra of electrons, calculated for $k_x = 0$, and various commensurate ratios $|\boldsymbol{G}|/|\boldsymbol{G}'| = p/q$ (controlled by the misalignment angle) are shown in Fig. 2. The hierarchy of "gaps" and bands in this spectrum, illustrated in the inset for $p/q = 0$, can be understood by using the $\boldsymbol{k}\cdot\boldsymbol{p}$ theory for the dispersion of electrons at a simpler fraction (such as $p/q = 0$; 1; or 1/2) and including incommensurability in the form of an additional perturbation. For example, the vicinity of $p/q = 0$ can be described by using the parabolic dispersion of electrons near the band edges of a spectrum evaluated for finite $\boldsymbol{G}'$ (see inset in Fig. 2), where as an additional slow-varying perturbation produces characteristic size-quantized levels near minima/maxima of the potential oscillating with wavevector $\boldsymbol{G}$. The same approach enables us to approximate the edges of the largest gaps in the spectrum, shown by black lines. Note that for all $k_x \neq 0$ such gaps appear at different energies, hence the overall fractal spectrum is not gapped.

Up to now, we have considered the effect of homogeneous strain on the form of moiré superlattices. Beyond that, any defect-induced smooth displacement field, $\boldsymbol{u}_\mathrm{d}(\boldsymbol{r})$, also affects the moiré superlattice geometry, and a smoothly varying local coordination between graphene and the substrate atoms would reflect such deformations in the Hamiltonian (eqn (2)) of graphene electrons. To follow how the actual lattice deformations can be related to the deformations of the moiré superlattice, we compare the points of constant phase (which are analogous to lattice "sites") in the Fourier harmonics included in eqn (2) with ($\boldsymbol{R}_\mathrm{d}$) and without ($\boldsymbol{R}$) deformations. Then, the "displacement field" in the moiré superlatice, $\boldsymbol{U}_\mathrm{d}(\boldsymbol{R}_\mathrm{d}) = \boldsymbol{R}_\mathrm{d} - \boldsymbol{R}$, describes the position of a particular moiré lattice "site" in the presence of a defect relative to its position in the absence of a defect. Using eqn (2),we find that

$$\boldsymbol{U}_\mathrm{d}(\boldsymbol{r}) = -\hat{M}\boldsymbol{u}_\mathrm{d}(\boldsymbol{r}), \qquad (4)$$

which enables us to match extended and topological defects in the top/bottom layer of the heterostructure to a partner defect in the moiré superlattice, which appears at a scale magnified by the matrix $\hat{M}$.

Below, we discuss in detail the two defect types shown in Fig. 3. One example illustrated in Fig. 3(a) is a wrinkle in the top graphene layer centred at the dashed white line along unit vector $\boldsymbol{l}_w$. The effect of such wrinkle on the moiré superlattice is characterized by a vector, $\boldsymbol{D}$, connecting moiré lattice "sites" on opposite sides

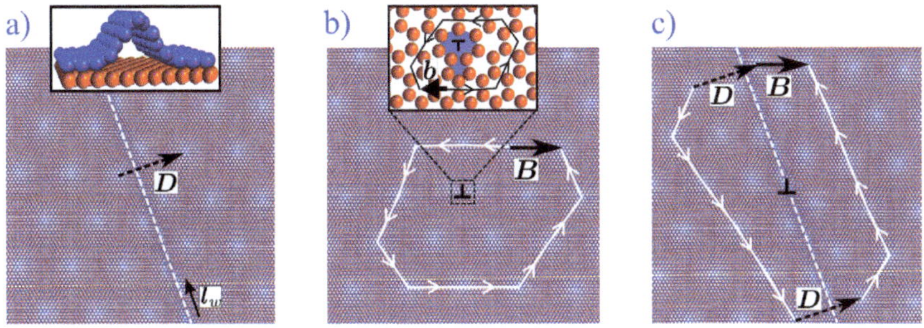

Fig. 3 The moiré superlattice with $\delta = 0.072$ and $\theta = 0$. (a) A wrinkle is present in the graphene layer. (b) A dislocation in the underlying material generates a closure failure of B around the loop (white) in the moiré superlattice. (c) Both a wrinkle and a dislocation are present. The vector D is used to connect the two sides of the wrinkle so that the presence of a dislocation is identified using the closure failure around a path in the superlattice (white).

of the wrinkle, measured at a distance larger than the wrinkle's width. Then, the shift

$$\boldsymbol{d} \equiv \int_D \mathrm{d}s \frac{\mathrm{d}\boldsymbol{u}_\mathrm{d}(\boldsymbol{r}(s))}{\mathrm{d}s}, \tag{5}$$

in atomic positions of carbon atoms with respect to their position in flat graphene can be found using

$$\boldsymbol{d} = -\hat{M}^{-1}(\boldsymbol{D} + n_0 A_0 + n_1 A_1), \tag{6}$$

where integers n_0 and n_1 account for the fact that $\boldsymbol{D}$ is only measured up to a vector from the moiré superlattice and should be used to choose the smallest possible $\boldsymbol{d}$. Eqn (6) is also valid if a discontinuity in the moiré superlattice is due to a step edge in the underlying material.[23] In this case $\boldsymbol{u}_\mathrm{d}(\boldsymbol{r})$ would include both the effect of curving the graphene flake over the step edge, as well as the shift in stacking characterised by the different atomic layers in the underlay.

In Fig. 3(b) we illustrate how a dislocation in the substrate, characterised by Burger's vector $\boldsymbol{b}$, is reflected by a dislocation in the moiré superlattice, with Burger's vector $\boldsymbol{B}$:

$$\boldsymbol{b} \equiv \oint \mathrm{d}s \frac{\mathrm{d}\boldsymbol{u}_\mathrm{d}(\boldsymbol{r}(s))}{\mathrm{d}s} = -\hat{M}^{-1}\boldsymbol{B}, \boldsymbol{B} \equiv \oint \mathrm{d}s \frac{\mathrm{d}\boldsymbol{U}_\mathrm{d}(\boldsymbol{r}(s))}{\mathrm{d}s}. \tag{7}$$

A dislocation in graphene with Burger's vector $-\boldsymbol{b}$ will have the same effect on the moiré superlattice. Interestingly, the moiré pattern analysis enables one to determine the presence of a dislocation even in a wrinkled heterostructure, as illustrated in Fig. 3(c).

In conclusion, we have shown that moiré superlattices characteristic for graphene heterostructures with hexagonal crystals, or graphene deposited on a substrate with a hexagonal surface layer, magnify the length-scale of deformations in either of the involved crystals, by generating deformations and defects in the moiré pattern. This result, discussed in detail for heterostructures of hexagonal crystals, remains qualitatively the same for other crystalline lattices. For

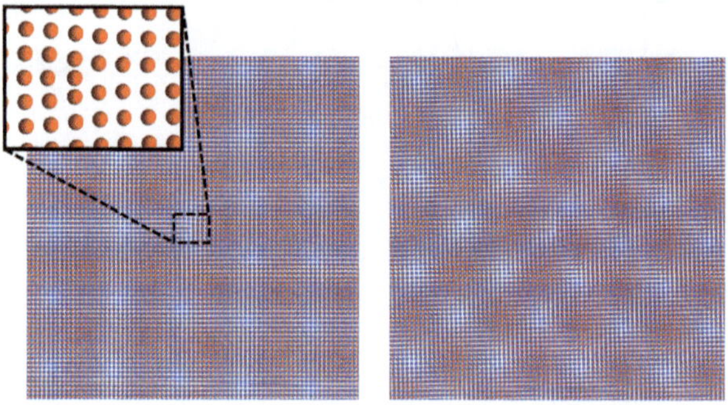

Fig. 4 Moiré pattern for square lattices. Moire patterns for square lattices with $\delta = 0.072$ and $\theta = 0$ (left) and $\delta = \theta = 0.05$ (right). The same dislocation is present in the underlying material in both images.

example, in Fig. 4, we show that a dislocation in one of two square lattice crystals in a heterostructure generates a clearly identifiable dislocation in the square moiré superlattice.

Acknowledgements

This work was supported by EC FP7 Graphene Flagship project CNECT-ICT-604391, ERC Grant Hetero2D, and the Royal Society Wolfson Research Merit Award. We are grateful to H. Schomerus and A. Geim for useful discussions.

References

1 F. R. S. Lord Rayleigh, *Philos. Mag.*, 1874, **47**, 81–94.
2 A. Righi, *Nuovo Cimento*, 1887, **21**, 203–228.
3 A. T. N'Diaye, S. Bleikamp, P. J. Feibelman and T. Michely, *Phys. Rev. Lett.*, 2006, **97**, 215501.
4 S. Marchini, S. Günther and J. Wintterlin, *Phys. Rev. B: Condens. Matter Mater. Phys.*, 2007, **76**, 075429.
5 P. Sutter, J. T. Sadowski and E. Sutter, *Phys. Rev. B: Condens. Matter Mater. Phys.*, 2009, **80**, 245411.
6 M. Sicot, P. Leicht, A. Zusan, S. Bouvron, O. Zander, M. Weser, Y. S. Dedkov, K. Horn and M. Fonin, *ACS Nano*, 2012, **6**, 151.
7 M. Batzill, *Surf. Sci. Rep.*, 2012, **67**, 83.
8 C. Berger, Z. Song, T. Li, X. Li, A. Y. Ogbazghi, R. Feng, Z. Dai, A. N. Marchenkov, E. H. Conrad, P. N. First and W. A. de Heer, *J. Phys. Chem. B*, 2004, **108**, 19912.
9 J. Xue, J. Sanchez-Yamagishi, D. Bulmash, P. Jacquod, A. Deshpande, K. Watanabe, T. Taniguchi, P. Jarillo-Herrero and B. J. LeRoy, *Nat. Mater.*, 2011, **10**, 282.
10 R. Decker, Y. Wang, V. W. Brar, W. Regan, H.-Z. Tsai, Q. Wu, W. Gannett, A. Zettl and M. F. Crommie, *Nano Lett.*, 2011, **11**, 2291.
11 Z. Y. Rong and P. Kuiper, *Phys. Rev. B: Condens. Matter Mater. Phys.*, 1993, **48**, 17427.

12 O. L. Blakslee, D. G. Proctor, E. J. Seldin, G. B. Spence and T. Weng, *J. Appl. Phys.*, 1970, **41**, 3373.
13 C.-H. Park, L. Yang, Y.-W. Son, M. L. Cohen and S. G. Louie, *Nat. Phys.*, 2008, **4**, 213.
14 M. Yankowitz, J. Xue, D. Cormode, J. D. Sanchez-Yamagishi, K. Watanabe, T. Taniguchi, P. Jarillo-Herrero, P. Jacquod and B. J. LeRoy, *Nat. Phys.*, 2012, **8**, 382.
15 C. Ortix, L. Yang and J. van den Brink, *Phys. Rev. B: Condens. Matter Mater. Phys.*, 2012, **86**, 081405.
16 M. Kindermann, B. Uchoa and D. L. Miller, *Phys. Rev. B: Condens. Matter Mater. Phys.*, 2012, **86**, 115415.
17 L. A. Ponomarenko, R. V. Gorbachev, G. L. Yu, D. C. Elias, R. Jalil, A. A. Patel, A. Mishchenko, A. S. Mayorov, C. R. Woods, J. R. Wallbank, M. Mucha-Kruczynski, B. A. Piot, M. Potemski, I. V. Grigorieva, K. S. Novoselov, F. Guinea, V. I. Fal'ko and A. K. Geim, *Nature*, 2013, **497**, 594.
18 C. R. Dean, L. Wang, P. Maher, C. Forsythe, F. Ghahari, Y. Gao, J. Katoch, M. Ishigami, P. Moon, M. Koshino, T. Taniguchi, K. Watanabe, K. L. Shepard, J. Hone and P. Kim, *Nature*, 2013, **497**, 598.
19 J. R. Wallbank, A. A. Patel, M. Mucha-Kruczyński, A. K. Geim and V. I. Fal'ko, *Phys. Rev. B: Condens. Matter Mater. Phys.*, 2013, **87**, 245408.
20 B. Hunt, J. D. Sanchez-Yamagishi, A. F. Young, M. Yankowitz, B. J. LeRoy, K. Watanabe, T. Taniguchi, P. Moon, M. Koshino, P. Jarillo-Herrero and R. C. Ashoori, *Science*, 2013, **340**, 6139.
21 G. André and G. Aubry, *Ann. Isr. Phys. Soc.*, 1980, **3**, 133.
22 B. Simons, *Adv. Appl. Math.*, 1982, **3**, 463.
23 J. Coraux, A. T. N'Diaye, C. Busse and T. Michely, *Nano Lett.*, 2008, **8**, 565.

Faraday Discussions

PAPER

Metal–semiconductor transition like behavior of naphthalene-doped single wall carbon nanotube bundles†

Fitri Khoerunnisa,*[a] Aaron Morelos-Gomez,[b] Hideki Tanaka,[c] Toshihiko Fujimori,[d] Daiki Minami,[d] Radovan Kukobat,[de] Takuya Hayashi,[e] Sang Young Hong,[f] Young Chul Choi,[f] Minoru Miyahara,[c] Mauricio Terrones,[g] Morinobu Endo[b] and Katsumi Kaneko*[d]

Received 30th May 2014, Accepted 27th June 2014

DOI: 10.1039/c4fd00119b

Naphthalene (N) or naphthalene-derivative (ND) adsorption-treatment evidently varies the electrical conductivity of single wall carbon nanotube (SWCNT) bundles over a wide temperature range due to a charge–transfer interaction. The adsorption treatment of SWCNTs with dinitronaphthalene molecules enhances the electrical conductivity of the SWCNT bundles by 50 times. The temperature dependence of the electrical conductivity of N- or ND-adsorbed SWCNT bundles having a superlattice structure suggests metal–semiconductor transition like behavior near 260 K. The ND-adsorbed SWCNT gives a maximum in the logarithm of electrical conductivity *vs.* T^{-1} plot, which may occur after the change to a metallic state and be associated with a partial unravelling of the SWCNT bundle due to an evoked librational motion of the moieties of ND with elevation of the temperature.

1 Introduction

Single wall carbon nanotubes (SWCNTs) have a wide range of electronic properties from semiconductor- to metal-like depending on the chiral indexes which describe the rolling up structure of graphene. As all carbon atoms of a SWCNT are exposed to the surface, the geometrical surface area is 2630 $m^2\ g^{-1}$ and SWCNTs

[a]*Department of Chemistry, Indonesia University of Education, Bandung 40154, Indonesia. E-mail: fitri.khoerunnisa@gmail.com*

[b]*Aqua Eco Center, Shinshu University, Nagano 380-8553, Japan*

[c]*Department of Chemical Engineering, Kyoto University, Nishikyo, Kyoto 615-8510, Japan*

[d]*Center for Energy and Environmental Science, Nagano 380-8553, Japan. E-mail: kkaneko@shinshu-u.ac.jp*

[e]*Department of Electrical Engineering, Shinshu University, Nagano 380-8553, Japan*

[f]*Research and Development Center, Hanwha Nanotech, Incheon 403-030, Republic of Korea*

[g]*Department of Physics, Pennsylvania State University, University Park, Pennsylvania 16802, USA*

† Electronic supplementary information (ESI) available. See DOI: 10.1039/c4fd00119b

are intrinsically surface sensitive.[1] Even exposure to an inert gas influence the electrical conductivity of a SWCNT with an atom collision-induced distortion in the graphene wall,[2] although adsorption of an electron donor or electron acceptor gases markedly changes the electrical conductivity of SWCNTs.[3–5] SWCNTs form an ordered bundle structure of a 2D hexagonal superlattice due to the inter-tube dispersion interaction.[6,7] Therefore, if we remove caps from the SWCNT by an oxidation-treatment, the open SWCNT bundles have three kinds of adsorption sites for molecules and atoms in the internal tube spaces, interstitial pore spaces and groove sites.[8] Here the internal tube spaces are in each open SWCNT, the interstitial pore spaces between three SWCNTs and the groove sites are at the area where two SWCNTs are arrayed on the external surface of the bundle. The internal tube spaces of the open SWCNT bundle are available for the adsorption of various molecules whose size is smaller than the effective size of the tube spaces. The in-pore superhigh pressure effect of the internal tube spaces of a SWCNT has been clearly demonstrated recently. Solid KI transforms into a high pressure phase at 1.9 GPa in the bulk; the high pressure phase transformation is induced by the introduction of KI nanocrystals into the carbon nanotube spaces even below 0.1 MPa.[9] It is known that solid sulfur having an eight member-ring structure, being a typical insulator, converts to 2D network structure consisting of 1D sulfur chains above 90 GPa. The introduction of sulfur atoms into the internal nanospaces of a SWCNT below 0.1 MPa gives an atomically 1D sulfur chain crystal of a metallic nature, indicating the intensive compression effect for sulfur atoms corresponding to 90 GPa at least.[10] As the interstitial nanopores offer the smallest pores and have the deepest interaction potential wells, accessible molecules can be strongly adsorbed there.[11] If a SWCNT is capped and closed, the internal spaces are not available for molecules and atoms, and thereby only the interstitial pores can accept atoms and molecules. When the interaction between the interstitial pore and atoms or molecules is strong enough to vary the SWCNT bundle structure, an intercalation-like structure similar to intercalation graphite[12] is formed. Br_2, K, TCNQ, and pentacene *etc.* are doped in the SWCNT bundles by vaporization of these intercalants in the gas phase at high temperature, giving the intercalation structures.[13,14] The intercalation structure formation changes the electrical conductivity of the SWCNT bundles remarkably. The highly crystalline SWCNTs give sharp X-ray diffraction patterns of the super lattice of the bundle,[6,7] although each SWCNT may perform a librational motion in the bundles and a weak inter-tube coupling can cause a slight change in the electronic structure around the Fermi level according to the theoretical studies.[15] Thus, the intercalation-type modification of the SWCNT bundles is effective for tuning the electronic properties as well as intercalation of graphite.

On the other hand, adsorption of C_{60} and polycyclic aromatic hydrocarbon (PAH) molecules on SWCNT bundles in the liquid phase induces their intercalation-like structure, where the charge–transfer interaction occurs between the PAH molecules and SWCNT.[16,17] In our previous study, a naphthalene-adsorbed SWCNT had an intercalation structure according to the X-ray diffraction pattern and the ultraviolet photoelectron spectroscopy shows an explicit increase of the density of state at the Fermi level which indicates a pseudo-metallic state.[18] In this study we modify SWCNT bundles with naphthalene and its derivatives of different electronic properties using liquid phase adsorption. The temperature dependence of the dc electrical conductivity of the naphthalene derivative (ND)-

adsorbed SWCNT was measured, suggesting a molecular motion-mediated unique temperature dependence of the electrical conductivity in addition to the Mott transition-like behavior which has been observed in transition metal oxides and disordered solids.

2 Experimental and computation procedures

SWCNTs produced from graphite by an arc-discharge method by Hanwha Nanotech Co. were used in this study. The SWCNTs were purified with 1 M HCl for 20 h at room temperature to remove metallic impurities and dried at 373 K after washing. Then the SWCNT samples were annealed at 1473 K in Ar for 30 min to remove amorphous carbons. The content of metal oxide impurities was 3 wt% as measured by thermal gravimetric analysis in a flow of O_2–N_2 mixed gas; the graphite remains according to X-ray diffraction measurement. The diameter of the SWCNTs are 1.3 to 1.5 nm and the bundle sizes are 20 nm to 30 nm on average according to the previous study.[19] Liquid phase adsorption of naphthalene (N) and its derivatives (ND) on the SWCNT bundle was carried out using the toluene solution at 298 K. Here, 1,5-dinitronaphthalene (DNN, $C_{10}H_6N_2O_4$) and 1,5-dihydroxynaphtalene (DHN, $C_{10}H_8O_2$) were used as the ND agent. Here DNN and DHN should work as an electron acceptor and an electron donor, respectively.[20] The dispersed SWCNT samples were filtrated and rinsed by toluene to remove any non-adsorbed N and ND molecules. The ND-adsorbed SWCNT samples were dried before characterization under vacuum at 323 K and 10 mPa for 24 h. The ND adsorption amount was determined from the concentration change of ND molecules in the solution using a UV-Vis-NIR spectrometer (JASCO; V-670). Raman spectra were obtained with 532 nm and 785 nm excitations, using a single-monochromator micro-Raman spectrometer in a back scattering configuration (inVia Raman Microscope, Renishaw). The surface coverage of SWCNTs with N or ND molecules was controlled to be 0.30 ± 0.10, which was determined by the surface area of SWCNTs from N_2 adsorption at 77 K and the area occupied by N or ND molecules using their molecular area. The optical absorption spectra of the SWCNT samples were collected using the UV-Vis-NIR spectrometer. The dc electrical resistance measurements of the SWCNT Bucky paper form were performed using the four probe method using a physical properties measurement system (model 600: Quantum Design) over the temperature range of 2 K to 340 K. The melting temperatures of DNN and DHN were measured using a DSC system with 10 K min^{-1} after calibration with In (DSC 8500: Perkin Elmer). The plausible configuration of DNN molecules in an interstice of the SWCNT array was simulated with a GCMC simulation using the structural parameters of the bundle estimated from X-ray diffraction experiment; in the simulation a (12,10) SWCNT of tube diameter of 1.495 nm was used. The simulation details are given in the ESI.† We computed the electronic energy levels of naphthalene and ND molecules and the density of states of the (17,0) SWCNT of tube diameter of 1.350 nm with *ab initio* density functional theory as implemented in the SIESTA code. The computational method is described in the ESI.† Here, both SWCNTs are semiconductive.

3 Results and discussion

The purified SWCNT is almost closed even after the above-mentioned purification, from the N_2 adsorption measurements at 77 K; the surface area of the

SWCNT is 330 $m^2 g^{-1}$, indicating that the SWCNT is closed.[19] All SWCNT samples are a mixture of metallic and semiconducting SWCNTs due to the feature of the G^- Raman band.[19] Fig. 1 shows SEM images of SWCNTs before the adsorption treatment and the DHN-adsorbed SWCNT bundles. We can see very clear SWCNT bundles and these SWCNT bundles are randomly entangled with each other before adsorption treatment. The SEM image of the DHN–SWCNTs is completely different from that of the SWCNTs before the treatment; the bundle size decreased markedly through the adsorption treatment, indicating the insertion of DHN molecules in the intertube spaces. Similar images were obtained for N- and DNN–SWCNTs. Fig. 2 shows the X-ray diffraction patterns of SWCNTs, N–SWCNTs, DNN–SWCNTs and DHN–SWCNTs. The diffraction pattern of the SWCNTs has sharp peaks coming from the well-ordered bundle structure of the hexagonal symmetry; the predominant peak at 2.72° is assigned to the 10 peak, the other peaks at 4.40°, 7.06° and 9.56° correspond to the reflections from 11, 20 and 21 superlattice planes of the bundles, respectively, although these positions are not corrected by the influence of the form factor. The inset shows the magnified X-ray diffraction patterns of the 10 peak for all samples. The sharp peak at 11.90° is caused by the reflection from the (002) plane of graphite as the impurity. The diffraction patterns consisting of the sharp peaks of the SWCNT dramatically change after the adsorption treatment with N and ND molecules. The intensity of all peaks coming from the ordered bundle structure decreases markedly, accompanying an explicit broadening. In particular, the 10 peak shows the greatest change, giving a lower diffraction angle shift corresponding to the expansion of the interlayer spacing by 0.12 nm for the DNN–SWCNTs, 0.063 nm for the DHN–SWCNTs and 0.10 nm for the N–SWCNTs. The magnified 10 peaks in the inset evidently show this change. Accordingly, the ordered SWCNT bundle structure should be partially broken to embrace the N or ND molecules, coinciding with SEM observation. Fig. 3 shows the optical absorption spectra of the adsorption-treated and non-treated SWCNT samples. The SWCNT sample is a mixture of metallic and semiconductive SWCNTs and we can observe the interband transitions due to metallic and semiconductive band structures. The spectrum of the SWCNTs has three broad bands due to the M_{11}, S_{11} and S_{22} transitions. Here, M_{11} is the transition between the first van Hove singularities of the metallic SWCNTs, whereas S_{11} and S_{22} are the transitions between the first and second van Hove singularities of the semiconductive SWCNTs, respectively.[21,22] The adsorption treatment induces an intensive hyperchromic effect for

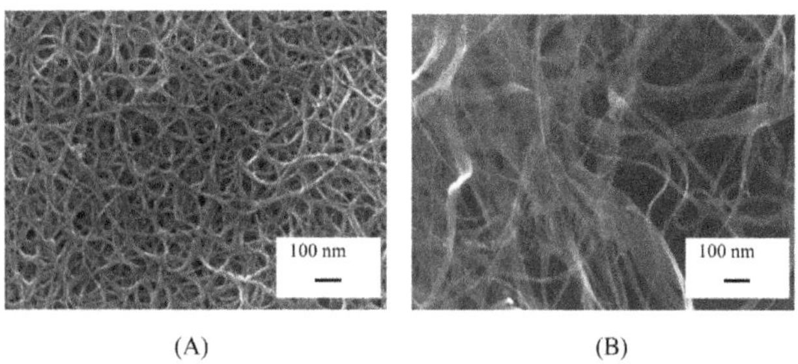

Fig. 1 SEM images of SWCNTs (A) and DHN-adsorbed SWCNTs (B).

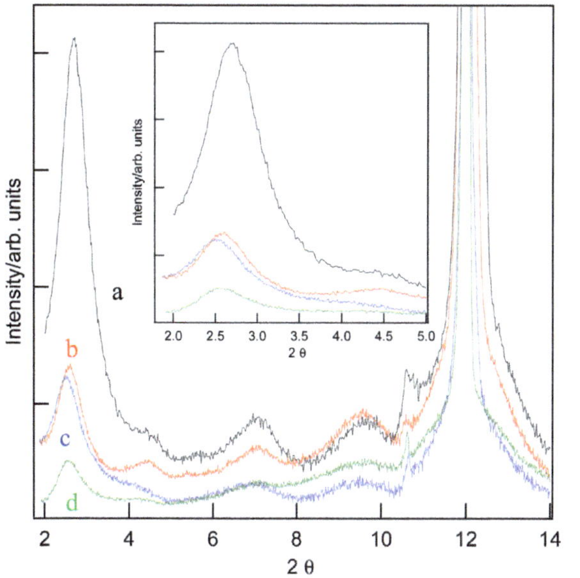

Fig. 2 X-ray diffraction patterns of SWCNT, N-adsorbed SWCNT and ND-adsorbed SWCNT bundles. Black (a): SWCNT, red (b): DHN–SWCNT, blue (c): DNN–SWCNT and green (d): N–SWCNT.

the M_{11}, S_{22} and S_{11} bands and an explicit red shift of the three bands, by 7–13 nm for M_{11}, 13–18 nm for S_{22}, 16–41 nm for S_{11}. The intensive hyperchromic effect and explicit red shift of the three bands clearly evidence the mixing of wave functions of the N or ND molecules and the SWCNTs regardless of the metallic or

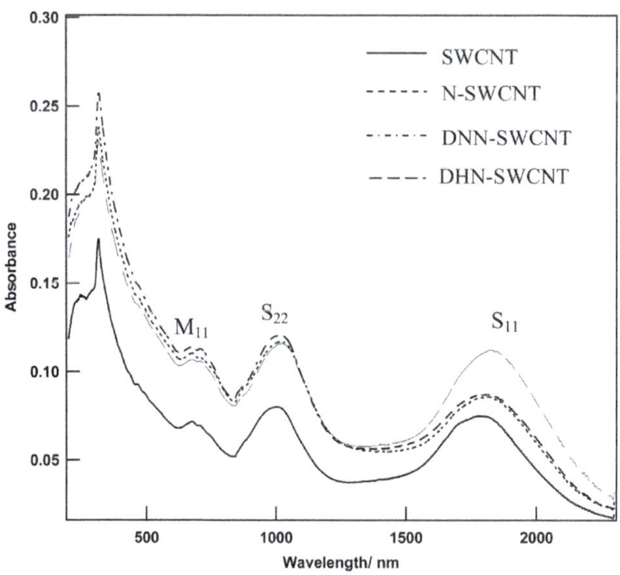

Fig. 3 Optical absorption spectra of the SWCNTs and N-, DNN- and DHN-adsorbed SWCNTs.

semiconductive nature of the SWCNTs; there should be a marked charge transfer interaction between naphthalene or ND and the SWCNTs. The naphthalene or ND adsorption treatment, however, did not give a marked change in the radial breathing mode (RBM) and G^--band frequencies:[23] G^+-band (cm^{-1}) by 532 nm excitation: SWCNT 1592; N–SWCNT 1594; DHN–SWCNT 1594; and DNN–SWCNT 1592. RBM (cm^{-1}) by 532 nm excitation: SWCNT 168; N-SWCNT 174; DHN–SWCNT 174; and DNN–SWCNT 170 (Fig. 2S(A)†). The higher frequency shift of the DNN–SWCNTs is slightly different from that of the N– and DHN–SWCNTs. We obtained similar results by excitation with 785 nm, as shown in Fig. 2S(B).† The Raman spectroscopic data are not clearly understood yet.

Fig. 4 shows the temperature dependence of the electrical conductivity of the SWCNT samples. The absolute electrical conductivity of the different SWCNT samples are different from each other. The DNN adsorption increases the electrical conductivity remarkably and the electrical conductivity is larger than that of SWCNTs by 50 times. N adsorption increases the electrical conductivity slightly, whereas DHN adsorption gives just a little decrease of the electrical conductivity. The extrapolation of the approximate linear plot above 125 K to the vertical axis gives the following electrical conductivity (see Fig. 5): SWCNT; 3.2×10^3 S m^{-1}, N–SWCNT; 5.1×10^3 S m^{-1}, DNN–SWCNT; 1.7×10^5 S m^{-1} and DHN–SWCNT; 2.8×10^3 S m^{-1}. Thus, the DNN adsorption treatment is quite efficient for enhancing

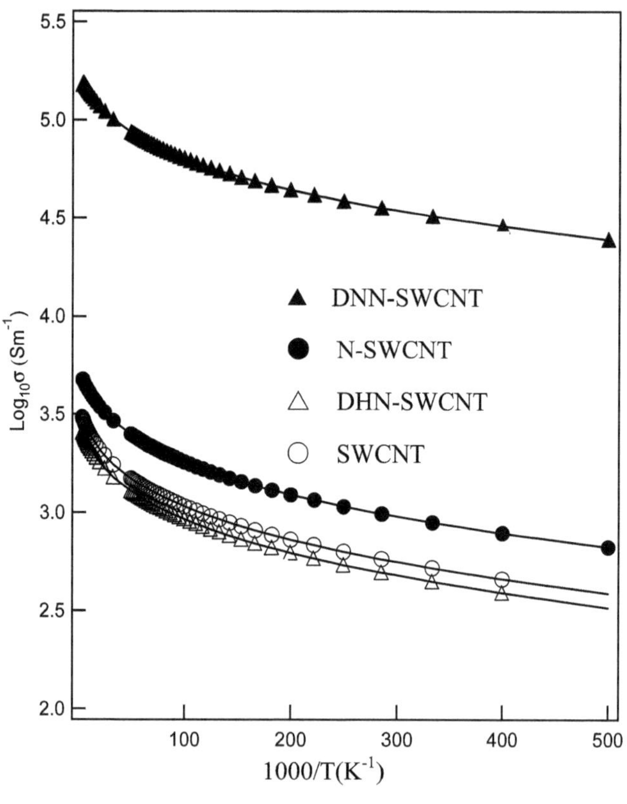

Fig. 4 The temperature dependence of the electrical conductivity σ of the SWCNT samples.

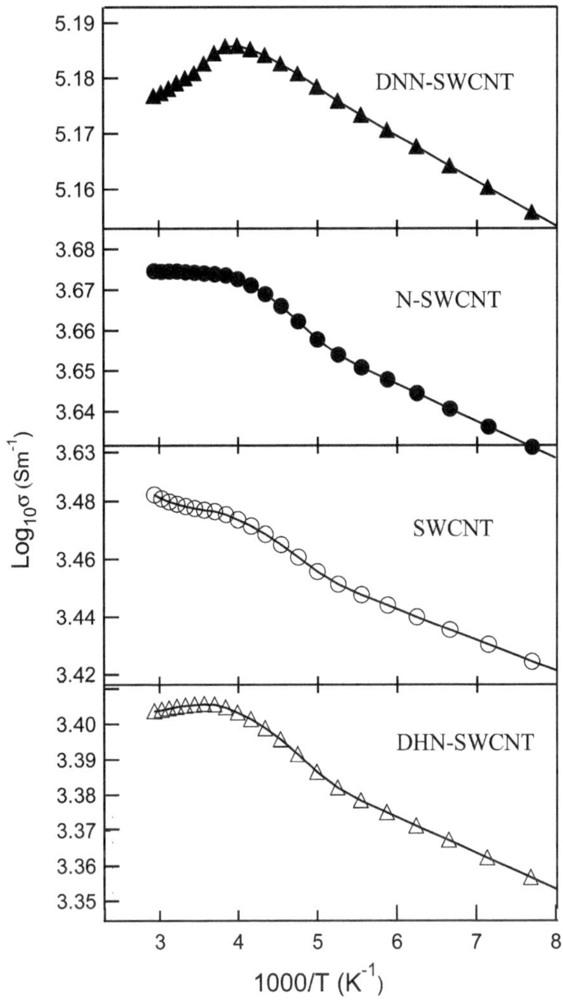

Fig. 5 Temperature dependence of the electrical conductivity σ at around ambient temperature

the electrical conductivity of the SWCNT bundles. The logarithm of electrical conductivity *vs.* T^{-1} plot is almost linear in the lower temperature region and deviates upward near an ambient temperature. The electrical conductivity change should be correlated with the electronic properties of the N, DNN and DHN molecules. It is assumed that a DNN molecule of an electron acceptor pulls an electron from the SWCNT to create a hole, being a charge carrier, to enhance the electrical conductivity, while a DHN molecule of an electron donor donates an electron to the SWCNT to annihilate a hole in the SWCNT. Thus, the concept of the electron donor and acceptor can explain qualitatively the electrical conductivity change through the adsorption treatments of the SWCNT bundles. However, the above explanation must be supported by a computational study on the electronic structure of the ND-adsorbed SWCNT. The preliminary computational results of the density of state of the semiconductive (17,0) SWCNT and the energy diagrams of three molecules are shown in Fig. 1S.† The HOMO level of DHN is slightly higher than the Fermi level, whereas the HOMO levels of naphthalene and DNN are lower than the Fermi level. On the other hand, the LUMOs of DNN and naphthalene are higher than the Fermi level by about 1 eV and 3 eV, respectively. In the case of charge–transfer complexes between donor and acceptor molecules, the LUMO–HOMO energy difference of 1–3 eV between acceptor and donor molecules induces a predominant charge–transfer interaction in the presence of sufficient orbital overlapping according to the molecular orbital study.[24] If we apply the charge–transfer interaction in the molecules to the molecule–SWCNT interaction, DNN whose LUMO is closer to the Fermi level of SWCNTs should induce a stronger charge transfer interaction than naphthalene. Consequently we conjecture that the adsorbed DNN molecules pull the electrons from SWCNT to create the holes more efficiently than naphthalene molecules, although a systematic study on the electronic interactions between ND or naphthalene and SWCNTs with the DFT computational method is necessary in future. The above distinct electrical conductivity changes of the SWCNT bundles with the adsorption of naphthalene or ND molecules suggest the formation of a partial intercalation structure between the SWCNT bundle and N or ND molecules. The interesting behavior is observed in the electrical conductivity near ambient temperature, as shown in Fig. 5. The electrical conductivity σ of the N–SWCNTs becomes almost constant above 250 K, being similar to a metallic conductor, whereas the log σ *vs.* T^{-1} plots of the DNN–SWCNTs and DHN–SWCNTs have a gentle maximum around 250 K and 270 K, respectively. Even the log σ *vs.* T^{-1} plot of the SWCNTs is slightly upward-deviating around 250 K. The log σ *vs.* T^{-1} plot of the N–SWCNTs agrees with the pseudo-metallization shown by the previous UV photoelectron spectroscopic study.[18] As the log σ *vs.* T^{-1} relation below the bending temperature is almost linear, the SWCNT samples are assumed to be semiconductors below the bending temperature. Consequently, phenomenologically the observed behavior in the N–SWCNTs suggests a metal–semiconductor transition which has been observed in transition metal oxides, covalent semiconductors containing impurities and disordered solids, which have localized electrons.[25,26] We infer that the unusual log σ *vs.* T^{-1} relations of DNN– and DHN–SWCNTs are also associated with the semiconductor–metal transition as well as that of the N–SWCNTs. Here we must remind readers that the electrical conductivity was measured using the SWCNT Bucky paper.

The SWCNT Bucky paper consists of many bundles and thereby there are two kinds of barriers for carrier transport. One is the SWCNT–SWCNT contact and another is the bundle–bundle contact. The electric charges must carry mainly through the metallic SWCNTs in the bundle and the charge carriers in the bundle should behave as localized ones, and then such localized carriers must transport the neighbor bundle through the contact. The Mott variable range hopping mechanism for transition metal oxides and disordered solids can be applicable to explain the electrical conduction behavior of the SWCNT bundle.[10,27–29] Snoussi *et al.* showed that the localization length is 4–11 nm, being the boundary size of the bundle for the SWCNT thin film using Mott theory.[27] Fig. 6 shows the 3D variable range hopping plots of the temperature dependence of the electrical conductivity of all SWCNT samples over the wide temperature range from 2 K to 340 K. The 3D variable range hopping plots are almost linear over most of the temperature range except for the higher temperature range. Here, the 3D variable

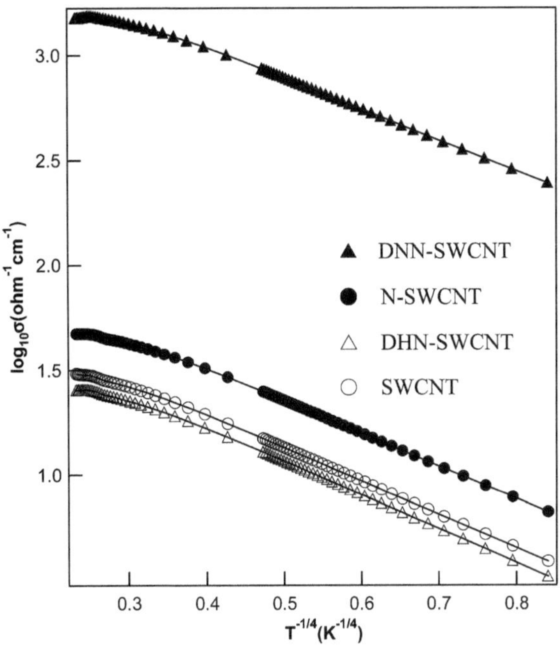

Fig. 6 The 3D variable range hopping plots. The electrical conductivity in the lower temperature range gives the best fit for the 3D plot.

range hopping plots give the best fit compared with the plots of other dimensions. Therefore, the electron conduction network should be three-dimensional, agreeing with the 3D entangled structure of the SWCNT bundles observed by SEM. The DNN molecules should inject the hole carriers into the SWCNTs by a charge transfer interaction and the fundamental charge transport mechanism in the DNN–SWCNT should be the same as for the other SWCNTs.

The well-known Mott transition of transition metal oxide crystals shows a vertical transition of the electrical conductivity from semiconductor to metal, while the transition is more gradual in the case of disordered solids.[25,26,30] The SWCNT bundles exhibit very sharp diffraction peaks coming from the superlattice structure, as already mentioned above; the superlattice system of the SWCNT bundle is much less crystalline compared with an ordered structure of a single crystalline transition metal oxide and the localized states in the SWCNT bundles should be not uniform. In particular, the superlattice structure of N or ND molecules-intercalated SWCNT bundles is more disordered, as shown by X-ray diffraction, leading to a widely distributed localized state for electrons. Hence, the Mott like transition of N and ND molecules-intercalated SWCNT bundles does not show a vertical change in the temperature dependence of the electrical conductivity.

In the next section we need to have an answer for the question of why the temperature dependence of the electrical conductivity of ND-adsorbed SWCNT bundles has a maximum, being quite different from the behavior of a well-known metal–semiconductor transition. We must notice that N-adsorbed SWCNT bundles do not have such a maximum in the log σ *vs.* T^{-1} plot. A great difference between the ND molecule and naphthalene is in the presence of the moieties; the DNN and DHN molecules have two nitro groups and hydroxyl ones. The framework of the naphthalene molecule has less motional freedom than the functional groups in the SWCNT bundle. The melting points of DNN and DHN are 505 K and 551 K, respectively. Confinement of naphthalene (N), methylnaphthalene (MN), methoxynaphthalene (MeN) and chloronaphthalene (ClN) in the silicious mesopores of SBA-15 of 5.5 nm lowers their melting temperature by 65–85 K.[31] As the confinement of DNN and DHN in the interstices of the SWCNT bundles is much more intensive, the depression of the melting temperature should be more than that in SBA-15. The depression of the melting temperature of DNN inserted in the SWCNT bundles by 200 K predicts an anomaly in the temperature dependence of the electrical conductivity around $T = 305$ K ($T^{-1} = 3.3 \times 10^{-3}$). Generally speaking, librational and/or rotational motions of moieties of the molecule should start below the melting temperature. The motion of the moieties on the pre-melting process is presumed to expand and fluctuate the intertube distances of the SWCNT bundles, bringing about a depression of the electrical conductivity. If librational or rotational motions of the nitro groups of the DNN molecule begin around 250 K, the maximum at 250 K can be understood. We can understand the maximum of $\log \sigma$ *vs.* T^{-1} plot of DHN–SWCNT in a similar way.

GCMC simulation shows the plausible intercalation configuration and mutual orientational structure of DNN molecules in the pried interstice of the SWCNT bundle estimated by X-ray diffraction data, as shown in Fig. 7. DNN molecules align along the interstice between the SWCNTs, which should be fit for a charge-transfer interaction. Thus, the DNN molecules can be intercalated in the SWCNT bundle. The moieties of the DNN molecule should do a librational or rotational motion around the temperature giving the maximum of $\log \sigma$ *vs.* T^{-1} plot, partially unravelling the ordered bundle structure and then the electrical conductivity drops above the temperature, as mentioned above. The DHN-adsorbed SWCNT is assumed to have a structure similar to that shown in Fig. 7. Also it is possible that the librational or rotational motion of the moieties of the naphthalene molecules may couple with the theoretically-predicted librational motion of the SWCNT bundles[10] to induce the observed drop in the electrical

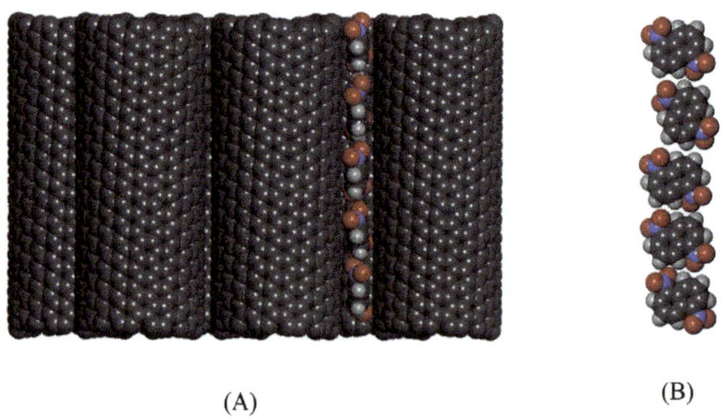

Fig. 7 Plausible intercalated configuration (A) and mutual molecular orientation (B) of DNN molecules in a pried interstice of the SWCNT array. Black; carbon, light grey; hydrogen, red; oxygen and purple; nitrogen.

conductivity. The newly observed molecular motion-aided electrical conductivity change in this work can be applicable to the design a new switching device.

We will compute the electronic structures with the DFT method and get the temperature dependence of the snapshots of ND molecules in the interstitial position of the SWCNT bundle with a GCMC simulation in a future study. Also we need to detect directly the above-mentioned molecular motion with DSC measurements and observe the ND molecular state with high resolution transmission electron microscopy in the future.

4 Conclusions

Naphthalene and its derivatives, 1,5-dinitronaphthalene and 1,5-dihydroxynaphthalene, are partially intercalated in the interstitial positions of the SWCNT bundles, bringing about charge–transfer interaction, though the intercalated SWCNT bundles have not necessarily well-ordered structure. The electrical conductivity of these naphthalene and its derivative-intercalated SWCNT bundles, except for in a higher temperature range, can be described by the variable range hopping mechanism. The naphthalene and naphthalene derivative-intercalated SWCNT bundles give a metal–semiconductor transition like behavior, but the change is not sharp because of less crystallinity of the superlattice structure of the SWCNT bundles. As 1,5-dinitronaphthalene or 1,5-dihydroxynaphthalene, as the intercalants, have functional groups which can induce a librational motion even in the interstitial position, the conduction path between the SWCNT bundles should be partially broken due to an unravelling of the bundle structure induced by the libration of the intercalant molecules. This mechanism gives an unusual temperature dependence of the electrical conductivity in addition to metal–semiconductor transition like behavior.

Acknowledgements

T. F., M. E., K. K. and M. T. were supported by Exotic Nanocarbons, Japan Regional Innovation Strategy Program by the Excellence, JST. This work was supported by Grant-in-Aid for Scientific Research (A) (24241038) and D. M. was partially supported by JST CREST "Creation of Innovative Functional Materials with Advanced Properties by Hyper-nano-space Design".

References

1 K. Kaneko, T. Itoh and T. Fujimori, *Chem. Lett.*, 2012, **41**, 466–475.
2 H. E. Romero, K. Bolton, A. Rosen and P. C. Eklund, *Science*, 2005, **307**, 89–93.
3 E. S. Snow, F. K. Perkins, E. J. Houser, S. C. Badescu and T. L. Reinecke, *Science*, 2005, **307**, 1942–1945.
4 A. Star, V. Joshi, S. Skarupo, D. Thomas and J. C. Gabriel, *J. Phys. Chem. B*, 2006, **110**, 21014–21020.
5 Y. Battie, O. Ducloux, P. Thobois, N. Dorval, J. S. Lauret, B. Attal-Tretout and A. Loiseau, *Carbon*, 2011, **49**, 3544–3552.
6 Y. Maniwa, H. Kataura, M. Abe, S. Suzuki, Y. Achiba, H. Kira and K. Matsuda, *J. Phys. Soc. Jpn.*, 2002, **71**, 2863–2866.

7 F. Khoerunnisa, T. Fujimori, T. Itoh, K. Urita, T. Hayashi, H. Kanoh, T. Ohba, S. Y. Hong, Y. C. Choi, S. J. Santosa, M. Endo and K. Kaneko, *J. Phys. Chem. C*, 2012, **116**, 11216–11222.
8 Y. Hattori, T. Ohba and K. Kaneko, *Comprehensive Inorganic Chemistry II: From Elements to Applications*, Elsevier, 2013, vol. 5, ch. 5.
9 K. Urita, Y. Shiga, T. Fujimori, Y. Hattori, H. Kanoh, T. Ohba, H. Tanaka, M. Yudasaka, S. Iijima, I. Moriguchi, F. Okino, M. Endo and K. Kaneko, *J. Am. Chem. Soc.*, 2011, **133**, 10344–10347.
10 T. Fujimori, A. Morelos-Gómez, Z. Zhu, H. Muramatsu, R. Futamura, K. Urita, M. Terrones, T. Hayashi, M. Endo, S. Y. Hong, Y. C. Choi, D. Tománek and K. Kaneko, *Nat. Commun.*, 2013, **4**, 2162–2169.
11 D. Noguchi, H. Tanaka, T. Fujimori, T. Kagita, Y. Hattori, H. Honda, K. Urita, S. Utsumi, Z.-M. Wang, T. Ohba, H. Kanoh, K. Hata and K. Kaneko, *J. Phys.: Condens. Matter*, 2010, **22**, 334207–334221.
12 H. Marsh, N. Murdie and I. A. S. Edwards, *Chem. Phys. Carbon*, 1987, **20**, 213–272.
13 R. S. Lee, H. J. Kim, J. E. Fischer, A. Thess and R. E. Smalley, *Nature*, 1997, **388**, 255–257.
14 T. Takenobu, T. Takano, M. Shiraishi, Y. Murakami, M. Ata, H. Kataura, Y. Achiba and Y. Iwasa, *Nat. Mater.*, 2003, **2**, 683–688.
15 Y.-K. Kwon and D. Tománek, *Phys. Rev. Lett.*, 2000, **84**, 1483–1486.
16 S. Gotovac, H. Honda, Y. Hattori, K. Takahashi, H. Kanoh and K. Kaneko, *Nano Lett.*, 2007, **7**, 583–587.
17 M. Arai, S. Utsumi, M. Kanamaru, K. Urita, N. Yoshizawa, D. Noguchi, K. Nishiyama, Y. Hattori, F. Okino, T. Ohaba, H. Tanaka, H. Kanoh and K. Kaneko, *Nano Lett.*, 2009, **9**, 3694–3698.
18 S. Gotovac-Atlagic, T. Hosokai, T. Ohba, N. Ueno, H. Kanoh and K. Kaneko, *Phys. Rev. B: Condens. Matter Mater. Phys.*, 2010, **82**, 075136.
19 F. Khoerunnisa, D. Minami, T. Fujimori, S. Y. Hong, Y. C. Choi, H. Sakamoto, M. Endo and K. Kaneko, *Adsorption*, 2014, **20**, 301–309.
20 H.-J. Shin, S. M. Kim, S.-M. Yoon, A. Benayad, K. K. Kim, S. J. Kim, H. K. Park, J.-Y. Choi and Y. H. Lee, *J. Am. Chem. Soc.*, 2008, **130**, 2062–2066.
21 H. Kataura, Y. Kumazawa, Y. Maniwa, I. Uemzu, S. Suzuki, Y. Ohtsuka and A. Achiba, *Synth. Met.*, 1999, **103**, 2555–2558.
22 R. B. Weisman and S. M. Bachilo, *Nano Lett.*, 2003, **3**, 1235–1238.
23 A. M. Rao, P. C. Eklund, S. Bandow, A. Thess and R. E. Smalley, *Nature*, 1997, **388**, 257–259.
24 K. Yakushi, M. Oguchi and H. Kuroda, *Bull. Chem. Soc. Jpn.*, 1979, **52**, 3180–3191.
25 N. F. Mott and E. A. Davis, *Electronic Processes in Non-Crystalline Materials*, Clarendon Press, Oxford, 1971.
26 N. F. Mott, *Metal-Insulator Transition*, Taylor & Francis, London, 1974.
27 K. Snoussi, A. Vakhshouri, H. Okimoto, T. Takenobu, Y. Iwasa, S. Maruyama, K. Hashimoto and Y. Hirayama, *Phys. Status Solidi C*, 2012, **9**, 183–186.
28 G. T. Kim, E. S. Choi, D. C. Kim, D. S. Suh and Y. W. Park, *Phys. Rev. B: Condens. Matter Mater. Phys.*, 1998, **58**, 16064–16069.
29 S. Luo, T. Liu, S. M. Benjamin and J. S. Brooks, *Langmuir*, 2013, **29**, 8694–8702.
30 P. A. Cox, *Transition Metal Oxides*, Clarendon Press, Oxford, 1995, ch. 5.
31 J. A. Lee, H. Rösner, J. F. Corrigan and Y. Huang, *J. Phys. Chem. C*, 2011, **115**, 4738–4748.

Faraday Discussions

PAPER

Chameleon-like behaviour of cyclo[*n*]paraphenylenes in complexes with C_{70}: on their impressive electronic and structural adaptability as probed by Raman spectroscopy†

Miriam Peña Alvarez,[a] Paula Mayorga Burrezo,[b] Takahiro Iwamoto,[c] Lili Qiu,[d] Miklos Kertesz,[d] Mercedes Taravillo,[a] Valentín G. Baonza,[a] Juan T. López Navarrete,*[b] Shigeru Yamago[c] and Juan Casado*[b]

Received 9th May 2014, Accepted 11th June 2014
DOI: 10.1039/c4fd00103f

A series of four 1 : 1 host–guest supramolecular complexes of [*n*]CPPs and C_{70} have been analyzed by Raman spectroscopy in the solid state and complemented with the analysis of their spectroscopic responses under mechanical and thermal stresses. By following the frequency behaviour of the G and RBM modes we have found that [10]CPP in the [10]CPP@C_{70} complex displays a more "ordered" structure. However, in [11]CPP@C_{70}, the nanoring becomes oval-shaped with closer contacts with the C_{70} poles and less conformational restriction in the flattened region. By mechanical and thermal stresses we are able to modify the lying conformation of [10]CPP@C_{70} towards a standing orientation. [11]CPP@C_{70} resists pressure changes, although it tends to shift from the standing to the lying orientation by heating. As for the crystal cell, the [*n*]CPPs occupy the residual empty spaces while the main crystallographic positions are reserved for C_{70}. These are new examples of the impressive adaptability of the [*n*]CPP molecules to different physico-chemical environments, a chameleon-like property which reveals the delicate equilibrium provided by cyclic conjugation and ring strain.

[a]*MALTA-Consolider Team, Department of Physical Chemistry, Chemistry Faculty, University Complutense of Madrid, 28040 Madrid, Spain*

[b]*Department of Physical Chemistry, University of Málaga, CEI Andalucía Tech, Campus de Teatinos s/n, 29071-Málaga, Spain. E-mail: casado@uma.es; Fax: +34952132000; Tel: +34952131868*

[c]*Institute for Chemical Research, Kyoto University, Uji 611-0011, Japan*

[d]*Department of Chemistry and Institute of Soft Matter, Georgetown University, 37th and O Streets, NW, Washington, D.C. 20057-1227, USA*

† Electronic supplementary information (ESI) available: Vibrational normal modes, detailed Raman spectra in the 1300 cm^{-1} region, and pressure behaviour of such regions are shown. Fullerene RBMs stress and temperature evolution for the different complexes are also presented. See DOI: 10.1039/c4fd00103f

Introduction

Cyclo[*n*]paraphenylenes (abbreviated as [*n*]CPP, see Fig. 1) are a class of conjugated pure-hydrocarbon molecules featuring a cyclic disposition of 1,4-substituted benzenes with the p_z carbon orbitals oriented towards the macroring center. Given their aesthetic shape, they have attracted the interest of chemists for a long time.[1] It was in 2008 when Jasti and Bertozzi[2] were able to synthesize the cyclo[*n*]paraphenylene compounds for the first time, [8]CPP, [12]CPP and [18]CPP, paving the way for the preparation of [*n*]CPP of almost all sizes, [6]- to [18]CPP, in a beautiful competition race mainly between the groups of Itami,[3] Yamago[4] and Jasti.[5] Very recently, the smallest member of the series, [5]CPP, has been successfully prepared almost simultaneously by Yamago and Jasti.[6]

The analogy between [*n*]CPPs and armchair single wall carbon nanotubes (SWCNT) is straightforward: the [*n*]CPP can be considered as the shortest version of arm-chair SWCNTs and, as such, can be considered as their molecular models. While the chemical lack of definition of SWCNTs is an insurmountable obstacle for the full understanding of their properties, the perfect knowledge of the chemical structures of the [*n*]CPP series allows researchers to evaluate the evolution of their electronic, structural and optical properties as a function of the [*n*]CPP size, thus permitting the first *oligomeric approach* to the physical properties of SWCNT. But [*n*]CPPs have been also used as templates for the chemical synthesis of SWCNT *via* a controlled vertical condensation of [*n*]CPP, as reported by Itami,[7] which is considered the first successful preparation of nanotubes by using an orthodox synthetic organic approach. In their own way, the [*n*]CPPs display very interesting optoelectronic properties as well thanks to the co-existence of rather unusual effects such as molecular strain and cyclic conjugation. Cyclic strain increases with the decrease of the number of benzenes[8] and produces in these rings some slight deformations from planarity and their partial benzo-quinonoidization.[9] On the other hand, cyclic conjugation, which is provided by these p_z orbitals contained in the circumferential molecular plane, further stabilizes the whole cyclic arrangement. As a result, these appealing molecules contain highly polarisable electronic shapes in the belt which originate other interesting chemical properties such as the ability to form host-guest π–π and van der Waals supramolecular complexes. These complexes are the focus of this article.[10]

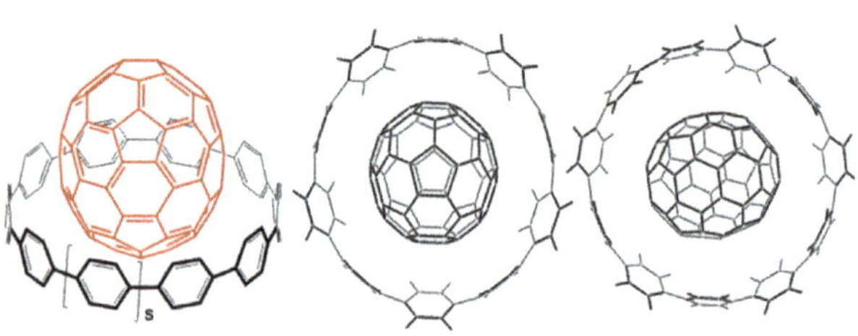

Fig. 1 Left: chemical structures of the [*n*]CPP compounds ($s = 0$ for [9]CPP, $s = 1$ for [10]CPP, $s = 2$ for [11]CPP and $s = 3$ for [12]CPP). Middle: lying disposition for [10]CPP@C_{70}. Right: standing disposition for [11]CPP@C_{70}.

We have recently reported on the dependence of the structural and electronic properties of the [*n*]CPPs (from [6]CPP to [12]CPP) with their size, and established robust relationships between quinonoidization, cyclic strain, cyclic conjugation and [*n*]CPP dimension, using Raman spectroscopy as the revealing tool.[11] Another structural feature of significance is their conformational effect, that makes these [*n*]CPPs irregular cylinders with the successive benzenes canted or staggered with opposite angles.[8,9] We have also addressed the structural flexibility of the whole nanohoop which, for the larger compounds, is able to undergo reversible ovallisation under strong mechanical stress.[11] The ability to re-adapt their shapes, their inner empty cavity, and the capacity to promote favourable surrounding π–π electronic interactions, are the basis for the formation of the 1 : 1 stoichiometric complexes.[10]

In this paper, we go a step further in these studies of [*n*]CPPs and address in detail the conformational and electronic effects of new [*n*]CPP complexes with a new guest, such as C_{70}.[12] The oval shape of this fullerene will allow us to get new insights on the connection between [*n*]CPP shape, its conformational plasticity and the electronic properties of the formed complexes. The whole study will be carried out by using Raman spectroscopy as the structural and electronic diagnostic tool. Raman spectroscopic properties will be investigated as a function of mechanical stress (pressure dependent measurements) and thermal stress (temperature dependent measurements). With this study we aim to provide new features of the particular physical and chemical properties of [*n*]CPP resulting from their impressive propensity to reshape their structures.

Experimental procedures and technical details

The syntheses of the [*n*]CPP and of their complexes with C_{70} have been published elsewhere.[12] The samples were studied as pure solids, and these were well characterized by different analytical and crystallographic techniques.

Raman measurements were carried out with a Senterra dispersive micro-Raman spectrometer from Bruker with a 785 nm laser as the excitation source. The Raman scattering radiation was collected in a back-scattering configuration with a standard spectral resolution of 3 cm^{-1}, a spatial resolution of 0.5 μm, and a spot size of about 3 μm.

High pressure studies were conducted in a sapphire anvil cell (SAC) with a culet diameter of 400 μm and a gold gasket.[13,14] No pressure transmitting medium was used and diamond chips were placed as the pressure calibrants. The Raman experiments were carried with the same Senterra microscope with 785 nm laser excitation. The recovered samples, after pressure cycles, were also analyzed, taking several spectra on different sample points to confirm the reproducibility on the same sample and to ensure whether the transformation of the samples is complete or not.

Laser excitation time dependent measurements were carried out in the micro-Raman spectrometer (BWTEK Voyage™ BWS435-532SY) with an excitation wavelength of 532 nm. These experiments were repeated at different temperatures ranging from 298 to 523 K. The [8]CPP sample was placed in a Pt crucible and situated inside the high temperature stage system (Linkam TS1500 with a T95 system controller). This system can heat within a temperature range from ambient up to 773 K, allowing us to observe and characterize samples through its quartz lid window. The high temperature stage system was equipped with a Pt/Rh thermocouple in direct contact with the ceramic heating element to detect and control the sample temperature. During the experiments, the temperature was stabilized within ±0.1 degrees. It was necessary to wait between 10 and 15 minutes for the sample to have a uniform temperature. Variable temperature Raman experiments on the complexes were carried out in the same Linkam TS1500 device but the excitation laser used was the 785 nm laser of the Senterra dispersive micro-Raman spectrometer from Bruker.

Vibrational Raman antecedents on [*n*]CPPs

The Raman spectra of [*n*]CPPs are characterized by the existence of four main Raman bands which constitute the fingerprint of their structure and electronic shapes.[11] For our discussion in this paper we will focus only on two: (i) the strongest band of the spectrum at 1580–1600 cm^{-1} which corresponds to the analogue G mode in SWCNTs (in our [*n*]CPP we call it the G mode as well) and is described as a CC stretching mode along the tangential direction of the tube. This G mode is composed of two sub-bands, the so-called G_1 and G_2 modes (see Fig. 2 and Fig. S1†). The frequency of the G_1 band in the [*n*]CPPs has been related to the degree of quinonoidization of the benzene units due to the cyclic strain. And (ii) the pseudo-radial breathing modes in our [*n*]CPPs which are related with the radial breathing vibrational mode of SWCNTs, a molecular motion that takes place along the radial direction of the tube and appears in the 400–100 cm^{-1} range. From a structural–spectroscopy relationship, the frequency of this pseudo-RBM band is quite sensitive to the cyclic radius (and therefore size) of the [*n*]CPP.

It is interesting to note the nature of the main Raman bands of the Raman spectrum of C_{70}, or those at: (i) 1564 cm^{-1} which is due to a tangential CC stretching mode or G-like mode; (ii) 456 cm^{-1} which is the radial breathing mode with has slightly more vibrational amplitudes along the *long* axial direction of the ellipsoid. And (iii) 256 cm^{-1} which also corresponds to a sort of breathing radial-like mode (squashing mode) featuring the main vibrational amplitudes in the atoms placed in the equatorial segment of C_{70}.[15] Fig. 2 depicts the normal modes associated to most of these relevant bands (see Fig. S2 of ESI† for further information).

Conformational studies

In this section we will address the analysis of the conformational properties of the [*n*]CPP as individual entities with the objective of understanding them before the CPP assemblies with C_{70} fullerene. To this end, we have studied the effect that simultaneous thermal and laser excitation stresses cause on the [*n*]CPP spectra. Fig. 3 shows the time evolution of the Raman spectra of [8]CPP, as a prototypical example, at 25 °C by exciting with the 2.33 eV laser. It must be indicated the same experiments with low energy Raman excitation at 1064 nm did not produce any spectral change.

Changes in the spectra are clearly visible resulting in a new peak at 1604 cm^{-1}. The time growth of the integrated area of this band (relative to that at 1582 cm^{-1}) allows us to obtain the rate constant for the process which, studied as a function of the temperature, gives us the activation barriers associated with the conversion

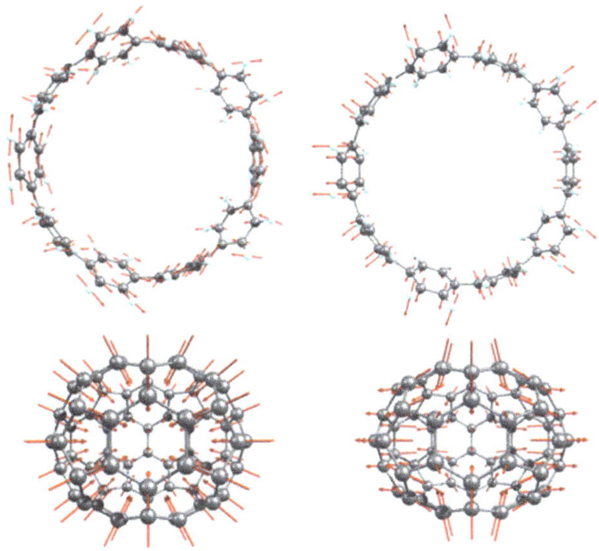

Fig. 2 Top: vibrational modes associated with the G_1 band of [10]CPP (left) with its pseudo radial breathing mode (right) from ref. 11. Bottom: radial breathing mode (left) and radial-like breathing mode (right, squashing mode) of C_{70}. In the experimental spectrum these modes are observed at 456 and 256 cm^{-1}, respectively.

towards the structure responsible for the new Raman component. For the growth of the 1604 cm^{-1} band a value of $E_a = (19.9 \pm 1.3)$ kJ mol^{-1} is determined. This activation energy is different depending on the [n]CPP size, for instance, in [12] CPP the strongest 1594 cm^{-1} band also evolves into the new band at 1604 cm^{-1} with a value of $E_a = (22.0 \pm 2.0)$ kJ mol^{-1}.[8,9] These experimental activation energies are in excellent agreement with those energy barriers calculated for the full rotation of a given benzene regarding their vicinal rings around the inter-ring C–C bonds. As a result, the energy barriers for the conversion into a species with a new band up-shifted (1604 cm^{-1} bands) with regards to the original ones is

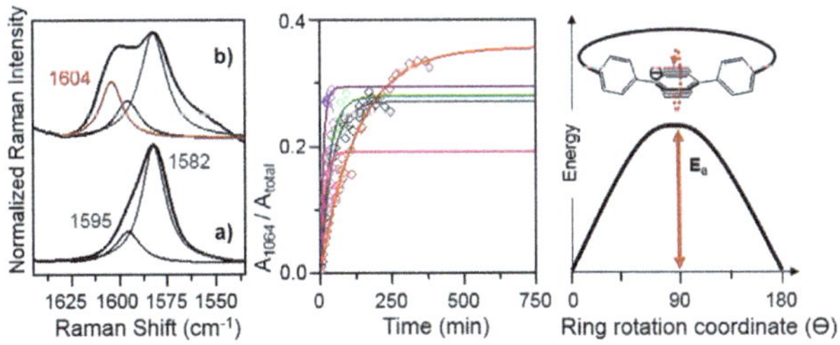

Fig. 3 532 nm Raman spectra of [8]CPP. (a) Deconvoluted spectrum at the start at 298 K; (b) deconvoluted spectrum recorded after exposure time to the laser radiation of about 300 min at 298 K. Middle: ratio between the area of the 1604 cm^{-1} band and the total area of these features as a function of the exposure time to the laser radiation at several temperatures; red (298.15 K), black (348.15 K), green (373.15 K), violet and pink (423.15 K). Right: representation of the ring rotation with its energy barrier.

consistent with the transformation into a more distorted or staggered conformation among the successive rings of the cycle, where these barriers might be viewed as the maximal energy costs for full rotation (Fig. 3). We speculate that the need for assistance by high energy radiation to develop the conformational effect can be explained by the formation after light absorption of an excited state closely related with the transient state of this internal rotation.

Complexation of [*n*]CPP with C_{70}: the 1 : 1 complexes

On the interpretation of the G modes of the [*n*]CPP

Fig. 4 and 5 display the Raman spectra of the 1 : 1 complexes, [10]CPP and [11]CPP; [9]CPP and [12]CPP, respectively, The solid state structures of [10]CPP and [11]CPP have been fully characterized by X-ray diffractometry in the solid state and by UV-Vis electronic absorption spectroscopy in solution.[12] For the shorter and larger complexes, no crystallographic data are available. At a first sight, in comparison with the respective [*n*]CPP, complexation induces much larger changes on the Raman spectra of [10]CPP and [11]CPP compared to those of [9]CPP and [12]CPP.

For [10]CPP@C_{70}, the G_1 mode frequency of [10]CPP changes from 1589 cm^{-1} to 1581 cm^{-1} in the complex. We have previously assigned the down-shift behaviour of the G_1 band in [*n*]CPP to the quinonoidization effect of the benzene rings that simultaneously brings the formation of less staggered structures between successive benzene rings (this effect is called below the "ordering" effect in the belt). A decrease in the number of benzene rings producing more effective cyclic conjugation gives also rise to a more quinonoid less staggered alternating ring sequence. A clear example of the connection between Raman down-shifts and quinonoidization is shown in the oxidation of a given cycloparaphenylene compound to its radical cation.[11] Therefore, the -8 cm^{-1} down-shift of the G_1 band on [10]CPP →[10]CPP@C_{70} can be attributed to the "ordering" effect on the

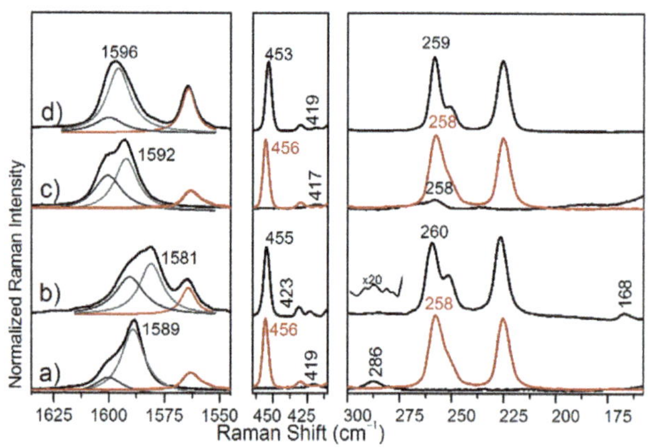

Fig. 4 Solid state 785 nm Raman spectra at 25 °C of: (a) [10]CPP (black) and C_{70} (red); (b) [10]CPP@C_{70}; (c) [11]CPP (black) and C_{70} (red); (d) [11]CPP@C_{70}. Thin lines correspond to the deconvoluted bands.

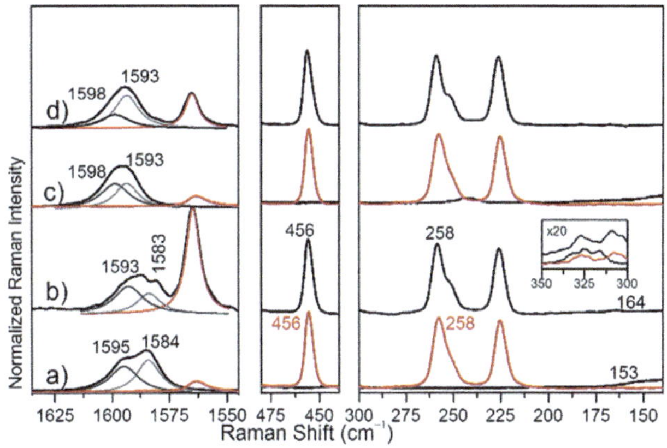

Fig. 5 Solid state 785 nm Raman spectra at 25 °C of: (a) [9]CPP (black) and C_{70} (red); (b) [9]CPP@C_{70}; (c) [12]CPP (black) and C_{70} (red); (d) [12]CPP@C_{70}. Thin lines correspond to the deconvoluted bands.

cycloparaphenylene structure as a result of the template effect that the C_{70} fullerene exerts on it, mediated by the attractive π–π forces established between both compounds in the complex. Solid state studies on [10]CPP@C_{70} reveal that the C_{70} fullerene occupies the internal cavity of the cycloparaphenylene with the CPP belt surrounding the shortest axis or equatorial segment of the fullerene ellipsoid, or *lying* disposition (see Fig. 1).[12] The size of this short axis of C_{70} is 0.712 nm which, compared with the diameter of [10]CPP, 1.380 nm, anticipates a favourable interaction between the two units with very suitable 0.33 nm van der Waals distances. As a result of the interaction on the circular equator, [10]CPP does not deform its cyclic shape which, together with the better alignment of the consecutive rings, favours cyclic electron delocalization as the origin of the Raman down-shift (lower torsional angles than in the pristine molecules, see Fig. S3a†).

The Raman behaviour in the case of the [11]CPP@C_{70} complex is different. The G_1 band, at 1592 cm^{-1} in the individual [11]CPP, up-shifts to 1596 cm^{-1}, by +4 cm^{-1}. In contrast to the case of [10]CPP@C_{70}, we should associate these changes also to the conformational effect, but in this case caused by a more staggered benzene ring distribution in the cycloparaphenylene (higher torsional angles between neighbouring benzenes, supported by the 1200 cm^{-1} spectral region behaviour, see Fig. S3b†). We have also shown that the formation of this more inter-ring distorted shape induces the diminishing of the quinoidal character, reverting towards a more aromatic form, justifying the up-shift of the G_1 band. The size of the C_{70} long axis is 0.796 nm while the [11]CPP diameter is 1.51 nm, thus it anticipates the two structures will interact by van der Waals interactions along the long C_{70} axes. Indeed, the solid state structure of [11]CPP@C_{70} has been resolved, in which C_{70} adopts a *standing* conformation (see Fig. 1) where the maximal coupling is through the two poles of C_{70} with the CPP rings of the belt all residing along the long axis of the ellipsoidal fullerene (this contrasts with the coupling through the equatorial part in [10]CPP@C_{70}).[12] As a result of this disposition, the [11]CPP undergoes a deformation of the cylindrical shape towards a more ellipsoidal flattened shape. This description is nicely in agreement with: (i) our previous publication[11] where we have described the great flexibility of the CPP compounds in a series of detailed studies as a function of variable pressure. We have shown that by applying mechanical stress to the samples these reversibly deformed into oval or flattened shapes. This mechanical deformation was followed by Raman spectroscopy which probed a frequency up-shift, in two linear regimes, with the increment of pressure such as observed here for the same G modes on [11]CPP → [11]CPP@C_{70}; (ii) the argument that in order to stabilize the standing conformation the benzenes placed in the equatorial side, given its unsuitable disposition for π–π coupling, would prefer to slightly rotate in order to promote energetically favourable hydrogen–C_{70} or H···π interactions[16] which might result again in a more staggered deposition for these benzenes also contributing to the Raman up-shift. This explanation would support an increment of entropy in the complexation reaction regarding the [11]CPP and C_{70} reactants, such as is deduced experimentally, thus allowing us to understand the entropy controlled regime for the exergonic formation reaction of this complex.[12]

On the interpretation of the [*n*]CPP RBMs

The low frequency region 500–100 cm^{-1} of the Raman spectra of the [*n*]CPPs is important because of the presence of the pseudo-RBM modes (see Fig. 2) whose frequencies also account for the alteration of their donut-like cylindrical shape with the formation of the complexes in the solid state. In [10]CPP and [11]CPP the pseudo-RBMs were assigned to the bands at 286 cm^{-1} and 258 cm^{-1} respectively, while the RBM modes were not detected in their Raman spectra.[11] After formation of the complexes in the low frequency region, the spectra are clearly dominated by the C_{70} Raman bands which mostly mask the Raman fingerprint of the cycloparaphenylene unit. In spite of this, three main findings need to be mentioned: (i) the pseudo-RBM modes, which seemingly disappear after complexation; (ii) in [10]CPP@C_{70} a completely new feature appears in the spectrum at 168 cm^{-1}; and (iii) in the region around 500–400 cm^{-1} there is a band at 423 cm^{-1} in [10]CPP@C_{70} which evolves from the [10]CPP one at 419 cm^{-1} and at 419 cm^{-1} in [11]CPP@C_{70} from that at 417 cm^{-1} in [11]CPP. It seems that the interaction in the complex drastically alters the dynamics of the pseudo-RBM modes provoking their disappearance. Simultaneously, the same electronic effect might give extra-intensity to the formerly weak (undetectable) RBM modes which emerge in the complex at 168 cm^{-1} (in the individual [*n*]CPP this mode would be expected to appear below 150 cm^{-1}). It might be argued that the slight symmetrical compression of [10]CPP around C_{70} could provoke a slight reduction of its diameter and, as a consequence, an increase of the RBM frequency given the reciprocal relationship between the RBM frequency and the diameter of the cylinder. Similar reasons can account for the frequency up-shifts of the 430–400 cm^{-1} bands from the unperturbed CPPs.

On the interpretation of the C_{70} fullerene modes

Complexation has a much smaller impact on the Raman frequencies of the C_{70} fullerene, given its rigidity, which contrasts with the elasticity of the CPPs. However, there are two interesting features in the spectra of the complexes compared to that of C_{70} alone: (i) the analogue radial breathing mode of C_{70} at 456

cm^{-1} (the main atomic displacements lie along the long axis and on the poles of the ellipsoid, see Fig. 2) changes by -1 cm^{-1} in [10]CPP@C_{70} and by -3 cm^{-1} in [11]CPP@C_{70}, the latter being the greatest in agreement with the formation of the standing complex where the strongest interaction between [11]CPP and C_{70} goes through these polar regions of the ellipse; and (ii) another radial breathing mode of the C_{70} is that at 256 cm^{-1} (squashing mode, with main atomic displacements on the equator of the ellipse, see Fig. 2) which appears at +2 cm^{-1} in [10]CPP@C_{70} and at +1 cm^{-1} in [11]CPP@C_{70}, the former being the greatest in agreement with the formation of the lying complex where the strongest interaction between [10] CPP and C_{70} goes through the equatorial belt of the fullerene. These are remarkable observations, since in SWCNTs peapods of C_{70}, not significant observations have been achieved on the C_{70} orientation changes along the SWCNT, due mainly to the lack of diameter, and thus of conformational, purity in the SWCNT samples. However, it is straightforward that C_{70} it is vibrationally affected by its orientation inside the tube cavity, as observed on CPPs.

The spectra of [9]CPP@C_{70} and [12]CPP@C_{70}

Their Raman spectra in Fig. 5 display small, but appreciable, differences regarding the superposition spectrum of the individual components. In spite of the fact that their innermost cavities are either too small or large to accommodate C_{70} ([9] and [12]CPP have diameters of 1.23 nm and 1.66 nm, respectively), the differences in the spectra seem to indicate some new insights. The G_1 and G_2 bands at 1584 cm^{-1} and 1595 cm^{-1} in [9]CPP move to 1583 cm^{-1} and 1593 cm^{-1} in [9]CPP@C_{70}. As in the [10]CPP@C_{70}, this down-shift emerges from a less inter-ring staggered conformer (see Fig. S4†), resulting from the template-ordering effect of C_{70} on the [9]CPP shape. The CC stretching modes of C_{70} fullerene are all up-shifted by 1–2 cm^{-1}. Although the internal volume of [9]CPP is unable to accommodate the C_{70} fullerene, it could be thought in a partial encapsulation, possibly through one of the two poles or through the corannulene segments of the fullerene "egg", forming a proto-complex. Due to the formation of this pseudo-complex, a new RBM band at 164 cm^{-1} (from that at 153 cm^{-1} in [9]CPP) is detected. Simultaneously, the C_{70} RBM bands (456 cm^{-1} and 256 cm^{-1}) change both by +0.5 cm^{-1}. The changes in the Raman spectrum of the [12]CPP@C_{70} from its components are certainly smaller in accordance with the mismatch of their relative sizes for the stabilization of any persistent concave–convex interaction in the internal cavity of the cycloparaphenylene.

Spectral responses of the [10]- and [11]CPP@C_{70} complexes under thermal and mechanical stresses

Variable pressure Raman experiments

Fig. 6 and 7 display the Raman spectra of [10]CPP@C_{70} and [11]CPP@C_{70} under pressure and on heating. Both thermal and mechanical treatments lead to rather different behaviours such as will be discussed below.

The application of pressure to [10]CPP@C_{70} induces an overall up-shift of the whole spectrum together with the typical spectral broadening as exerted by the increment of the intermolecular contacts in the pressed sample. The G_1 band up-

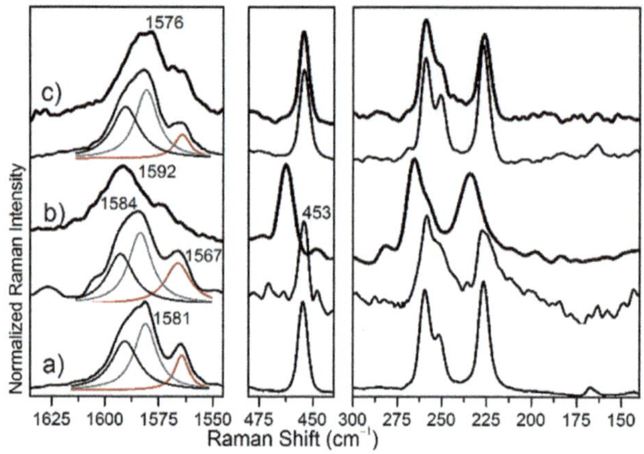

Fig. 6 785 nm Raman spectra of [10]CPP@C_{70}: (a) at room temperature and pressure; (b) thick black line corresponds to the spectrum at high pressure (≈2.0 GPa) and thin black line corresponds to the recovered spectrum after pressure cycle; both spectra taken at 25 °C; (c) thick black line corresponds to the spectrum at high temperature (140 °C) and thin black line corresponds to the recovered spectrum after thermal cycle. Both spectra are taken at room pressure. Thin lines correspond to the deconvoluted bands.

shifts from 1581 cm^{-1} at room pressure to 1592 cm^{-1} at 2 GPa, a significant shift which might be justified, not only because of the increasing frequency effect of pressure, but also because of the additional flattening of the [10]CPP structure towards a more oval shape.[11] This change is accompanied by a similar +12 cm^{-1} displacement (from 1564 cm^{-1} at room pressure to 1576 cm^{-1} at 2 GPa) of the tangential CC stretching mode of fullerene. The C_{70} bands associated to the RMB modes also experience big up-shifts such as are displayed in Fig. 6. After pressure release the spectrum does not recover its original aspect, hence the G_1 band

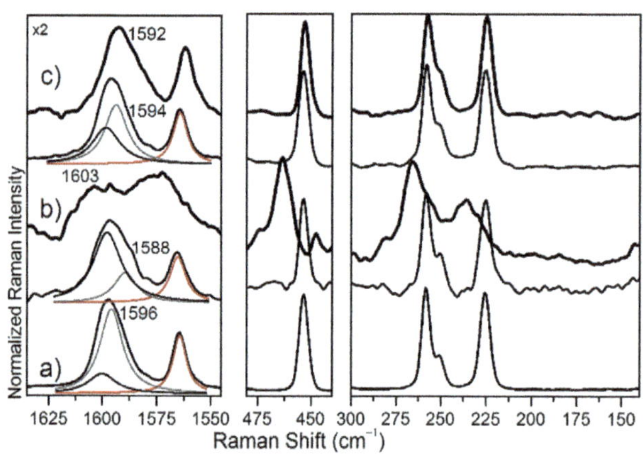

Fig. 7 785 nm Raman spectra of [11]CPP@C_{70}: (a) at room temperature and pressure; (b) thick black line corresponds to the spectrum at high pressure (≈2.0 GPa) and thin black line corresponds to the recovered spectrum after pressure cycle, both spectra taken at 25 °C; (c) thick black line corresponds to the spectrum at high temperature (140 °C) and thin black line corresponds to the recovered spectrum after thermal cycle. Both spectra are taken at room pressure. Thin lines correspond to the deconvoluted bands.

maximum is at 1584 cm^{-1} while the C_{70} tangential mode band is displaced to 1567 cm^{-1}.

In the case of the [11]CPP@C_{70}, the application of pressure again produces an up-shift of its G mode frequency from 1596 cm^{-1} to 1603 cm^{-1} which is smaller than in [10]CPP@C_{70} (+11 cm^{-1}) corroborating the connection between frequency up-shifts and structural flattening as this varies from sample to sample. The smaller change in [11]CPP@C_{70} comes from the fact that the [11]CPP structure is already partially flattened in the standing complex at room pressure so that the space for further deformations is smaller. This could be due to the fact that the π–π contacts in the molecular region around the equatorial segment are not optimal leaving the possibility of a further ovallization. Similarly to [10]CPP@C_{70}, the C_{70} fullerene bands undergo ≈+10 cm^{-1} frequency displacements with pressure. The back to room pressure spectrum has suffered a ignificant down-shift on the CPPs' G bands, confirming the formation of a more staggered form to achieve better π–π contacts with C_{70}, but the low frequency C_{70} remains as in the pristine complex indicating the prevalence of the complex in its standing orientation.

The fact that [10]CPP@C_{70}, does not recover the original aspect would be indicating that the [10]CPP could have slightly rotated from the lying disposition towards a situation closer, but not identical, to the standing form. After pressure release this high pressure structure would result in a meta-stable kinetically blocked species. Fig. 8 represents the evolution of the main Raman band frequencies of [10]CPP@C_{70} with pressure. In general, the G bands of the cycloparaphenylenes follow a different variation pattern compared to that of fullerene. These G bands of [10]CPP linearly shift up to a critical pressure, 0.6 GPa, at which they stay constant with the increment of pressure up to 1.6 GPa from which a second linear region is detected. However, the tangential CC stretching mode of fullerene follows a continuous linear regime in the whole range of pressure. This behavior on the cycloparaphenylene fragment could be related to a re-orientational effect from the lying to standing shape in the first linear regime, while for the second linear segment, at higher pressures, the CPP undergoes a flattening process (see Fig. S5 and Fig. S7† for further information). The existence in the recovered spectrum of an up-shift (from 1581 cm^{-1} to 1584 cm^{-1}) tells us about the reorientational movement in the complex which is the same effect found

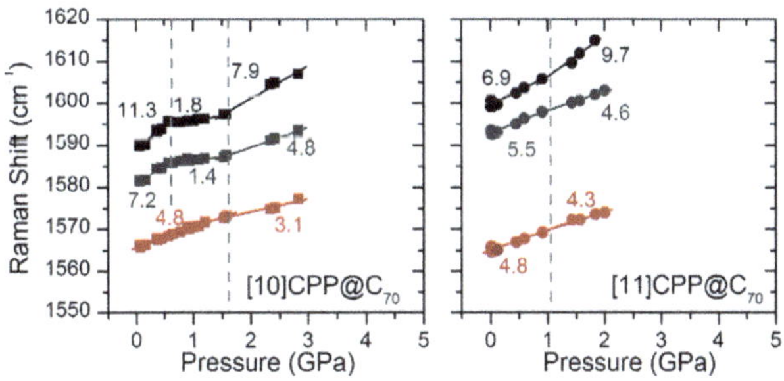

Fig. 8 Evolution of the Raman frequencies of the G modes of the [10]CPP@C_{70} and [11]CPP@C_{70} samples as a function of pressure. Inserted values correspond to the pressure coefficients of each linear trend expressed in units of cm^{-1} GPa^{-1}. Black and grey dots correspond to the G_1 and G_2 bands of the [*n*]CPPs, respectively, while red points correspond to the 1565 cm^{-1} band of C_{70} (see Fig. S7† for the C_{70} RBMs).

going on from a lying form in [10]CPP@C_{70} (1581 cm^{-1}) to the standing shape in [11]CPP@C_{70} (1592 cm^{-1}). In addition, the combination of both effects, reorientation plus flattening, imparts the observed irreversibility of the whole cycle. In the case of [11]CPP@C_{70}, the differentiation of the two frequency-pressure regimes is much less clear and the almost continuous behaviour before and after 1 GPa should be related to the predominance of the flattening process in the whole cycle process.

The second aspect of relevance is that while from complex to complex at room conditions the main Raman spectral changes are observed for the CPP bands, here in the experiments at variable pressure, the C_{70} fullerene frequencies are those mainly altered (see Fig. S8†). This could reveal that in the solid the relevant crystallographic positions are occupied by C_{70} while the [*n*]CPPs might mostly fill the empty space among them, with the result that the main transmitters of the pressure in the solid are the C_{70} molecules which thus undergo the greater frequency changes.

These changes of the Raman frequencies of the [*n*]CPP in the complex by pressure in terms of re-orientational and flattening arguments are in agreement with the Raman data on the literature reported for analogous binary systems of C_{70} inside SWCNTs, or peapods.[17,18]

Variable temperature Raman experiments

Fig. 4 and 5 also display the spectra within the thermal stress cycle. The spectrum of [10]CPP@C_{70} at 140 °C is slightly different to that at room conditions, with the G band displaced −2 cm^{-1}. The formation process of this complex is exothermic meaning that heating would produce the rupture of the interaction in the complex leading to exactly the inverse effect of the G frequencies. Thus we argue that temperature only imparts a bulk effect with no effect on the molecular structure of the complex, an interpretation corroborated by the spectrum when the complex is taken back to room temperature which is exactly identical to the original one. In contrast, for [11]CPP@C_{70} it turns out that the G mode band is scarcely changed at 140 °C (−0.5 cm^{-1}) while the recovered spectrum at room temperature shows the G band at 1592 cm^{-1}, which is unexpectedly similar to that of [11]CPP alone, 1592 cm^{-1}. A possible explanation is that heating would rotate the cycloparaphenylene from the standing situation towards the lying around the equatorial belt of the fullerene where the π–π distances increase, the π–π interaction is apparently interrupted and the [11]CPP starts to behave as an isolated entity (again this would be a kinetically blocked state in the solid after thermal template).

As for the C_{70} bands, temperature scarcely affects them by −1 cm^{-1} in all cases in the range of temperature analyzed, an observation that contrasts with their significant changes in the pressure experiments.

Structural responses of the [9]- and [12]CPP@C_{70} complexes under mechanical and thermal stresses

Fig. 9 displays the Raman spectra of [9]CPP@C_{70} under pressure. Given the small volume of the innermost cavity of [9]CPP, we have inferred from the Raman

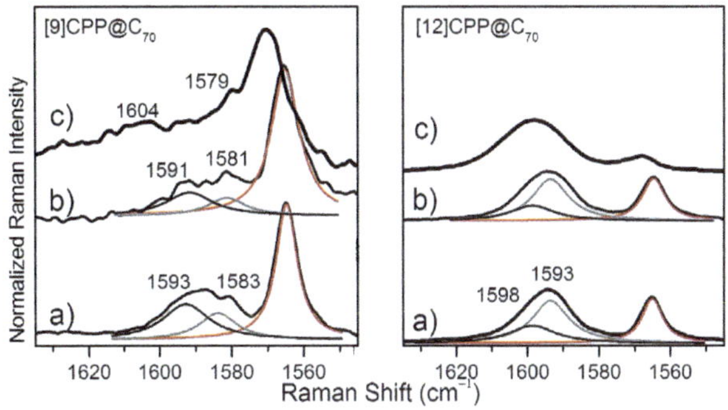

Fig. 9 Left: 785 nm Raman spectra of [9]CPP@C_{70}: (a) at room temperature and pressure; (b) black line corresponds to the spectrum at high pressure (≈2.0 GPa); and (c) the recovered spectrum after pressure cycle. Both spectra taken at 25 °C. Thin lines correspond to the deconvoluted bands. Right: the same for [12]CPP@C_{70}.

changes in the complex the formation of a species of proto-complex. The Raman spectrum of [9]CPP@C_{70} as a function of pressure in the initial stages of the compression displays the typical frequency up-shift inherent to the increment of intermolecular contacts; however, at around ≈2 GPa, the spectrum evolves by showing a dominant low frequency component at 1579 cm^{-1} together with another feature at 1604 cm^{-1}. These changes at high pressure persist after pressure release, the G bands down-shift 2 cm^{-1}, which might be interpreted by arguing that the application of pressure further helps the formation of the supramolecular complex, a compound that it is not formed directly from solution. Nonetheless, the binary system formed from solution is in a suitable disposition to progress with pressure towards the final complex where it is likely that the [9]CPP more effectively surrounds the corannulene fragments of the fullerene.

In the case of [12]CPP@C_{70}, the increment of pressure up to ≈2 GPa scarcely affects the frequencies of cycloparaphenylene unit, likely due to the moderate pressure increment. Again the largest changes are detected for the fullerene bands in agreement with these molecular entities supporting the transmission of the mechanical stress.

Conclusions

The electronic and structural properties of host–guest supramolecular 1 : 1 complexes of [n]CPPs and C_{70}, [n]CPP@C_{70} with $n = 9$, 10, 11 and 12, have been thoroughly analyzed by using Raman spectroscopy. We have found that the vibrational Raman spectra, and in particular the bands associated with the CC stretching tangential G modes and with the radial breathing modes of both [n]CPP and C_{70}, contain the relevant electronic and structural information resulting from complexation. Depending on their sizes, the [n]CPPs have been shown to be able to adapt their structures to form stable couplings with C_{70}: in [10]CPP@C_{70} the fullerene is surrounded by the [10]CPP by its equatorial part, or lying orientation, a situation that we now further characterize as having a less staggered structure for the CPP. In [11]CPP@C_{70} the fullerene is wrapped up through its long ellipsoidal axis, or standing orientation, where the nanoring becomes oval-shaped with close contacts in the C_{70} poles and less conformational restriction in the flattened region.

Under mechanical stress, the largest frequency changes are found for the fullerene bands revealing that the [*n*]CPPs, as far as the unit cell of the solid is concerned, occupy the empty spaces, with the C_{70} molecules in the prevalent sites. In addition, pressure produces the modification of the lying structure of the [10]CPP@C_{70} complex towards a more standing-like conformation followed by the flattening of its structure, a dual process that overall renders an irreversible structural mechanical cycle. [11]CPP@C_{70}, however, resists in its standing position the applied pressure, and it is by heating that the CPP ring seems to rotate towards a lying disposition by finally recovering its "isolated" spherical aspect. All in all, these data display the great adaptability of the [*n*]CPPs to the shape of C_{70} and also to the crystalline structure in the fullerene solid. C_{70} molecules are encapsulated in SWCNTs forming the well-known peapods where the fullerene molecules might adopt both lying and standing positions, existing in many cases as mixtures of both. Here, in the analysis of the [*n*]CPP as models of SWCNTs, the characterization by Raman of these well-defined complexes might help us to understand the complexity of these peapods between fullerene and SWCNT.

The main message of this study is the impressive structural versatility and adaptability of the [*n*]CPP molecules to different physico-chemical environments, a chameleon-like property provided by the conjunction of unusual effects such as cyclic conjugation and ring strain cylindrical shape which form a delicate ground electronic state equilibrium easily modifiable through external physical and chemical inputs.

Acknowledgements

Financial support from MINECO of Spain, CTQ2012-33733, CTQ2012-38599-C02-02, CSD2007-00045, Junta de Andalucía (Project P09-FQM-4708) and Comunidad de Madrid (S2009/PPQ-1551) are acknowledged. M.P.A. is grateful to the Spanish MEC for an FPU grant.

Notes and references

1 V. C. Parekh and P. C. Guha, *J. Indian Chem. Soc.*, 1934, **11**, 95.

2 R. Jasti, J. Chattacharjee, J. B. Neaton and C. R. Bertozzi, *J. Am. Chem. Soc.*, 2008, **130**, 17646.

3 H. Omachi, S. Matsuura, Y. Segawa and K. Itami, *Angew. Chem., Int. Ed.*, 2010, **49**, 10202; H. Takaba, H. Omachi, Y. Yamamoto, J. Bouffard and K. Itami, *Angew. Chem., Int. Ed.*, 2009, **48**, 6112; Y. Segawa, S. Miyamoto, H. Omachi, S. Matsuura, P. Senel, T. Sasamori, N. Tokitoh and K. Itami, *Angew. Chem., Int. Ed.*, 2011, **50**, 3244; H. Omachi, Y. Segawa and K. Itami, *Acc. Chem. Res.*, 2012, **45**, 1378; Y. Ishii, Y. Nakanishi, H. Omachi, S. Matsuura, K. Matsui, H. Shinohara, Y. Segawa and K. Itami, *Chem. Sci.*, 2012, **3**, 2340.

4 E. Kayahara, T. Kouyama, T. Kato, H. Takaya, N. Yasuda and S. Yamago, *Angew. Chem., Int. Ed.*, 2013, **52**, 13722; T. Iwamoto, Y. Watanabe, Y. Sakamoto, T. Suzuki and S. Yamago, *J. Am. Chem. Soc.*, 2011, **133**, 8354; S. Yamago, Y. Watanabe and T. Iwamoto, *Angew. Chem., Int. Ed.*, 2010, **49**, 757; E. Kayahara, Y. Sakamoto, T. Suzuki and S. Yamago, *Org. Lett.*, 2012, **14**, 3284.

5 J. Xia and R. Jasti, *Angew. Chem., Int. Ed.*, 2012, **51**, 2474; T. J. Sisto, M. R. Golder, E. S. Hirst and R. Jasti, *J. Am. Chem. Soc.*, 2011, **133**, 15800; T. J. Sisto and R. Jasti, *Synlett*, 2012, **23**, 483; E. S. Hirst and R. Jasti, *J. Org. Chem.*, 2012, **77**, 10473; R. Jasti and C. R. Bertozzi, *Chem. Phys. Lett.*, 2010, **494**, 1; J. Xia, J. W. Bacon and R. Jasti, *Chem. Sci.*, 2012, **3**, 3018; M. R. Golder, B. M. Wong and R. Jasti, *Chem. Sci.*, 2013, **4**, 4285; A. V. Zabula, A. S. Filatov, J. Xia, R. Jasti and M. A. Petrukhina, *Angew. Chem., Int. Ed.*, 2013, **52**, 5033.

6 E. Kayahara, V. K. Patel and S. Yamago, *J. Am. Chem. Soc.*, 2014, **136**, 2284; P. J. Evans, E. R. Darzi and R. Jasti, *Nat. Chem.*, 2014, **6**, 404.

7 H. Omachi, T. Nakayama, E. Takahashi, Y. Segawa and K. Itami, *Nat. Chem.*, 2013, **5**, 572.

8 Y. Segawa, H. Omachi and K. Itami, *Org. Lett.*, 2010, **12**, 2262.

9 See papers in ref. 2–5 where the molecular structures of some [*n*]CPPs are resolved by X-ray diffraction. In addition: S. Taubert, D. Sundholm and F. Pichierri, *J. Org. Chem.*, 2010, **75**, 5867 U. H. F. Bunz, S. Menning and N. Martín, *Angew. Chem., Int. Ed.*, 2012, **51**, 7094; S. Schrettl and H. Frauenrath, *Angew. Chem., Int. Ed.*, 2012, **51**, 6569.

10 T. Iwamoto, Y. Watanabe, T. Sadahiro, T. Haino and S. Yamago, *Angew. Chem., Int. Ed.*, 2011, **50**, 8342.

11 M. Peña Alvarez, P. Mayorga Burrezo, M. Kertesz, T. Iwamoto, S. Yamago, J. Xia, R. Jasti, J. T. López Navarrete, M. Taravillo, V. G. Baonza and J. Casado, *Angew. Chem., Int. Ed.*, 2014, **53**, 7033.

12 T. Iwamoto, Y. Watanabe, H. Takaya, T. Haino, N. Yasuda and S. Yamago, *Chem.–Eur. J.*, 2013, **19**, 14061.

13 V. G. Baonza, M. Taravillo, A. Arencibia, M. Cáceres and J. Núñez, *J. Raman Spectrosc.*, 2003, **34**, 264.

14 E. del Corro, J. González, M. Taravillo, E. Flahaut and V. G. Baonza, *Nano Lett.*, 2008, **8**, 221516.

15 G. Sun and M. Kertesz, *J. Phys. Chem. A*, 2002, **106**, 6381; V. Schettino, M. Pagliai and G. Cardini, *J. Phys. Chem. A*, 2002, **106**, 1815; D. Jing and Z. Pan, *European Journal of Mechanics, A: Solids*, 2009, **28**, 948.

16 S. Tsuzuki, K. Honda, T. Uchimaru, M. Mikami and K. Tanabe, *J. Am. Chem. Soc.*, 2002, **124**, 104–112.

17 C. Caillier, D. Machon, A. San-Miguel, R. Arenal, G. Montagnac, H. Cardon, M. Kalbac, M. Zukalova and L. Kavan, *Phys. Rev. B: Condens. Matter Mater. Phys.*, 2008, **77**, 125418.

18 P. M. Rafailov, C. Thomsen and H. Kataura, *Phys. Rev. B: Condens. Matter Mater. Phys.*, 2003, **68**, 193411; A. N. Khlobystov, R. Scipioni, D. Nguyen-Manh, D. A. Britz, D. G. Pettifor, G. A. D. Briggs, S. G. Lyapin, A. Ardavan and R. J. Nicholas, *Appl. Phys. Lett.*, 2004, **84**, 772.

Faraday Discussions

PAPER

Role of edge geometry and chemistry in the electronic properties of graphene nanostructures

Shintaro Fujii,†*[a] Maxim Ziatdinov,†[b] Misako Ohtsuka,[a] Koichi Kusakabe,[c] Manabu Kiguchi[a] and Toshiaki Enoki*[a]

Received 18th April 2014, Accepted 3rd July 2014
DOI: 10.1039/c4fd00073k

The geometry and chemistry of graphene nanostructures significantly affects their electronic properties. Despite a large number of experimental and theoretical studies dealing with the geometrical shape-dependent electronic properties of graphene nanostructures, experimental characterisation of their chemistry is clearly lacking. This is mostly due to the difficulties in preparing chemically-modified graphene nanostructures in a controlled manner and in identifying the exact chemistry of the graphene nanostructure on the atomic scale. Herein, we present scanning probe microscopic and first-principles characterisation of graphene nanostructures with different edge geometries and chemistry. Using the results of atomic scale electronic characterisation and theoretical simulation, we discuss the role of the edge geometry and chemistry on the electronic properties of graphene nanostructures with hydrogenated and oxidised linear edges at graphene boundaries and the internal edges of graphene vacancy defects. Atomic-scale details of the chemical composition have a strong impact on the electronic properties of graphene nanostructures, *i.e.*, the presence or absence of non-bonding π states and the degree of resonance stability.

Introduction

The electronic properties of graphene are crucially dependent on the geometrical shape and chemistry of its edge boundaries. Indeed, several experimental studies have demonstrated the dramatic influence of the crystallographic orientation of graphene one-dimensional edges (*i.e.*, zigzag and armchair [Fig. 1a]) on the electronic structure of graphene. According to current

[a]*Department of Chemistry, Tokyo Institute of Technology, 2-12-1 Ookayama, Meguro-ku, Tokyo 152-8551, Japan. E-mail: fujii.s.af@m.titech.ac.jp; tenoki@chem.titech.ac.jp*

[b]*Department of Organic and Polymeric Materials, Tokyo Institute of Technology, Ookayama, Meguro-ku, Tokyo 152-8552, Japan*

[c]*Graduate School of Engineering Science, Osaka University, 1-3 Machikaneyama-cho, Toyonaka, Osaka 560-8531, Japan*

† These authors made equal contributions to the work.

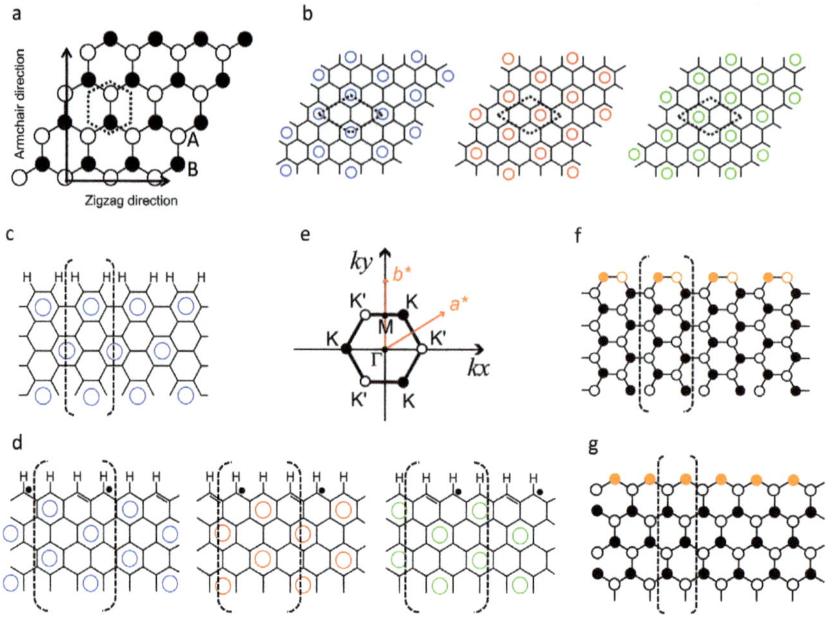

Fig. 1 (a) Graphene honeycomb bipartite lattice structure, where the white and black circles denote A and B sublattice sites, respectively. The unit cell is indicated by a dotted hexagon. The two typical crystallographic directions of the armchair and zigzag directions are indicated by arrows. (b) Chemical structures of infinite graphene, which are described by resonance of three Clar formulas. The sextet consisting of six π electrons is represented by a circle. The unit cell is indicated by the dotted lines. (c) Unique Clar formula of the hydrogenated armchair edge. (d) Clar formulas of the mono-hydrogenated zigzag edge, which are represented by three degenerated formulas. The black dot denotes an unpaired electron. (e) First Brillouin zone of graphene. Reciprocal lattice vectors are denoted by **a*** and **b***. $K = 2\pi/a(1/3, 1/\sqrt{3})$, $K' = 2\pi/a(2/3, 0)$, $\mathbf{\Gamma} = (0, 0)$. (f) Bipartite model of the armchair edge, where open and filled circles denote the A and B sublattice sites, respectively. Secondary carbon sites belonging to A or B sublattices at the boundary are coloured orange. (g) Bipartite model of the zigzag edge. Secondary carbon sites at the boundary are denoted by orange circles. In (c), (d), (f), and (g) the unit cells are denoted by dotted brackets.

understanding, the zigzag edge of graphene supports the localised π state (edge state) on its boundary and gives rise to intra-valley scattering of extended electronic states. The edge state, which has a large local density of states in the edge region and is spin polarised, plays an important role in giving electronic, magnetic and chemical activities to zigzag edges. In contrast, its armchair counterpart does not possess the edge state and thus leads to inter-valley scattering of charge carriers.[1–15] The inter-valley scattering causes the formation of a standing wave, resulting in energetic stability in armchair edges. The electronic properties of graphene edges are also significantly affected by how foreign chemical species bond to the edge carbon atoms.[6,16–30] It is therefore appealing to tune the electronic, magnetic and chemical properties of graphene-based nanostructures by changing the chemistry and geometrical shape of the graphene edge. In general, the chemistry of graphene nanostructures is poorly characterised on the atomic scale, largely due to the difficulty in controlling the chemistry at the edges and because of the poor chemical sensitivity of local probe methods such as scanning transmission electron microscopy and scanning probe microscopy, although the situation is improving rapidly with increasing interest in graphene materials.[10–15,31] In order to better understand the relationships between chemical and electronic properties in graphene nanostructures and to bring them to their possible practical applications, a number of studies including ours have succeeded in providing structural and electronic characterisations of representative and realistic graphene nanosystems in chemically and structurally well-defined samples.[6,25–30] Here we present our findings from scanning probe microscopy and first-principles characterisation of graphene nanostructures with different geometrical shapes and chemical compositions.

Effect of geometry at the edge

Graphene is a zero-gap semiconductor and its low-energy π state is characterised by electron–hole symmetry and linear conduction and valence bands within the nearest-neighbour tight-binding approximation.[11,13,31] The unit cell of graphene contains two physically non-equivalent carbon atoms denoted as A and B atoms (Fig. 1a). From a chemical perspective, graphene is a large polyaromatic hydrocarbon, and it can be represented by the resonance of three degenerate Clar formulas with fully benzenoid characters[32] (Fig. 1b). Thus graphene is essentially an inert material, but the introduction of boundaries into the bulk graphene drastically affects the chemical reactivity and the electronic properties around the boundaries. We start by introducing the most typical graphene boundaries of a one-dimensional linear edge. When the graphene is cut along fundamental crystallographic directions and hydrogen atoms are attached to each edge carbon atom, two main types of linear edges, called armchair and zigzag shapes, are generated, as shown in Fig. 1c and d, respectively. The local chemical structure around the armchair-shaped edge can be represented by a unique Clar formula[32] with a thermodynamically stable (*i.e.*, fully benzenoid) character. The non-uniform sextet distribution (localisation of the sextet) is in sharp contrast to that of bulk graphene, in which the sextet migrates over the honeycomb lattice to form a uniform π-electron distribution.[31] In contrast, the local chemical structure around the zigzag-shaped edge is described by the resonance of three degenerated formulas with chemically reactive radical characters at the edge sites (*i.e.*, non-Kekulé type Clar formulas). In addition to the radical character at the edge site, the sextet migration in the resonance leads to uniform sextet distribution, as in the case of bulk graphene. The unit cell of the degenerated Clar formulas around the zigzag-shaped edge contains three edge sites, and an unpaired electron is localised at one of these three edge sites.

We can understand the edge-geometry-induced modification of the electronic features on the basis of the physics of the two-dimensional graphene-bipartite lattice, in which electrons move with their behaviour as that upon a massless Dirac fermion, as given by the Weyl equation;

$$\hat{H} = v_{\mathrm{F}}\sigma\mathbf{p}, \tag{1}$$

where v_F, σ and **p** are the Fermi velocity, the Pauli matrix of the pseudo-spin, and the momentum, respectively.[13] The pseudo-spin appears upon from the equivalence of the structural degree of freedom 2 in the A/B bipartite lattice to the spin degree of freedom ↑↓. Two sets of linear conduction and valence bands (Dirac cones) are created at K and K′ points in the Brillouin zone, due to the biparticity of the lattice (Fig. 1e). The fully benzenoid structure of the armchair edge originates from the formation of a standing wave as a consequence of electron wave interference, which is due to the inter-valley scattering at the armchair edge, in which the K and K′ points participate. In contrast, the presence of unpaired electrons is caused by the broken symmetry of the pseudo-spin of the massless Dirac fermion at the zigzag edges (*i.e.*, the broken symmetry between the A and B sublattices in the bipartite system).[13] As shown in the bipartite models in Fig. 1f and g, the local sublattice symmetry at the graphene boundaries consisting of secondary carbon sites is broken for the zigzag edge, whereas the A and B sublattices are always paired for the armchair boundary. Fig. 2a and b show band structures for graphene nanoribbon models with hydrogenated armchair and zigzag edges, respectively. The gap opening for the armchair model and the flat bands near the Brillouin-zone boundary for the zigzag model (see the band dispersion at $(2/3)\pi < \boldsymbol{k} < \pi$), respectively, manifest the benzenoid and edge-localised radical characters,[1,2] as predicted in the corresponding Clar formulas. It should be noted here that beyond the two typical armchair- and zigzag-shaped edges, local probe analysis on chiral-shaped edges prepared by the unzipping of carbon nanotubes has been reported recently.[25,33]

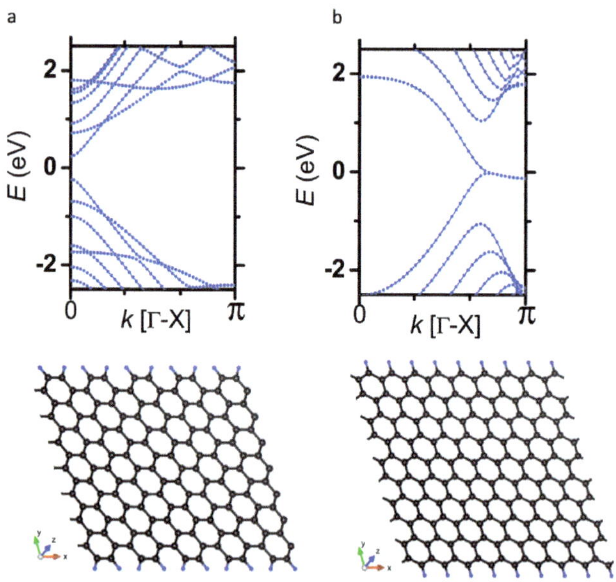

Fig. 2 LDA-calculated band structures and graphene nanoribbon models with the (a) armchair edge and (b) zigzag edge. Black and blue balls in the models represent carbon and hydrogen atoms, respectively. The *xyz* axis is shown at the bottom left corner in each model. In all the figures, DFT calculations were performed in a similar manner as described in ref. 6, 27 and 28 unless otherwise noted.

Discussion

The chemistry at the edge has a remarkable impact on the electronic structure near the graphene edge. Understanding the chemistry has been a long-standing problem, and it is mostly discussed based on theoretical models.[16–20] However, it is still a challenging task to realize the chemical modification of a graphene edge and to tune the electronic properties in a controlled manner. In response to the challenge, several local probe analyses of chemically-modified graphene edges in combination with bottom-up syntheses and nano-fabrication techniques have been reported recently.[6,15,21–27,29] We too have focused on scanning probe characterisation of chemically modified graphene edges in order to study the effect of the chemistry on the electronic properties of graphene.[6,22,24,25] In the following, we discuss the effect of hydrogenation and oxidation on the electronic structure of graphene edges by referring to the model structures of zigzag and armchair edges described above (Fig. 1 and 2).

Effect of chemistry at the edge

Hydrogenated graphene edges with atomically smooth boundaries contained within nano-sized pits were prepared by atomic hydrogen etching of single vacancies in the topmost graphene layer of graphite.[6] To simulate scanning probe microscopic (STM) images and corresponding tunnelling spectroscopy data, electronic calculations based on density functional theory (DFT) were performed. High-resolution STM characterisations and DFT simulations demonstrate that atomically well-defined mono-hydrogenated zigzag edges are preferentially formed at the periphery of the nanoholes with mono-atomic depth, as shown in Fig. 3a and b. As predicted in the DFT simulation, the low-energy STM contrast with bright spots at the zigzag-shaped edge is dominated by an edge-localised non-bonding π state at the boundary (Fig. 3b and c), which reflects the π-radical character (Fig. 1d) and is due to the higher local density of states (LDOS) originating from the flat band (Fig. 2b). The edge-localised state decays into bulk quickly to form an $a \times a$ periodic pattern (a: lattice constant (0.246 nm) in the graphene-bipartite lattice (Fig. 3b]). The edge-localised non-bonding state, which is also called an edge state,[1,2] can be a source of electronic and magnetic activity, and thus the potential application of the edge-localised non-bonding state functionalised in graphene materials has long been discussed in the nanocarbon community.[10–15] The results of our scanning tunnelling spectroscopy on the hydrogenated zigzag edge suggests that a spin gap is formed as a result of the ferromagnetic ordering of the magnetic moment along the zigzag edge sites,[6] which puts forward the potential electronic and magnetic functionalisation of graphene-based materials with controlled edge-chemistry. It is noteworthy that the size of the spin gap is sensitive to the details of the edge preparation and hence its chemistry. Indeed, the spin gap reported for the oxidised edges of unzipped carbon nanotubes, reconstructed edges in chemical vapour deposition-grown graphene ribbons, and hydrogen-etched edges in graphite nanoholes is 0.03–0.05 eV,[34] 0.05–0.08 eV,[33] and 0.16 eV,[6] respectively.

Hydrogenated armchair-shaped edges were prepared by the hydrogenation of the graphene edges on graphite. The electronic state around the armchair-shaped edge is characterised by extended π states with a superperiodic pattern with

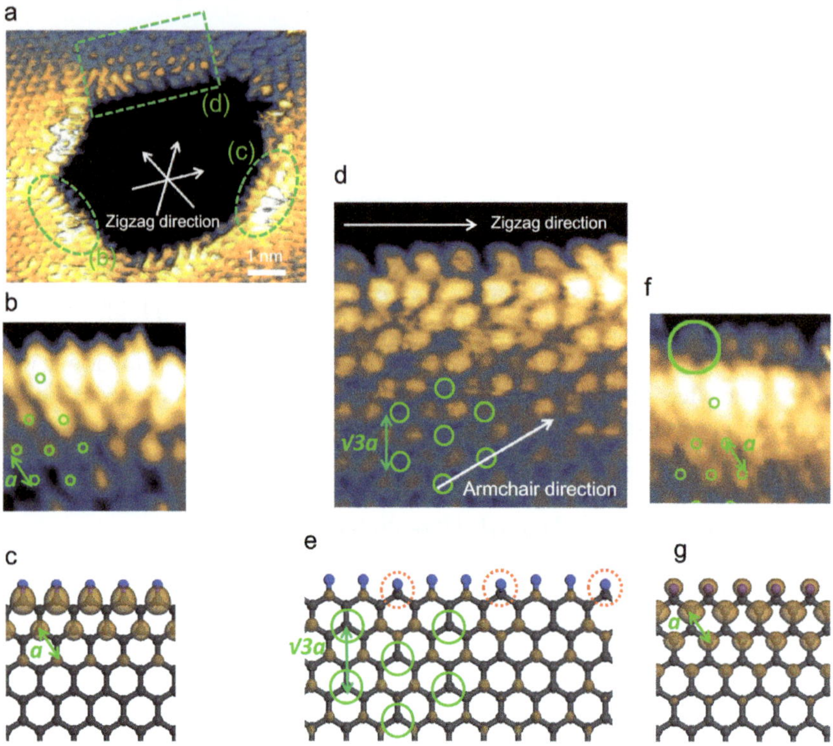

Fig. 3 (a) Experimental ultrahigh vacuum (UHV)-STM height image of a nanohole with mono-atomic depth prepared by atomic hydrogen etching of the single atomic vacancies in the graphene layer on graphite (imaging conditions: $V_s = 0.1$ V, $I = 0.9$ nA, scale bar = 1 nm). Arrows indicate the zigzag directions of the graphene lattice. The nanohole edges are aligned in the zigzag directions. Three types of nanohole edges (b)–(d) are indicated by the dotted lines. (b–g) STM images and local density approximation (LDA) simulations for the mono-hydrogenated zigzag edge (b and c, respectively), the partially di-hydrogenated zigzag edge (d and e, respectively), and the fully di-hydrogenated zigzag edges (f and g, respectively) (imaging conditions: $V_s = 0.1$ V, $I = 0.7$ nA, imaging area = 1.35×1.35 nm^2). The images correspond to the edges (b), (c) and (d) indicated in (a). The observed periodicities in the STM images are schematically represented by green circles: $a \times a$, $\sqrt{3}a \times \sqrt{3}a$, and $a \times a$ for (b), (d), and (f), respectively. The zigzag and armchair directions are indicated by arrows in (d). The di-hydrogenated site is indicated by dotted circles in (e). The hydrogen atom in the C–H participating in hyperconjugation is encircled by a green line in (f). Black and blue balls in the models represent carbon and hydrogen atoms, respectively. The isosurface obtained by integrating the local density of states within a bias range from 0.1 to 0 V in the simulation and those in Fig. 4, 7, 9, 15 and 16 are coloured orange.

$\sqrt{3}a \times \sqrt{3}a$ periodicity (Fig. 4a and b), which is associated with the non-uniform sextet distribution (sextet localisation, Fig. 1c) and can be interpreted as a result of inter-valley scattering of the low-energy electron waves and the resultant interference patterns with a charge-density wave (CDW) formation around the edge.[7–9]

In addition to the conventional mono-hydrogenated zigzag edge, we found a new type of zigzag edge on the periphery of the nanohole, which was characterised by the disappearance of the edge-localised state and by the formation of a

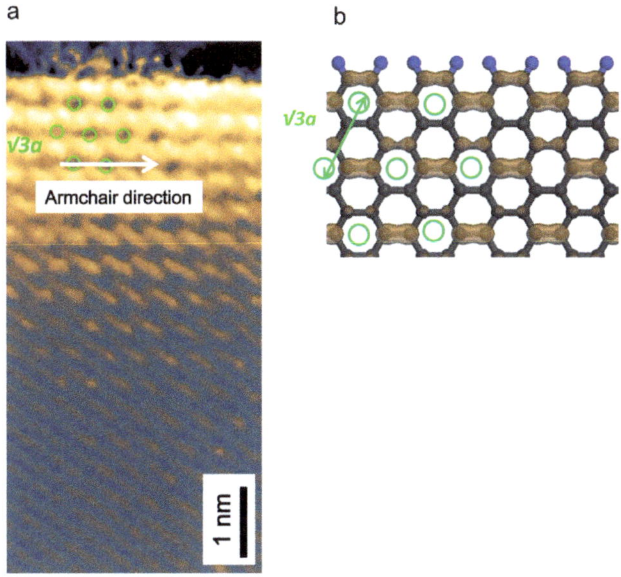

Fig. 4 Experimental UHV-STM height image of the hydrogenated armchair edge and the LDA simulation (imaging conditions: $V_s = 0.1$ V, $I = 0.7$ nA, scale bar = 1 nm). The observed $\sqrt{3}a \times \sqrt{3}a$ periodicity in the STM image is extended into the interior and is schematically represented by green circles. An arrow indicates the armchair direction. Black and blue balls in the models represent carbon and hydrogen atoms, respectively.

superperiodic pattern with $\sqrt{3}a \times \sqrt{3}a$ periodicity (Fig. 3d). The structure of the superperiodic pattern mimics that around the armchair-shaped edges with a thermodynamically stable character, which implies non-uniform sextet distribution and inter-valley scattering around the edge.

To understand these two main features, we have to take into account the chemistry at the zigzag edge. Systematic DFT simulations show that the edge-localised state can be removed when every third zigzag-edge site is di-hydrogenated while the other two sites remain mono-hydrogenated (Fig. 3e). The DFT simulation reproduces the absence of the localised state at the edge sites, as well as the presence of the extended superperiodic $\sqrt{3}a \times \sqrt{3}a$ patterns (*i.e.*, CDW formation) around the edge boundary in the STM image (Fig. 3e). A previous DFT study of a variety of hydrogenated graphene edges already demonstrated that the partially di-hydrogenated zigzag edge is energetically stable.[20] The partially di-hydrogenated zigzag edge can be represented by a Kekulé-type Clar formula, in which radical characters of the original mono-hydrogenated edge are quenched by the attachment of an additional (second) hydrogen atom to every third carbon site (Fig. 5a). The non-uniform sextet migration gives the localised character of the sextet distribution (Fig. 5b), which can be associated with the superperiodic distribution of the π electrons. The DFT-calculated band structure confirms the disappearance of the flat bands around the Fermi energy E_F and the opening of the gap (Fig. 6a). The local sublattice symmetry of the partially di-hydrogenated zigzag boundary can be intuitively understood from the effective bipartite model, in which the number of the A and B sublattice sites at the boundary is balanced (Fig. 5c).

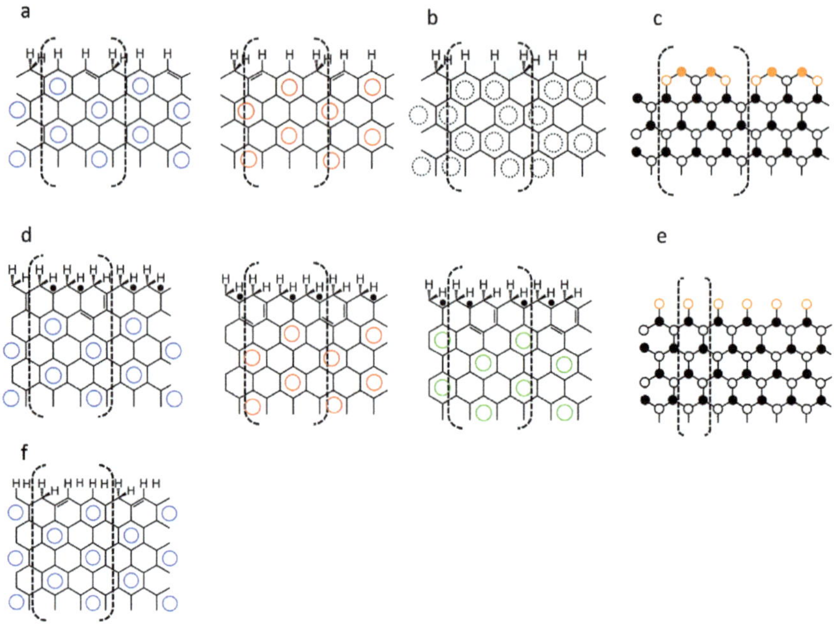

Fig. 5 (a) The two degenerated Clar formulas of the partially di-hydrogenated zigzag edge. (b) The superposition of the sextet positions in the two formulas in (a). (c) The bipartite model of the partially di-hydrogenated edge. The secondary carbon sites at the boundary are denoted by orange circles. (d) The three degenerated Clar formulas of the fully di-hydrogenated edge. The black dot denotes an unpaired electron. (e) The bipartite model of the fully di-hydrogenated edge. The primary carbon sites at the bearded boundary are denoted by orange circles. (f) One of the Clar formulas of the fully di-hydrogenated edge where the CH_2 moiety is involved in a hyperconjugation interaction with the bulk graphene π-electron network. The unit cells are denoted by dotted brackets.

In addition to the two main types of hydrogenated zigzag edges, some fraction (less than 5%) of the short zigzag edges with atomically smooth boundaries features an unconventional π-state distribution, in which one can recognise a weak amplitude at the edge sites (Fig. 3f). We attributed this structure to fully di-hydrogenated zigzag edges (also known as Klein edges[35,36]). The observed STM image and DFT simulation show a reasonable match (Fig. 3f and g). In a manner similar to that of the mono-hydrogenated zigzag edge, the fully di-hydrogenated edge can be represented by the resonance of three degenerated Clar formulas with radical characters at the bearded edge sites (Fig. 5d). The unit cell of the degenerated Clar formulas contains three edge sites, and two unpaired electrons are localised at two of the three edge sites. In a crude discussion, the degree of the radical character is two times higher for the di-hydrogenated edge than for the mono-hydrogenated edge (*i.e.*, the unit cell contains an unpaired electron, which is shared by three edge-carbon sites for the mono-hydrogenated edge, whereas two unpaired electrons are shared by three edge-carbon sites for the di-hydrogenated edge). The higher degree of the radical character, that is, the larger density of states of the edge-localised state, is evident in the DFT band structures of the graphene nanoribbon models. Fig. 6b shows band structures for graphene nanoribbon models terminated by the di-hydrogenated zigzag edge. In contrast to

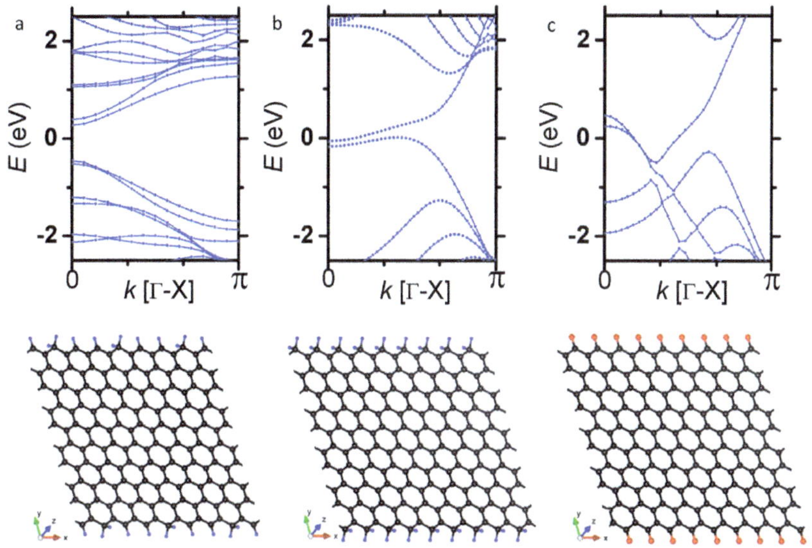

Fig. 6 LDA-calculated band structures and graphene nanoribbon models with (a) the partially di-hydrogenated zigzag edge, (b) the fully di-hydrogenated zigzag edge, and (c) the carboxylated zigzag edge. The LDA approximation was used for (a) and (b) and the PBE0 approximation was used for (c).[26] Black, red, and blue balls in the models represent carbon, oxygen, and hydrogen atoms, respectively. The *xyz* axis is shown at the bottom left corner in each model.

the mono-hydrogenated zigzag edge, flat bands appear at $0 < \boldsymbol{k} < (2/3)\pi$, indicating that the density of states is approximately two times larger than that of the mono-hydrogenated edge ($(2/3)\pi < \boldsymbol{k} < \pi$). The bipartite model suggests that the local sublattice symmetry is broken at the fully di-hydrogenated edge where the boundary is terminated by one of the sublattices (Fig. 5e).

The di-hydrogenated zigzag edge can gain energetic stability *via* a hyper-conjugation-type interaction.[36,37] The hyperconjugation originates from the interaction of the electrons in a σ bond (C–H bond at the di-hydrogenated site) with an adjacent p_z orbital in graphene, leading to an extended π-electron network. In the corresponding Clar formulas, two unpaired electrons participate in the graphene π-electron network by forming additional aromatic sextets, which leads to resonance stabilisation (Fig. 5f). The interaction between the electrons in the C–H bond and p_z orbital causes a finite increase in the amplitude of the localised state at the di-hydrogenated moiety in the DFT simulation (Fig. 3g). We believe that the bearded π state in the STM image in Fig. 3f is a strong indication of the hyperconjugation-type interaction.‡

To further investigate the effect of chemistry on the electronic properties of graphene edges, we investigated the effect of oxidation. The oxidised structure is

‡ On the basis of a simple valence bond theory, the electronic state around the partially hydrogenated edge could be represented by a superposition of the three resonant structures (states). It should be noted here that, for more precise discussion, we have to consider weighting of each of the three states in the superposition depending on the eigenvalue of each state if available.

commonly present in real-world graphene materials that are in contact with an ambient environment during their preparation process. In our studies, the oxidised graphene edges with atomically smooth edge boundaries were prepared by the electrochemical oxidation of a graphitic surface.[26] The electrochemical oxidation preferentially proceeds at the graphene edges and leads to fragmentation of the graphitic surface (Fig. 7a). A closer examination of the atomic-scale STM images of the nano-fragment revealed that not only ragged edge boundaries but also atomically sharp edge boundaries that terminated along a zigzag direction were present on the periphery of the nanographene fragments (Fig. 7b). In sharp contrast to the conventional hydrogenated graphene edge, electrochemically oxidised edges (carboxylated edges in this study) with higher thermodynamic stability exhibited significantly modulated π states near the Fermi level (Fig. 6c), which is characterised by a quasi-localised edge state with a spatially extended character in both the experimental data and the DFT simulation (Fig. 7c). Although the mono-hydrogenated zigzag edge features strongly

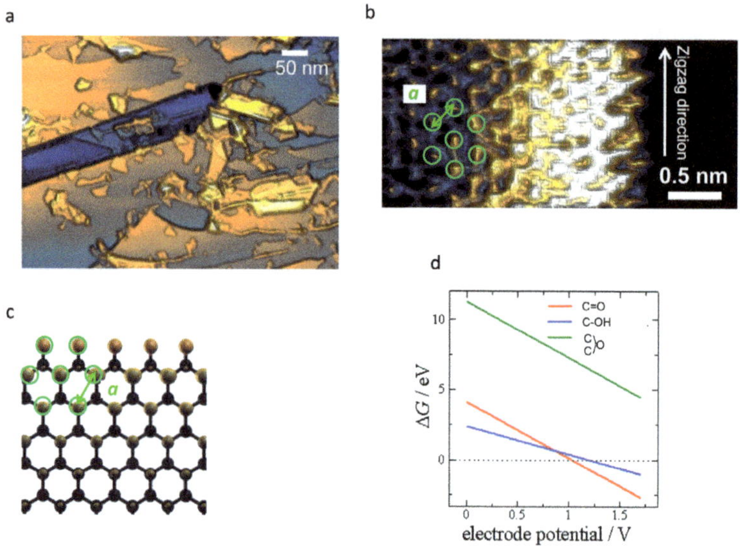

Fig. 7 (a) Experimental ambient STM height image of the graphite surface after 10 cycles of a potential sweep from 1.0 to 1.9 V at 20 mV s^{-1} in a 0.1 M H_2SO_4 solution at room temperature (imaging conditions: $V_s = 0.8$ V, $I = 0.4$ nA, scale bar = 50 nm). (b) High-resolution STM-current image of an oxidised zigzag edge of the nanographene fragment with an atomically sharp zigzag edge boundary. The edge-localised state represents large amplitudes on one of the graphene sublattice sites with $a \times a$ periodicity ($a = 0.24$ nm) that decay into the interior from the edge boundary. The arrow indicates the zigzag direction of the graphene lattice (imaging conditions: $V_s = 0.2$ V, scale bar = 0.5 nm). (c) PBE0 simulation for the carboxylated zigzag edge, in which the $a \times a$ periodicity is superimposed. Black and red (located at the edge) balls represent carbon and oxygen atoms, respectively. The hybrid PBE0 calculations were performed using the Quantum Espresso code with norm-conserving pseudopotentials and a wave function cutoff energy of 40 Ry. The Brillouin zone integration was performed using a uniform 4 × 4 × 1 Monkhorst-type k-point grid. (d) LDA-based free energy of the oxidised zigzag graphene nanoribbon models terminated by carboxyl, alcohol, and ether groups *versus* the electrochemical potential, which are denoted by red, blue, and green lines, respectively.

edge-localised π states at several sublattice sites from the boundary, the quasi-localised state in the oxidised edge decays into bulk up to 2 nm. This modulation can be understood in terms of the participation of the oxygen π-electrons into the graphene π-electron system and the change in the effective boundary conditions of the π-electron network, as predicted previously.[38] The carboxylated edge can be represented by the resonance of three degenerated Clar formulas with radical characters at the oxygen edge sites (Fig. 8a). The π-electron network and corresponding bipartite model (Fig. 8b) are similar to those of the fully di-hydrogenated edge (*i.e.*, Klein type) in terms of the effective geometry of the network.

A remarkable difference between the two Klein-type edges of the di-hydrogenated edge and the carboxylated edge is the energy dispersion of the low-energy π state (flat bands near E_F, Fig. 6c). The oxygen atoms at the bearded site of the carboxylated edge with higher electronegativity induces band bending *via* the charge–transfer effect,[22,26] which causes a quasi edge-localised state with a spatially extended character. Calculation of the free energy (eqn (2)) of the expected hydrolysis reaction in our electrochemical oxidation process (eqn (3)), as described below, revealed that the observed carboxylated edge was in fact the most favourable edge under electrochemical potentials from 0.8 to 1.7 V (*vs.* RHE, reversible hydrogen electrode, Fig. 7d).

$$\Delta G = E_{C_xO_yH_z} + (2y - z)\left(E_{H_3O^+} + RT\ln a\left(H_3O^+\right) + E_F - F\Delta\phi\right) - xE_C - (3y - z)E_{H_2O} \quad (2)$$

$$xC + (3y - z)H_2O \rightarrow C_xO_yH_z + (2y - z)(H_3O^+ + e^-) \quad (3)$$

here $E_{C_xO_yH_z}$ is the DFT-calculated total energy of the graphene nanoribbon model with the chemical composition of $C_xO_yH_z$. The values E_C, $E_{H_3O^+}$, and E_{H_2O} are the total energies of bulk carbon, H_3O^+, and H_2O, respectively, F is the Faraday constant, $\Delta\phi$ is the electrochemical potential, and a is the activity. The oxidised

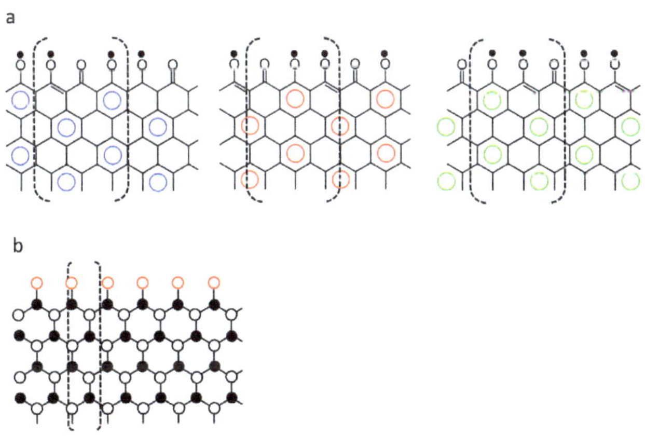

Fig. 8 (a) Three degenerated Clar formulas and (b) bipartite model of the carboxylated zigzag edge. The black dot denotes an unpaired electron in (a). Primary carbon sites at bearded oxygen sites at the boundary are denoted by red circles in (b). The unit cells are denoted by dotted brackets.

linear zigzag edge is stable in air and could be one of the possible designs to functionalize graphene nanostructures based on the edge-related unique electronic states.§

Effect of chemistry at point defects

Although the number of studies dealing with the chemistry of one-dimensional linear edges has continuously increased in recent years, there is a significant lack of experimental and theoretical studies on the chemistry of point defects (*i.e.*, zero-dimension vacancy-internal edges). In a manner similar to that of the one-dimensional edge, the geometry and chemistry of the zero-dimensional vacancy-internal edges could affect the electronic properties.

A majority of the atomic-scale characterisations of point defects have focused on several model systems, such as pristine and/or reconstructed pristine vacancy defects.[11] In contrast to the case of the one-dimensional edge, adsorption of foreign chemical species on vacancy-internal edges induces large out-of-plane deformation of the graphene plane around the defects, which would lead to the formation of several metastable structures, as we will see in the following discussion.

Hydrogenated single vacancies were prepared by exposing non-passivated single vacancies (Fig. 9a) to an atomic hydrogen environment.[28] After the hydrogenation process, the single vacancies showed significant changes in their electronic state near the vacancy sites, which are characterised by two main types of electronic states (Fig. 9b and c).

The non-passivated single vacancy has three under-coordinated carbon atoms with σ-dangling bonds. Each under-coordinated carbon can in principle accommodate up to two hydrogen atoms. To simulate the possible structures of the hydrogenated vacancies, systematic DFT calculations were performed, in which a variety of hydrogenated structures are evident (Fig. 10a). Free-energy analysis of the possible metastable structures reveals that there are two preferential hydrogenated vacancy structures, within our preparation conditions, V(111) and V(211) (Fig. 10b). The free energy G_H is calculated as

$$G_H = (E_{Vn1,n2,n3} - E_V - n_H \times E_H) - n_H \times \mu_H \quad (4)$$

where $E_{Vn1,n2,n3}$, E_V, E_H are the total energies of a hydrogenated atomic vacancy (the indices $n1$, $n2$, and $n3$ denote the number of hydrogen atoms attached to each of the vacancy's under-coordinated carbon atoms), the non-passivated vacancy, and an isolated hydrogen atom, respectively, whereas n_H stands for the total number of hydrogen atoms in the vacancy and μ_H stands for the chemical potential. One can see a good match between the experimental STM data (Fig. 9b and c) and the DFT simulations (Fig. 9d and e). The preferable hydrogenated vacancy structures are characterised by the emergence and disappearance of the vacancy-internal edge localised π state (a flat band near E_F) for fully mono-

§ In the hydrolysis process, which involves oxygen evolution (*i.e.*, $2H_2O \rightarrow O_2 + 4H^+ + 4e^-$), the effective local pressure (concentration) of the oxygen molecules at the graphene–liquid interface is as high as 10^{-2} Pa, which could accelerate the oxidation process.

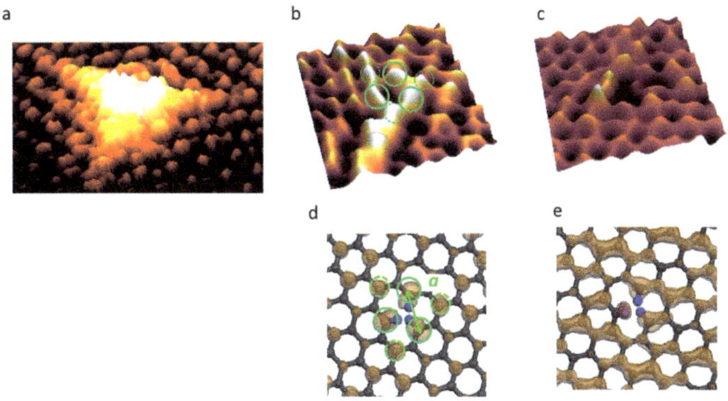

Fig. 9 (a) Experimental UHV-STM height image showing a non-passivated mono-atomic vacancy in the topmost graphite layer with three-fold symmetry (imaging conditions: $V_s = 0.1$ V, $I = 0.4$ nA, imaging area: 2.2×2.2 nm^2). (b)–(e) UHV-STM height images ((b), (c)) and LDA simulations ((d), (e)), respectively, for the two main types of hydrogenated vacancies with and without the vacancy-internal-edge-localised state, where the three edge sites are mono-hydrogenated in the former, while one of the edge sites is di-hydrogenated with others being mono-hydrogenated in the latter. The three vacancy-internal-edge sites with higher local densities of states are marked by green circles in (b) and (d) (imaging conditions: $V_s = 0.1$ V, $I = 0.4$ nA). The mono-hydrogenated vacancy in (d) is characterized by an $a \times a$ periodic pattern (a: lattice constant (0.246 nm)) on one of the graphene-sublattice-sites, while the partially di-hydrogenated vacancy in (e) does not indicate such a periodic pattern. Black and blue balls represent carbon and hydrogen atoms, respectively.

hydrogenated (V(111)) and partially di-hydrogenated (V(211)) vacancies, respectively (Fig. 11a and b).

Lieb's theorem predicts the total spin of the ground state of the Hubbard model in graphene-bipartite lattices.[39] The total spin S of the ground state is $S = (N_A - N_B)/2$ where N_A and N_B are the numbers of carbon-sites belonging to A and B sublattices, which indicates that the edge-localised non-bonding state appears if the sublattice symmetry is broken. The sublattice symmetry of the fully mono-hydrogenated and partially di-hydrogenated edges can be respectively estimated to be imbalanced and balanced (Fig. 11c and d), *i.e.*,

$$N_{\text{bulk}}(\text{C}_\text{A} + \text{C}_\text{B}) - \text{C}_{\text{A\_vacancy}} = (N_{\text{bulk}} - 1)(\text{C}_\text{A} + \text{C}_\text{B}) + \text{C}_\text{B} \text{ for V(111)}$$

$$(N_{\text{bulk}} - 1)(\text{C}_\text{A} + \text{C}_\text{B}) + \text{C}_\text{B} - \text{C}_{\text{B\_CH}_2} = (N_{\text{bulk}} - 1)(\text{C}_\text{A} + \text{C}_\text{B}) \text{ for V(211)}$$

where N_{bulk} is the number of A–B paired bulk carbons of C_A and C_B, $\text{C}_{\text{A\_vacancy}}$ is the A sublattice of the vacancy site, and $\text{C}_{\text{B\_CH}_2}$ is the sublattice of the di-hydrogenated carbon site. These findings again confirm that the chemistry at the graphene edge is inevitably linked to the unique π state and related electronic functionalities in the graphene materials.

There has been much interest in a point defect-induced magnetism in graphene materials.[11] The defect induced magnetism has been addressed in a considerable number of theoretical works, in which non-passivated single vacancy and partially hydrogenated single vacancies (*i.e.*, V(000),[11,17,18] V(100),[17] V(111)[17] and V(110)[18] in our notation) are used as model systems. The V(111) model has been considered to be the most unlikely case in the previous paper[17]

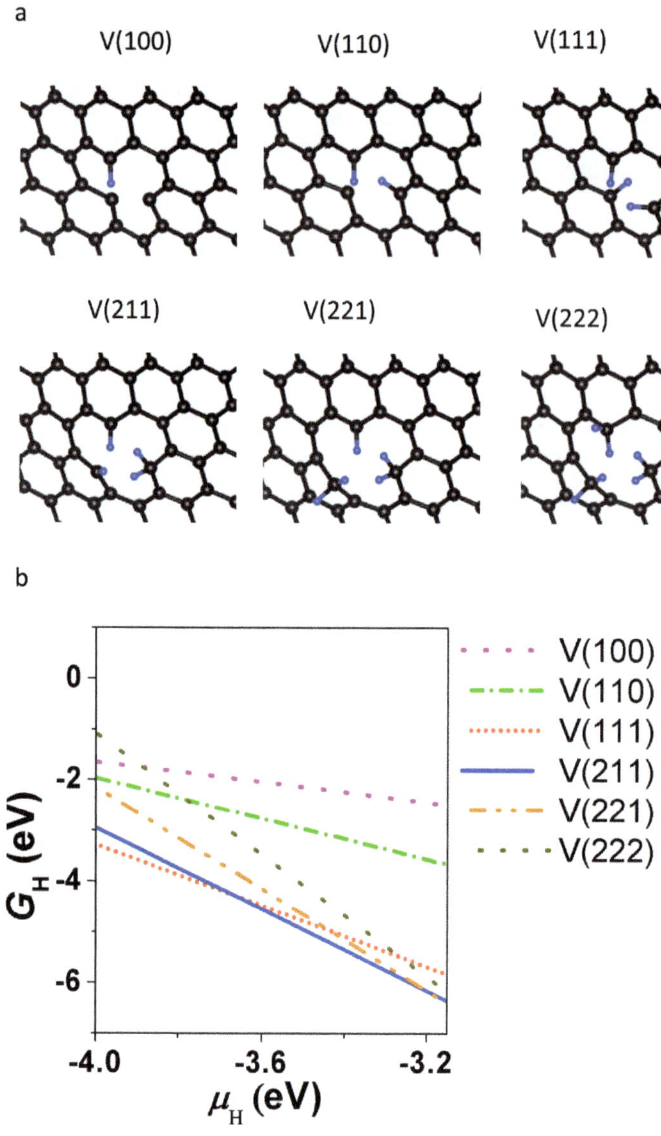

Fig. 10 (a) Possible hydrogenated vacancy structures obtained by geometrical optimisation of mono-atomic vacancies with 1–6 hydrogen atoms in a graphene supercell ($8a \times 8a$, a = graphene lattice constant). The initial atomic configurations were generated with the hydrogen atoms on the top and/or bottom of the vacancy-internal edge sites. The atomic positions were then relaxed to the closest local minimum of the LDA total energy. The notation of V($n1$, $n2$, $n3$) represents the chemical compositions of the hydrogenated mono-atomic vacancies. The indices $n1$, $n2$, and $n3$ denote the number of hydrogen atoms attached to each of three vacancy's under-coordinated carbon sites. Black and blue balls in the models represent carbon and hydrogen atoms, respectively. (b) Free energy of the hydrogenated vacancy, G_H, *versus* chemical potential, μ_H, of the atomic hydrogen gas phase for the models in (a). The potential window is estimated from partial pressure of the atomic hydrogen gas and the temperature range of 800–1000 °C which corresponds to the experimentally measured temperatures (900 ± 100 °C) during the sample preparation process.

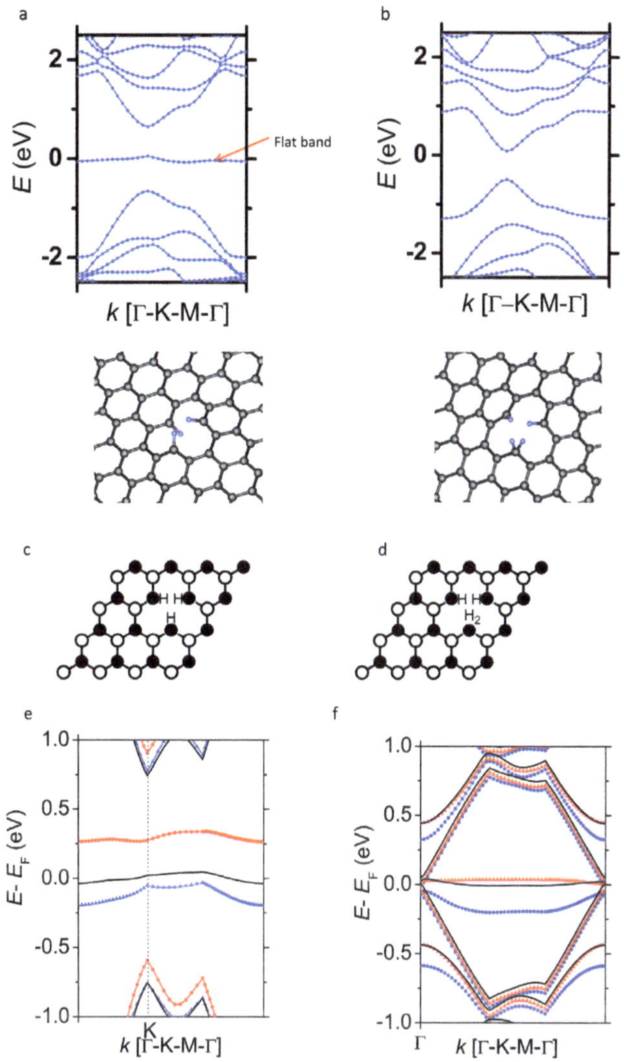

Fig. 11 (a) and (b) LDA-calculated band structures and graphene supercell models of the energetically favourable hydrogenated vacancies V(111) and V(211), respectively (see the caption in Fig. 10). Here we used a relatively small graphene supercell ($4a \times 4a$, a = graphene lattice constant), which indicates that basic physics (*i.e.*, the emergence and disappearance of the flat band, the charge density distribution pattern of the low-energy π state, and the optimised geometries of the hydrogenated vacancy structures) is almost unchanged regardless of the cell sizes. For band structures of larger graphene supercell calculations ($8a \times 8a$, a = graphene lattice constant), see ref. 28. The flat band in (a) is indicated by an arrow. Black and blue balls in the models represent carbon and hydrogen atoms, respectively. (c) and (d) The bipartite models for V(111) and V(211), respectively. The CH_2 site in (d) has no free p_z orbital and thus is regarded as a vacancy. The white and black circles denote A and B sublattice sites, respectively. (e) and (f) Spin-GGA-calculated band structures of V(111) for graphene supercells ($4a \times 4a$ (e) and $9a \times 9a$ (f), a = graphene lattice constant). The blue circles and red triangles denote bands of α-spin and β-spin states, respectively. The black line is a band for nonmagnetic state. The Dirac point is located at the K point in the Brillouin zone (BZ) in (e). The dotted line is for the eyes to indicate band points at the K point. In (f), the Dirac point is at the Γ point, which is due to the BZ folding.

possibly due to undesirable steric hindrance effects among the hydrogen atoms at the vacancy. Thus the point defect-induced magnetism for the V(111) model has not been well addressed yet. In our previous paper,[6] the plane wave-based LSDA calculations within the Quantum ESPRESSO code[40] suggested that the magnetic moment associated with the point defect of V(111) vanishes at the sufficiently large supercell size. In the current work we address the magnetic properties of the V(111) complex by adapting the strictly localized basis sets as implemented in the SIESTA code.[41] The generalized gradient approximation (GGA) exchange–correlation density functional[42] was employed together with a double-ζ plus polarization basis set, norm-conserving pseudopotentials, and a mesh cutoff of 200–400 Ry. For the Brillouin zone integrations a $10 \times 10 \times 1$ Monkhorst–Pack k-mesh[43] was used. The spin-polarized calculations show that the defect-induced π-localized state of the V(111) complex has a magnetic moment of *ca.* 1 μ_B, which is coupled to the magnetic moments in the neighboring supercells *via* the ferrimagnetic superexchange interaction for supercell models of $4a \times 4a$ and $9a \times 9a$ (a = graphene lattice constant). The calculated band structures (Fig. 11e and f) exhibit a spin gap formation of *ca.* 320 and 70 meV at the Dirac point for the $4a \times 4a$ and $9a \times 9a$ models, respectively. The spin-polarized states are *ca.* 50 and 70 meV more stable than the nonmagnetic ones for the $4a \times 4a$ and $9a \times 9a$ models, respectively. In a striking contrast to the V(000) and V(100) models in which magnetic moments vanish along with increasing the supercell size (*i.e.*, decreasing defect density), the V(111) models preserve a magnetic moment of 1 μ_B at the sufficiently large supercell sizes of $8a \times 8a$, $9a \times 9a$ and $12a \times 12a$ (*i.e.*, at low defect densities). This result within DFT calculations of the single defect models in periodic supercells has important implications for defect-induced magnetism based on chemical modification as it provides a possible route to design a magnetic state in graphene even at a relatively low density of the vacancy defects.

Typically, the non-passivated single vacancy model has been used to explain observed single-vacancy-related phenomena such as the emergence of the vacancy-internal-edge-localised π state and its interaction with conduction electrons (*e.g.*, the Kondo effect and magnetic ordering associated with the localised π state[11,44]). As we have seen above, small variations in the atomic details at the linear and vacancy-internal edges, which is how chemical species are attached to the edges, have a considerable impact on the electronic structure. This is a natural consequence of the chemistry at the graphene edge in realistic samples; however, it has not been considered seriously, which is most likely due to the apparent complexity in edge-chemical compositions. In striking contrast to the chemical modification of monovalent hydrogen atoms, oxidation of the vacancy-internal edges introduces apparent complexity into the system because of the variation in the oxygen bonding patterns; nevertheless, we could treat this issue and consider realistic chemistry at the edge with the aid of DFT simulations.

Oxidised vacancy defects were prepared by chemical and thermal reduction of single-layered oxidised graphene on Au(111).[27] After the removal of the oxygen functional groups on the oxidised graphene, the remaining oxidised vacancy-defects featured vacancy-internal-edge-localised states with an apparent size of several nanometers (Fig. 12a). Interestingly, some fraction of the defects exhibited a dynamic ON–OFF switching of the vacancy-internal-edge-localised states, which can be induced by scanning probe manipulation. During raster scanning from the

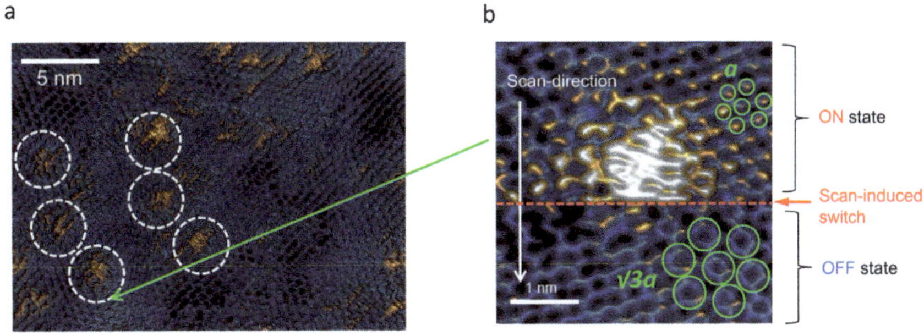

Fig. 12 (a) Current mapping image of the reduced-oxidised graphene on Au(111), where oxidised vacancy defects exhibit a vacancy-internal-edge-localised state, as indicated by the dotted circles. The defect exhibited reversible ON–OFF switching during the repeated imaging. (b) One of the switching events of the defect in (a). During the imaging from top to bottom, the vacancy-internal-edge-localised state switched from ON to OFF. The switching position is indicated by the dotted line (imaging conditions: $V_s = 2$ mV). The low-energy charge density patterns around the vacancy defects for the ON and OFF states are characterised by the $a \times a$ and $\sqrt{3}a \times \sqrt{3}a$ periodicities, respectively, which are schematically superimposed on the image.

top to the bottom of the image, the vacancy-internal-edge-localised state is manipulated from ON to OFF when the scanning probe is scanned across the defect site (Fig. 12b). The switching proceeds reversibly (*i.e.*, from ON to OFF and from OFF to ON). To identify the oxidised defects subject to the reversible switching, DFT modelling was performed in a manner similar to the hydrogenated vacancy modelling. Here we consider oxidised mono- and di-vacancy models to reproduce the reversible ON–OFF switching of the vacancy-internal-edge-localised state and corresponding reversible structural change between two metastable configurations. The systematic DFT simulation within the oxidised vacancy models revealed that only a limited number of models can explain the reversible switching of the vacancy-internal-edge-localised state. Transition-state calculations based on the linear synchronous transit/quadratic synchronous transit (LST/QST) algorithm indicated that Model 2 of the oxidised di-vacancy with an activation barrier of <1 eV is the most probable among the considered structures within the LDA calclations[27] (Fig. 13). The ON–OFF switching is caused by the change in the oxygen-bonding patterns, where two carbonyl groups are changed into two ether (epoxy) groups in a concerted fashion (Fig. 14a). The cooperative reaction essentially reduces the reaction barrier height between the two states (<1 eV), which act as a moderate barrier to separate the two states and enables the reaction to be manipulated (*i.e.*, ON–OFF switching of the edge-localised states) by a scanning probe.

The ON–OFF switching of the defect-localised states can be intuitively understood in terms of the sublattice symmetry. Based on the simple consideration that the π electrons in the ether group have nothing to do with the graphene π-electron network, the carbonyl group adds an additional π electron to the graphene π system and the epoxy group removes two π electrons from the neighbouring sublattice sites. Bipartite models of the ON and OFF states are shown in Fig. 14b. The sublattice symmetry of the ON and OFF states can be respectively described to be imbalanced and balanced, *i.e.*,

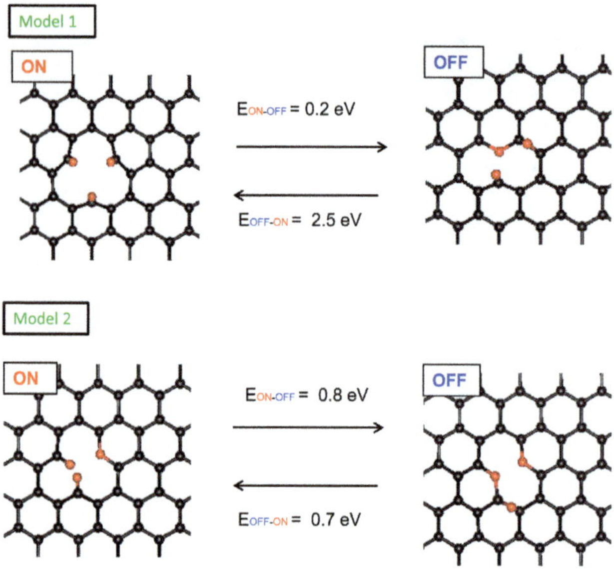

Fig. 13 Pairs of oxidised vacancy models with the reversible switching of the vacancy-internal-edge-localised state. The ON state was characterised by quasi-flat band(s) crossing the Fermi level, whereas the OFF states did not exhibit such flat bands in the LDA-calculated band structures (not shown here). Rectangular graphene supercells containing 72 carbon atoms (1.476 × 1.278 nm) were used for the calculations. The activation barriers for the ON-to-OFF/OFF-to-ON transitions were 0.2/2.5 eV and 0.8/0.7 eV for Models 1 and 2, respectively. The smaller activation barrier (0.2 eV) for the ON-to-OFF transition in Model 1 was not substantially enough to separate the two states at room temperature.

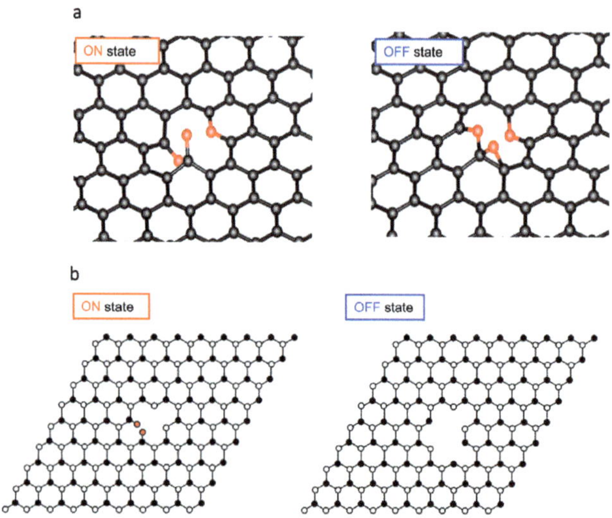

Fig. 14 (a) Top perspective of the two configurations of the ON and OFF states in Model 2 (Fig. 13) with and without an edge-localised π state. (b) Bipartite models for ON and OFF states. The carbonyl group has a free p_z orbital and thus is regarded as an additional A sublattice site, which is represented by a red circle. The white and black circles denote A and B sublattice sites, respectively.

$$N_{bulk}(C_A + C_B) - (C_A + C_B)_{\_divacancy} + 2C_{A\_carbonyl} = (N_{bulk} - 1)(C_A + C_B) + 2C_A$$

for the ON state

$$N_{bulk}(C_A + C_B) - (C_A + C_B)_{\_divacancy} - (C_A + C_B)_{\_epoxy} = (N_{bulk} - 2)(C_A + C_B)$$

for the OFF state where N_{bulk} is the number of the A–B-paired bulk carbons of C_A and C_B, $-(C_A + C_B)_{\_divacancy}$ stands for the removal of the A–B-paired sublattice sites due to di-vacancy-generation, $+C_{A\_carbonyl}$ is the addition of the A sublattice site corresponding to the attachment carbonyl group, and $-(C_A + C_B)_{\_epoxy}$ stands for the removal of the A–B-paired sublattice sites due to epoxy adsorption.

DFT-calculated band structures confirm that the ON state is characterised by quasi-flat dual bands (*i.e.*, imbalance = $+2C_A$) crossing the Fermi level, whereas the OFF state does not have a flat band (*i.e.*, imbalance = 0, Fig. 15). DFT simulation indicates that the ON state has the edge-localised π state strongly localised

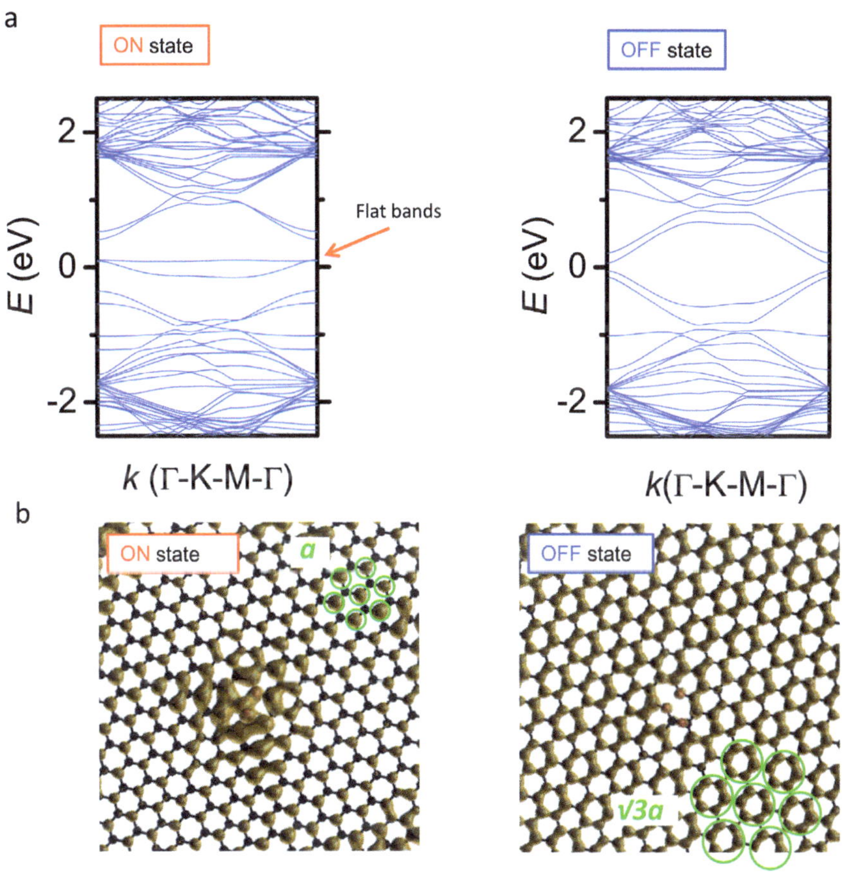

Fig. 15 (a) LDA-calculated band structures of ON and OFF states. The flat band of the ON state is indicated by an arrow. (b) LDA simulations of the ON and OFF states, which reproduce the $a \times a$ and $\sqrt{3}a \times \sqrt{3}a$ periodicities around the defect and the ON–OFF switching of the vacancy-internal edge. A graphene supercell ($9a \times 9a$, a = graphene lattice constant) was used for the calculations.

at the vacancy-internal edge sites, which decay into the bulk with a finite amplitude at one of the graphene sublattice sites with the $a \times a$ periodicity, whereas the edge-localised π state is quenched for the OFF state, in which an extended π state propagates around the defect region to form a superperiodic pattern of $\sqrt{3}a \times \sqrt{3}a$ periodicity (Fig. 15). A closer examination of the simulated STM-isosurface reveals that the decaying π-state distribution with the $a \times a$ periodicity for the ON state and the extended π state with $\sqrt{3}a \times \sqrt{3}a$ periodicity for the OFF state correlate well with the experimentally observed patterns. The triangular pattern of the edge-localised state in the ON state is in qualitative agreement with the experimental data, whereas the complete disappearance of the edge-localised state in the OFF state is evident from both theoretical and experimental results.

The switching of the electronic activity promotes further arguments. As is well-known, the effective Hamiltonian of graphene, *i.e.*, the low-energy effective Dirac field theory (Weyl equation of (eqn (1)) or the $k \cdot p$ theory) and the tight-binding Hamiltonian for the π electron system, is classified by its time-reversal symmetry and the particle conservation.[45] The topological number appears as a winding number, representing the topological charge. The topological quantum number is insensitive to disorder (the deformation of graphene plane), indemnifying a defined state (ON or OFF) against loss of the information. Then, the switching is naturally realized by chemistry-induced modulation only, where the charge topological number (the particle number) is switched. The action is defined in the microscopic scale and thus it is expected to be controllable and more stable than a method using bias controlled charge modulation. All of examples shown in this paper prove that the chemical control of "edge mode" enhances the possibility to tune the electronic, magnetic, and chemical properties of graphene and related nanostructures in various preferable manners.

Stability of the edge mode in nanoscale graphene corners

We have previously studied the electronic properties of the edge-localized state in the quasi-isolated linear edges with several nano-meter lengths. Here we focus on the properties of the edge mode along the nanoscale corner edges in graphene. We are particularly interested in the corner structures in graphene nanosized two-dimensional holes. The nanohole structures have attracted much interest because of the possibility to introduce effectively edge structures into bulk graphene to modify its electronic properties. Of specific interest are the so-called graphene nano-mesh and antidote lattices. It has been argued that lattice scale irregularities in the nanoholes can significantly affect the spin transport through the antidote lattices.[46] Therefore, atomic-scale experimental data on the properties of nanoholes could be crucial for future spintronic devices. We have recently succeeded in preparing nanosized hole-structures with atomically smooth edges by a combination of Ar ion sputtering and subsequent atomic hydrogen-etching of the graphitic surface[6] (Fig. 16a and b). The hole-edges are terminated along zigzag directions and the emergence of the edge-localized state at the hole-edges is characterised as an increased LDOS around the nanohole (Fig. 16a). The hole structure possess corners where two linear hole-edges are interconnected at the

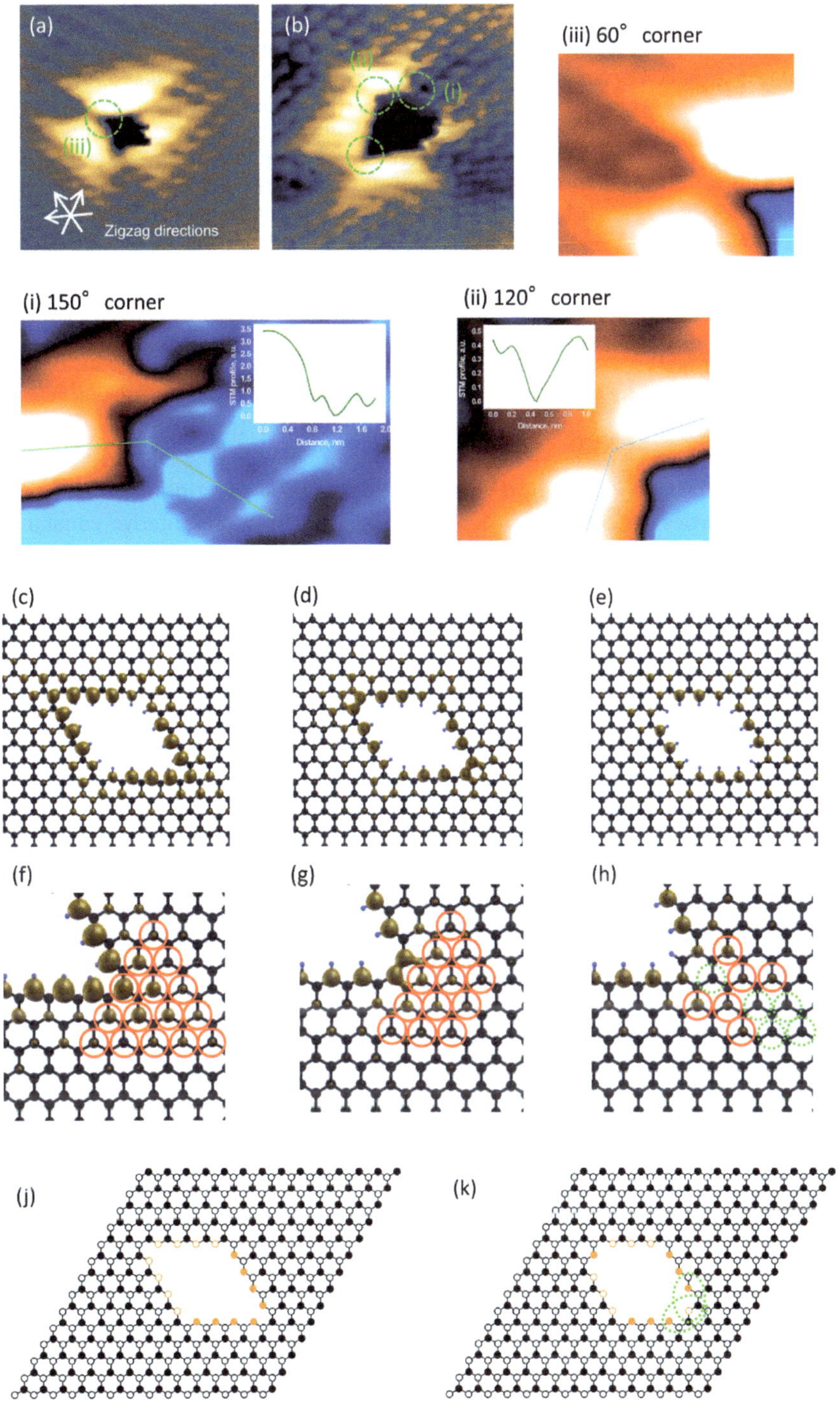

Fig. 16 (a) and (b) Representative STM images of nanoholes with atomically smooth boundaries. Images size 6 × 6 nm^2. Imaging conditions: V_s = 0.35 V, I = 0.3 nA. In (a), arrows indicate the zigzag directions of the graphene lattice. (i)–(iii) Zoomed-in images of 150°- (i), 120°- (ii), and 60°- (iii) corner edges. The positions of (i)–(iii) are represented by dotted circles in (a) and (b). STM cross-sectional profiles along the green solid line in (ii) and

(iii) are shown in the insets. (c)–(e) LDA-calculated LDOS isosurfaces for three possible structures of the 60° corner edge (see the text for details). (f)–(h) Zoomed-in images of the isosurface at the corners of (c)–(e), respectively, in which inner carbon sites with higher LDOS are marked by red circles, while carbon sites with reduced or vanished LDOS (*i.e.*, node sites) are indicated by dotted circles. Black and blue balls represent carbon and hydrogen atoms, respectively. DFT calculations are performed in a similar manner as described in ref. 6. (j) Bipartite models for (c) and (d). (k) Bipartite models for (e). The dotted circle in (k) is one example of a local pairing of A(B) and B(A) sublattices at the edge sites. Secondary carbon sites at the hole-edges are denoted by orange circles. The white and black circles denote A and B sublattice sites, respectively.

corner sites and thus a simple question arises: do the low energy edge-localized states appear at the site that interconnects two edges? To answer this question, we analyzed the distribution of the edge-localized states in the corners of the nanoholes. We are particularly interested in the monohydrogenated edges, whose properties can be well compared with the effective mass description of the electronic properties of the graphene corner edges.[47] The representative constant current STM images of the nanoholes with the zoomed in areas of 150°, 120°, 60° nanoscale edge corners are shown in Fig. 16i–iii. The individual bright spots associated with the edge state, as described earlier, can be distinguished at the edges of these nanoholes at lower bias voltages and higher tunnelling currents. DFT structural realizations were performed using the SIESTA code[41] with a double-ζ plus polarization basis set, norm-conserving pseudopotentials, a mesh cutoff of 200 Ry and 2 × 2 × 1 Monkhorst–Pack *k*-mesh. All atoms were fully relaxed until the forces were less than 0.01 eV Å^{-1}. Simulations of the STM isosurface were obtained for the relaxed atomic positions using the Quantum Espresso code[40] at the LDA level. The electron–ion interaction was described by the projector augmented-wave method.[48] The wave function cutoff energy was set at 40 Ry. The Brillouin zone integration was performed using a uniform 4 × 4 × 1 Monkhorst-type *k*-point grid. We assume a quasi-flat geometry of the edges, noting that the small out-of-plane bending of the monohydrogenated edge does not affect the appearance of the edge state in the constant current STM image as discussed previously.[6] We therefore work within the framework of the Tersoff–Hamann approximation,[49,50] where the STM data at relatively low bias voltages represents contour maps of constant surface LDOS at the Fermi energy. We found a significant decrease in the LDOS at all the edge corners (Fig. 16i and ii), which can be interpreted as a disappearance of the edge-localized states in these corners. The absence of the edge state in the 150° and 120° corners is in a good agreement with theoretical predictions based on the effective mass approximation[47] and DFT theory.[51] On the other hand, absence of the edge state features in the 60° corner edge (Fig. 16iii) looks surprising and deviates from predictions of the effective mass scheme. To understand this discrepancy we first point out that in a non-relaxed structure of a 60° corner, two hydrogen atoms are located very close to each other. We therefore examined more realistic chemistry at the corners and calculated the charge density distribution in three possible structural modifications of the corner edge, namely: (1) a structure where two hydrogen atoms move apart from each other, yielding relatively large displacement (>0.2 Å) of the carbon atoms to which they are attached (Fig. 16c and f), (2) a structure where the absence of the two hydrogen atoms under consideration results in a partial reconstruction of the dangling bonds in the corner edge (Fig. 16d and g), and (3) a structure with an extra carbon atom, terminated with hydrogen, in the corner (Fig. 16e and h). The first corner structure shows a profound signal in the corner, which is reduced in the second structure due to a partial break of the bipartite symmetry upon reconstruction of dangling bonds. On the other hand, the third structure is characterized by the complete disappearance of the edge-localized states in the corner. A closer examination of the spatially decaying behaviour of the edge localized states (Fig. 16f–h) revealed that a reduced or vanished LDOS at the inner carbon sites is apparent (see the dotted circle in Fig. 16h), which is reproduced well in the experiment (Fig. 16iii). To understand the disappearance of the edge-localized states, we examine more closely the balance between the two sublattices in the corner (Fig. 16j and k). It is well known that the edge-localized states originate from the local imbalance between two sublattices at the edges (see Fig. 1), which can be seen as a larger LDOS at four and three zigzag sites that are aligned linearly along the hole-edges for the models in Fig. 16c–e, respectively. In contrast to the linearly aligned zigzag-edge sites, the extra carbon site at the corner in the model in Fig. 16e is isolated and belongs to the different sublattice compared to the rest of the hole-edge sites (see Fig. 16k). Thus the isolated carbon site belonging to A(B) sublattice locally makes a pair with an adjacent hole-edge site belonging B(A) sublattice (see the dotted circle in Fig. 16k). The resultant local sublattice symmetry is balanced in a similar manner to that of the armchair edge (see Fig. 1). Since the A–B paired armchair edge does not support the edge states, the local paring of the A–B sublattice explains the disappearance of the edge state at the corner structure in Fig. 16iii and h.

Conclusions

Combined scanning probe and first-principles characterisation of a chemically modified graphene nanostructure make it possible to identify the exact chemistry of the graphene edges on the atomic scale and their related electronic activities. The chemistry-dependent electronic properties can be basically understood on the basis of the interplay of edge geometry and chemistry, which induces the modulation of a graphene-π electron network, as can be intuitively seen in the corresponding Clar formulas. Depending on the number of hydrogen atoms attached to the linear zigzag edge, the edge-localised π-state with the large densities of states near E_F (*i.e.*, radical character) switches back and forth, which leads to ON–OFF switching of the electronic activity. The attachment of oxygen atoms at the linear zigzag edges causes the drastic change in the effective boundary conditions of the π-electron network, which induces a quasi-localised edge state with a spatially extended character. In a similar manner as the hydrogenated linear edges, the electronic properties of the hydrogenated vacancy-defects are subject to significant modulation in response to chemical details at the internal defect-edges. The defect-localized π-state at the fully mono-hydrogenated vacancy can be quenched by addition of a hydrogen atom at the defect-internal edge, which inevitably links to the electronic and magnetic activities. We provided experimental evidence of dynamical switching functionality of the edge-localised π state at the oxidised point defects, which is induced by dynamic changes in the bonding configuration of the oxygen atoms (*i.e.*, ether/carbonyl forms). In addition to the number of foreign chemical species attached to the edge, their bonding configurations have in fact a strong impact on the electronic properties.

This study reveals that precise control of the chemistry at the graphene edge is crucial for understanding the unique defect-localised π state in graphene materials, which could be an important aspect of fabricating future graphene-based electronic devices.

Acknowledgements

This work was supported by Grants-in-Aid for Scientific Research (no. 20001006, no. 23750150, no. 25790002 and no. 261075260) from the Ministry of Education, Culture, Sports, Science and Technology of Japan.

Notes and references

1 M. Fujita, K. Wakabayashi, K. Nakada and K. Kusakabe, Peculiar localized state at zigzag graphite edge, *J. Phys. Soc. Jpn.*, 1996, **65**, 1920–1923.

2 K. Nakada, M. Fujita, G. Dresselhaus and M. S. Dresselhaus, Edge state in graphene ribbons: nanometer size effect and edge shape dependence, *Phys. Rev. B: Condens. Matter Mater. Phys.*, 1996, **54**, 17954–17961.

3 S. E. Stein and R. L. Brown, π-Electron properties of large condensed polyaromatic hydrocarbons, *J. Am. Chem. Soc.*, 1987, **109**, 3721–3729.

4 Y. Kobayashi, K. Fukui, T. Enoki, K. Kusakabe and Y. Kaburagi, Observation of zigzag and armchair edges of graphite using scanning tunneling microscopy and spectroscopy, *Phys. Rev. B: Condens. Matter Mater. Phys.*, 2005, **71**, 193406.

5 Y. Niimi, T. Matsui, H. Kambara, K. Tagami, M. Tsukada and H. Fukuyama, Scanning tunneling microscopy and spectroscopy studies of graphite edges, *Appl. Surf. Sci.*, 2005, **241**, 43–48.

6 M. Ziatdinov, S. Fujii, K. Kusakabe, M. Kiguchi, T. Mori and T. Enoki, Visualization of Electronic States on Atomically Smooth Graphitic Edges with Different Types of Hydrogen Termination, *Phys. Rev. B: Condens. Matter Mater. Phys.*, 2013, **87**, 115427.

7 H. Yang, A. J. Mayne, M. Boucherit, G. Comtet, G. Dujardin and Y. Kuk, Quantum Interference Channeling at Graphene Edges, *Nano Lett.*, 2010, **10**, 943–947.

8 C. Park, H. Yang, A. J. Mayne, G. Dujardin, S. Seo, Y. Kuk, J. Ihm and G. Kim, Formation of Unconventional Standing Waves at Graphene Edges by Valley Mixing and Pseudospin Rotation, *Proc. Natl. Acad. Sci. U. S. A.*, 2011, **108**, 18622–18625.

9 K. Sakai, K. Takai, K. Fukui, T. Nakanishi and T. Enoki, Honeycomb superperiodic pattern and its fine structure near the armchair edge of graphene observed by low-temperature scanning tunneling microscopy, *Phys. Rev. B: Condens. Matter Mater. Phys.*, 2010, **81**, 235417.

10 C. N. R. Rao, A. K. Sood, K. S. Subrahmanyam and A. Govindaraj, Graphene: The New Two-Dimensional Nanomaterial, *Angew. Chem., Int. Ed.*, 2009, **48**, 7752–7777.

11 O. V. Yazyev, Emergence of Magnetism in Graphene Materials and Nanostructures, *Rep. Prog. Phys.*, 2010, **73**, 056501.

12 X. Jia, J. Campos-Delgado, M. Terrones, V. Meunier and M. S. Dresselhaus, Graphene Edges: A Review of Their Fabrication and Characterization, *Nanoscale*, 2011, **3**, 86–95.
13 T. Enoki, Role of Edges in the Electronic and Magnetic Structures of Nanographene, *Phys. Scr., T*, 2012, **146**, 014008.
14 K. P. Loh, Q. Bao, P. K. Ang and J. Yang, The Chemistry of Graphene, *J. Mater. Chem.*, 2010, **20**, 2277–2289.
15 L. Chen, Y. Hernandez, X. Feng and K. Müllen, From Nanographene and Graphene Nanoribbons to Graphene Sheets: Chemical Synthesis, *Angew. Chem., Int. Ed.*, 2012, **51**, 7640–7654.
16 J. M. Carlsson, F. Hanke, S. Linic and M. Scheffler, Two-Step Mechanism for Low-Temperature Oxidation of Vacancies in Graphene, *Phys. Rev. Lett.*, 2009, **102**, 166104.
17 A. A. El-Barbary, R. H. Telling, C. P. Ewels, M. I. Heggie and P. R. Briddon, Structure and energetics of the vacancy in graphite, *Phys. Rev. B: Condens. Matter Mater. Phys.*, 2003, **68**, 144107.
18 P. O. Lehtinen, A. S. Foster, Y. Ma, A. V. Krasheninnikov and R. M. Nieminen, Irradiation-Induced Magnetism in Graphite: A Density Functional Study, *Phys. Rev. Lett.*, 2004, **93**, 187202.
19 L. R. Radovic and B. Bockrath, On the Chemical Nature of Graphene Edges: Origin of Stability and Potential for Magnetism in Carbon Materials, *J. Am. Chem. Soc.*, 2005, **127**, 5917–5927.
20 T. Wassmann, A. P. Seitsonen, A. M. Saitta, M. Lazzeri and F. Mauri, Structure, Stability, Edge States, and Aromaticity of Graphene Ribbons, *Phys. Rev. Lett.*, 2008, **101**, 096402.
21 H. Zheng and W. Duley, First-Principles Study of Edge Chemical Modifications in Graphene Nanodots, *Phys. Rev. B: Condens. Matter Mater. Phys.*, 2009, **78**, 045421.
22 G. Lee and K. Cho, Electronic Structures of Zigzag Graphene Nanoribbons with Edge Hydrogenation and Oxidation, *Phys. Rev. B: Condens. Matter Mater. Phys.*, 2009, **79**, 165440.
23 A. P. Seitsonen, A. M. Saitta, T. Wassmann, M. Lazzeri and F. Mauri, Structure and Stability of Graphene Nanoribbons in Oxygen, Carbon Dioxide, Water, and Ammonia, *Phys. Rev. B: Condens. Matter Mater. Phys.*, 2010, **82**, 115425.
24 T. Wassmann, A. P. Seitsonen, A. M. Saitta, M. Lazzeri and F. Mauri, Clar's Theory, π-Electron Distribution, and Geometry of Graphene Nanoribbons, *J. Am. Chem. Soc.*, 2010, **132**, 3440–3451.
25 X. Zhang, O. V. Yazyev, J. Feng, L. Xie, C. Tao, Y.-C. Chen, L. Jiao, Z. Pedramrazi, A. Zettl, S. G. Louie, *et al.*, Experimentally Engineering the Edge Termination of Graphene Nanoribbons, *ACS Nano*, 2013, **7**, 198–202.
26 M. Ohtsuka, S. Fujii, M. Kiguchi and T. Enoki, Electronic State of Oxidized Nanographene Edge with Atomically Sharp Zigzag Boundaries, *ACS Nano*, 2013, **7**, 6868–6874.
27 S. Fujii and T. Enoki, Rearrangement of π-Electron Network and Switching of Edge-Localized π State in Reduced Graphene Oxide, *ACS Nano*, 2013, **7**, 11190–11199.
28 M. Ziatdinov, S. Fujii, K. Kusakabe, M. Kiguchi, T. Mori and T. Enoki, Direct imaging of monovacancy-hydrogen complexes in a single graphitic layer, *Phys. Rev. B: Condens. Matter Mater. Phys.*, 2014, **89**, 155405.

29 L. Talirz, H. Söde, J. Cai, P. Ruffieux, S. Blankenburg, R. Jafaar, R. Berger, X. Feng, K. Müllen, D. Passerone, *et al.*, Termini of Bottom-Up Fabricated Graphene Nanoribbons, *J. Am. Chem. Soc.*, 2013, **135**, 2060–2063.
30 H. Huang, D. Wei, J. Sun, S. L. Wong, Y. P. Feng, A. H. C. Neto and A. T. S. Wee, Spatially Resolved Electronic Structures of Atomically Precise Armchair Graphene Nanoribbons, *Sci. Rep.*, 2013, **2**, 983.
31 S. Fujii and T. Enoki, Nanographene and Graphene Edges: Electronic Structure and Nanofabrication, *Acc. Chem. Res.*, 2013, **46**, 2202–2210.
32 E. Clar, *The Aromatic Sextet*, Wiley, London, 1972.
33 C. Tao, L. Jiao, O. V. Yazyev, Y.-C. Chen, J. Feng, X. Zhang, R. B. Capaz, J. M. Tour, A. Zettl, S. G. Louie, *et al.*, Spatially Resolving Edge States of Chiral Graphene Nanoribbons, *Nat. Phys.*, 2011, **7**, 616–620.
34 M. Pan, E. C. Girao, X. Jia, X. S. Bhaviripudi, Q. Li, J. Kong, V. Meunier and M. S. Dresselhaus, Topographic and Spectroscopic Characterization of Electronic Edge States in CVD Grown Graphene Nanoribbons, *Nano Lett.*, 2012, **12**, 1928–1933.
35 D. J. Klein, Graphitic Polymer Strips with Edge States, *Chem. Phys. Lett.*, 1994, **217**, 261–265.
36 K. Kusakabe and M. Maruyama, Magnetic Nanographite, *Phys. Rev. B: Condens. Matter Mater. Phys.*, 2003, **67**, 092406.
37 Y. Li, Z. Zhou, C. R. Cabrera and Z. Chen, Preserving the Edge Magnetism of Zigzag Graphene Nanoribbons by Ethylene Termination: Insight by Clar's Rule, *Sci. Rep.*, 2013, **3**, 2030.
38 Y.-W. Son, N. Park, S. Han and J. Yu, Magnetic Ordering at the Edges of Graphitic Fragments: Magnetic Tail Interactions between the Edge-Localized States, *Phys. Rev. B: Condens. Matter Mater. Phys.*, 2005, **72**, 174431.
39 E. H. Lieb, Two theorems on the Hubbard model, *Phys. Rev. Lett.*, 1989, **62**, 1201.
40 P. Giannozzi, *et al.*, QUANTUM ESPRESSO: a modular and open-source software project for quantum simulations of materials, *J. Phys. Condens. Matter.*, 2009, **21**, 395502.
41 J. M. Soler, E. Artacho, J. D. Gale, A. García, J. Junquera, P. Ordejón and D. Sánchez-Portal, The SIESTA method for *ab initio* order-N materials simulation, *J. Phys.: Condens. Matter*, 2002, **14**, 2745.
42 J. P. Perdew, K. Burke and M. Ernzerhof, *Phys. Rev. Lett.*, 1996, **77**, 3865.
43 H. J. Monkhorst and J. D. Pack, *Phys. Rev. B: Solid State*, 1976, **13**, 518.
44 J.-H. Chen, L. Li, W. G. Cullen, E. D. Williams and M. S. Fuhrer Tunable, Kondo effect in graphene with defects, *Nat. Phys.*, 2011, **7**, 535–538.
45 A. Kitaev, Periodic table for topological insulators and superconductors, *AIP Conf. Proc.*, 2009, **1134**, 22–30, arXiv:0901.2686[cond-mat.mes-hall].
46 M. Vanević, V. M. Stojanović and M. Kindermann, Character of electronic states in graphene antidot lattices: Flat bands and spatial localization, *Phys. Rev. B: Condens. Matter Mater. Phys.*, 2009, **80**, 045410.
47 Y. Shimomura, Y. Takane and K. Wakabayashi, Electronic States and Local Density of States in Graphene with a Corner Edge Structure, *J. Phys. Soc. Jpn.*, 2011, **80**, 054710.
48 P. E. Blochl, Projector augmented-wave method, *Phys. Rev. B: Condens. Matter Mater. Phys.*, 1994, **50**, 17953.

49 J. Tersoff and D. R. Hamann, Theory and Application for the Scanning Tunneling Microscope, *Phys. Rev. Lett.*, 1983, **50**, 1998–2001.
50 J. Tersoff and D. R. Hamann, Theory of the Scanning Tunneling Microscope, *Phys. Rev. B: Condens. Matter Mater. Phys.*, 1985, **31**, 805–813.
51 N. T. Cuong, M. Otani and S. Okada, Absence of edge states near the 120° corners of zigzag graphene nanoribbons, *Phys. Rev. B: Condens. Matter Mater. Phys.*, 2013, **87**, 045424.

Faraday Discussions

PAPER

Maximum aromaticity or *maximum pentagon separation*; which is the origin behind the stability of endohedral metallofullerenes?†

Antonio Rodríguez-Fortea* and Josep M. Poblet

Received 6th May 2014, Accepted 19th June 2014
DOI: 10.1039/c4fd00096j

Two different interpretations have been recently proposed to rationalize the stabilization of some hosting cages in endohedral metallofullerenes as a consequence of the larger localization of the negative charge on pentagonal rather than on hexagonal faces. We try to figure out the physical origin that mainly governs the stability of charged fullerenes; is it aromaticity or electrostatics?

1 Introduction

The hollow space inside fullerenes can host a wide variety of metal atoms and clusters giving rise to the so-called endohedral metallofullerenes (EMFs).[1–5] In 1999, Dorn and co-workers reported the characterization of $Sc_3N@I_h(7)\text{-}C_{80}$,[6] which is the third most abundant fullerene after C_{60} and C_{70}. Since then, a large number of EMFs with a great diversity of metal clusters such as M_3N, M_2, M_2C_2, M_4O_3, M_2S, *etc.* (M = group 3 or rare earth metal) have been isolated and characterized (Fig. 1).[7–11] The electronic structure of EMFs can be easily rationalized from an *ionic model*, which considers that there is a formal transfer of electrons from the internal cluster to the carbon cage. Using this idea, in 2005 Poblet and co-workers formulated a simple orbital rule to identify the most suitable hosting cages in EMFs.[12] For nitride EMFs, with six electrons formally transferred from the cluster to the cage, $(M_3N)^{6+}@(C_{2n})^{6-}$, a suitable fullerene host must have three low-lying unoccupied molecular orbitals and a sizeable energy gap between the LUMO+3 and the LUMO+2.[12,13] The rule holds not only for six, but also for other numbers of electrons transferred.[14] This molecular orbital rule, although useful to

Departament de Química Física i Inorgànica, Universitat Rovira i Virgili, Marcel·lí Domingo s/n, 43007 Tarragona, Spain. E-mail: antonio.rodriguezf@urv.cat; josepmaria.poblet@urv.cat

† Electronic supplementary information (ESI) available: Additional plots of the relative energies with respect to (i) the electrostatic contributions to the energy; (ii) the additive local aromaticity (ALA) for different C_{2n} families. Correlations between ALA and the number of pyracylene motifs are also shown. See DOI: 10.1039/c4fd00096j

identify the best cage, did not provide the final explanation for the selection of a particular cage to host a specific metal cluster. In 2008, Alcamí and co-workers analysed the IPR (Isolated Pentagon Rule) and non-IPR isomers of C_{60} and C_{70} and showed that the relative stability of a charged fullerene is mainly determined by the charge distribution in the fullerene and the subsequent Coulomb repulsion between the charges of equal signs.[15] In line with this work, in 2010 we gave physical support to the orbital rule and showed that the structures with the largest separation between pentagons, the faces that mostly localize the negative charge in EMFs, minimize the Coulomb interaction, which is the major factor in determining the stability of charged fullerenes, *i.e.* the so-called *maximum pentagon separation rule*.[16] We also showed that the cages with the most separated pentagons are those that present low numbers of pyracylene motifs (Scheme 1).

Recently, Solà and co-workers, following our work and also based on the fact that charge in EMFs is preferentially accumulated on pentagonal faces, have proposed a *maximum aromaticity criterion* as a guiding principle for the most suitable hosting cages in EMFs.[17] In particular, they have defined an additive local aromaticity (ALA) index based on one structural indicator of aromaticity (HOMA), which is found to be maximum for those cages in EMFs.

In the present work, we discuss the strengths and weaknesses of the two interpretations given for the same physical outcome, *i.e.* the stabilization of some hosting cages due to the larger localization of the negative charge on pentagons rather than on hexagons. We find that the structural additive local aromaticity

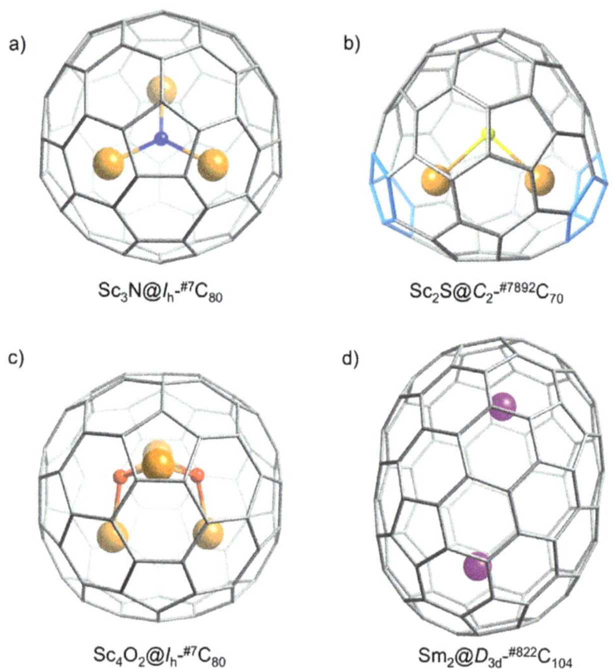

Fig. 1 Structures of some representative nitride, sulfide, oxide and M_2 EMFs. Color codes: Sc, orange; Sm, purple; O, red; N, blue; S, yellow; and C, grey. Adjacent pentagons are highlighted in blue in the non-IPR structure.

index proposed by Solà and co-workers correlates with the number of pyracylene motifs in the cage. Therefore, ALA can be seen as a continuous descriptor of the topology of the fullerene that has to be maximized so as to minimize the number of pyracylenes in charged fullerenes. Our final goal in this work is to answer the following question: what is behind the stabilization of charged fullerenes; aromaticity or electrostatics?

2 Computational methodology

Semi-empirical calculations at the AM1 level were performed using the Gaussian-09 code.[18] DFT calculations were carried out using the ADF-2012 program.[19,20] The exchange and correlation functionals of Becke and Perdew (BP86) were used.[21,22] Relativistic corrections were included by means of the zero-order regular approximation (ZORA). Triple-ζ polarization basis sets were employed to describe the valence electrons of the C atoms. Frozen cores consisting of the 1s shell were described by means of single Slater functions. Charges have been computed using the Mulliken partition analysis. Throughout this paper we identify individual isomeric cages by their symmetry and for non-IPR cages by the number assigned by the spiral algorithm,[23] whereas the truncated numbering system that counts only IPR isomers is used for IPR cages.

3 Results and discussion

3.1 The *maximum pentagon separation rule*: minimization of the electrostatic energy in EMFs

We recently found that for charged fullerenes, and by extension EMFs, (i) there is a preferential location of the charge density on the carbon atoms that belong to the pentagonal faces, caused by their higher pyramidalization angles; and (ii) those cages with the maximum separation among the vicinal pentagons, *i.e.* lowest inverse pentagon separation index (IPSI), show the lowest energies. We thus inferred that the electrostatic term is likely the dominant factor that determines the stability of charged fullerenes and EMFs. The Coulomb repulsion decreases when the pentagons are separated maximally; or put in another words, the cages with the most appropriate distribution of pentagons, *i.e.* maximally separated, are those that allow the pentagons to accumulate the largest amount of charge. We therefore formulated the so-called *maximum pentagon separation* (MPS) *rule* as "The electron transfer from the internal cluster to the fullerene host preferentially adds electrons to the pentagons; therefore, the most suitable carbon cages are those with the largest separation among the 12 pentagons".[16] The molecular electrostatic potentials (MEP) computed at DFT (BP86/TZP) level for some hexa- and tetra-anionic cages provided additional evidence of the preferential charge localization on pentagons.[16] Moreover, the good correlation between the stability of the fullerenes and the electrostatic contribution to the bond energy, according to the partition scheme proposed by Morokuma,[24] supports the idea that the electrostatic term must be the dominant factor that governs the relative stability of the charged fullerene isomers. We also realized that those cages that encapsulate clusters in EMFs show the lowest number of pyracylene motifs (Scheme 1), which is the structural arrangement of pentagons that provides the shorter pentagon–pentagon distances in IPR structures. Thus,

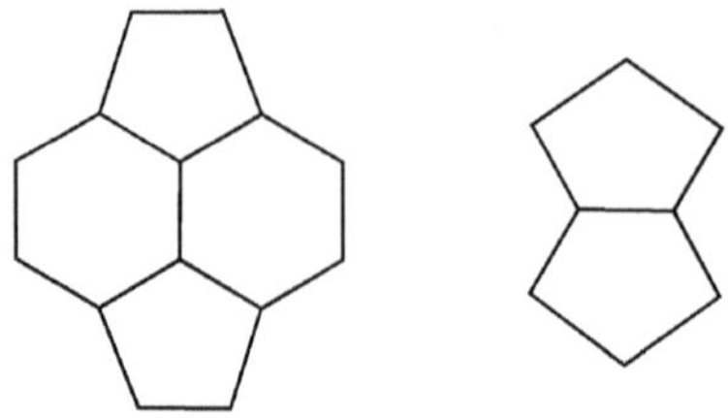

Pyracylene Pentalene

Scheme 1 Representation of the pyracylene and pentalene motifs.

we found a very simple and useful criterion to discard fullerene cages that are not suitable to encapsulate metal clusters.

Solà and co-workers, in a recent work,[17] proposed an alternative guiding principle to predict the most suitable hosting cages in EMFs based on a *maximum aromaticity criterion*. At this point, we would like to remark that these authors misinterpreted our results and conclusions regarding the MPS rule. As stated previously, cages with most separated pentagons and lowest IPSI values allow largest accumulation of charge on pentagons. Coulomb repulsion precludes charge localization on nearby isolated pentagons. The concentration of the charge on the symmetrically distributed pentagons is favoured from a pure electrostatic point of view. The larger amount of charge on pentagons may cause, as noted by Solà and co-workers, the cages with lowest IPSI showing larger repulsion between pentagons, if the repulsion term is computed in a rather approximated manner (see ESI† and caption of Fig. 2). However, the larger the charge is located on pentagons, the smaller it is on hexagons, and for middle- and large-size fullerenes ($2n > 50$) these latter faces are much more abundant than the former ones ($2n/2$–10 hexagons *vs.* 12 pentagons). Therefore, the much smaller repulsion between hexagons would cause the *total* Coulomb repulsion as well as the electrostatic energy to be minimized in the structures with most separated pentagons (Fig. 2).

A very good correlation is found between the relative energies of the hexa- and tetra-anions with respect to the electrostatic contribution to the bond energy, within the scheme proposed by Morokuma, for the large set of C_{2n} families

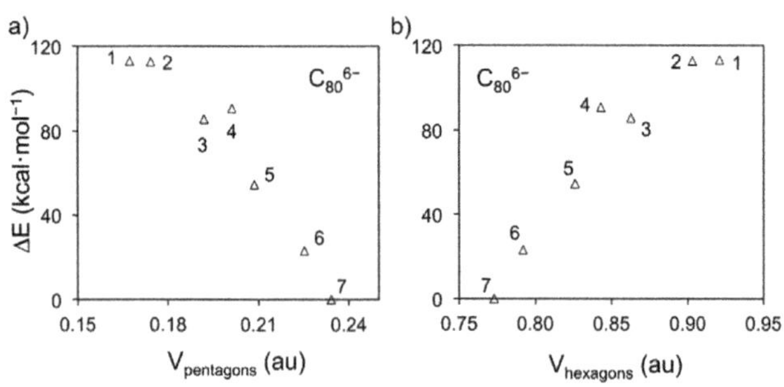

Fig. 2 Plots of the relative energies of IPR C_{80}^{6-} isomers with respect to I_h-$^{\#7}C_{80}$ *versus* an approximated estimation of the Coulomb repulsion (a) between pentagons, $V_{pentagons}$; (b) between hexagons, $V_{hexagons}$. See ESI† for details.

analysed here (Fig. 3a and ESI†). Such an electrostatic term corresponds to the classical electrostatic interaction between the unperturbed charge distributions of the fragments, C atoms in the present case, as they are brought together at their final positions, giving rise to an overall density that is simply a superposition of fragment densities, without taking into account any kind of relaxation of the orbitals.[20] Even though some correlation may seem to be present, much more scattered values are observed when the same two variables are plotted for C_{2n} neutral cages (see Fig. 3b and ESI†), which confirms the relevance of electrostatics on the stabilization of charged fullerenes and EMFs.

3.2 The additive local aromaticity is related to the topology of the fullerene

Aromaticity is a qualitative concept that has been used among chemists for many years. Several criteria for aromaticity have been used so far, based on (i) structural; (ii) energetic; and (iii) magnetic properties, which provide equivalent predictions when prototypical systems are considered. Contradictions and paradoxes might appear for other less usual molecules. Stanger pointed out that the fundamental problem in defining and quantifying aromaticity is the lack of "aromaticity units".[25] Fullerenes, in general, cannot be defined as aromatic, non-aromatic or antiaromatic molecules. Moreover, it may happen that the different criteria make controversial predictions for a given fullerene. A large number of works have been devoted to this subject and it is not our intention to mention them, but to draw the reader's attention to such a delicate issue.[25–29] The additive local aromaticity (ALA) is defined by Solà and co-workers as the sum of the local aromaticities of all the rings in the fullerene structure.[17] The local aromaticities can be computed using any of the indices that have been defined so far according to the aforesaid criteria. Indices based on the study of electron delocalization are found to be the most reliable descriptors of aromaticity.[30] The much simpler structural-based index HOMA (harmonic oscillator model of aromaticity), with a very low computational cost, has been extensively used, although it may show some limitations.[30,31]

In order to show the relationship between the ALA index and the topology of the fullerenes, we have first carried out a detailed analysis for the prototypical IPR

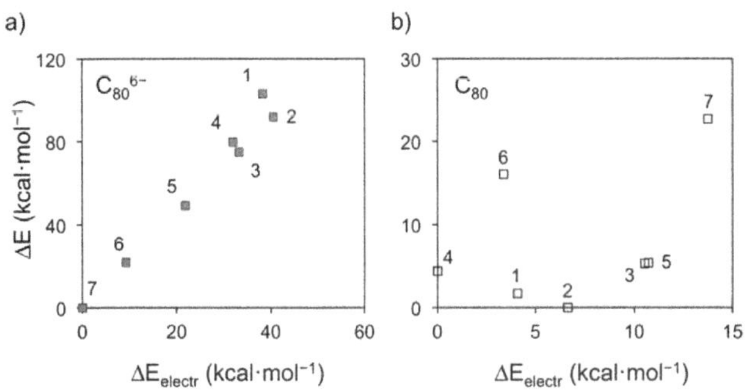

Fig. 3 Plot of the relative energies of IPR C_{80} isomers *versus* the relative electrostatic contributions to the bond energy according to the partition of Morokuma for (a) hexa-anions; (b) neutral cages. Both the energies and the electrostatic contributions are relative to the lowest-energy isomer.

C_{80} family. We have computed the ALA index for the seven isomers in different charged states (neutral and hexa-anions) at different levels of theory (see ESI†). Table 1 shows the ALA values for the seven hexa-anions and neutral isomers obtained for the optimized structures at two different computational levels, semiempirical AM1 and DFT BP86/TZP.

As pointed out previously,[17] hexa-anions are more aromatic than neutral species, according to the values of the ALA index computed at the semiempirical AM1 level (Table 1). However, we did observe the opposite behaviour when the fullerene structures were optimized at the more accurate DFT BP86/TZP level. The reason for such discrepancies can be found in the definition of the HOMA descriptor,[29] which is very sensitive to the C–C bond distances (R_i in eqn (1), n is the number of bonds in the ring).

$$\mathrm{HOMA} = 1 - \frac{257.7}{n}\sum_{i=1}^{n}(1.388 - R_i)^2 \quad (1)$$

Therefore, the following question may arise: are the neutral IPR C_{80} isomers less or more aromatic than their hexa-anions? The geometry-based criterion given by the HOMA descriptor does not answer clearly this question and other indicators (magnetic, electronic) should be used instead as well as measures of global aromaticities. It is important to remark that regardless of the charge state and the computational level used to optimize the geometry, the sorting order given by ALA is essentially the same for the seven IPR isomers of C_{80}: #7 > #6 > #5 > #3 ~ #4 > #1 ~ #2, which might indicate that this index could be related to the topology of the cages. In fact, ALA correlates with IPSI and with the number of pyracylene motifs (see ESI†). Those cages with lowest IPSI, *i.e.* maximum separated pentagons, show the largest ALA values whatever the charge state considered. The correlation improved when the number of pyracylenes was considered (Fig. 4 and ESI†). But, what is at the origin of these correlations? An inspection of the definition of HOMA, eqn (1), helped to answer this query. The larger the standard deviation of the C–C distances in the fullerene, R_i, with respect to the optimal distance of 1.388 Å, the more negative (less positive) the HOMA value, and therefore the smaller the ALA index. Although the pyracylene [6,6] bonds show distances very similar to 1.388 Å and provide a significant contribution to the HOMA value, there is an important dispersion of the C–C distances in the rest of the bonds within the

Table 1 ALA values for the seven hexa-anions and neutral IPR isomers of C_{80} computed at semiempirical AM1 and DFT BP86/TZP levels

	Hexa-anions		Neutral	
Isomer #	AM1	BP86/TZP	AM1	BP86/TZP
1	11.31	5.55	5.25	15.28
2	10.93	5.71	5.47	12.85
3	11.79	4.60	6.48	14.40
4	11.87	4.64	8.87	15.01
5	13.15	6.56	7.60	15.33
6	14.47	8.28	10.11	16.81
7	15.63	9.63	10.95	19.67

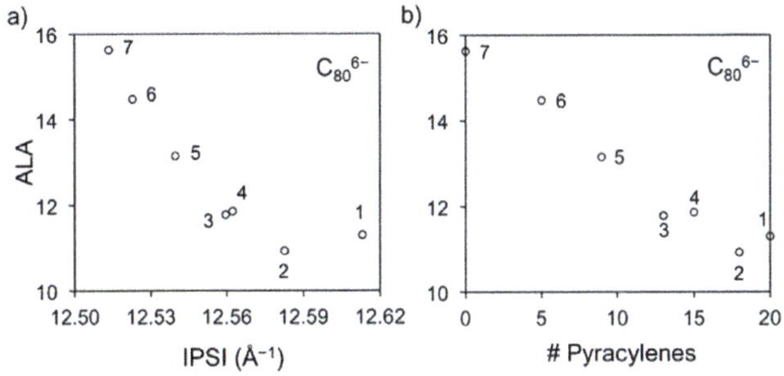

Fig. 4 ALA index for the seven IPR C_{80} isomers with respect to (a) the IPSI; and (b) the number of pyracylene motifs of the cages.

pyracylene motifs, with some bond lengths as large as 1.50 Å. These longer bonds make significant negative contributions to the HOMA of the polygons and, subsequently, lead to a smaller ALA index (Fig. 5), which is the sum of the HOMA for all the pentagonal and hexagonal faces. Therefore, the larger the number of pyracylene motifs, the larger the dispersion in the C–C bonds of the fullerene, as shown in the histogram of Fig. 5b, and the smaller the corresponding ALA index.

Acceptable correlations between the ALA index, based on the HOMA descriptor, and the number of pyracylene motifs for larger C_{2n} families were also observed (see ESI†). Thus, the ALA index might be seen as a *continuous* descriptor of the topology of fullerenes. The continuous nature of this descriptor makes it correlate somewhat better with the relative energies of the charged fullerenes than the *discrete* number of pyracylenes. However, this last descriptor has the advantage that it is much more intuitive and easier to know with a simple inspection of the fullerene topology, without the necessity of making any kind of computation.

3.3 The case of non-IPR fullerenes

The MPS rule also holds for non-IPR isomers with the same number of adjacent pentagon pairs (APPs), for which their IPSI can be directly compared. The electrostatic gain due to the larger charge localization on [5,5] bonds counterbalances

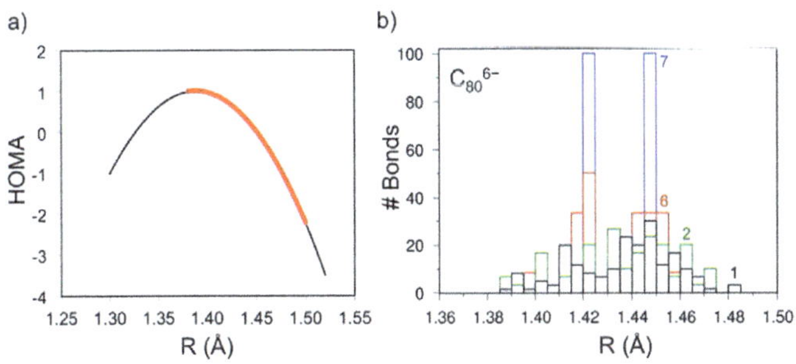

Fig. 5 (a) Representation of HOMA with respect to the C–C bond distance, eqn (1), as if the n sides of the polygon had identical distances, $R_i = R$. The range of C–C distances in fullerenes is highlighted in red. (b) Histogram of the C–C bond distances for IPR isomers #1 (black), #2 (green), #6 (red) and #7 (blue) of the C_{80} family.

the strain caused by the presence of adjacent pentagons and, consequently, many non-IPR structures are observed as EMFs.[32] This usually happens for middle-size cages whose IPR isomers contain a significant number of pyracylene units, as for example C_{84}. The main drawback of the IPSI is that it does not allow comparisons between families with different numbers of adjacent pentagons pairs (APPs). However, we were able to compare IPR and non-IPR structures when we considered the electrostatic contributions to the bond energy, which showed very good correlations with respect to the relative energies, as can be seen in Fig. 6 for the C_{84} IPR and non-IPR hexa-anions with one adjacent pentagon pair (APP1). The same plot for the neutral isomers shows, again, much more scattered points.

As pointed out previously,[17] the ALA index correlates with the relative stabilities of the hexa-anions for the whole set of structures in a given C_{2n} family, regardless of the number of APPs. Hence, if it is seen as a topological parameter, it is more useful than IPSI because, for example, it allows comparison between IPR and non-IPR isomers. The two largest non-IPR cages characterized by X-ray crystallography so far,[33,34] C_s-$^{\#39663}C_{82}$ and C_s-$^{\#51365}C_{84}$, have structures that are intimately related to I_h-$^{\#7}C_{80}$, the prototype for the encapsulation of nitrides and other clusters.[35] Therefore, these structures are topologically favoured, with few pyracylene motifs, and large ALA values that can compete with the 9 and 23 IPR isomers of C_{82} and C_{84}, respectively, with a much larger number of pyracylenes.

3.4 Can the stability of fullerenes be seen as an interplay between strain energy and aromaticity?

We have shown that the ALA index, based on the HOMA criterion, could be seen as a continuous topological index that allows the comparison of the whole set of the C_{2n} isomers. The ALA correlates with the relative stabilities of C_{2n} hexa-anions, but provides scattered plots when the relative energies of neutral isomers are considered. Solà and co-workers concluded that there is 'an interplay between strain energy and aromaticity' in fullerenes. Reducing the strain energy is the main factor for determining the stability of neutral empty fullerenes, whereas maximizing the aromaticity is the main stabilizing force for highly charged fullerenes or EMFs.[17] If ALA, based on HOMA, were a measure of aromaticity for

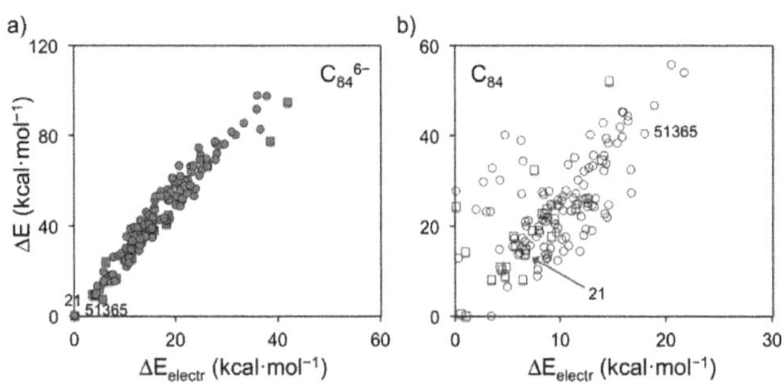

Fig. 6 Plot of the relative energies of IPR (squares) and non-IPR APP1 (circles) C_{84} isomers *versus* the relative electrostatic contributions to the bond energies for (a) hexa-anions; and (b) neutral cages. The two lowest-energy hexa-anions, IPR #21 and non-IPR #51365 are highlighted. Both the energies and the electrostatic contributions are relative to the lowest-energy isomer.

the neutral IPR isomers of C_{80}, cages #6 and #7 would be most aromatic structures (see Table 1). According to Raghavachari's neighbour indexes,[23] the least strained C_{80} cages are also #6 and #7. Therefore, the most aromatic and least strained IPR #6 and #7 isomers should be the most stable neutral cages in striking contrast with the theoretical prediction and the experimental isolation of structures #2 and #1.[36–38]

At this point, it is interesting to remark the results obtained by those authors for the C_{72} family.[17] They found that the good correlation that appeared for the relative energies of the hexa-anions with respect to ALA was progressively lost when passing to tetra-anions, di-anions and neutral species, for which the plot showed completely scattered points. Interestingly, the correlation is somewhat recovered when tetra-cations were considered. We have checked that such a correlation remains for hexa-cations (Fig. 7a), with the cages that have the most separated pentagons and a low number of pyracylenes appearing as the ones with the lowest energies. If the ALA were a measure of the aromaticity of fullerenes, we would state that the most aromatic hexa-cations are those with the lowest energies and that aromaticity is more important than strain in the above-mentioned interplay. However, this conclusion would be difficult to understand because hexa-cations should be, according to their negative ALA values, much less aromatic than neutral species (Fig. 7a and ESI†). On the other hand, the correlation between the electrostatic contribution to the bond energy and the relative energies is better than for the neutral cages, even though it is not as good as for hexa-anions (Fig. 7b). Thus, the relevance of electrostatics seems to be recovered again when moving from neutral cages to hexa-cations.

3.5 Is the stability of the endohedral metallofullerenes cycloadducts related to aromaticity?

Recent results have shown that the thermodynamics of the Bingel–Hirsch addition on non-IPR EMFs is governed by the additive local aromaticity of the rings of the final products.[39] To verify whether this result is a general trend for other cycloaddition reactions on EMFs, we have analysed the Diels–Alder reaction and the 1,3-dipolar cycloaddition of azomethine ylides, the so-called Prato reaction,

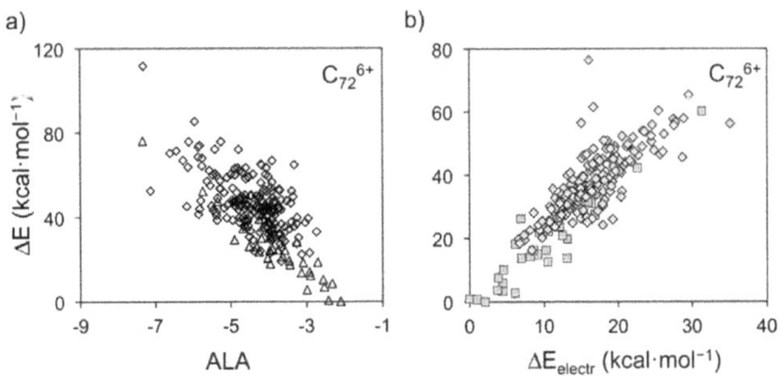

Fig. 7 Plot of the relative energies of IPR (squares), non-IPR APP1 (circles), APP2 (triangles) and APP3 (diamonds) C_{72} hexa-cations *versus* (a) the ALA index calculated with the HOMA descriptor; (b) the relative electrostatic contributions to the bond energies. Both the energies and the electrostatic contributions are relative to the lowest-energy isomer.

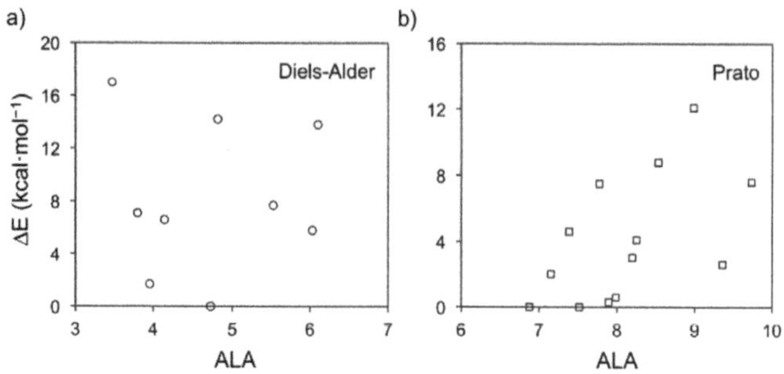

Fig. 8 Relative energies of the different regioisomers of the Diels–Alder (a) and Prato (b) reactions on $Sc_3N@D_{5h}$-$^{\#6}C_{80}$ with respect to the ALA index.

on $Sc_3N@D_{5h}$-$^{\#6}C_{80}$, where several regioisomers are possible. According to the ALA index, there is a decrease in the aromatic character of each monoadduct with respect to the pristine $Sc_3N@D_{5h}$-$^{\#6}C_{80}$ because the bond where the addition takes place is significantly enlarged up to around 1.60 Å, with the corresponding decrease of the HOMA of the rings that share this bond (Fig. 5a) and the total ALA. In sharp contrast to what was observed for the Bingel–Hirsch monoadducts, the relative energies of the different Diels–Alder and Prato regioisomers do not correlate with their ALA indexes, as shown in Fig. 8. The rather small structural changes that take place around the bond that is functionalized are not sufficient to justify the relative energies of the monoadducts.

Finally, we analyzed the eighteen different possible Bingel–Hirsch monoadducts on $Sc_3N@D_3$-$^{\#6140}C_{68}$ and found no clear correlation between their relative energies and ALA indexes (see ESI†). Most of the points, 14 out of 18, are mainly scattered as found for the Diels–Alder and Prato adducts, except the four closed-cage regioisomers, which show very low values for the ALA index because the functionalized bond can be as large as 1.70 Å (see ESI†). Whether this long bond should be included in the computation of the ALA index for the closed-cage adducts is in our opinion not clear.

4 Conclusions

In this work, we have confirmed again that the *maximum pentagon separation rule*, which provides a guideline for the structures of EMFs or charged fullerenes, is mainly governed by electrostatics. Moreover, we have found that IPR and non-IPR cages with different numbers of adjacent pentagon pairs can be directly compared when considering their *electrostatic energies*.

Is the *maximum aromaticity criterion* related to the MPS rule? Is the ALA, based on HOMA, a proper measure of aromaticity for these systems? Since the HOMA descriptor depends importantly on the structure, controversial results may emerge depending on the computational level. For example, the ALA index is not able to answer clearly whether IPR C_{80} neutral cages are more or less aromatic than hexa-anions. As usually happens when measuring aromaticity, different criteria would be needed. Based on the good correlations that exist between ALA and the number of pyracylene motifs, we propose ALA as a *continuous* topological parameter that allows comparisons between IPR and non-IPR structures with different numbers of APPs. On the other hand, the interpretation that the stabilization of fullerenes is an interplay between strain energy and aromaticity seems oversimplified and, at least, does not hold for the IPR C_{80} family, for which neutral cages #6 and #7 should be observed instead of cages #2 and #1. Finally, we have found that if ALA were a measure of aromaticity, the aromaticity would not be the driving force for the stabilization of many cycloadducts on EMFs. Therefore, what is the answer to the question posed in the title? We have shown that the electrostatic term is an important contribution to the stability of EMFs or charged fullerenes, but we cannot completely discard that aromaticity is also playing a role because other descriptors, more reliable than HOMA, and measures of global aromaticities should confirm the results and explain the inconsistencies reported here. Nevertheless, we believe that it is reasonable to think that in charged fullerenes the electrostatic term has to play an important role, whereas the not clearly defined and difficult to determine aromaticity and strain energy terms should participate in the stabilization of both neutral and ionic systems.

Acknowledgements

This work is supported by the Spanish Ministerio de Economía y Competitividad (Project no. CTQ2011-29054-C02-01) and by the Generalitat de Catalunya (2014SGR199 and XRQTC).

Notes and references

1 X. Lu, L. Feng, T. Akasaka and S. Nagase, *Chem. Soc. Rev.*, 2012, **41**, 7723–7760.
2 A. A. Popov, S. F. Yang and L. Dunsch, *Chem. Rev.*, 2013, **113**, 5989–6113.
3 D. M. Rivera-Nazario, J. R. Pinzon, S. Stevenson and L. A. Echegoyen, *J. Phys. Org. Chem.*, 2013, **26**, 194–205.
4 A. Rodriguez-Fortea, A. L. Balch and J. M. Poblet, *Chem. Soc. Rev.*, 2011, **40**, 3551–3563.
5 J. Y. Zhang, S. Stevenson and H. C. Dorn, *Acc. Chem. Res.*, 2013, **46**, 1548–1557.
6 S. Stevenson, G. Rice, T. Glass, K. Harich, F. Cromer, M. R. Jordan, J. Craft, E. Hadju, R. Bible, M. M. Olmstead, K. Maitra, A. J. Fisher, A. L. Balch and H. C. Dorn, *Nature*, 1999, **401**, 55–57.
7 M. N. Chaur, F. Melin, A. L. Ortiz and L. Echegoyen, *Angew. Chem., Int. Ed.*, 2009, **48**, 7514–7538.
8 N. Chen, M. N. Chaur, C. Moore, J. R. Pinzon, R. Valencia, A. Rodriguez-Fortea, J. M. Poblet and L. Echegoyen, *Chem. Commun.*, 2010, **46**, 4818–4820.
9 L. Dunsch, S. F. Yang, L. Zhang, A. Svitova, S. Oswald and A. A. Popov, *J. Am. Chem. Soc.*, 2010, **132**, 5413–5421.
10 X. Lu, T. Akasaka and S. Nagase, *Acc. Chem. Res.*, 2013, **46**, 1627–1635.
11 S. Stevenson, M. A. Mackey, M. A. Stuart, J. P. Phillips, M. L. Easterling, C. J. Chancellor, M. M. Olmstead and A. L. Balch, *J. Am. Chem. Soc.*, 2008, **130**, 11844–11845.
12 J. M. Campanera, C. Bo and J. M. Poblet, *Angew. Chem., Int. Ed.*, 2005, **44**, 7230–7233.
13 R. Valencia, A. Rodriguez-Fortea and J. M. Poblet, *Chem. Commun.*, 2007, 4161–4163.

14 R. Valencia, A. Rodriguez-Fortea and J. M. Poblet, *J. Phys. Chem. A*, 2008, **112**, 4550–4555.
15 H. Zettergren, M. Alcamí and F. Martin, *ChemPhysChem*, 2008, **9**, 861–866.
16 A. Rodriguez-Fortea, N. Alegret, A. L. Balch and J. M. Poblet, *Nat. Chem.*, 2010, **2**, 955–961.
17 M. Garcia-Borràs, S. Osuna, M. Swart, J. M. Luis and M. Solà, *Angew. Chem., Int. Ed.*, 2013, **52**, 9275–9278.
18 M. J. Frisch, G. W. Trucks, H. B. Schlegel, G. E. Scuseria, M. A. Robb, J. R. Cheeseman, G. Scalmani, V. Barone, B. Mennucci, G. A. Petersson, H. Nakatsuji, M. Caricato, X. Li, H. P. Hratchian, A. F. Izmaylov, J. Bloino, G. Zheng, J. L. Sonnenberg, M. Hada, M. Ehara, K. Toyota, R. Fukuda, J. Hasegawa, M. Ishida, T. Nakajima, Y. Honda, O. Kitao, H. Nakai, T. Vreven, J. Montgomery, J. A. Montgomery, J. E. Peralta, F. Ogliaro, M. Bearpark, J. J. Heyd, E. Brothers, K. N. Kudin, V. N. Staroverov, R. Kobayashi, J. Normand, K. Raghavachari, A. Rendell, J. C. Burant, S. S. Iyengar, J. Tomasi, M. Cossi, N. Rega, J. M. Millam, M. Klene, J. E. Knox, J. B. Cross, V. Bakken, C. Adamo, J. Jaramillo, R. Gomperts, R. E. Stratmann, O. Yazyev, A. J. Austin, R. Cammi, C. Pomelli, J. W. Ochterski, R. L. Martin, K. Morokuma, V. G. Zakrzewski, G. A. Voth, P. Salvador, J. J. Dannenberg, S. Dapprich, A. D. Daniels, Ö. Farkas, J. B. Foresman, J. V. Ortiz, J. Cioslowski and D. J. Fox, Gaussian, Inc, Wallingford CT, 2009.
19 E. J. Baerends, D. E. Ellis and P. Ros, *ADF 2012.01*, Department of Theoretical Chemistry, Vrije Universiteit, Amsterdam.
20 G. te Velde, F. M. Bickelhaupt, E. J. Baerends, C. F. Guerra, S. J. A. v. Gisbergen, J. G. Snijders and T. Ziegler, *J. Comput. Chem.*, 2001, **22**, 932–967.
21 A. D. Becke, *Phys. Rev. A*, 1988, **38**, 3098–3100.
22 J. P. Perdew, *Phys. Rev. B: Condens. Matter Mater. Phys.*, 1986, **33**, 8822–8824.
23 P. W. Fowler and D. E. Manolopoulos, *An Atlas of Fullerenes*, Oxford University Press, Oxford, 1995.
24 K. Morokuma, *J. Chem. Phys.*, 1971, **55**, 1236–1244.
25 A. Stanger, *Chem. Commun.*, 2009, 1939–1947.
26 M. Bühl and A. Hirsch, *Chem. Rev.*, 2001, **101**, 1153–1183.
27 Z. F. Chen and R. B. King, *Chem. Rev.*, 2005, **105**, 3613–3642.
28 Z. F. Chen, J. I. Wu, C. Corminboeuf, J. Bohmann, X. Lu, A. Hirsch and P. V. Schleyer, *Phys. Chem. Chem. Phys.*, 2012, **14**, 14886–14891.
29 T. M. Krygowski and M. K. Cyranski, *Chem. Rev.*, 2001, **101**, 1385–1419.
30 F. Feixas, E. Matito, J. Poater and M. Solà, *J. Comput. Chem.*, 2008, **29**, 1543–1554.
31 M. Garcia-Borràs, S. Osuna, M. Swart, J. M. Luis and M. Solà, *Chem. Commun.*, 2013, **49**, 1220–1222.
32 Y. Z. Tan, S. Y. Xie, R. B. Huang and L. S. Zheng, *Nat. Chem.*, 2009, **1**, 450–460.
33 B. Q. Mercado, C. M. Beavers, M. M. Olmstead, M. N. Chaur, K. Walker, B. C. Holloway, L. Echegoyen and A. L. Balch, *J. Am. Chem. Soc.*, 2008, **130**, 7854–7855.
34 T. Zuo, K. Walker, M. M. Olmstead, F. Melin, B. C. Holloway, L. Echegoyen, H. C. Dorn, M. N. Chaur, C. J. Chancellor, C. M. Beavers, A. L. Balch and A. J. Athans, *Chem. Commun.*, 2008, 1067–1069.

35 M. Mulet-Gas, A. Rodriguez-Fortea, L. Echegoyen and J. M. Poblet, *Inorg. Chem.*, 2013, **52**, 1954–1959.
36 F. Furche and R. Ahlrichs, *J. Chem. Phys.*, 2001, **114**, 10362–10367.
37 F. H. Hennrich, R. H. Michel, A. Fischer, S. Richard-Schneider, S. Gilb, M. M. Kappes, D. Fuchs, M. Burk, K. Kobayashi and S. Nagase, *Angew. Chem., Int. Ed. Engl.*, 1996, **35**, 1732–1734.
38 M. L. Sun, Z. Slanina, S. L. Lee, F. Uhlik and L. Adamowicz, *Chem. Phys. Lett.*, 1995, **246**, 66–72.
39 M. Garcia-Borràs, S. Osuna, M. Swart, J. M. Luis, L. Echegoyen and M. Solà, *Chem. Commun.*, 2013, **49**, 8767–8769.

Faraday Discussions

PAPER

Nitrogen segregation in nanocarbons

C. P. Ewels,*[a] D. Erbahar,[ab] Ph. Wagner,[a] X. Rocquefelte,[a] R. Arenal,[cd] P. Pochet,[e] M. Rayson,[f] M. Scardamaglia,[h] C. Bittencourt[h] and P. Briddon[ag]

Received 16th May 2014, Accepted 13th June 2014

DOI: 10.1039/c4fd00111g

We explore the behaviour of nitrogen doping in carbon nanomaterials, notably graphene, nanotubes, and carbon thin films. This is initially *via* a brief review of the literature, followed by a series of atomistic density functional calculations. We show that at low concentrations, substitutional nitrogen doping in the sp^2-C graphenic basal plane is favoured, however once the nitrogen concentration reaches a critical threshold there is a transition towards the formation of the more thermodynamically-favoured nitrogen terminated 'zigzag' type edges. These can occur either *via* formation of finite patches (polycyclic aromatic azacarbons), strips of sp^2 carbon with zigzag nitrogen edges, or internal nitrogen-terminated hole edges within graphenic planes. This transition to edge formation is especially favoured when the nitrogen can be partially functionalised with, *e.g.* hydrogen. By comparison with available literature results, notably from electron energy loss spectroscopy and X-ray spectroscopy, the current results suggest that much of the nitrogen believed to be incorporated into carbon nanoobjects is instead likely to be present terminating the edges of carbonaceous impurities attached to nanoobject's surface. By comparison to nitrogen-doped tetrahedrally amorphous carbon, we suggest that this transition at around 10–20% nitrogen concentration and above towards sp^2 coordination *via* internal nitrogen-terminated edge formation may be a general property of nitrogen-doped carbon materials.

[a]*IMN, CNRS UMR6502, Universit de Nantes, 44300 Nantes, France. E-mail: chris.ewels@cnrs-imn.fr; Tel: +33 (0)2 40 37 64 07*

[b]*Physics Department, Gebze Institute of Technology, Gebze, Kocaeli 41400, Turkey*

[c]*Laboratorio de Microscopias Avanzadas (LMA), Instituto de Nanociencia de Aragon (INA), Universidad de Zaragoza, Calle Mariano Esquillor, 50018 Zaragoza, Spain*

[d]*Fundacion ARAID, 50018 Zaragoza, Spain*

[e]*Univ. Grenoble Alpes, INAC-SP2M, L Sim and CEA, INAC-SP2M, Atomistic Simulation Lab., F-38000 Grenoble, France*

[f]*Department of Chemistry, University of Surrey, Guildford, Surrey, UK*

[g]*School of Natural Sciences, University of Newcastle upon Tyne, Newcastle upon Tyne, UK*

[h]*Chemistry of Interaction Plasma Surface (ChIPS), University of Mons, Belgium*

1 Introduction

Heteroatom doping has long been of interest as a way to modify the physical, chemical and electronic properties of carbon materials.[1–5] With its five valence electrons compared to carbon's four, and very similar atomic radius, nitrogen is a natural choice for doping nanocarbons.[5–8] Nitrogen doping of carbon nanomaterials has been explored as a way to electronically dope carbon, chemically activate the carbon surface and change solubility, and improve the electrical and thermal conductivity of carbon nanotubes (CNTs)[9–13] with the resulting materials showing promise for oxygen reduction reactions[14,15] energy-storage,[16,17] gas[18] and bio[19] sensors, and in the synthesis of 3D macrostructures.[20]

Nitrogen-doped single walled nanotubes (SWNTs) generally exhibit morphologies similar to their undoped counterparts, namely straight non-buckled nanotube walls.[21,22] However nitrogen-doped multi-walled nanotubes (MWNTs) exhibit very distinct morphologies, significantly different from their undoped counterparts. These tubes always have a so-called "bamboo" structure, where the nanotube contains a regularly spaced array of internal carbon walls, or 'herringbone' structures where the tubes are constructed from a stacked-cone carbon arrangement. Nitrogen-doped nanotubes are less stable than their pure carbon counterparts, breaking easily under the electron beam in the transmission electron microscope (TEM)[23,24] and oxidising at lower temperatures than undoped tubes.[25] Similar behavior under the electron beam has also been observed in N-doped graphene compared to pristine graphene.[23,26]

Nitrogen doping in nanocarbons exhibits similar spectral signatures whether in graphene,[27,28] nitrogen-doped carbon thin films,[29] SWNTs[23,30] (see Fig. 1) or nanodiamonds.[31] The nitrogen N1 is seen to split into two distinct peaks at $\approx$398 and 402 eV, with a third intermediate peak at $\sim$400 eV seen in MWNTs.[27,32] The peak at $\sim$398 eV is commonly associated with pyridinic nitrogen, defined as a nitrogen atom with two aromatic carbon neighbours. This can occur next to a vacancy, or at an edge site, for example at an open nanotube tip[7] (see Fig. 2). The peak at $\sim$400 eV is commonly assigned to pyrrolic nitrogen,[32,33] substitutional nitrogen atoms in a region of defective non-aromatic lattice (such as neighbouring pentagons and/or heptagons)[24] or sp cyanide (triple bonded to carbon).[34,35] Finally the peak at $\sim$401–402 eV is normally assigned to substitutional (graphitic) nitrogen.[36]

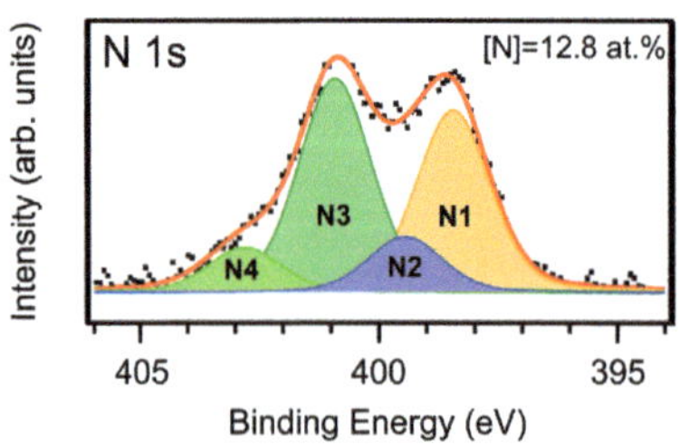

Fig. 1 N1s core level spectra from nitrogen implanted CVD few-layer graphene; experimental data (dotted line), peaks resulting from a least-square fitting procedure (continuous red line). A Shirley-type background was subtracted (adapted from ref. 27).

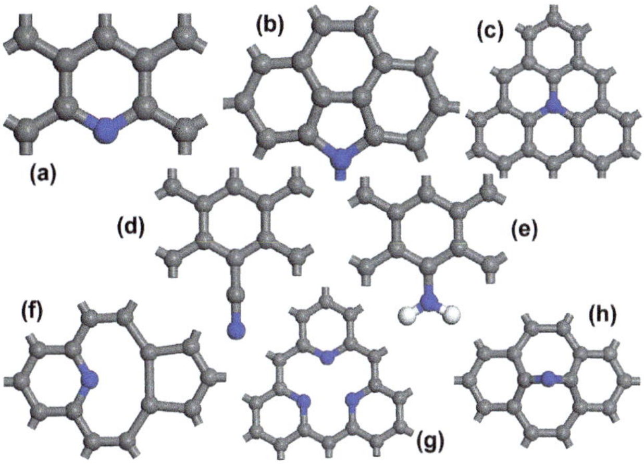

Fig. 2 Possible bonding configurations for N in graphitic networks, (a) pyridinic-like N, (b) pyrrolic-type nitrogen (other pyrrolic configurations are possible provided the nitrogen remains sp^3 coordinated), (c) graphitic substitutional nitrogen, (d) Nitrile $-C{\equiv}N$, (e) $-NH_2$, (f) pyridinic N-vacancy complex, (g) pyridinic N_3-vacancy, (h) interstitial nitrogen. Reproduced from ref. 7.

Other macroscopic spectroscopic techniques as Raman spectroscopy have also been successfully employed for the study of doped sp^2-carbon nanomaterials.[37,38] In N-doped SWNTs, the shift and broadening of characteristic Raman peaks provides information about the doping degree of these materials.[39,40] We focus in this current article on X-ray and electron spectroscopy results.

2 Substitutional nitrogen in bilayer graphene

The current investigation was initiated by the fact that experiments seem to suggest fundamental limits to nitrogen doping concentrations in CNTs. In SWNTs concentrations appear limited to ~1% at most,[23,24,30] whereas in MWNTs average concentrations can reach up to ~15–20%.[41,42] While local concentrations can reach 25–30%,[43] it has been shown with electron tomography that the nitrogen is not homogeneously distributed in this case and instead the sample contains nitrogen-rich non-graphitic regions.[42] To determine a theoretical maximum attainable upper limit for nitrogen substitution in an sp^2-layered carbon system, we performed a series of density functional calculations.

We constructed supercells containing AB-bilayer graphene (see Computational method below), then substituted random carbon atoms with nitrogen, imposing the rule that neighbouring atoms could not be substituted. The systems were then fully geometrically optimized. This process was repeated for a number of different nitrogen configurations, at a range of different nitrogen concentrations from 0–50%. Substituting carbon with nitrogen disrupts the surrounding electronic π-cloud, and at a certain point the surrounding carbon atoms are no longer stable in an sp^2 configuration, and instead have a tendency to cross-link with the neighbouring layer forming sp^3-bonded carbon. The natural limit of this is at 50% nitrogen incorporation, at which point every other carbon atom has been nitrogen substituted. The remaining carbon atoms all cross-link between the two layers, resulting in a fully sp^3-coordinated system with interior atoms sp^3-linked carbon atoms, and the external faces terminated with pyramidal nitrogen atoms (see Fig. 3).

Our results are shown in Fig. 3, where we can see three distinct regimes. Below ~10% nitrogen, no cross-linking between the layers occurs, and the system remains fully sp^2-coordinated. In the range 10–20% nitrogen the first interlayer C–C bonds form. Above 20–25% the cross-linking between carbon atoms in adjacent layers increases significantly, and the layered sp^2-carbon structure breaks down until, at 50% nitrogen, all the carbon atoms are sp^3-coordinated. Thus, these findings could also explain the different morphologies existing in N-doped MWNTs, which, as mentioned above, are associated with the different local amounts of N incorporated on these NTs.

Thus this result defines an upper limit for local substitutional nitrogen concentration in a sp^2-C layered system at around ~10–20%. At first sight this would appear a highly promising result, as it lies in a similar range as the upper doping limit observed in N-doped MWNTs.[44,45] It also coincides with the composition at which CN_x thin films have been shown to become ultra-hard,[46] whose TEM microstructures show heavily cross-linked layers and inter-layer spacings consistent with these calculated here. It is close to the threshold limit nitrogen concentration observed in N-implanted graphene,[27] and also matches early PM3 calculations of nitrogen-doped graphene monolayers which found sheet corrugation begins (corresponding to sp^2-C breakdown) at around 20% [N].[47]

However in fact, this predicted transition from sp^2-C towards sp^3-C at ~20% [N] is actually a misleading coincidence, as we will now show.

2.1 Nitrogen stabilisation of sp^2 carbon

Unfortunately the sp^2/sp^3 carbon ratio in nitrogen-doped carbon nanomaterials as a function of nitrogen concentration is rarely discussed, since typically the XPS C1s peaks are broad and difficult to deconvolute.[27] We therefore instead turn to

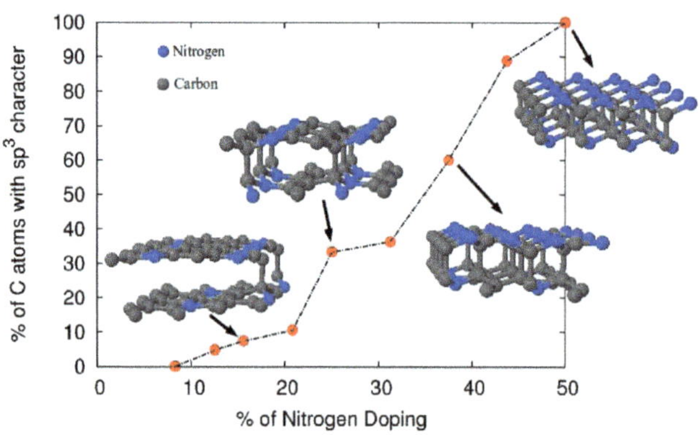

Fig. 3 Fraction of carbon atoms in sp^3 configuration in bilayer graphene as a function of substitutional nitrogen doping percentage, after full structural optimisation. sp^3 coordination in all cases consists of C–C cross-linking between layers. Ball and stick model structures of selected data points indicated by arrows are included.

the literature for nitrogen-doped tetrahedrally-amorphous carbon (ta-C). Previous theoretical works also indicated that N incorporation favors the formation of reactive C-sp^3 in adjacent sites.[48] Furthermore, this prediction has been also used to support the hypothesis that C-sp^3 atoms act as cross-linking sites.[46] However, experimentally it has been found that high nitrogen concentration CN_x films are composed of ~100% sp^2-configurations.[49,50] Fig. 4 shows two plots from different sample preparations. As can be seen, the predicted behavior here is the exact *inverse* of our finding above, namely, in the range 1–10% or 10–20% depending on the sample, the system undergoes a transition from primarily sp^3-coordinated carbon to sp^2-coordination, *i.e.* increasing the nitrogen concentration above this threshold appears to *stabilize* sp^2-coordination. We therefore need a different explanation of nitrogen behaviour at higher concentrations.

Our calculations thus far have assumed essentially homogenous samples with only substitutionally-bonded nitrogen bonding species. However as discussed above, XPS studies clearly see nitrogen in other configurations, notably pyridinic. This is the configuration shown in Fig. 2a, a nitrogen atom with only two neighbours, *i.e.* nitrogen adjacent to either an external or internal edge (such as vacancies).

Variable nitrogen concentration, [N], studies of these two peaks show the substitutional nitrogen 400.5 eV peak remains constant independent of [N], whereas the pyridinic 398.2 eV peak intensity is proportional to nitrogen concentration.[45] Studies of N-doped ultra-hard thin films grown by magnetron sputtering found a rapid increase in elastic recovery behaviour as [N] passed from 5 to 15%, with a corresponding rapid and continuous increase in the pyridinic 398 eV peak intensity as compared to the substitutional 400 eV peak.[52] Similar behavior was also seen in our recent X-ray photoelectron spectromicroscopy studies of the nitrogen ion implantation of nanotubes, where the pyridinic/substitutional ratio in randomly aligned multi-walled carbon nanotubes increased with [N].[53]

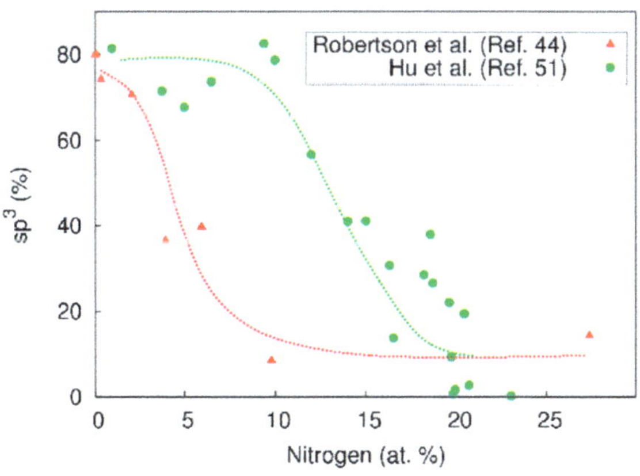

Fig. 4 sp^3-C fraction in nitrogen-doped tetrahedrally-amorphous carbon (ta-C) thin films as a function of nitrogen concentration as determined by EELS spectroscopy, showing a marked transition towards a high sp^2-fraction when the nitrogen concentration passes a threshold. Data from ref. 44 and 51. Data from ref. 51 is upshifted here by 10% as compared to that in the original reference, due to assumed mislabelling in the original published plot.

Similarly when implanting individual few-layer graphene sheets, we found that while the concentration of substitutional nitrogen increases by only about 1.6% between 5 and 15 min of nitrogen ion bombardment, the pyridinic concentration almost doubles from 2.5 to 4.7%.[27] Whilst these latter cases can be understood *via* the formation of carbon vacancies and voids due to the sputtering process, it would be desirable to see if there is a thermodynamic driving force towards pyridinic nitrogen formation at higher nitrogen concentrations. A way to achieve this would be through segregation, with primary possible segregation locations being edges (internal or external), grain boundaries, or new phase formation.

3 Nitrogen segregation at boundaries

Our calculations for a single substitutional nitrogen atom (N_{subst}) in a large 287 atom graphene sheet (0.3% N) show little change in bond length for the nitrogen compared to carbon, consistent with previous calculations[14,54] (shifts of less than 0.04 Å during geometrical optimisation after substitution). Our calculated formation enthalpy for substitutional nitrogen is +0.21 eV, using the formalism

$$E_F(N_{subst}) = E_{TOT} - n_c E_F(\text{graphene}) - 1/2E_F(N_2) \quad (1)$$

where E_{TOT} is the total system energy for the nitrogen substituted graphene containing n_C carbon atoms, E_F(graphene) is the energy of one carbon atom in an infinite graphene sheet, $E_F(N_2)$ is the total energy for an N_2 molecule. By decreasing the hexagonal cell size we increase the nitrogen concentration while maintaining a homogenous nitrogen distribution. The trend in formation enthalpy is shown in Fig. 5; it can be seen that once a 2 × 2 8-atom cell is reached, the formation enthalpy for substitutional nitrogen has risen to +0.50 eV.

While this result precludes stabilization *via* simple nitrogen segregation to high density substitutional regions (at least, on the same sub-lattice), its primary

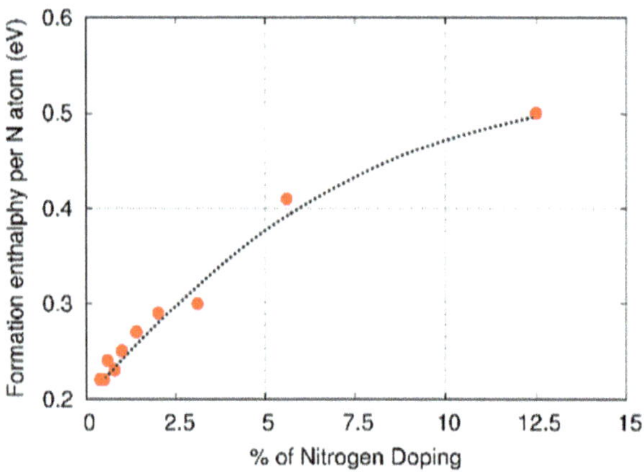

Fig. 5 Calculated formation enthalpy for homogeneously (hexagonally) distributed substitutional nitrogen in pristine infinite graphene, as a function of nitrogen concentration.

purpose is to provide formation enthalpies for comparison with alternative segregation routes.

There are three primary segregation possibilities for nitrogen in sp^2-carbon lattices: segregation *via* simple aggregation, segregation at boundaries (grain boundaries, sheet edges), and segregation *via* the formation of new phases. The first of these is effectively excluded above, so we next turn to segregation at sheet edges.

There are two primary cutting directions in the hexagonal graphene lattice, the $\langle 1\bar{1}00 \rangle$ and the $\langle 2\bar{1}\bar{1}0 \rangle$, creating so-called *armchair* and *zigzag/Klein* edges respectively. All intermediate orientations can be described in terms of alternating sections of these edge orientations,[55] in general called *chiral* edges. For a more detailed discussion of edge structure and termination see ref. 56.

In the following section we construct orthorhombic supercells with ~50 Å width graphene nanoribbons, since these have been shown to be sufficiently wide that the edges are decoupled.[56] Similar convergence criteria and cell dimensions are used to our previous studies with hydrogen,[56] hydroxyl,[57] sulphur and halogen[58] terminated edges, however we quote here formation enthalpies per nitrogen atom rather than per unit length to remain comparable to the bulk defect energies above.

In the case of nitrogen-terminated edges there are very few possible structural configurations, once nitrogen pairs and undercoordinated nitrogen atoms are excluded. We have considered here the two primary $\langle 1\bar{1}00 \rangle$ and $\langle 2\bar{1}\bar{1}0 \rangle$ edge structures, along with their formation enthalpies (Fig. 6). There are two key observations we draw, firstly that the armchair $\langle 1\bar{1}00 \rangle$ edge structure is extremely unstable (E_F = +4.62 eV per nitrogen atom), and secondly the surprisingly high stability of the zigzag $\langle 2\bar{1}\bar{1}0 \rangle$ edge structure (formation enthalpy of only +0.81 eV per N atom). Thus this result suggests that nitrogen-terminated edges are likely to be exclusively zigzag oriented.

Unfortunately, the few experimental works on this topic to date have investigated flat and homogeneous regions relatively far from the edge of the graphene layer.[23,26,28] However we note that this situation of exclusively nitrogen-terminated zigzag edges is in line with predictions and observations in h-BN.[60,61]

This result is also entirely consistent with the nitrogen-terminated vacancy shown in Fig. 2g. This is effectively the smallest possible zone of internal zigzag edge of the type discussed here, with three pyridinic nitrogen atoms surrounding a single carbon vacancy. Our calculated formation enthalpy for this defect in the centre of a very large 284 C atom supercell (*i.e.* at 1.0% [N]) is +0.80 eV per N atom, matching our low density edge value found here.

The +0.81 eV per N formation enthalpy for the zigzag edge shown above is close to the formation enthalpy for substitutional nitrogen in the 10–20% concentration range in the basal plane (+0.4 to +0.6 eV, see Fig. 5), but still remains higher. Thus simple segregation into zigzag nitrogen-terminated edges in this concentration range is not a candidate thermodynamic driving force for the observed behavioural change. However there are a number of ways that the formation enthalpy of such edges can be decreased further.

The first is to explore increasing effective edge density; in practice this could represent either a higher density of nitrogen terminated vacancy loops in the sheet, or increasingly small nitrogen-terminated polycyclic aromatic carbon patches (azacarbons). For our simulations we increase the edge site density by

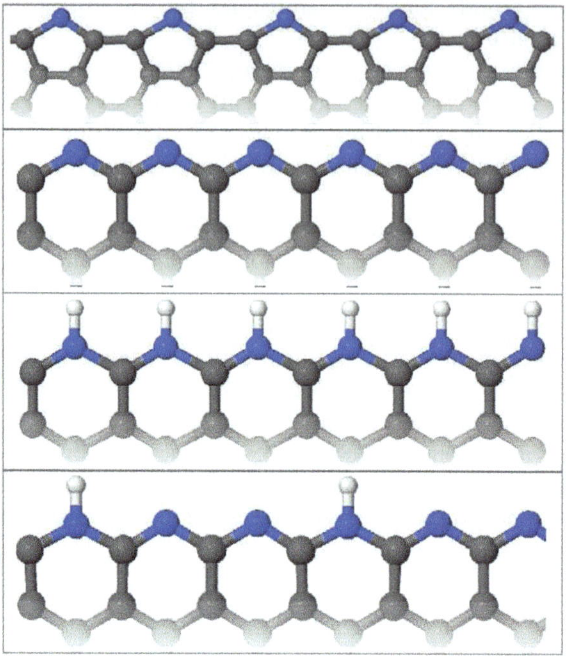

Fig. 6 Nitrogen-terminated edge structures (a) unstable $\langle 1\bar{1}00\rangle$ arm-chair structure (formation enthalpy of +4.62 eV per N), (b) stable $\langle 2\bar{1}\bar{1}0\rangle$ zigzag nitrogen-terminated edge (formation enthalpy of +0.81 eV per N), (c) fully hydrogenated $\langle 2\bar{1}\bar{1}0\rangle$ zigzag nitrogen-terminated edge (formation enthalpy of +0.23 eV per N), and significantly more stable than all of the others, (d) partially hydrogenated $\langle 2\bar{1}\bar{1}0\rangle$ zigzag nitrogen-terminated edge (formation enthalpy of −0.10 eV per N), a nitrogen analogue of the z_{211} hydrogenated zigzag edge.[59]

decreasing the width of nanoribbon we have simulated. The resulting variation in formation energy is shown in Fig. 7.

While decreasing the spacing between edges significantly decreases their formation enthalpy at high edge densities, at nitrogen concentrations in the 10–20% range the enthalpy decrease is relatively minor.

Another way to stabilize zigzag nitrogen-terminated edges is through interlayer interaction, for example if nitrogen-terminated azacarbons are attached to the surface of graphene or carbon nanotubes; preliminary calculations (not reported here) suggest this stabilizes such nitrogen-terminated edges significantly.

3.1 Hydrogen stabilisation of nitrogen-terminated edges

Finally we consider another novel way to further stabilize nitrogen-terminated zigzag edges, through modifying the edge chemistry. Noting that nitrogen is isoelectronic with (CH), we can draw analogies between nitrogen-terminated zigzag edges and hydrogenated zigzag edges.

DFT studies of hydrogen terminated zigzag edges demonstrated that the stability of single hydrogenated edges can be greatly increased by introducing an additional hydrogen to every third edge carbon atom. This converts the atom to sp^3 coordination and stabilizes the edge, opening up the band gap and removing the unstable dispersed Fermi level state that localizes along a singly hydrogenated

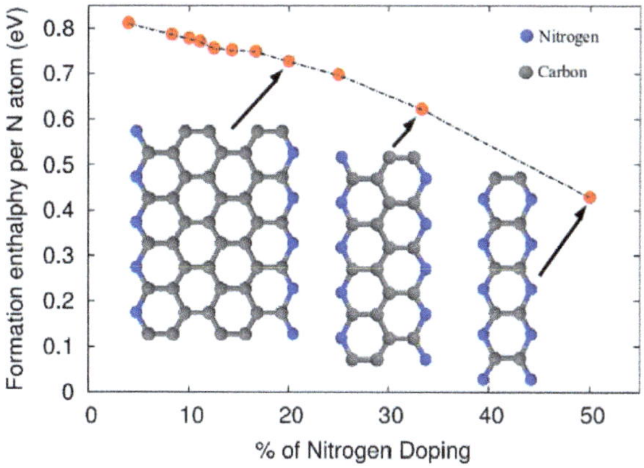

Fig. 7 Calculated effect of decreasing ribbon width on formation enthalpy of nitrogen terminated $\langle 2\bar{1}\bar{1}0 \rangle$ zigzag edges in graphene. Formation enthalpy drops from +0.81 eV per N at [N] = 4% to +0.46 eV per N at [N] = 50%.

edge.[59] A similar effect was noted for the thermodynamically preferred reconstructed Klein edges.[56]

By analogy, hydrogenating every third nitrogen atom along a nitrogen-terminated zigzag edge should also stabilize that edge structure. We have tested this effect with the nitrogen edge-terminated graphene nanoribbons given in Fig. 7, both with a fully hydrogenated $(NH)_3$ edge, and with every third nitrogen hydrogenated $N_2(NH)$. The results are shown with filled circles in Fig. 8. The stabilizing effect on the edges is remarkable, dropping the formation energies at even low edge concentrations.

Just as for the hydrogenated carbon edge, the greatest stabilization comes from hydrogenating every third nitrogen atom, which goes so far as to render the formation enthalpy of the edge negative. Unlike the hydrogenated edge,[56] there is still significant stabilization upon total hydrogenation of the nitrogen edge. For the isolated edges at low % N, the formation enthalpy per N atom drops from +0.79 eV per N to +0.23 eV per N for the fully hydrogenated edge, and −0.10 eV per N for the edge with every third nitrogen hydrogenated. These formation energies mean that the binding enthalpy of $1/2H_2$ to an unfunctionalised zigzag nitrogen-terminated edge is a massive 2.67 eV, *i.e.* the unfunctionalised nitrogen zigzag edge will bc highly chemically reactive.

Whilc thc formation enthalpies of these edge structures are now comparable and lower than those for substitutional nitrogen, a better measure in the presence of hydrogen is obtained by considering a molecular hydrogen gas atmosphere around the graphene edge. The calculated total edge formation energy E_{edge} can be adjusted using the hydrogen chemical potential μ_{H_2} and the hydrogen edge density ρ_H, to give a first approximation to the relative Gibb's free energy of the edge, G_{edge},

$$G_{edge} = E_{edge} - \rho_H \mu_{H_2}/2. \qquad (2)$$

The hydrogen chemical potential μ_{H_2} depends on the pressure and temperature of the system.[59] As an indication, at ambient conditions the chemical

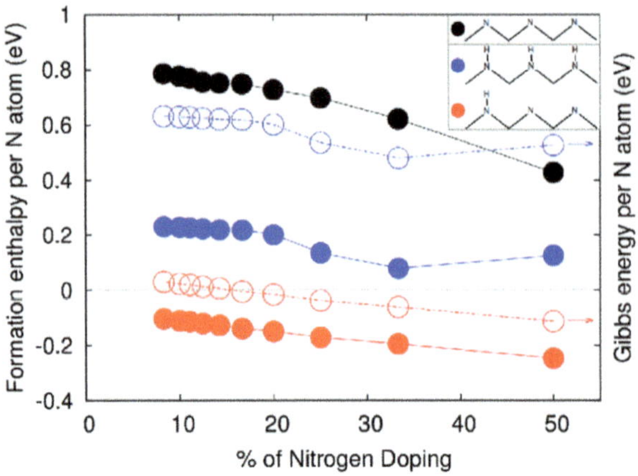

Fig. 8 Effect of total $(NH)_3$ (blue) and partial $(N_2(NH))$ (red) hydrogenation on the stability of nitrogen terminated $\langle 2\bar{1}\bar{1}0 \rangle$ zigzag edges in graphene (for edge structures see Fig. 6C and D respectively). The formation enthalpy of pure N edges (black filled circles) is included for comparison. Formation enthalpies (filled circles) and Gibbs free energies (empty circles) are shown. A chemical potential of 0.8 eV is used for H_2 when calculating the Gibbs energies (see text).

potential μ_{H_2} at 300 K and partial H_2 pressure in air of $P_{H_2} \approx 10^{-4}$ mbar gives $\mu_{H_2} \approx -0.4$ eV.[62] For both decreasing P_{H_2} and increasing temperatures, the chemical potential decreases. Low temperature and higher pressure conditions ($\mu_{H_2} \approx -1.2$ to -0.5 eV) are currently of interest for cost efficient CVD graphene growth on metal surfaces.[63–66] In Fig. 8 the empty circles indicate the Gibbs free energy of formation for the fully and partially hydrogenated nitrogen zigzag edges, taking a typical hydrogen chemical potential of -0.8 eV.

This shows that the different nitrogen terminated zigzag edges, from non- to fully hydrogenated, span a similar energy range to the formation enthalpy for substitutional nitrogen at varying densities.

Clearly there are further terms that should be incorporated into the Gibb's free energy (such as edge entropy), and it is difficult to extrapolate a general argument much further than this, since thereafter experimental specifics such as growth conditions and gas atmosphere become predominant. However broadly, these edge results confirm our hypothesis that as nitrogen concentration increases in sp^2-nanocarbon samples, there should be a thermodynamic transition in stability between substitutional nitrogen, in favour of segregation to nitrogen terminated zigzag edges, almost certainly with some degree of heteroatom functionalisation. In sp^3-like ta-C, this effect will tend to favour the formation of nitrogen-terminated regions of sp^2-C, favouring a transition from sp^3 to sp^2 as observed. The formation energies of the hydrogenated nitrogen-terminated edges are in the right range. Notably, if feed gas chemical potential is important in determining the precise concentration at which this stability transition between substitutional and edge nitrogen occurs, this may also help to explain the different nitrogen concentrations in Fig. 4 at which the ta-C undergoes transition from sp^3 to sp^2, if this is dependent on the growth conditions.

4 Discussion

To summarise thus far, the calculations suggest that substitutional nitrogen should only be stable in sp^2 carbon layers at relatively low concentrations. At high concentrations ($\sim$15–20% and above) the presence of the nitrogen would cause the collapse of the π-cloud and breakdown of the sp^2 network *via* sp^3-cross linking between layers. However before this occurs, at intermediate concentrations (5–15%) it becomes thermodynamically favoured for nitrogen to segregate to nitrogen-terminated zigzag edges, for example at edges of sp^2-C patches or around holes in sp^2-C sheets. We emphasise that this is a relatively general thermodynamic analysis aimed at identifying trends in driving forces and thermodynamic concentration limits. Clearly experimental growth and dynamic implantation studies require additional input on the reaction kinetics and external environmental factors.

The model fits well with our recent HRTEM/EELS observations of nitrogen-doped single-walled carbon nanotubes (SWNTs)[24] (see Fig. 9). Bulk to nanometer-resolved EELS measurements of nitrogen-doped SWNTs typically show a pyramidal N1s spectra centred around 407 eV, resembling Fig. 9C(i and ii).[30,67] EELS simulations suggest that this signal comes from the σ^* response of pyridinic nitrogen atoms,[24] and spatially mapping the EELS signal shows this signal comes primarily from amorphous and disordered regions in the sample, and amorphous material on the nanotube surfaces. In fact, atomic resolution EELS studies of pristine tube surfaces find very low nitrogen concentrations (typically much less than 1%), and EELS spectra of individual substitutional defects shows a very different spectral form (Fig. 9C(iii and iv)). In these cases there is a large, sharper peak centred around 400–403 eV, also ascribed primarily to σ^* bonding, but of substitutional nitrogen. The observed peak splitting in Fig. 9C(iii) is explained *via* local lattice asymmetry in the σ bonds when the substitutional nitrogen sits next to a lattice defect such as in Fig. 9E.[24]

Thus in the case of nitrogen-doped SWNTs it appears that the non-local (down to nanometer-resolved) spectroscopic picture of nitrogen behavior is misleading and does not represent nitrogen incorporation in the nanotube itself. Instead the HRTEM, EELS and modeling results point towards a picture whereby nitrogen-doped SWNTs have very low concentrations of substitutional nitrogen doping. Thereafter the nitrogen level in the sample is increased *via* the presence of pyridinic nitrogen in the surface material attached to the tube, likely in the form of nitrogen-terminated zigzag edges of polycyclic carbon patches.

This strongly resembles the behavior of oxygen impurities and associated functionalisation for single-walled[68] and multi-walled nanotubes,[69–72] carbon fibres,[73] and graphene oxide.[74–76] These studies demonstrate that many of the observed oxygen-derived functional groups previously assigned to these nanocarbon objects are actually associated with oxidative debris; highly oxidized polycyclic aromatic carbon patches attached to the surface of the nanobject, which can often be removed *via* base washing, an approach which would be interesting to try with the nitrogen-doped samples. In the case of graphene oxide it has been shown that the oxidative debris is inherently electroactive, and may in fact account for the majority of the electroactivity which makes graphene oxide an interesting material for energy storage applications such as supercapacitors.[76] It

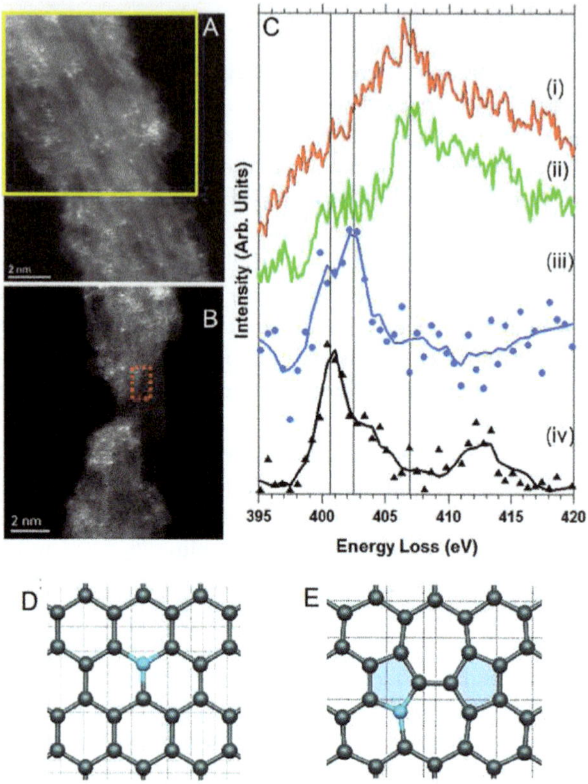

Fig. 9 Spatially-resolved EELS from nitrogen containing SWNTs and associated debris. (A) and (B) HAADF images of a bundle of single-walled nanotubes covered by an amorphous material. (C) N1s ELNES spectra obtained from (i) a cumulative EELS spectrum over the area in A marked by the yellow rectangle, (ii) sum of 10 spatially localised EELS spectra from the top-left corner of the red dotted rectangle in B. (i) and (ii) correspond to typical 'bulk' EELS spectra for nitrogen-doped SWNTs. In contrast, (iii) shows the spectrum obtained from a single pixel (non-averaged) at a substitutional nitrogen atom in the wall of a SWNT as shown in D, and (iv) a similar spectrum from substitutional nitrogen neighbouring a topological defect, as shown in E ("pyrrolic" N). These local substitutional defects clearly show a markedly different peak profile to the bulk measurements, in agreement with our proposition that the majority of nitrogen in the sample is not embedded in the nanotube wall but is instead in the intervening "amorphous" material. (Figure adapted from ref. 24.)

seems possible that the enhanced supercapacitive performance seen for nitrogen-doped nanotubes[5,7,16,17,77,78] for example, may also be strongly affected by the presence of nitrogen-terminated polycyclic fragments attached to the material surface.

There are also structural implications for the morphology of heavily nitrogen-doped MWNTs. Addition of nitrogen to the feedstock gas in CVD nanotube growth modifies the classic MWNT geometry, from walls aligned parallel to the nanotube axis to either a compartmentalised structure ('bamboo' nanotubes), or a structure where the nanotube walls are inclined at a fixed angle to the nanotube axis ('herringbone' nanotubes), similar to a series of stacked cones (with or without a hollow core, see Fig. 10). This second structure has potentially very different physical and chemical behavior to conventional parallel wall tubes, not

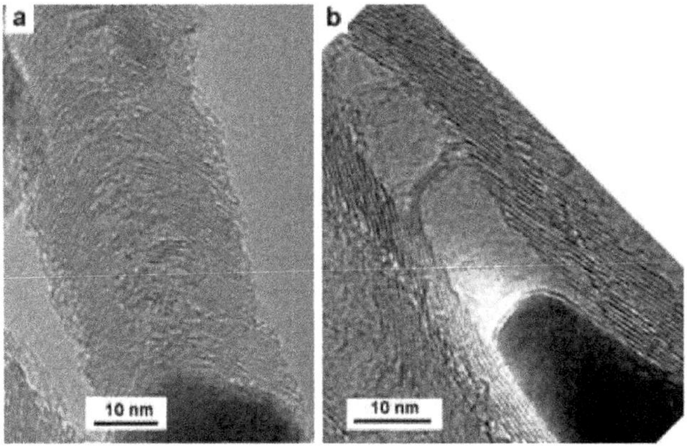

Fig. 10 HRTEM showing typical stacked-cup structure of nitrogen-doped herringbone multi-walled carbon nanotubes (figure from ref. 79).

least in terms of surface activity. Whereas conventional tubes surface terminate with a graphene basal plane, the graphene planes in herringbone tubes terminate at the tube surface, and therefore (excepting the case of layer folding) the nanotube surface consists of basal plane edges. Our current results suggest a mechanism of nitrogen stabilization of edge sites and may be relevant in the structural stability and subsequent surface chemistry of herringbone tubes.

The current article has focused on the interplay between substitutional and pyridinic-type nitrogen. The chemistry of the nitrogen–carbon system is more complex and richer than this, for example we have not considered here the capability of nitrogen to stabilize regions of high curvature within an sp^2-carbon lattice *via* the formation of pentagons[46,52] and pentagon pairs,[80] or the interaction between nitrogen and carbon nanomaterial catalysts, and resultant effects on the growth kinetics and subsequent morphology. Nonetheless we hope to have identified some key general points in nitrogen behaviour in nanocarbons. The first of these is the general observation of a thermodynamic driving force towards segregation at $\langle 2\bar{1}\bar{1}0\rangle$ zigzag edges in sp^2-C networks above a certain nitrogen concentration, resulting in N-edge terminated PAHs and hole formation, and stabilising N-edge terminated sp^2 material in nitrogen-doped carbon thin films. The second is that much of the nitrogen observed in 'N-doped' nanocarbons may instead be present in amorphous and polycyclic carbon species attached to the nanocarbon surface, notably as pyridinic nitrogen at edge sites in such material. This attached material may be responsible for some of the improved electroactive behaviour observed in nitrogen-doped nanocarbons. We have also tried to highlight the complementarity between atomically resolved EELS, time resolved XPS and DFT calculations, in unravelling the complex and fascinating interactions between nitrogen and carbon.

5 Computational method

Spin polarized density functional (DFT) calculations using LDA-PW92 (ref. 81) were performed as implemented in the AIMPRO code.[82–84] The calculations were carried out using supercells, fitting the charge density to plane waves with an energy cutoff of 150 Hartrees. Relativistic pseudopotentials generated by

Hartwigsen, Goedecker and Hutter were used.[85] 22 and 40 independent Gaussian functions were used as basis sets for carbon and nitrogen respectively, 12 for hydrogen. Periodic boundary conditions have been applied by using supercells. Supercell sizes have been checked and chosen to be sufficiently large (vacuum distance between all 1-D and 2-D structures larger than 15 Å) to avoid interactions. The bilayer graphene systems were arranged with AB stacking, with cells containing up to 64 atoms depending on system symmetry. A fine k-point grid was chosen (from $2 \times 2 \times 1$ to $8 \times 8 \times 1$ for bilayer graphene depending on cell size, $8 \times 1 \times 1$ for graphene nanoribbons), and electronic level occupation was obtained using a Fermi occupation function with $kT = 0.04$ eV. Absolute energies are converged in the self-consistency cycle to better than 10^{-5} Ha. Atomic positions and lattice parameters were geometrically optimized until the maximum atomic position change in a given iteration dropped below 10^{-4} a_0 (a_0: Bohr radius).

Acknowledgements

We would like to acknowledge TUBITAK-2219 Abroad Post-Doctoral Research Funding Programme (Ref. no. 1059B191301289) for supporting DE. CB and MS are supported by the Belgian Fund for Scientific Research (FSR-FNRS) under contract 'CHEMOGRAPHENE' (no. 2.4577.11). CPE, PW acknowledge the French SPRINT project ANR-10-BLAN-0819. RA acknowledges the European Union Seventh Framework Program under Grant Agreement 312483 – ESTEEM2 (Integrated Infrastructure Initiative – I3), and all authors the COST network MP0901 NanoTP. We thank Odile Stephan, Katia March, Matthieu Kociak and Annick Loiseau for stimulating discussions. CE and CB thank PHC Tournesol 31160RC for funding. R.A. acknowledges funding from the Spanish Ministerio de Economia y Competitividad (FIS2013-46159-C3-3-P).

References

1 Y. Ma, A. Foster, A. V. Krasheninnikov and R. Nieminen, *Phys. Rev. B: Condens. Matter Mater. Phys.*, 2005, **72**, 205416.
2 C. Zhou, J. Kong, E. Yenilmez and H. Dai, *Science*, 2000, **290**, 1552–1555.
3 Y. K. Yap, *B-C-N Nanotubes and Related Nanostructures*, Springer, 2009.
4 M. Monthioux, *Introduction, Carbon Meta-Nanotubes: Synthesis, Properties and Applications*, John Wiley & Sons, 2011.
5 P. Ayala, R. Arenal, A. Loiseau, A. Rubio and T. Pichler, *Rev. Mod. Phys.*, 2010, **82**, 1843–1885.
6 Q. Jiao, L. Hao and Y. Zhao, *Carbon*, 2013, **61**, 647–649.
7 C. P. Ewels and M. Glerup, *J. Nanosci. Nanotechnol.*, 2005, **5**, 1345.
8 P. Ayala, R. Arenal, M. Rummeli, A. Rubio and T. Pichler, *Carbon*, 2010, **48**, 575–586.
9 H. R. Barzegar, E. Gracia-espino, T. Shari, F. Nitze and T. Wagberg, *J. Phys. Chem. C*, 2013, **117**, 25805–25816.
10 A. Lopez-Bezanilla, *J. Phys. Chem. C*, 2014, **118**, 1472–1477.
11 A. P. Alegaonkar, A. Kumar, S. H. Patil, K. R. Patil, S. K. Pardeshi and P. S. Alegaonkar, *J. Phys. Chem. C*, 2013, **117**(51), 27105–27113.

12 M. A. Kanygin, O. V. Sedelnikova, I. P. Asanov, L. G. Bulusheva, A. V. Okotrub, P. P. Kuzhir, A. O. Plyushch, S. A. Maksimenko, K. N. Lapko, A. A. Sokol, O. A. Ivashkevich and P. Lambin, *J. Appl. Phys.*, 2013, **113**, 144315.
13 V. Krstic, G. L. J. A. Rikken, P. Bernier, S. Roth and M. Glerup, *Europhys. Lett.*, 2007, **77**, 37001.
14 A. Zhao, J. Masa, W. Schuhmann and W. Xia, *J. Phys. Chem. C*, 2013, **117**, 2428324291.
15 Y. Tang, S. C. Burkert, Y. Zhao, W. A. Saidi and A. Star, *J. Phys. Chem. C*, 2013, **117**, 25213–25221.
16 R. Mi, H. Liu, H. Wang, K.-W. Wong, J. Mei, Y. Chen, W.-M. Lau and H. Yan, *Carbon*, 2014, **67**, 744–752.
17 S. Hussain, R. Amade, E. Jover and E. Bertran, *J. Mater. Sci.*, 2013, **48**, 7620–7628.
18 J.-J. Adjizian, R. Leghrib, A. A. Koos, I. Suarez-Martinez, A. Crossley, P. Wagner, N. Grobert, E. Llobet and C. P. Ewels, *Carbon*, 2014, **66**, 662–673.
19 Q. Sheng, R. Liu and J. Zheng, *Bioelectrochemistry*, 2013, **94**, 39–46.
20 C. Shan, W. Zhao, X. L. Lu, D. J. O'Brien, Y. Li, Z. Cao, A. L. Elias, R. Cruz-Silva, M. Terrones, B. Wei and J. Suhr, *Nano Lett.*, 2013, **13**, 5514–5520.
21 R. Droppa Jr, P. Hammer, A. C. M. Carvalho, M. C. dos Santos and F. Alvarez, *J. Non-Cryst. Solids*, 2002, **299**, 874–879.
22 M. Glerup, J. Steinmetz, D. Samaille, O. Stephan, S. Enouz, A. Loiseau, S. Roth and P. Bernier, *Chem. Phys. Lett.*, 2004, **387**, 193.
23 T. Susi, J. Kotakoski, R. Arenal, S. Kurasch, H. Jiang, V. Skakalova, O. Stephan, A. V. Krasheninnikov, E. I. Kauppinen, U. Kaiser, *et al.*, *ACS Nano*, 2012, **6**, 8837–8846.
24 R. Arenal, K. March, C. P. Ewels, X. Roquefelte, M. Kociak, A. Loiseau and O. Stephan, *Nano Lett.*, 2014, DOI: 10.1021/nl501645g.
25 C. J. Lee, S. C. Lyu, H.-W. Kim, J. H. Lee and K. I. Cho, *Chem. Phys. Lett.*, 2002, **359**, 115.
26 J. C. Meyer, S. Kurasch, H. J. Park, V. Skakalova, D. Knzel, A. Gro, A. Chuvilin, G. Algara-Siller, S. Roth, T. Iwasaki, U. Starke, J. H. Smet and U. Kaiser, *Nat. Mater.*, 2011, **10**, 209–215.
27 M. Scardamaglia, B. Aleman, M. Amati, C. Ewels, P. Pochet, N. Reckinger, J. F. Colomer, T. Skaltsas, N. Tagmatarchis, R. Synders, L. Gregoratti and C. Bittencourt, *Carbon*, 2014, **73**, 371.
28 U. Bangert, W. Pierce, D. M. Kepaptsoglou, Q. Ramasse, R. Zan, M. H. Gass, J. A. Van den Berg, C. B. Boothroyd, J. Amani and H. Hofsass, *Nano Lett.*, 2013, **13**, 4902–4907.
29 D. Marton, K. J. Boyd, A. H. al Bayati, S. S. Todorov and J. W. Rabalais, *Phys. Rev. Lett.*, 1994, **73**(1), 118.
30 H. Lin, R. Arenal, S. Enouz-Vedrenne, O. Stephan and A. Loiseau, *J. Phys. Chem. C*, 2009, **113**, 9509–9511.
31 R. Arenal, O. Stephan, P. Bruno and D. M. Gruen, *Appl. Phys. Lett.*, 2009, **94**, 111905.
32 H. Wang, T. Maiyalagan and X. Wang, *ACS Catal.*, 2012, **2**, 781–794.
33 J. Casanovas, J. M. Ricart, J. Rubio, F. Illas and J. M. Jime, *J. Am. Chem. Soc.*, 1996, **7863**, 8071–8076.
34 J. M. Ripalda, E. Roman, N. Diaz, L. Galan, I. Montero, G. Comelli, A. Baraldi, S. Lizzit, A. Goldoni and G. Paolucci, *Phys. Rev. B: Condens. Matter Mater. Phys.*, 1999, **60**, 3705–3708.

35 I. Shimoyama, G. Wu, T. Sekiguchi and Y. Baba, *Phys. Rev. B: Condens. Matter Mater. Phys.*, 2000, **62**, 6053–6056.
36 C. Ronning, H. Feldermann, R. Merk, H. Hofsass, P. Reinke and J. U. Thiele, *Phys. Rev. B: Condens. Matter Mater. Phys.*, 1998, **58**(4), 2207.
37 A. Jorio, M. S. Dresselhaus, R. Saito and G. Dresselhaus, *Raman Spectroscopy in Graphene Related Systems*, Wiley-VCH Verlag, 2011.
38 E. S. Reich, C. Thomsen and J. Maultzsch, *Carbon Nanotubes: Basic Concepts and Physical Properties*, Wiley-VCH, 2004.
39 I. Maciel, N. Anderson, M. Pimenta, A. Hartschuh, H. Quian, M. Terrones, H. Terrones, J. Campos-Delgado, A. Rao, L. Novotny and A. Jorio, *Meteorit. Planet. Sci.*, 2008, **7**, 878.
40 F. Villalpando-Paez, A. Zamudio, A. Elias, H. Son, E. Barros, S. Chou, Y. Kim, H. Muramatsu, T. Hayashi, J. Kong, H. Terrones, G. Dresselhaus, M. Endo, M. Terrones and M. Dresselhaus, *Chem. Phys. Lett.*, 2006, **424**, 345–352.
41 M. Glerup, M. Castignolles, M. Holzinger, G. Hug, A. Loiseau and P. Bernier, *Chem. Commun.*, 2003, 2542.
42 I. Florea, O. Ersen, R. Arenal, D. Ihiawakrim, C. Messaoudi, K. Chizari, I. Janowska and C. Pham-Huu, *J. Am. Chem. Soc.*, 2012, **134**, 9672–9680.
43 M. Castignolles, Ph.D. thesis, Universit Montpellier II, France, 2004.
44 J. Robertson and C. A. Davis, *Diamond Relat. Mater.*, 1995, **4**, 441–444.
45 S. Souto, M. Pickholz, M. C. dos Santos and F. Alvarez, *Phys. Rev. B: Condens. Matter Mater. Phys.*, 1998, **57**(4), 2536.
46 H. Sjöström, S. Stafström, M. Boman and J.-E. Sundgren, *Phys. Rev. Lett.*, 1995, **75**, 1336–1339.
47 M. dos Santos and F. Albarez, *Phys. Rev. B: Condens. Matter Mater. Phys.*, 1998, **58**, 13918.
48 S. Stafstrom, *Appl. Phys. Lett.*, 2000, **77**, 3941–3943.
49 W. J. Gammon, D. I. Malyarenjo, O. Kraft, G. L. Hoatson, A. C. Reilly and B. C. Holloway, *Phys. Rev. B: Condens. Matter Mater. Phys.*, 2002, **66**, 153402.
50 W. J. Gammon, G. L. Hoatson, B. C. Holloway, R. L. Vold and A. C. Reilly, *Phys. Rev. B: Condens. Matter Mater. Phys.*, 2003, **68**, 195401.
51 J. Hu, P. Yang and C. M. Lieber, *Phys. Rev. B: Condens. Matter Mater. Phys.*, 1998, **57**, R3185.
52 N. Hellgren, M. Johansson, E. Broitman, L. Hultman and J.-E. Sundgren, *Phys. Rev. B: Condens. Matter Mater. Phys.*, 1999, **59**, 5162.
53 M. Scardamaglia, M. Amati, B. Llorente, P. Mudimela, J.-F. Colomer, J. Ghijsen, C. Ewels, R. Snyders, L. Gregoratti and C. Bittencourt, *Carbon*, 2014, **77**, 319–328.
54 J.-Y. Yi and J. Bernholc, *Phys. Rev. B: Condens. Matter Mater. Phys.*, 1993, **47**, 1708–1711.
55 P. S. Branicio, M. H. Jhon, C. K. Gan and D. J. Srolovitz, *Modell. Simul. Mater. Sci. Eng.*, 2011, **19**, 054002.
56 P. Wagner, V. V. Ivanovskaya, M. Melle-Franco, B. Humbert, J.-J. Adjizian, P. R. Briddon and C. P. Ewels, *Phys. Rev. B: Condens. Matter Mater. Phys.*, 2013, **88**, 094106.
57 P. Wagner, C. P. Ewels, V. V. Ivanovskaya, P. R. Briddon, A. Pateau and B. Humbert, *Phys. Rev. B: Condens. Matter Mater. Phys.*, 2011, **84**, 134110.

58 P. Wagner, C. P. Ewels, J.-J. Adjizian, L. Magaud, P. Pochet, S. Roche, A. Lopez-Bezanilla, V. V. Ivanovskaya, A. Yaya, M. Rayson, P. Briddon and B. Humbert, *J. Phys. Chem. C*, 2013, **117**, 26790–26796.
59 T. Wassmann, A. P. Seitsonen, A. M. Saitta, M. Lazzeri and F. Mauri, *Phys. Rev. Lett.*, 2008, **101**, 096402.
60 C. Jin, F. Lin, K. Suenaga and S. Iijima, *Phys. Rev. Lett.*, 2009, **102**, 195505.
61 E. Machado-Charry, P. Boulanger, L. Genovese, N. Mousseau and P. Pochet, *Appl. Phys. Lett.*, 2012, **101**, 132405.
62 M. W. Chase, *J. Phys. Chem. Ref. Data*, 1998, **9**, 1–1951.
63 X. Li, W. Cai, J. An, S. Kim, J. Nah, D. Yang, R. Piner, A. Velamakanni, I. Jung, E. Tutuc, S. K. Banerjee, L. Colombo and R. S. Ruoff, *Science*, 2009, **324**, 1312–1314.
64 L. Gao, W. Ren, J. Zhao, L.-P. Ma, Z. Chen and H.-M. Cheng, *Appl. Phys. Lett.*, 2010, **97**, 183109.
65 K. Celebi, M. T. Cole, K. B. K. Teo and H. G. Park, *Electrochem. Solid-State Lett.*, 2011, **15**, K1–K4.
66 A. T. Murdock, A. Koos, T. B. Britton, L. Houben, T. Batten, T. Zhang, A. J. Wilkinson, R. E. Dunin-Borkowski, C. E. Lekka and N. Grobert, *ACS Nano*, 2013, **7**, 1351–1359.
67 S. Enouz-Vdrenne, O. Stphan, M. Glerup, J.-L. Cochon, C. Colliex and A. Loiseau, *J. Phys. Chem. C*, 2008, **112**, 16422–16430.
68 L. Shao, G. Tobias, C. G. Salzmann, B. Ballesteros, S. Y. Hong, A. Crossley, B. G. Davis and M. L. H. Green, *Chem. Commun.*, 2007, 5090–5092.
69 R. Verdejo, S. Lamoriniere, B. Cottam, A. Bismarck and M. Shaffer, *Chem. Commun.*, 2007, 513–515.
70 S. Fogden, R. Verdejo, B. Cottam and M. Shaffer, *Chem. Phys. Lett.*, 2008, **460**, 162–167.
71 C. Salzmann, S. Llewellyn, G. Tobias, M. Ward, Y. Huh and M. Green, *Adv. Mater.*, 2007, **19**, 883–887.
72 Z. Wang, M. D. Shirley, S. T. Meikle, R. L. Whitby and S. V. Mikhalovsky, *Carbon*, 2009, **47**, 73–79.
73 Z. Wu, C. U. Pittman Jr and S. D. Gardner, *Carbon*, 1995, **33**, 597–605.
74 J. P. Rourke, P. A. Pandey, J. J. Moore, M. Bates, I. A. Kinloch, R. J. Young and N. R. Wilson, *Angew. Chem., Int. Ed.*, 2011, **50**, 3173–3177.
75 A. F. Faria, D. S. T. Martinez, A. C. M. Moraes, M. E. H. Maia da Costa, E. B. Barros, A. G. Souza Filho, A. J. Paula and O. L. Alves, *Chem. Mater.*, 2012, **24**, 4080–4087.
76 A. Bonanni, A. Ambrosi, C. K. Chua and M. Pumera, *ACS Nano*, 2014, **8**, 4197–4204.
77 L. G. Bulusheva, E. O. Fedorovskaya, A. G. Kurenya and A. V. Okotrub, *Phys. Status Solidi B*, 2013, **250**, 2586–2591.
78 R. A. Fisher, M. R. Watt and W. Jud Ready, *ECS J. Solid State Sci. Technol.*, 2013, **2**, M3170–M3177.
79 M. Monthioux, L. Noe, L. Dussault, J.-C. Dupin, N. Latorre, T. Ubieto, E. Romeo, C. Royo, A. Monzon and C. Guimon, *J. Mater. Chem.*, 2007, **17**, 4611–4618.
80 C. Ewels, *Nano Lett.*, 2006, **6**, 890–895.
81 J. Perdew and P. Wang, *Phys. Rev. B: Condens. Matter Mater. Phys.*, 1992, **45**, 13244–13249.

82 P. R. Briddon and R. Jones, *Phys. Status Solidi B*, 2000, **217**, 131–171.
83 M. J. Rayson and P. R. Briddon, *Phys. Rev. B: Condens. Matter Mater. Phys.*, 2009, **80**, 205104.
84 P. R. Briddon and M. J. Rayson, *Phys. Status Solidi B*, 2011, **248**, 1309–1318.
85 C. Hartwigsen, S. Goedecker and J. Hutter, *Phys. Rev. B: Condens. Matter Mater. Phys.*, 1998, **58**, 3641.

Faraday Discussions

DISCUSSIONS

Doping and Theory: general discussion

Philip Davies, Irena Kratochvílová, Chris Ewels, Tariq Aqeel, Nazario Martin, Vladimir Fal'ko, Toshiaki Enoki, Antonio Rodriguez-Fortea, Katsumi Kaneko, Ayse Turak, Carlo Buono, Valeri Kovalenko, Pulickel Ajayan, Malcolm Heggie, Varsha Khare, Andreas Hirsch, Thurid Gspann, Alan Windle, Pedro Costa, Milo Shaffer, Juan Casado, Mary Chan-Park and Rosa Busquets

DOI: 10.1039/c4fd90044h

Toshiaki Enoki opened the discussion of the paper by Vladimir Fal'ko: The out-of-plane phonon mode is very soft and affected by the inter-layer interaction. Accordingly, moiré pattern, wrinkles, *etc.* in multi-layer graphene affect the out-of-plane phonon mode. It is interesting to investigate the correlation between the mutual angle in the moiré; pattern and the phonon mode is interesting. I would like to have information on this if the author has investigated it.

Vladimir Fal'ko answered: The effect of the moiré pattern on such modes is that phonon lines in the Raman spectrum of graphene acquire satellites, corresponding to the Bragg scattering of a gamma-point phonon by the reciprocal lattice vectors of the moiré superlattice. Such an effect has been seen in a pair of slightly misaligned graphene sheets.

Pulickel Ajayan said: Does your technique allow observation and understanding of larger area mismatch effects at low magnifications?

Vladimir Fal'ko responded: In principle, yes.

Pulickel Ajayan remarked: Are you able to observe such defects even at lower magnifications, using your interference techniques?

Vladimir Fal'ko responded: The spacial resolution needed to observe the dislocation is determined by the period of the moiré pattern, which is much longer than the actual period of the crystalline lattice. Hence, fine (atomic-scale) resolution in STM is not needed.

Pedro Costa remarked: What would be the effect of grain boundaries given their pronounced existence in CVD grown graphene? How extended would their influence be felt (*i.e.* relaxed *versus* strained areas along the boundary line)?

Fig. 1 Moiré pattern visualization of a non-basal edge dislocation. (a) Non-basal dislocation in 6/8 form at centre of graphene. (b) Graphene, (c) non-basal dislocation in a (002) twist boundary obtained by superposition of (a) with (b) at a small angle.

Nazario Martin commented: Did you observe different electronic densities on the valley and top positions of the moiré pattern?

Vladimir Fal'ko answered: For neutral graphene, no. However, when graphene is doped, added charges would, certainly, be distributed inhomogeneously across the moiré supercell. Such inhomogeneity would be characteristic for moiré patterns in graphene grown on metal surfaces, where large charge transfer takes place, between graphene and the metal.

Nazario Martin said: In this case, is there an electron transfer between the boron nitride and the graphene?

Vladimir Fal'ko responded: There is no charge transfer between graphene and hexagonal boron nitride (hBN), since hBN has a huge band gap.

Malcolm Heggie commented: This is a beautiful theoretical paper,[1] which ends by working through what appears to be a novel mechanism for visualizing dislocations. In 2002, I published on the non-basal (5/7) dislocations discussed (and another 6/8 version) with Chris Ewels[2] making regrettably little reference to the literature. Helped by the comments of a reviewer of a subsequent paper on buckling and folding,[3] I researched the original studies dislocations in graphite, which informed the conference dinner lecture I delivered to the Extended Defects in Semiconductors at Brighton in 2010. It turned out that the magnifying effect of moiré patterns for visualizing dislocations was already well developed in the 1950s, enabling the low resolution (*ca.* 1 nm) electron microscopes of the day to image dislocations.[4,5] Below I reproduce the three slides I used to explain the observation of the non-basal dislocation in graphite, using isotropic elasticity theory to generate the 6/8 dislocation of Fig. 1(a). The interference pattern of Fig. 1(c) gives the highly magnified representation of the dislocation, which Cosma et al. report[1] in their Fig. 3b. In 1959, Dawson and Follett reported[6] the experimental measurement of one of the sets of interference fringes shown here. They inferred the existence of the non-basal dislocation in graphite and were the first to speculate on the 5/7 core structure, to the best of my knowledge.

1 D. A. Cosma, J. R. Wallbank, V. Cheianov and V. I. Fal'ko, *Faraday Disc.*, 2014, **173**, DOI:10.1039/C4FD00146J.
2 C. P. Ewels, M. I. Heggie and P. R. Briddon, *Chem. Phys. Lett.*, 2002, **351**, 178.
3 M. I. Heggie, I. Suarez-Martinez, C.R. Davidson, G. L. Haffenden, *J. Nucl. Mat.*, 2011, **413**, 150.

4 H. Hashimoto and R. Uyeda, *Acta Cryst.*, 1957, **10**, 143.
5 G. A. Bassett, J. W. Menter and D. W. Pashley, *Proc. Roy. Soc. A*, 1958, **246**, 345.
6 I. M. Dawson and E. A. C. Follett, *Proc. Roy. Soc. A,* 1959, **253**, 390.

Toshiaki Enoki remarked: The orbital susceptibility of a single graphene sheet is $-\infty$ when the magnetic field is applied perpendicular to the sheet. The inter-sheet interaction modifies the orbital susceptibility. The presence of inter-sheet interaction with two sheets being mutually rotated is expected to have the mutual angle dependence in the orbital susceptibility. I would like to have information on this if the author has investigated it.

Vladimir Fal'ko answered: We studied orbital magnetism of electrons affected by the moiré superlattice. There might be a strong signal (sharp peak in susceptibility) in the case when chemical potential is near the edge of the first moiré miniband, whereas a peak in diamagnetic susceptibility exactly at the edge between conduction and valence bands of graphene is preserved.

Nazario Martin commented: In our paper on graphene on ruthenium,[1] doped with TCNQ we find electron transfer, we get spin orientated electrons on the TCNQ molecules on the graphene forming domains on the surface, thus producing magnetism.

1 M. Garnica, D. Stradi, S. Barja, F. Calleja, C. Díaz, M. Alcamí, N. Martín, A. L. Vázquez de Parga, F. Martín and R. Miranda, *Nature Physics*, 2013, **9**, 368.

Mary Chan-Park opened the discussion of the paper by Katsumi Kaneko: Would there be a difference in the behavior if you used purely metallic or purely semi-conducting nanotubes?

Katsumi Kaneko responded: As we do not measure the electrical conductivity of pure metallic or semiconductive SWCNT, I cannot reply with a concrete answer yet. My assumption is as follows. We use the bucky paper form sample for the electrical conductivity measurements. The intertube electrical contact is an important key in both the metallic and semiconductive SWCNT bundles, although the energy barrier of the electrical contact depends on the metallic or semiconductive state. Even metallic SWCNT bundles should have the conduction energy barrier for intertube hopping in the lower pressure region, showing a semiconductive behavior. Then, the temperature dependence of the electrical conductivity similar to semiconductor SWCNT bundles should occur. However, the changing temperature from semiconductive to metallic behavior depends on the metallic or semiconductive SWCNT.

Ayse Turak asked: Given that the DHN and DNN are roughly similar in size, do you have an explanation for why there is greater disorder in the DHN carbon nanotubes than that observed for the DNN ones?

Katsumi Kaneko answered: At first I must make an apology for a mistake in the figure caption of Figure 2 in the paper; b and d must be exchanged with each other. Doping naphthalene whose size is the smallest decreases the X-ray intensity more markedly than doping of DNN or DHN; the molecular size of DHN is

0.17 nm^3, being larger than that of naphthalene, while that of DNN is 0.19nm^3. One plausible reason is that N-molecules can be inserted even in the central bundle sites, whereas DNN and DHN molecules are doped in the interstitial positions near the bundle surface. However, still we have no clear evidence on this explanation.

Ayse Turak said: There appears to be little correlation between the superlattice disorder and the conductivity, given that DNN and N have roughly the same disorder with different effects on the conductivity. Can you comment on the conductivity mechanism for the tube bundles and why the conductivity is uncorrelated to the disorder.

Katsumi Kaneko replied: The electrical conductivity is governed by charge carrier number and mobility. The mobility is deeply associated with the lattice ordering, because the charge carriers are scattered by the disordered sites. The most striking effect in the doping of naphthalene and the derivative molecules comes from controlling of the charge carrier number. The electron acceptor molecules can create positive hole carriers more efficiently and thereby DNN and N increase the charge carrier number, while DHN annihilate the positive holes. If the mobility is constant, the electrical conductivity is in the order of DNN, N, and DHN. However, the mobility must be influenced by the crystallinity. Then we need to take into account the crystalline structure, as you suggested. In this case, we do not determine the carrier number and mobility separately, and then we cannot give an explicit explanation so far. However, the charge carriers in the SWCNT bundle migrate the conducting paths of the best electronic contacting place, which is not necessarily associated with the superlattice structure. The electrical conductivities of the N and ND doped SWCNT, probably, are not clearly correlated with the super lattice structure.

Chris Ewels commented: You describe very nicely the effect of naphthalene and its derivatives on the intra-nanotube bundle behaviour. Do you have an idea of how it might be affecting *inter*-nanotube bundle behaviour?

Katsumi Kaneko replied: Naphthalene molecules can occupy the interstitial pore space of the carbon nanotube bundle, giving rise to stable charge transfer interactions. The naphthalene molecules in the bundle cannot be desorbed even on heating at 473 K under UHV conditions. On the contrary, naphthalene molecules are physically adsorbed on the external surface of the bundle, giving rise to almost no direct affect on the inter-nanotube bundle behavior. However, the doping of naphthalene molecules in the interstitial pore spaces induces debundling as shown in the SEM image. The doping of the naphthalene in the interstitial spaces affects the inter-nanotube bundle behavior too.

Juan Casado asked: What would you expect if using flexible molecules instead of naphthalene?

Katsumi Kaneko responded: Doping of a flexible molecule having a conjugated electron system should give an interesting result. I expect that flexible molecules could achieve a stable accommodation configuration in the SWCNT bundle. It should be a quite interesting future research subject.

Nazario Martin remarked: How did you determine that it was a charge–transfer interaction as opposed to simple π–π interactions? Probably both supramolecular interactions are occurring simultaneously.

Katsumi Kaneko answered: We observed the Raman peak-shifts of RBM and G bands and a marked hyperchromic effect in the optical absorption. It is well-known that the observed Raman shifts come from the charge-transfer interaction and the hyperchromic effect is caused by the mixing of the electron orbitals between electron donor and acceptor molecules. Thus we conclude that the charge transfer interaction is dominant. As we did not take into account the π–π interactions, we need to analyze our data with the relevance to the π–π interactions also in the future study.

Nazario Martin said: Did you try to use other stronger electron donors or acceptors? If so, is this behaviour a general effect?

Katsumi Kaneko replied: Yes we doped with TCNQ as a stronger electron acceptor. However, TCNQ did not give a marked effect compared with naphthalene and its derivative. Probably the TCNQ molecule is too large to be stably accommodated in the interstitial spaces of the nanotube bundle. If we anneal the TCNQ-doped nanotube bundles, for example, possibly we can observe a stronger effect. This kind of experiment is necessary in the future.

Toshiaki Enoki asked: The Mott transition is governed by the competition between the kinetic energy and the electron correlation. Meanwhile variable range hopping takes place in the disordered metal, in which serious randomness makes electrons localized. There is inconsistency between the explanation of the Mott transition and the variable range hopping, which is confirmed in the experiment. The explanation on the conductivity is wrong. The author should revise the explanation.

Katsumi Kaneko replied: One of the reasons why I attend this Faraday discussion is to get a more essential understanding on this unique bahavior of the temperature dependence of the electrical conductivity. The comment is really appreciated and I will correct the explanation after this comment. I can plan to do a future study on the right direction and hope to have future collaborative work.

Malcolm Heggie remarked: Can you fit TCNQ in between your carbon nanotubes?

Katsumi Kaneko answered: The molecular size of TCNQ is larger than naphthalene molecules and we need to search for the optimum occlusion conditions for TCNQ molecules. As TCNQ has a planar structure, the SWCNT bundle and TCNQ molecules should form an assembly structure; the assembly is expected to show attractive properties.

Nazario Martin commented: As soon as you use a benzene ring fused to the TCNQ molecule, it becomes highly distorted. You could use some DCNQI derivatives which are planar electron acceptors.

Katsumi Kaneko replied: The comment is very helpful for the next stage of the study. We want to use DCNQI derivatives instead of TCNQ to obtain the doped SWCNT bundles of splendid electrical conductivity, if we can get hold of the DCNQI derivatives.

Tariq Aqeel communicated: Is the distribution of the doped nitrogen atom in the carbon structure uniform or random?

Katsumi Kaneko communicated in reply: We observed explicit effects on the physical properties of the SWCNT bundles by doping naphthalene and the derivative molecules, as shown in my presentation. However, our doping method is still not established for the ordered SWCNT bundles in which doped molecules are uniformly distributed. We must try to challenge this subject.

Chris Ewels opened the discussion of the paper by Juan Casado: Your studies are performed in the solid state. Do you expect to see a difference in the C70 orientation when the CPP/C70 hybrids are not in interaction with each other, *i.e.* there is no stacking interaction between neighbouring C70 molecules?

Juan Casado answered: The complexes are formed in solution and then precipitated. I do not expect changes from solution to the solid state. All Raman studies presented in the paper under discussion were conducted in the solid state.

Andreas Hirsch asked: Did you look at the bond lengths as the molecules become more quinoidal?

Juan Casado responded: Yes, however, while the intra-ring bond lengths become quinoidal in the sense that those bonds parallel to the inter-ring bonds become reinforced, those intermolecular bonds are weakened. So it is quinoidized at the level of the ring but not for the intra-ring C–C bonds. This is a rather different situation to that when we speak about quinoidization in linear p-conjugated molecules.

Malcolm Heggie said: You remark that the bond between benzene rings gets longer but it is still quinoidal. Does this mean you believe that the bonds would have been longer but for this quinoidal effect?

Juan Casado answered: What I mean is that this is not the same effect as in linear conjugated oligomers and polymers where quinoidization simultaneously reinforces the intrannular bonds and the interring bonds. Here because of the cyclic structure, the only way to accommodate the cyclic strain and provide the needed curvature is by enlarging the inter-ring C–C bonds at the same time that locally the rings become quinoidal in the sense that parallel bonds are strengthened. For the carbon atoms connecting the rings, they are pyramidalized and acquire a sort of radical character.

Katsumi Kaneko queried: The CPP crystals have a soft nature for the adsorption of water and alcohol molecules according to the *in situ* synchrotoron X-ray diffraction measurements. I want to know whether an explicit distortion occurs on inclusion of C70 molecules in the lattice?

Juan Casado responded: Regarding the crystal structure of the complex, it seems that the structure is not much altered regarding the pristine CPP. This is discussed in the paper.

Antonio Rodriguez-Fortea said: Have you considered performing simulations to support your assertions about the changes from the standing disposition towards the lying due to the temperature?

Juan Casado answered: These theoretical studies are not the focus of the paper. However, they have been already done in a former publication[1] but at the level of single molecules in a vacuum, but no dynamic simulations.

1 T. Iwamoto, Y. Watanabe, H. Takaya, T. Haino, N. Yasuda and S. Yamago, *Chem. Eur. J.*, 2013, **19**, 14061–14068.

Antonio Rodriguez-Fortea remarked: Shouldn't you observe a rotation of the benzene ring in the cyclo[*n*]paraphenylenes at ambient temperature as the energy barriers are rather small?

Juan Casado replied: This is a unresolved question. The only proof we have is that we need simultaneous excitation to promote full rotation, that is, to pass the energy barrier.

Malcolm Heggie asked: I remember the days in Sussex after the discovery of fullerenes when Humberto and Mauricio Terrones would always be asking what would happen to such-and-such a material or molecule if it were pyrolysed. Do you think there could be interesting pyrolysis products from your cyclo[*n*] paraphenlylenes?

Juan Casado replied: probably we would obtained decomposition and a chemical behaviour typical of carbon materials. However, controlled heating of the sample allows for the growth of a SWCNT by coupling one CPP on top of the other.

Malcolm Heggie remarked: If you took one of your complexes with C_{70} and heated it, could you get C_{70} with a skirt?

Juan Casado responded: You mean a chemically attached skirt. Possibly but it is hard to predict.

Malcolm Heggie commented: You argue that the complexes you discuss are good models for peapod carbon structures, but surely these peapod structures are easily studied directly in high resolution electron microscopy?

Juan Casado answered: Yes, this is actually done in C70@SWCNT peapods. However to see the transition from lying to standing and *vice versa* you need either heating or pressure and these are nicely coupled to a Raman spectrometer, but not that much to high resolution electron microscopy.

Rosa Busquets remarked: have the authors thought if NMR could provide any useful information?

Juan Casado responded: Yes, you can check the references on the paper to see a complete study of the NMR in solution from the monomers to the complexes. However, the studies using pressure are in the solid state and are not feasible to follow with NMR. Here is where Raman spectroscopy is particularly useful.

Tariq Aqeel said: Have you tried labelling some of the carbon atoms at different locations in the C70, C60 (fullerenes) and in the cyclo[*n*]paraphenylene ([*n*]CPP) compounds, during preparation, to understand the deformation and the orientation changes that take place on the fullerenes and the CPPs during the heat pressure studies?

Juan Casado responded: These studies are interesting but we do not need a description of the vibrational nature of the G mode in the discussion. So the labelling studies are not commented on here.

Vladimir Fal'ko commented: Concerning the discussion of the effect of isotopes of carbon in nanotubes, I recall a joint theory-experiment paper on C-13 isotopic marking of CNTs.[1] In that work, the effect of isotopes on phonons was studied, including their appearance in Raman spectroscopy.

1 F. Simon, V. Zólyomi, R. Pfeiffer, H. Kuzmany, J. Koltai, and J. Kürti, *Physical Review B*, 2010, **81**, 125434.

Malcolm Heggie remarked: There are many reports that a sulfur : graphite composite, obtained by grinding them together can be superconducing.[1]

1 Y. Kopelevich, R. R. da Silva, J. H. S. Torres, S. Moehlecke, M. B. Maple, *High-temperature local superconductivity in graphite and graphite–sulfur composites, Physica C*, 2004, 77, 408–410.

Malcolm Heggie remarked: My student, Jean-Jo Adjizian did some nice DFT calculations that showed that sulfur was extraordinarily stable attached to a zig zag edge.[1]

1 J.-J. Adjizian, C. D. Latham, S. Öberg, P. R. Briddon, M. I. Heggie, *DFT study of the chemistry of sulfur in graphite, including interactions with defects, edges and folds, Carbon*, 2013, **62**, 256–262.

Alan Windle answered: Thank you for drawing attention to our work. The fact that sulphur's attachment to a graphene edge is predicted to be likely underlines its role as a promotor in the edge growth of graphene structures. However, in our process we have a much more complicated situation. It appears that the nanotubes grow from the surface of nano-scaled iron particles, so sulphur is playing its role on the surface of iron, and as a further challenge we have active carbon atoms present, presumably to replace the edge sulphur with further carbon. I believe that modelling will contribute greatly to our understanding over the next decade, as compared with theory it is generally robust under increasing complexity.

Malcolm Heggie commented: Jean-Jo's DFT calculations showed that sulfur can sit in a divacancy, *i.e.* as if it was replacing a carbon in a split-vacancy (*loc. cit.*)

Malcolm Heggie said: Interestingly, sulfur in diamond also takes up the split-vacancy configuration, probably for the same reason - to generate enough space to locate the much larger sulfur atom. However its doping properties change.[1]

1 J. M. Baker, J. A. van Wyk, J. P. Goss and P. R. Briddon, *Electron paramagnetic resonance of sulfur at a split-vacancy site in diamond, Phys. Rev. B*, 2008, **78**, 235203.

Pulickel Ajayan commented: There have been a lot of studies on coal, which contains quite a bit of sulfur - maybe there are reports discussing sulfur doping in graphitic substrates in the literature on coal and perhaps we need to look into those.

Malcolm Heggie replied: Literature on graphitisation is summarised in a DFT study of the chemistry of sulfur in graphite, including interactions with defects, edges and folds.[1] Coal is probably too complex a system to allow discussion of 'doping'.

1 J.-J. Adjizian, C. D. Latham, S. Öberg, P. R. Briddon, M. I. Heggie, *Carbon*, 2013, **62**, 256–262.

Malcolm Heggie remarked: It should be worth referring to Jean-Jo's paper on sulfur in graphite, but we have not seen any specific change in structure of the graphite with sulfur, just that the temperatures at which graphitisation processes occur are reduced.

Katsumi Kaneko enquired: I have never studied role of sulfur in the synthesis of carbon nanotubes. However recently we prepared an atomically 1D sulfur chain crystal inside a SWCNT and DWCNT[1] and showed that the 1D sulfur chain crystal is extremely stable even in the atmosphere below 470 K. I assume that sulfur has a self-locking function against reaction with oxygen in the atmosphere due to a specific chemical interaction with the edge carbon atoms. The interaction between sulfur and the edge carbon may be quite important. Is there any possibility that such a specific interaction is associated with the efficient catalytic effect for the carbon nanotube synthesis?

1 T. Fujimori, A. Morelos-Gómez, Z. Zhu, H. Muramatsu, R. Futamura, K. Urita, M. Terrones, T. Hayashi, M. Endo, S. Y. Hong,Y. C. Choi, D. Tománek and K. Kaneko, *Nature Comm.*, 2013, **4**, 2162.

Alan Windle replied: Our view is that sulphur plays two roles in the very rapid growth of carbon nanotubes at 1000 °C plus. One is that it plays a role in helping the nanotube to break free from the metal particle, a surfactant role, otherwise the particles simply become encapsulated in a spherical graphene coating, and thus are poisoned as far as their role as CNT catalysts is concerned. We observe many such particles in the absence of sulphur. The other role is an active role in helping the addition of new carbon atoms to the growing end of the nanotube in contact with the metal, perhaps the sulphur aids the breaking of the carbon–metal bond ready for its replacement by a new carbon atom. There have been several studies of the role of sulphur both in poisoning and activating transition metal catalysts used for hydrocarbon reactions. (Especially for molybdenum.) However, the degree to which this chemistry can be applied exactly to the particular circumstances of carbon nanotube synthesis in the presence of sulphur remains unclear. I wonder whether the most useful advances in our understanding of this very complex situation are, over the next decade, more likely to come from computational modelling.

Malcolm Heggie addressed Nazario Martin: You mention that carbon nanotubes (CNTs) are p-type. Is this the well known result that CNTs are p-type under atmosphere through electrochemical doping and surface reactions?

Nazario Martin responded: To answer in a simple way this question is not easy. Of course, the redox properties of SWCNTs depend on their nature (diameter, chirality, *etc.*). In order to get a better idea of the redox properties of SWCNTs, a comparison with the well-known C60 could be done. The redox potential values for the first oxidation and first reduction of C60 with the respective values for SWCNTs shows that SWCNTs behave, in general, as both better electron acceptors and donors than C60, but while the electron accepting properties are rather comparable, oxidation of C60 is energetically more demanding than p-doping of SWCNTs. For more details, please see ref. 1. Furthermore, semiconducting SWCNTs behave typically as p-type materials due to the presence of chemical species, mainly adsorbed oxygen, acting as week p-type dopants.

1 F. Paolucci *et al.*, in *Carbon Nanotubes and Related Structures*, ed. D. M. Guldi and N. Martin, Wiley, 2010, ch. 3, pp. 53–75.

Pulickel Ajayan remarked: There has been a lot of work done on point defects such as vacancies and interstitials in graphite - what are the similarities and differences of these defects in graphene compared to those in graphite?

Malcolm Heggie replied: On the whole there are great similarities. In graphene of course there are no interlayer interactions possible (hence the spiro-interstitial is not relevant and only the adatom (which equates to the metastable grafted interstitial in graphite) exists as the ground state for an extra atom. For mulitlayer graphenes (bilayer...few layer) there should be quite close similarities (and of course it is often mentioned that ten layers approximates graphite well). The detailed differences concern stacking - graphite has two inequivalent atomic sites and graphene only one (if it is infinite). Generally differences between defects at these sites and differences between graphite and graphene are in the region of 0–

0.2 eV for defects with formation energies normally more than 5 eV. It is an irony that most publications of the theory of the vacancy in graphite before 2004 were in fact graphene calculations. I can recommend the work of Drs Christopher Latham and Thomas Trevethan within my group[1-6] for the detail, as well as the prodigious outputs of Arkady Krashenninikov and the Helsinki group.[7]

1 T. Trevethan, C. D. Latham, M. I. Heggie, P. R. Briddon and M. J. Rayson, *Vacancy diffusion and coalescence in graphene directed by defect strain fields*, *Nanoscale*, 2013, **6**, 2978-2986.
2 C. D. Latham, M. I. Heggie, M. Alatalo, S. Öberg and P. R. Briddon, *The contribution made by lattice vacancies to the Wigner effect in radiation-damaged graphite*, *Journal of Physics: Condensed Matter*, 2013, **25**, 135403.
3 C. D. Latham, M. I. Heggie, J. A. Gámez, I. Suárez-Martínez, C. P. Ewels, and P. R. Briddon, *The di-interstitial in graphite*, *Journal of Physics: Condensed Matter*, 2008, **20**, 395220.
4 R. H. Telling, A. El-Barbary, C. P.Ewels and M. I. Heggie, *Wigner defects bridge the graphite gap*, *Nature Materials*, 2003, **2**, 333–337.
5 C. P. Ewels, R. H. Telling, A. El-Barbary, M. I. Heggie and P. R. Briddon, *Metastable Frenkel pair defect in graphite: source of Wigner energy*, *Physical Review Letters*, 2003, **91**, 025505.
6 A. A. El-Barbary, R. H. Telling, C. P. Ewels, M. I. Heggie and P. R. Briddon, *Structure and energetics of the vacancy in graphite*, *Physical Review B*, 2003, **68**, 144107.
7 G A. Gulans, A. V. Krasheninnikov, M. J. Puska, and R. M. Nieminen, *Bound and free self-interstitials in graphite and bilayer graphene: a computational study*, *Phys. Rev. B*, 2011, **84**, 024114.

Pulickel Ajayan addressed Malcolm Heggie and Vladimir Fal'ko: Are there experimental results to back up these studies on adatoms/doping effects?

Vladimir Fal'ko replied: There is no discussion of adatoms or doping in this paper.

Malcolm Heggie replied: Our own work has focused on S and B, with these papers[1,2] giving some background to experimental situation. There is also much work on radiation damage in graphene and graphite, notably Kirkland coworkers[3] in Oxford, Krasheninnikov[4,5] and coworkers in Helsinki and Germany, and many more.

1 J.-J. Adjizian, C. D. Latham, S. Öberg, P. R. Briddon and M. I. Heggie, *DFT study of the chemistry of sulfur in graphite, including interactions with defects, edges and folds*, *Carbon*, 2013, **62**, 256–262.
2 I. Suarez-Martinez, A. A. El-Barbary, G. Savini and M. I. Heggie, *First principles simulations of boron diffusion in graphite*, *Physical Review Letters*, 2007, **98**, 015501.
3 A. W. Robertson, C. S. Allen, Y. A. Wu, K. He, J. Olivier, J. Neethling, A. I. Kirkland and J. H. Warner, *Spatial control of defect creation in graphene at the nanoscale*, *Nature Comms.*, 2012, **3**, 1144.
4 O. Lehtinen, S. Kurasch, A. V. Krasheninnikov, U. Kaiser and K. Nordlund, *Atomic scale study of the life cycle of a dislocation in graphene from birth to annihilation*, *Nature Comms*, 2013, **4**, 2098.
5 A. V. Krasheninnikov and K. Nordlund, *Ion and electron irradiation-induced effects in nanostructured materials*, *J. Applied Physics*, 2010, **107**, 071301.

Philip Davies opened the discussion of the paper by Toshiaki Enoki: Have you investigated the stability of the different graphene edges against attack by oxygen or water?

Toshiaki Enoki replied: We have investigated the adsorption of oxygen to the graphene edges not by STM/STS observation but by NEXAFS, electrical conductivity and TPD. According to our experimental results, zigzag edges, which are thermodynamically unstable, are selectively etched away under the presence of oxygen at high temperature, resulting in the edge geometry change from zigzag to armchiar edges, the latter of which are thermodynamically stable. This oxygen etching process should be tracked by STM/STS observations in the future. We have not investigated the reaction with water. The interaction with oxygen and water are interesting issues which should be investigated by STM/STS observations together with DFT calculations, as we use graphene or carbon nanomaterials in the ambient atmosphere, in which oxygen and water are included.

Philip Davies remarked: How did you create the carbonic and hydroxyl groups that you compared with TPD?

Toshiaki Enoki answered: We created carbonyl groups using electrochemical oxidation of graphene edges, and confirmed the presence of the groups by STM observations. We also observed ether, epoxy and carbonyl groups bonded to the edge carbon atoms of point defects created in the reduced graphene oxide. Different from the present work, we have been working on experiments of NEXAFS and TPD. By means of these measurements, we found these oxygen-including functional groups. Depending on the heat-treatment temperature, the functional groups vary from carboxyl, carbonyl, hydroxyl, to hydrogen as the temperaturere is elevated.

Chris Ewels asked: Could you extract further information from your STM studies of large hole edges by performing them at different bias voltages, in order to sample deeper into the gap? In particular, do you think you could distinguish different sp_3-coordinated bonding at the edge this way?

Toshiaki Enoki responded: Depending on the functional groups to be bonded to the edge carbon atoms, edge-related modification of the electronic structure appears not in a narrow energy range in the vicinity of the Fermi level but is extended into a wide energy range. In this case, investigation in a wide range of bias voltage is required. One example is the case of a carbonyl group. The participation of oxygen $2p_z$ orbital into the π state works to split the edge state into two states accompanied by the charge transfer effect. Then we should investigate the modification of the electronic structure in a wide range of bias voltage. The sp_3-coordinated bonding is created when two hydrogen atoms are bonded to an edge carbon atom. In this case the 1s orbitals of two hydrogen atoms bonded to the edge carbon atom participate in the π state. The edge state has an energy dispersion in a range of *ca.* 0.5 eV. Accordingly, we need to investigate the edge-related modification in a wide range of bias voltage.

Alan Windle said: You relate the additional groups at the edge of the ribbons to the electrical conductivity along the ribbon. My question is: for the ribbons you showed, *e.g.* such as that in Fig 1f in your paper, what proportion of the current will be carried by the edges and what proportion by the ribbon itself. Are, for example, the edges dominant as charge carriers? The ribbon which has zigzag edges, if wrapped round and joined to make a single walled nanotube would make an armchair nanotube (standard nomemclature) which would show metallic conductivity, whereas the ribbon with armchair edges, if similarly converted into a single walled nanotube, would form a zigzag nanotube which would only show semi-conductivity. So I am interested in your view as to the importance of non-edge conductivity in your ribbons.

Toshiaki Enoki answered: The edge geometry dependence of the electrical conductivity is one of the interesting points in graphene nanostructures. It is completely different from that in carbon nanotubes. In an armchair edged graphene nanoribbon, the boundary condition at the edges gives a standing wave accompanied by the formation of a $\sqrt{3} \times \sqrt{3}$ superlattice. This can be explained in terms of the formation of a charge density wave, in the language of physics, or the localization of Clar's aromatic sextets, in the language of chemistry. The formation of a standing wave gives rise to an energy gap, resulting in the change in the electronic structure from semimetallic in infinite graphene to semiconductive in an armchair edged graphene nanoribbon. Accordingly, the conductivity becomes low by cutting graphene into graphene nanoribbons along the armchair direction. In contrast, there is no standing wave in a zigzag edged graphene nanoribbon, and instead a nonbonding state (edge state or zero mode) localized in the zigzag edges is created. The edge state or zero mode does not carry electrical current. On the other hand, the delocalization of sextets, which makes a standing wave absent, gives no energy gap, resulting in a conducting feature in the electron transport in a zigzag edged graphene nanoribbon similar to infinite graphene. We investigated the electrical conduction of graphene nanostructures depending on the edge geometry, using STM and AFM.[1] Our work confirmed the edge geometry dependence as given above. Another issue which should be addressed is the spatial distribution of the current density along the direction perpendicular to the ribbon axis. The current density becomes largest in the center of the graphene nanoribbon and it becomes smaller as we go from the center to the edge irrespective of the edge geometry.

1 S. Fujii and T. Enoki, *Angew. Chem. Int. Ed.*, 2012, **51**, 7236–7241.

Malcolm Heggie said: There was a question about whether the electron states of graphene nanoribbons (GNRs) can be related to those in graphene in the same way as they can be for carbon nanotubes (CNTs). The answer is broadly yes for those states which do not interact strongly with edge states. Just as CNT states are obtained in a reasonable approximation as graphene states subject to the cyclic boundary condition around the circumference, so GNR states are determined by graphene states subject to a boundary condition corresponding to a vacuum (*i.e.* effectively a high potential which means the wave function must go to zero). Both kinds of states are standing waves. The approximations ignore $\sigma : \pi$ interactions and any deviations from the ideal graphene bond length.

Irena Kratochvílová commented: What happens with the states of electrons at the surface/bulk in the ribbon if you deposit electropositive or electronegative groups on the edges?

Toshiaki Enoki answered: The electronic state of the edges depends not only on charge transfer but also on orbital hybridization between the edge carbon atoms and the functional groups to be bonded to the edge carbon atoms. This means that the electronegativity is only one of the factors which determine the electronic structures. In the case of the carbonyl group, oxygen $2p_z$ orbital participates in the carbon π state. This induces the charge transfer from carbon to oxygen, the latter of which is more electronegative than the former. We have not strong charge transfer for a hydroxyl group. In the case of monohydrogenated edges, charge transfer is weak. For the dihydrogenated zigzag edge, two hydrogen atoms participate in the carbon π state due to hyperconjugation.

Milo Shaffer queried: To what extent does the nature of the underlying graphitic layer(s) effect the STM observations that you make?

Toshiaki Enoki answered: We have employed the uppermost graphene layers of graphite and also graphene sheets on a Au substrate in our microprobe experiments. In the work of hydrogenated graphene defects on graphite, we confirmed that the results are not seriously affected by the interaction with the underlying graphite. The DFT results on thermodynamic stability and presence/absence of localized states in the different H-passivated states of single vacancies were also tested against the model of double-layer graphene and no significant difference between mono-layer and bi-layer cases were found. Therefore, we expect that our major results on the connection between the vacanciy's chemical composition and its electronic structure will be valid for both free-standing monolayer graphene and the top graphene layer of multilayer graphene structures.

Carlo Buono enquired: It is possible, by Bader analysis,[1] to quantify the charge transfer from graphene carbon to hydrogen and oxygen atoms. If the electron transfer takes place from carbon to the electronegative oxygen, could you explain what happens in presence of hydrogen which has electron donor properties? How can the presence of both functional groups (hydrogen and oxygen atoms) influence the density of state (DOS) of a functionalised mono-vacancy? An example of carbonyl and hydrogen functional groups at graphene defect sites are given from water dissociation at the mono-vacancy reported in the work of Cabrera-Sanfelix *et al.*[2]

1. R. F. W. Bader, *Atoms in Molecules–A Quantum Theory*, Oxford University Press, Oxford, 1990.
2. P. Cabrera-Sanfelix and G. R. Darling, *J. Phys. Chem. C*, 2007, **111**, 18258–18263.

Toshiaki Enoki responded: Yes we estimated the charge transfer between carbon atoms in graphene and oxygen/hydrogen atoms bonded to the edge carbon atoms using quantum mechanical calculations. The oxygen atom of the carbonyl-bonded edge and the hydrogen atom of monohydrogenated edge have $-1.49e/-0.19e$ and $+0.04e/+0.13e$, respectively, for the Bader/Lowdin charges. The transferred charges are shared by the carbon atoms in graphene with the carbon atoms directly bonded to the edge functional groups being affected most seriously. The theoretical results are confirmed by STM experiments. The paper by P. Cabrera-Sanfelix and G. R. Darling is interesting, and we should confirm the theoretical result of water adsorption at a point defect by microprobe experiments in future work.

Andreas Hirsch opened the discussion of the paper by Antonio Rodriguez-Fortea: Another aromaticity descriptor is the homo shift. Have you considered other descriptors?

Antonio Rodriguez-Fortea replied: In the present work, we have only considered HOMA as aromaticity descriptor. Solà and co-workers[1] checked for some IPR hexaanions that aromaticity trends given by the HOMA index calculated from the AM1 optimized geometries correlated to those obtained with the electronic multicenter index (MCI) at the BP86/cc-pVDZ DFT level, which is an aromaticity descriptor based on the wavefunction (and therefore much more expensive from the computational point of view).[2]

1 M. Garcia-Borràs, S. Osuna, M. Swart, J. M. Luis and M. Solà, *Angew. Chem., Int. Ed.*, 2013, **52**, 9275–9278.
2 P. Bultinck, R. Ponec and S. Van Damme, *J. Phys. Org. Chem.* 2005, **18**, 706–718.

Chris Ewels remarked: Regarding your electrostatic model deriving the thermodynamic drive for isolated pentagons without the presence of cations within the cage. I could imagine that for a large fullerene without cations it might be advantageous to form a high symmetry, roughly spherical cage, where the pentagons are separated to a maximum degree. However if two cations were included, and showed strong electrostatic interaction with the negatively charged pentagons, there might instead be more of a driving force to form an extended 'buckytube' type structure with pentagon localisation at the ends around the two cations. So my question is really, to what extent do you think that inclusion of the cations will influence your findings?

Antonio Rodriguez-Fortea replied: The interaction of the internal metal cations or clusters with the formally negatively charged fullerene cages is also a point that has to be considered to fully understand the relative energies between the different isomers. The maximum pentagon separation rule tells us which are the most stable negative cages, provided that the charge transfer is significant, and it is able to explain the experimental results, that is, the cages that encapsulate the characterized endohedral metallofullerenes (EMFs). For those EMFs in which the interaction between the cluster and the cage is important, because the cluster does not fit in the cage or the internal cations would be not far from each other, the relative stabilities of EMFs can be different to those between the corresponding negative anions. In these cases, a cage that is not the most stable anion, but among the most stable ones, can lead to the most stable EMF. For the example of large fullerenes that is commented in the question, if the charge transfer is significant, the relative energies between the anionic fullerenes will not differ so much with respect to those between the EMFs.

Valeri Kovalenko (on behalf of his own and **Ayrat Khamatgalimov**) remarked: It has been 30 years since the discovery of fullerenes. A huge volume of information on fullerene synthesis, structure and derivatives, both within and outside the cage, has been obtained. However, in our opinion, theoretical studies still progress very slowly and this in turn limits the widespread application of fullerenes. To consciously control the synthesis of fullerenes, one needs a theory, but to date

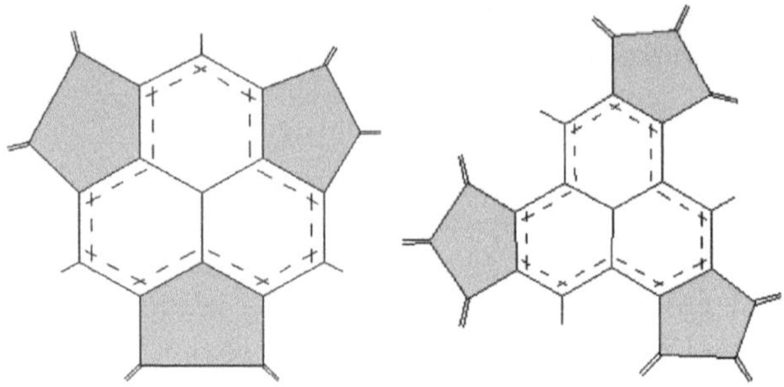

Fig. 2 Phenalenyl-radical substructures that most often occur in molecules of higher fullerenes.

there are only some approximations to it, *i.e.* attempts to find criteria, use descriptors, *etc.* We believe that for the most part they are all steps to a good future theory. That is what we have heard today in the form of an extramural discussion between two approaches, devoted to the problems of stability of fullerenes. At our institute we are working on developing our own empirical approach supported by quantum-chemical calculations.[1,2] We have already demonstrated the following: (1) structural formulas of the molecules for a series of isomers of IPR and non-IPR fullerenes C_{66}, C_{68}, C_{72}, C_{74}, C_{76} and up to C_{86}; wherein the electron density distribution is given in the form of the traditional three types of bonds: single, double, and pi-delocalized in hexagon. (2) Two main causes of instability of fullerene molecules. On one hand, it is the local excess strain caused by ineffectual topology of pentagons and hexagons when there are a lot of condensed hexagons. On the other hand, it is a radical nature of the molecule. The number of such radical fullerenes is very large (Fig. 2). All of these structural features of the molecules have been shown by us for some isomers of various fullerenes.[3-10]

The most important factor accompanying our approach is full correspondence of our descriptions and predictions with already existing or later appearing experimental results. One of them is particularly impressive. It is obvious that fullerene-radicals are unstable mainly because of their high reactivity. Thus, it is natural to expect that adding a corresponding number of electrons can stabilize them. Diener and co-workers have recently proposed a method of obtaining fullerenes in anionic form.[11] It was found that the insoluble residue referred to simply as soot contains more than 50% of higher fullerenes! It is not difficult to understand that radical fullerenes having 2 or more unpaired electrons readily polymerize transitioning into an insoluble form for IPR fullerenes. Moving them into an ionized form is obviously accompanied by depolymerization and accordingly their transition into a soluble form.

Returning to the report under discussion, we note that all fullerene molecules shown in Figure 2 of the paper are just radical fullerenes. Indeed, spin density is distributed along the periphery of the radical substructure, which necessarily involves pentagons (this is also a similarity of our considerations to those set out in this report). That is exactly where an additional negative charge is located after the addition of a corresponding number of electrons. That charge orients the metal cation inside the cage[6] (Fig. 3). For exohedral derivatives obtained in the radical addition reactions[12,13] we have shown the most "attractive" position on

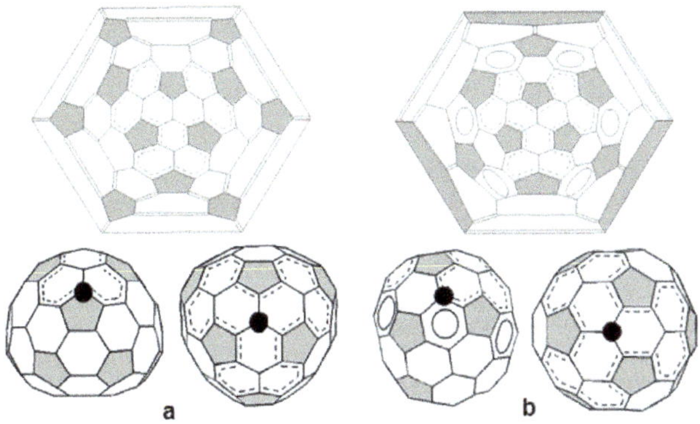

Fig. 3 Schlegel diagrams (upper part) with distributed bonds types: single, double, delocalized in hexagons (circles), and radical substructures (dotted lines), and the positions of endohedral metal atoms[6], depicted by black circle, two views (bottom part)) near radical substructures (dotted lines) in isomer 7 (C3v) (a) and 8 (C3v) (b) of fullerene C_{82}. Adapted with permission from Ref. 6. Copyright 2011 American Chemical Society.

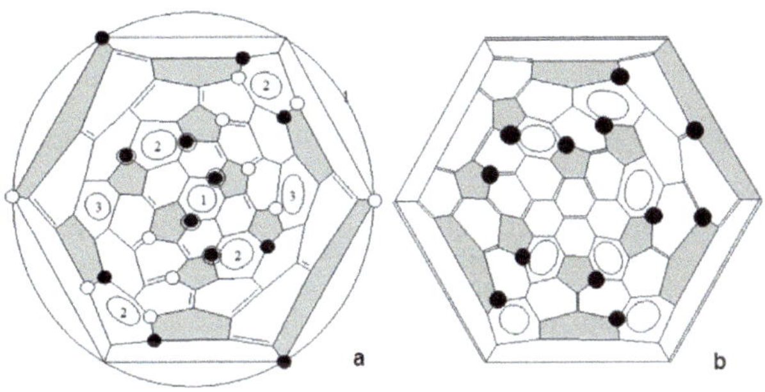

Fig. 4 Bonds distribution in the molecule of the isomer 22 (D2) of C_{84} fullerene (a) and 32 (C1) (b) of C_{90} fullerene with addends positions in adducts 22(D2)(C2F5)12 and 22(C1)(CF3) 16 (a) (black and white circles, respectively) and 32(C1)(CF3)12 (b) (black circles). Numeration of equivalent hexagons is according to molecule symmetry.

the fullerene sphere: they are hexagons with π-delocalized bonds (Fig. 4),[11] as well as the above-mentioned radical substructure.[10]

It is only left to add that there is another good correspondence between our data and those obtained by the authors of the paper being discussed. It was found that the sum of a number of π-delocalized hexagons Hd (our data) plus a number

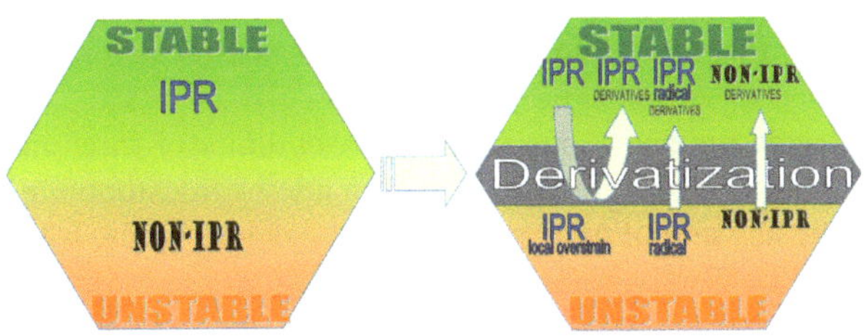

Scheme 1

of pyracylene motifs (mentioned by the authors), in the report, is a constant, *e.g.* for all 24 IPR isomers of C_{84} it is equal to 18! The fewer the number of pyracylene motifs, the more the number of π-delocalized hexagons Hd! Scheme 1 shows a transformation of connection "IPR-Stability" starting at very beginning of the fullerene *era* up to now. In summary, theoretical approaches developed by different researchers should gradually gather into a single channel - the theory of the molecular structure of fullerenes and their derivatives.

1 V. I. Kovalenko, A. R. Khamatgalimov, *Regularities in the molecular structure of stable fullerenes*, *Russ. Chem. Rev.* ,2006, **75**, 981–988.
2 V. I. Kovalenko, A. R. Khamatgalimov, *Electronic Structure and Stability of Higher Fullerenes*, in *NATO Security through Science Series A: Chemistry and Biology. Hydrogen Materials Science and Chemistry of Carbon Nanomaterials*, Springer, Dordrecht, The Netherlands, 2007, pp. 437–441.
3 A. R. Khamatgalimov, V. I. Kovalenko, *The Structure of Fullerene C_{66}, which Does not Obey the Rule of Isolated Pentagons, and Endohedral Metallofullerene $Sc_2@C_{66}$: Quantum-Chemical Calculations*, *Russ. J. Phys. Chem. A*, 2008, **82**, 1164–1169.
4 A. R. Khamatgalimov, V. I. Kovalenko, *Electronic structure and stability of fullerene C_{76} Isolated-Pentagon-Rule isomers*, *Fuller. Nanotub. Car. Nanostruct.*, 2015, **23**, 148–152.
5 A. R. Khamatgalimov, V. I. Kovalenko, *Electronic structure and stability of C_{80} fullerene IPR isomers*, *Fuller. Nanotub. Car. Nanostruct.*, 2011, **19**, 599–604.
6 A. R. Khamatgalimov, V. I. Kovalenko, *Electronic structure and stability of fullerene C_{82} IPR isomers*, *J. Phys. Chem. A*, 2011, **115**, 12315–12320.
7 A. R. Khamatgalimov, A. V. Luzhetskii, V. I. Kovalenko, *Unusual pentagon and hexagon geometry of three isomers (No 1, 20, and 23) of fullerene C_{84}*, *Int. J. Quant. Chem.*, 2008, **108**, 1334–1339.
8 A. R. Khamatgalimov, V. I. Kovalenko, *Deformation and Thermodynamic Instability of a C_{84} Fullerene Cage*, *Russ. J. Phys. Chem. A*, 2010, **84**, 721–726.
9 A. R. Khamatgalimov, V. I. Kovalenko, *Electronic structure and stability of C_{86} fullerene Isolated-Pentagon-Rule isomers*, *Int. J. Quant. Chem.*, 2011, **111**, 2966–2971.
10 V. I. Kovalenko, A. R. Khamatgalimov, *Open-shell fullerene C_{74}: phenalenyl-radical substructures*, *Chem. Phys. Lett.*, 2003, **377**, 263–268.
11 J. W. Raebiger, J. M. Alford, R. D. Bolskar, M. D. Diener, *Carbon*, 2011, **49**, 37–46.
12 N. B. Tamm, L. N. Sidorov, E. Kemnitz, S. I. Troyanov, *Chem. Eur. J.*, 2009, **15**, 10486–10492.
13 I. E. Kareev, A. A. Popov, I. V. Kuvychko, N. B. Shustova, S. F. Lebedkin, V. P. Bubnov, O. P. Anderson, K. Seppelt, S. H. Strauss, O. V. Boltalina, *J. Am. Chem. Soc.*, 2008, 13471–13489.
14 V. I. Kovalenko, R. A. Tuktamysheva, A. R. Khamatgalimov, *Electronic Structures of Some of C_{84} Fullerene Isomers and the Structures of Their Perfluoroalkyl Derivatives*, *Russ. J. Phys. Chem. A*, 2014, **88**, 103–107.

Antonio Rodriguez-Fortea asked: As far as I understood from the comment by delegate V. Kovalenko, the stability of IPR endohedral metallofullerenes (EMFs) might be also explained as considering the radical substructures or structural motifs within a cage. It is not surprising and already known that cages such as icosahedral C_{80}, which are not observed as empty cages, should be radical species considered as neutral and the electron transfer from the internal metal clusters render them very high stable EMFs with rather large HOMO–LUMO gaps. However, I guess that many cages and not only those that are seen to encapsulate clusters may have this radicaloid behavior in their neutral states.

Malcolm Heggie quiered: Your model invokes units of three neighbouring hexagons, which being like the polycyclic aromatic hydrocarbon are π radicals. When a fullerene has several such units, surely the radical nature will be removed through interaction of the molecular orbitals of these units and/or through Jahn Teller distortion?

Valeri Kovalenko answered: Being spatially separated, fullerene polyradicals are treated as a set of monoradicals like most typical organic polyradicals (for example, see ref. 1 and references therein). In addition when we consider the experimental data; they are in accordance with our model. Jahn–Teller distortions seem not to be an effect that works due to the low geometry configuration of the substructures of fullerene molecules. That prevents degeneracy of the ground state (may be excluding the C_{80} (Ih) molecule?); I suppose further investigation will cast more light upon the question.

1 A. Rajca *et al.*, *J. Am. Chem. Soc.*, 2013, **135**, 18205.

Malcolm Heggie asked: So, in your model, all these neutral molecules with several C_{13} units will have half-full degenerate molecular orbitals?

Malcolm Heggie questioned: Different theoretical methods, *e.g.* DFT with different functionals, can vary bond lengths by a matter of percents, so in the descriptor for aromaticity based on bond lengths, the reference bond length of 138.8 pm would need to be rescaled for each different method, would it not?

Antonio Rodriguez-Fortea replied: The reference length of 138.8 pm for the C–C bonds that appears in the definition of the HOMA descriptor (eqn 1 in the paper), also called optimal aromatic bond length, is not based on computations. The HOMA model is based on the assumption that the harmonic oscillator energy of extension or compression of a bond depends on the force constants, which, in turn, are dependent on the bond lengths. The optimal bond length is the length of a bond for which equal energy inputs are required to extend it to the length of the single bond or to compress it to the length of the double bond. The C–C bond lengths in buta-1,3-diene (taken from the electron diffraction data), 1.467 and 1.349 Å for single and double bonds, respectively, were taken as reference values, assuming the relationship between force constants for double and single bonds to be 2 : 1.[1] When people estimate the aromaticity of systems with C–C bonds with the HOMA descriptor, the value of 138.8 pm is usually considered, regardless of the theoretical method used to optimize the structure.

1 Krygowski *et al.*, *Chem. Rev.*, 2014, **114**, 6383–6422.

Nazario Martin said: In the case of La@C_{82}, this is a paramagnetic molecule - what do you conclude about the nature of this molecule in your model?

Antonio Rodriguez-Fortea replied: The maximum pentagon separation (MPS) rule predicts that the cage that encapsulates in La@C_{82} is among the IPR cages that show the lowest IPSI values for trianionic $C_{82}{}^{3-}$ species, although the applicability of the rule is in the limit since there is a formal electron transfer of three electrons. All the $C_{82}{}^{3-}$ isomers are paramagnetic, as is La@C_{82}. Moreover, when computing at DFT level the spin density distribution in La@C_{82}, it is seen that it is delocalized on the carbon cage as a consequence of the electron transfer. Therefore, the ionic model of charge transfer is completely applicable to the paramagnetic La@C_{82} and the MPS rule works rather well even if the charge transfer is not very high.

Nazario Martin asked: What does your model say about $Li^+@C_{60}$, X^-?

Antonio Rodriguez-Fortea answered: The maximum pentagon separation rule is valid to explain the relative stabilities of endohedral metallofullerenes (EMFs) provided that the formal electron transfer from the internal metal atoms or clusters to the cage is significant. The ideal situation is for trimetallic nitride M_3N EMFs (or other clusters such as Sc_4O_3, *etc.*), for which there is a transfer of six electrons. It also works pretty well for four electron transfers (M_2C_2, M_2O, M_2S, where M is always a trivalent metal). However, when the electron transfer is small, as for $Li@C_{60}$ with only one electron transfer, electrostatic effects are not so important and cannot reverse the relative stabilities of the neutral cages. The cage in $Li@C_{60}$ is icosahedral IPR C_{60}, the most stable (and observed!) C_{60} neutral empty cage.

Andreas Hirsch opened the discussion of the paper by Chris Ewels: If you dissolved your carbon nitrides in water you should be able to prove your results experimentally quite easily.

Chris Ewels answered: Thanks, this goes well with a previous comment, it's something we will definitely try.

Nazario Martin said: It would be better to protonate the sample first - this would protonate the nitrogens and removing the salts in water will become easier.

Chris Ewels responded: Thanks, this is a good suggestion, I'll discuss this with my experimental colleagues.

Milo Shaffer remarked: We also found that N-doping during CVD growth changed the morphology.[1,2] In our case, the walls of the resulting tubular structures were highly crystalline, and only slightly herringbone (in other words the graphitic layers were only a few degrees from parallel to the axis). By EELS, the nitrogen was also concentrated in the core where the planes terminated/spanned. The EELS spectra evolved with time, indicating the ejection of nitrogen from the carbon structure and the formation of nitrogen gas. I tended to assume that the lack of nitrogen signal from the outer edges of these herringbone fibres was due to similar beam damage effects, accelerated by exposure to high vacuum (unlike the core where nitrogen gas accumulates, at least until the shell is punctured by the electron beam). Do you see similar beam effects, or do you have a different model for the outer surface termination?

1 Koziol *et al.*, *Adv. Mat.*, 2005, **17**(6), 760–763.
2 Ducati *et al.*, *Small*, 2006, **2**, 774–784.

Chris Ewels responded: Since developing the ideas presented in this Faraday paper, we haven't done any detailed atomic resolution EELS studies of herringbone tubes. With the earlier work performed by Odile Stephan and colleagues on bamboo tubes,[1,2] they saw very high interior nitrogen concentrations much as you describe (particularly located on the interior compartment walls, but also the internal wall) and in private discussions they said that they also saw rapid evolution of the N1s EELS signal suggesting amorphisation/phase change under the beam in these zones, also consistent with your observations.

1 S. Trasobares, O. Stéphan, C. Colliex, W. K. Hsu, H. W. Kroto, *Compartmentalized CNx nanotubes: Chemistry, morphology and growth*, *J. Chem. Phys.*, 2002, **116**(20), 8966.
2 S. Enouz, J. L. Bantignies, M. R. Babaa, L. Alvarez, P. Parent, F. Le Normand, O. Stéphan, P. Poncharal, A. Loiseau, B. Doyle, *Spectroscopic Study of Nitrogen Doping of Multiwall Carbon Nanotubes*, *J. Nanoscience and Nanotechnology*, 2007, 7, 1–4.

My feeling is that there are distinct differences between structure and resultant beam behaviour between inner and outer regions of such tubes. The inner zones have such high nitrogen concentrations, up to 30%, that this cannot be considered as nitrogen doped layered carbon but must be effectively a new phase(s), amorphous or crystalline. The outer layers show very low nitrogen levels but are relatively crystalline and layered in the microscope, consistent with the segregation model proposed in the current paper for layered sp2-carbons. You may well be correct concerning nitrogen loss from the outer surface, this would not be inconsistent with a model where the exposed outer edges in a herringbone structure are stabilised and terminated with nitrogen during growth. This would also match with other herringbone HRTEM images I have seen, where it is sometimes possible to see 'looping back', *i.e.* adjacent layers bonding to each other along the exposed outer edges, implying lack of dangling bonds and associated functionalisation in these cases. This might be a beam induced effect in this case.

Even if not evaporated by the beam, the atomic nitrogen concentration of a nitrogen terminated edge at the herringbone surface would not be that high, and it may be difficult to distinguish the signal from background "noise" coming from the much higher concentrations in the cores (and possibly also nitrogen in any surface amorphous material). I would also expect however the inner core regions, with their very high nitrogen concentrations, to be more beam sensitive. It may be that alternative less destructive characterisation techniques (possibly TERS?) might be useful to fully characterise herringbone surface functionalisation.

Thurid Gspann addressed Chris Ewels and Alan Windle: Throughout the conference, we have had quite a few comments regarding the function of sulphur, nitrogen and oxygen and whether they have mostly similar or very different effects on the structural growth and synthesis of CNTs. For paper 2 the comment was made that in the floating catalyst synthesis at least the primary roles regarding process and spinability are very different (see previous comment), sulphur acts as a growth accelerating agent, while neither nitrogen nor oxygen can fulfil that role. However, looking at figure 10 in Chris Ewels paper describing the effect of nitrogen regarding the structural features of the grown CNT, we did observe very similar deformations if the system is overloaded with sulphur, see Fig. 5. So, regarding the formation of the CNT structure, the role of nitrogen and sulphur are possibly indeed akin.

Chris Ewels answered: I suspect we need to distinguish (at least) two distinct roles for heteroatom impurities in CNT growth, namely the direct chemical impact on the morphology and behaviour of the CNT tube itself, and the effect on the catalyst particle and growth kinetics. Paper 2 gave excellent insight into the second of these, with the model for sulphur build-up, poisoning, and island

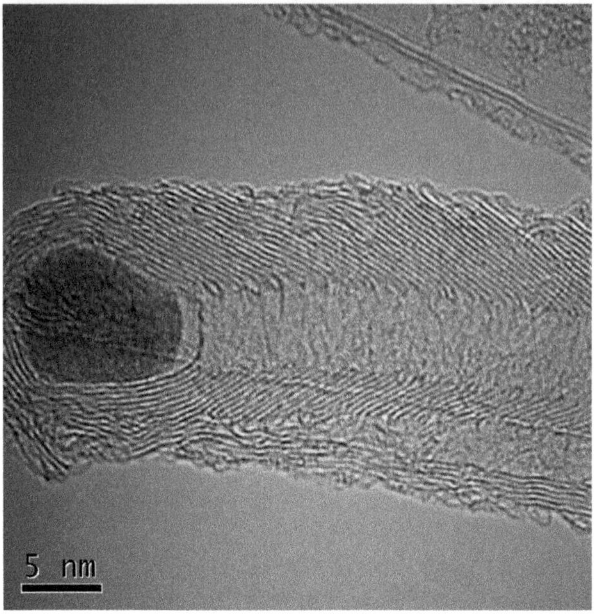

Fig. 5 Herringbone structure observed as defective tube in CNT fibres grown by the floating catalyst method as described in paper 2, when the system gets overloaded with sulphur. Although sulphur is necessary as a growth accelerator in our process, too much of it leads to structural deformations, partially very similar to the structural effects due to nitrogen as described in paper 12.

formation on the catalyst surfaces. I suspect that nitrogen also plays a key role interacting with the catalyst, probably inhibiting certain diffusion pathways, changing stabilisation of different phases, and possibly also facets on the catalyst particle, which in turn will affect the nanotube nucleation and ejection processes. This isn't something I covered at all in Paper 12 (and is quite hard to say meaningful things about *via* modelling). However for these processes, I would imagine quite strong differences in the behaviour of nitrogen, sulphur and oxygen, for example. Concerning the morphology and carbon chemistry however, I could imagine more similarities between dopants. Paper 12 concentrates on the stability of nitrogen at edge sites in sp2-carbon, but edge stabilisation has also been shown theoretically for both sulphur and oxygen.[1] Notably sulphur has been shown to catalyse unzipping of high curvature sp2 sheets resulting in sulphur terminated edges.[2] This seems very consistent with your observation of herringbone tube formation in conditions of high sulphur concentration, and it would be interesting to find a way to characterise the tube surface to confirm whether the available edge sites are sulphur terminated.

1 P. Wagner, C. P. Ewels, J.-J. Adjizian, L. Magaud, P. Pochet, S. Roche, A. Lopez-Bezanilla, V. V. Ivanovskaya, A. Yaya, M. Rayson, P. Brisson and B. Humbert, *Band Gap Engineering via Edge Functionalisation of Graphene Nanoribbons*, *J. Phys. Chem. C*, 2013, **117**(50), 26790–26796.
2 J.-J. Adjizian, C. D. Latham, S. Oberg, P. R. Briddon and M. I. Heggie, *DFT study of the chemistry of sulfur in graphite, including interactions with defects, edges and folds*, *Carbon*, 2013, **62**, 256–262.

Malcolm Heggie asked: If you use segregation to explain the herringbone/stacked cone structures, what is the explanation for bamboo structures?

Chris Ewels replied: That's very difficult to say. There are clearly multiple effects at work here. The growth process from the metal nanoparticle catalyst is changed in the presence of nitrogen; notably the orientation of the graphitic layers with respect to the catalyst particle surface changes between "conventional" axis-parallel multi-walled tubes, herringbone and bamboo structures. My instinct is that it's a careful balance of impurity, hydrogen and carbon concentration and temperature, and that the nitrogen is significantly modifying different diffusion pathways and surface reactions on the metal catalyst. Notably the chemical composition in bamboo tubes is very different, the surface layers show very low doping, likely comparable with SWNT doping processes discussed in the paper, whereas the interior cavities and interior dividing walls can show nitrogen concentrations up to 25-30% in which case these can no longer be considered as doped graphene but instead new phases. It is an area for further investigation.

Pulickel Ajayan said: How did you use EELS to interpret the location of the nitrogen in the lattice? Also can you comment on the doping of graphene with nitrogen - does it still produce an ordered structure?

Chris Ewels replied: Full details are given in our recent Nano Letters publication.[1] We used a combination of HRTEM HAADF imagery (comparing image intensity with simulations of substitutional doping), and then simulated EELS spectra of various different configurations of nitrogen in a graphitic lattice and correlated these with the experimental spectra we observed. The EELS was performed with a STEM in spectrum-image mode (*i.e.* a spatial grid of points were scanned, producing an EELS spectrum at each pixel), and the only place nitrogen was detected in the EELS spectra corresponded with our HAADF observations. Some subtleties were also identified, notably an explanation for a secondary peak identified in the C1s spectrum in the presence of some nitrogen defects, which is also seen at defect sites associated with hydrogen and oxygen impurities. This peak corresponds to assymetry in the sigma-bonding of carbon atoms neighbouring the defects (this can be identified from the modelling).

Concerning nitrogen doping in graphene this remains an area of open and very interesting research, notably the recent findings of preferential substitutional nitrogen doping on only one of the two carbon sublattices. There are different models in the literature for this (invoking Friedel oscillations, kinetics of the growth process),[2] and I am waiting to see what consensus is reached. I think that care should be taken in drawing comparisons between graphene doping and other carbon nanoobjects, particularly in these cases where graphene is grown on metallic substrates.

1 R. Arenal *et al.*, *Nano Letters*, 2014, DOI:10.1021/nl501645g.
2 L. Zhao, R. He, K. T. Rim, T. Schiros, K. S. Kim, H. Zhou, *et al.*, *Science*, 2011, **333**, 999–1003.

Varsha Khare remarked: In graphene produced by a solution method I do see phosphorous substitution, have you tried doping with phosphorous?

Chris Ewels replied: We have looked at phosphorus doping in carbon nanotubes and also fullerenes, both experimentally and theoretically. In nanotubes the doping levels appear very low indeed and very hard to detect, even harder to characterise.[1] Other groups have also looked at phosphorus doping in CNTs and tend to find co-doping is easier to achieve.[2] This is an area of ongoing study in the literature, notably it is not clear whether segregation processes may also be important for phosphorus.[3]

As well as for the more obvious size-difference problems (nitrogen can in principle substitute easily for carbon as they have almost identical atomic radii, whereas phosphorus is significantly larger), I think phosphorus doping behaviour will also be different to nitrogen due to the strong tendency of phosphorus to oxidise to give phosphene oxide. An example in a related material is hetero-fullerenes. The only studies to date that I know of, of phosphorus doping in fullerenes, found a very strong C59PO signal despite in principle having an oxygen free system,[4] and our calculations show oxidation of phosphorus to be highly exothermic. In the case of hetero-fullerenes this switches the doping from n- to p-type.[5] Phosphorus oxidation was also consistent with our measurements for phosphorus doped tubes.[1] Given the extremely high reactivity, it would be necessary to provide a good explanation why any phosphorus present in a given system was not oxidised I think, particularly if exposed to air.

1 V. Krstic *et al.*, *ACS Nano*, 2010, **4**(9), 5081.
2 Koos *et al.*, *Carbon*, 2012, **50**(8), 2816.
3 G. Kucukayan-Dogu *et al.*, *J. Phys. Chem. C*, 2013, **117**(46), 24554–24560.
4 C. Möschel and M. Jansen, *Z. Anorg. Allg. Chem.*, 1999, **625**, 175; S. Glenis *et al.*, *Chem. Mater*, 1994, **6**, 1850; N. Felder *et al.*, *J. Non-Cryst. Solids*, 2004, **333**, 301.
5 C. P. Ewels *et al.*, *Phys. Chem. Chem. Phys.*, 2008, **16**, 2145–2148.

Varsha Khare commented: The result may depend on the environmental conditions, - I haven't seen much oxidation in my systems. By changing the environment, we might be able to prevent the oxidation of phosphorous.

Chris Ewels replied: See my previous response.

Antonio Rodriguez-Fortea commented: You have used spin-polarized LDA. This method is known to delocalize spin density more than GGA or hybrid functionals. Have you checked the effects of such larger delocalization on the formation enthalpies per N?

Chris Ewels answered: We did perform test calculations on some of the structures using GGA. The aim of the calculations in the article are to provide a qualitative guide to the relative thermodynamic stability of different nitrogen configurations, I would be wary about treating the calculated values too quantitatively. I would assume that the largest effect would be to substitutional nitrogen configurations where the additional valence electron compared to carbon is comparatively delocalised.

Antonio Rodriguez-Fortea commented: The formation enthalpy per N atom wouldn't change so much and the relative values for the different edge structures wouldn't be significantly affected.

Faraday Discussions

PAPER

Spectroscopic and atomic force studies of the functionalisation of carbon surfaces: new insights into the role of the surface topography and specific chemical states†

Carlo Buono, Philip R. Davies,* Robert J. Davies,‡ Thomas Jones, Jiří Kulhavý, Ryan Lewis, David J. Morgan, Neil Robinson and David J. Willock

Received 8th April 2014, Accepted 12th May 2014
DOI: 10.1039/c4fd00061g

The utility of carbon materials in applications as diverse as drug delivery and photocatalysis is often undermined by the complexity of their surface chemistry; different sources of carbon give rise to a varied mixture of functional groups and hence different properties. Considerable efforts have been made to identify specific groups at these surfaces and elucidate the complex interactions that take place but even on materials such as graphene and carbon nanotubes there remains uncertainty about the nature of the components present and their role in the nucleation of other functional materials at the surface. The present study uses highly ordered pyrolytic graphite (HOPG) as a model on which the fundamental properties of specific functional groups and their interactions with deposited nanoparticles can be characterised. We have shown that treatment of HOPG surfaces with low concentrations of hydrochloric acid results in significant topographic changes to the surface and a low concentration of oxygen containing species. From selective derivatization and a comparison of their XP spectra, the latter can be unambiguously identified as surface hydroxyls. DFT calculations have shown that these groups are stable in close proximity to each other. Heating to 573 K leads to conversion of the hydroxyls to mixture of two states, one of which is identified as a ketone whilst the other is proposed to be an ether. Gold deposition on the surface from aqueous solutions of chloroauric acid is shown to be strongly influenced by the nature of the oxygen species present.

Introduction

In addition to their unique electronic properties carbon nano-materials have advantageous physical properties that have been utilised in a variety of different

Cardiff Catalysis Institute, School of Chemistry, Cardiff University, Cardiff, CF10 3AT, UK. E-mail: daviespr@cardiff.ac.uk; Fax: +44 (0)29 2087 4030; Tel: +44 (0)29 2087 4072

† Electronic supplementary information (ESI) available. See DOI: 10.1039/c4fd00061g

‡ Present address: Imperial College, London SW7 2AZ, UK.

applications. Their high surface area, high physical strength and low weight makes them ideal support surfaces for uses ranging from nano-oscillators to heterogeneous catalysts. A common factor in all such applications is the need to control the interface between the carbon allotrope involved and the components that need to bind to it. The nature of that interface will be dependent on the structure and functionality of the carbon surface and many applications of carbon materials apply empirically developed treatments to modify the surfaces concerned.[1,3–7] Carbon nanotubes are a good example, modifications to their surface chemistry being used to improve their dispersal in polymers and solvents as well as to tailor their electronic properties.[6,8]

Typically, surface modifications involve the use of strong oxidants to create, or functionalise, defect sites; sonication in 8 M HNO_3 for 1 hour[7] or refluxing in 2.6 M HNO_3 for 48 h[9] being representative examples. However, determining the effects of such treatments is not always straightforward and there remains considerable uncertainty about the functional groups present at carbon surfaces; reviews by Boehm[10] and Rodríguez-reinoso[11] for example highlight the difficulty of unravelling this chemistry. Methods used to identify functional groups include temperature programmed desorption (TPD) and infrared spectroscopy (IR) but the former is a rather indirect probe and IR is only reliable for the identification of functional groups with strong transition dipoles. It is possible for minority species to dominate the infrared spectra whilst the majority state is undetected. A relevant example is the work of Goncalves *et al.*[12] who studied the attachment of gold nanoparticles to oxygen functionalised graphene. Although they were able to demonstrate the importance of the oxygen functionality at the surface the authors were unable to distinguish between the roles of the different possible oxygen groups in the attachment of the gold.

X-ray photoelectron spectroscopy (XPS) is a surface sensitive and quantitative tool that has great potential in the characterisation of functional groups at carbon surfaces[13–15] but because of the huge complexity of the materials involved there remains uncertainty about species assignments in XPS. One approach to address this complexity is the selective chemical derivatization of functional groups discussed by Everhart & Reilley in 1981 (ref. 16) and Batich in 1988.[17] The method, which has previously been applied to the study of the functionality of polymer and carbon nanotube surfaces,[18–20] is semi-quantitative and involves the selective reaction of specific functional groups with a distinctive marker species, in this case fluorine, which is easily detectable in XP spectra.

Our approach to this complex area has been to focus on surfaces that are as reproducible as possible and HOPG is an ideal starting point, being readily available in well characterised orientations and quality. HOPG is also an ideal substrate for comparison with high quality DFT modelling, enabling us to probe the way in which functional groups are formed at HOPG surfaces. In our initial investigations[2] we explored changes in the topography of the surface after treatment with water and dilute acids under ambient conditions (a parallel investigation into HOPG under UHV conditions is also underway).[21] We have shown that dilute HCl and HNO_3 cause areas of local delamination at the surface of the HOPG and speculate that these are due to the formation of local defects with associated oxygen containing functional groups that reduce the interlayer bonding. In the present paper, we explore the identity of the functional groups formed by the acids and the effects of heat treatment on the surfaces and on the deposition of gold nanoparticles from aqueous solution. The experimental protocol we have adopted mimics that of the preparation of gold supported catalysts for the hydrochlorination of ethyne and we hope eventually to explain the effects that acid washing has on the performance of these and other catalysts.[4,5,22]

Experimental

HOPG samples (~10 mm square, ZYH quality, supplied by NT-MDT ltd, Europe) were cleaned before each experiment by peeling off the top few graphene monolayers using adhesive tape. Acid treatment involved placing a 100 μl droplet of fresh diluted HCl (Fischer Chemicals Ltd, 99%) and leaving to stand for 30 minutes before drying under a pure helium stream. The acid wet the surface very effectively and so larger droplets were not needed to provide complete coverage of the crystal surfaces. Gold deposition was achieved in a similar fashion: a 100 μl droplet of a gold precursor solution ($HAuCl_4 \times H_2O$, Sigma-Aldrich, 99.999%) was placed on the pre-treated sample and left for 1 minute. The sample was subsequently dried under a stream of helium for 2 minutes.

Ultra-pure water (UPW; Millipore) was used for dilution of the acids before surface treatment and also for control experiments that replicated dosing and analysis conditions. Selected samples were heated to 423 K and 573 K for 1 hour after surface treatment and allowed to cool to ambient temperature before analysis.

Derivatization reagents trifluoroacetic anhydride (TFAA, 99%), trifluoroethyl hydrazine (TFH, 70 wt% in water), triflouroethanol (TFE, 99%) and *N,N*′-di-*tert*-butylcarbodiimide (DTBC, 99%) were purchased from Sigma-Aldrich and used as received. Derivatization reactions were carried out in a sealed glass chamber pumped with a rotary vane pump to a base pressure of $\sim10^{-2}$ mbar. (Initial attempts at derivatization using a stainless steel vacuum chamber were abandoned because subsequent analysis by XPS showed transport of copper from the metal vacuum seals to the HOPG surfaces. No contamination issues occurred in the glass chamber.) 2 ml of TFAA and TFH liquids were frozen and placed into a reservoir of liquid nitrogen inside the chamber; this enabled the evacuation of the chamber without pumping away the reactant. TFE on the other hand was dosed from a 2 ml mixture of TFE, pyridine and DTBC in a 9 : 4 : 8 ratio, respectively, according to the procedure reported by Everhart & Reilley.[16] HOPG samples were placed into the chamber avoiding contact with the reservoir and walls. The chamber was evacuated until the liquid nitrogen reservoir had evaporated, at which point it was isolated from the pump and allowed to equilibrate for ~24 hours. The chamber was opened to the air and the samples typically analysed by XPS within a few hours of removal. However, control experiments in which the derivatised samples were allowed to stand in air for up to 7 days showed no significant change in the surface chemistry.

It is worth noting here that the concentration of different oxygen components in the XP spectra of the treated HOPG surfaces were slightly variable from sample to sample (±20%) meaning that quantitative estimates between samples is difficult. However, binding energies of the components within the spectra were reproducible from sample to sample.

XP spectra were recorded with a Kratos Axis Ultra-DLD photoelectron spectrometer using a monochromatic Al Kα X-ray source and the "hybrid spectroscopy" mode, resulting in an analysis area of 700 × 400 μm at a pass-energy of 40 eV for high resolution scans and 160 eV for survey scans. The XPS data was analysed using CasaXPS[23] with all binding energies referenced to the C(1s) peak at 284.5 eV. Curve fits were made using Gaussian line profiles and surface concentrations were calculated using the method described in detail by Carley *et al.*[24,25] AFM images were acquired on a Bruker Veeco Multimode probe in tapping mode using silicon nitride tips. AFM images were analysed using WSxM software.[26]

The periodic density functional theory (DFT) calculations presented here were carried out with the VASP code.[27–30] Convergence was found for the bulk structure of graphite using a 500 eV energy cut off for the plane-wave basis and a 5 × 5 × 1 Monkhorst–Pack *k*-point mesh.[31] We use the PBE generalised gradient approximation[32] functional with Grimme's parameterised dispersion interaction[33] included (PBE-D2) in order to model the van der Waals interactions which are the major force operating between graphitic layers. The core 1s states for C and O are represented using the projector augmented wave (PAW) approach while the ultrasoft pseudopotential is employed for H atoms.[34,35] The cell dimensions were optimised by systematic variation of the equivalent *a* and *b* lattice vectors followed by variation of the *c*-parameter. This process gave a lower cell energy than the alternative of simple cell volume scaling and resulted in hexagonal cell parameters of $a = b = 2.465$ Å and $c = 6.297$ Å. The *c*-lattice vector is controlled by the interlayer spacing and we were only able to obtain a stable minimum with respect to lattice expansion in this direction when the D2 corrected functional is employed. The pure graphite structure is hexagonal with an ABA... stacking sequence so that the *c*-vector represents two inter-layer separations. The inter-layer separation found here (3.149 Å) is shorter than that observed experimentally[36,37] (3.354 Å) by 6% although it is in line with earlier simulations using similar functionals[38] (3.21 Å).

The graphite $(10\bar{1}5)$ plane was used to model step edge features and the convergence of results with respect to supercell size and slab thickness was assessed using the calculated edge energies (see eqn (5)) for hydrogenated step edges. Slabs containing two graphite layers and doubled in the direction parallel to the edge were constructed so that the opposite edges of graphite sheets on different sides of the slab were separated by at least four aromatic rings.

Results and discussion

XPS and AFM studies of acid treated HOPG

AFM imaging shows that HOPG surfaces treated with 0.2 M HCl under ambient conditions become decorated with raised approximately circular features 50–100 nm in diameter and ~15–20 nm in height above the surrounding lattice, Fig. 1. We have previously demonstrated that these features are due to local areas of delamination, possibly due to the functionalization of defects sites and the reduction in inter-layer bonding and shown that the size of the islands is related to acid concentration.[2] Control experiments using ultra-pure water, in contrast,

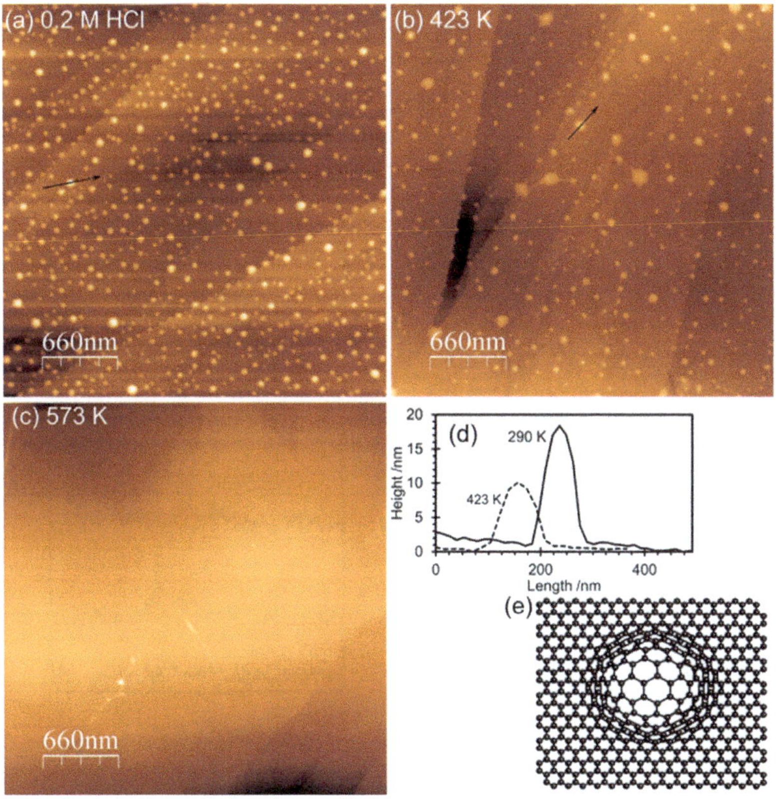

Fig. 1 AFM images of the acid treated surfaces: (a) HOPG surface treated with 0.2 M HCl for 30 minutes; (b) after heating to 423 K for 1 hour; (c) after heating to 573 K for 1 hour; (d) line profiles from the AFM images; (e) model illustrating the local delamination proposed to be the cause of the features in the AFM images.

showed neither the local delamination effects nor any significant change in the species present at the surface.

The corresponding XP spectra are shown in Fig. 2, and reveal a single peak in the O(1s) region at 532.6 eV with a calculated surface concentration of 1.1×10^{15} cm^{-2} corresponding to ~0.27 monolayers. Changes in the C(1s) spectrum are too small to be detected against the background but there are small increases in the Cl(2p) spectra (~3.5×10^{13} cm^{-2}) and, perhaps surprisingly, in the N(1s) region with the development of a peak at 400.2 eV with a surface concentration corresponding to 1.6×10^{14} cm^{-2}. The N(1s) binding energy of 400.2 eV is relatively low, more indicative of an amine than a nitrate.

Heating the sample in stages causes a decrease in the *number* of features in the AFM images with an accompanying decline in the overall *volume* of the features. By 573 K virtually all of the features in the AFM images have disappeared. In the XP spectra the peaks in the Cl(2p) and N(1s) region vanish by 573 K whereas the single peak in the oxygen region splits into two with binding energies of 531.5 and 533.3 eV. Repeated experiments suggest only a small reduction in surface oxygen concentration occurs on heating from 290 K to 573 K.

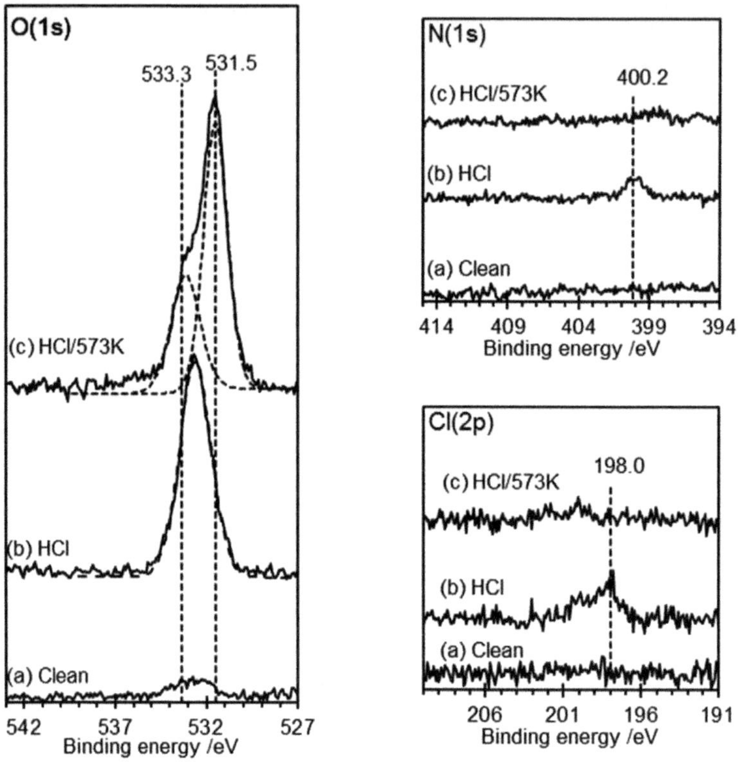

Fig. 2 XP spectra of HOPG after treatment with 0.5 M HCl and subsequent heating to 573 K. (a) Clean surface; (b) after treatment with 0.5 M HCl for 30 minutes; (c) after heating to 573 K for 1 hour and cooling to ambient.

Identification of functional groups at acid treated HOPG surfaces by selective derivatisation

Identifying hydroxyl groups at carbon surfaces. Trifluoroacetic anhydride (TFAA, $CF_3C(O)OC(O)CF_3$) reacts selectively with hydroxyl groups at carbon surfaces with a 1 : 1 stoichiometry and high selectivity, generating an adsorbed fluoroacetate and an acetic acid that desorbs or reacts with other hydroxyl groups to give the same product:[17,39]

$$CF_3C(O)OC(O)CF_3\ (g) + OH\ (a) \rightarrow CF_3C(O)O\ (a) + CF_3CO_2H\ (g) \qquad (1)$$

HOPG surfaces, freshly cleaved in air, always show a small concentration ($\sim 1 \times 10^{14}$ cm^{-2}) of oxygen dominated by a peak at 532.6 eV. Exposure of this surface to TFAA leads to a similarly weak peak in the F(1s). When the HOPG surface treated with 0.5 M HCl was exposed to TFAA the strong O(1s) peak at 532.7 eV is broadened with the development of a shoulder to higher binding energy; curve-fitting of the spectra show that the resulting envelope is consistent with peaks at 532.6 eV and 534.0 eV with area ratios of ~3 : 1. In the F(1s) region a strong peak at 688.8 eV indicates the presence of fluorine with a surface concentration that is a little less than twice that of the oxygen peak at 534.0 eV, Fig. 3.

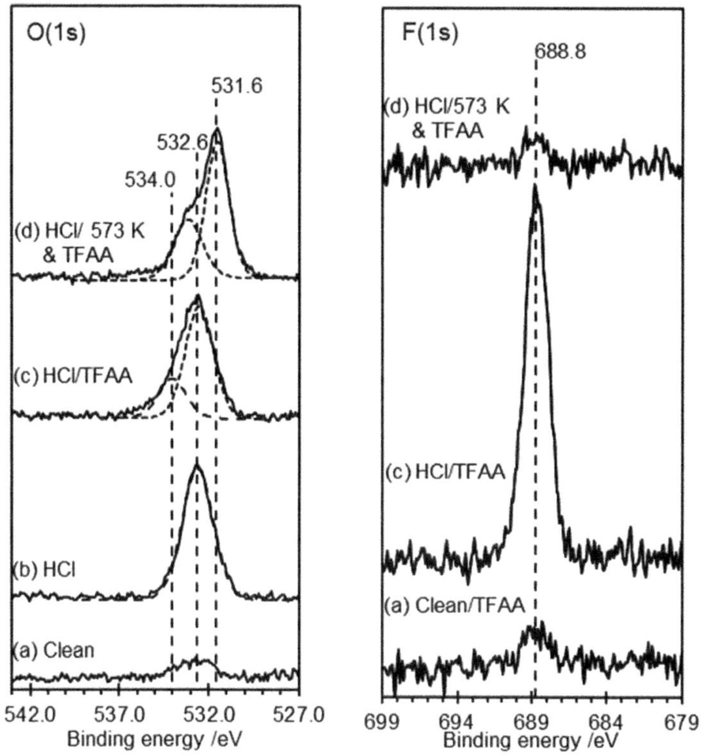

Fig. 3 XP spectra showing O(1s) and F(1s) regions of HOPG after treatment with HCl and TFAA. (a) Clean HOPG exposed to TFAA; (b) (in O(1s) spectra only for comparison), HOPG treated with 0.5 M HCl for 30 minutes; (c) surface in (b) after exposure to TFAA; (d) after heating (b) to 573 K for 1 hour followed by exposure to TFAA.

The data suggest that the HCl treatment of graphite surfaces generates a high concentration of hydroxyl species, with a characteristic binding energy of ~532.7 eV. The fluoroacetate produced from the reaction of the TFAA with this hydroxyl will have two distinct oxygen species, a carbonyl and a methoxy type functionality (C–O–CF_3). We show below that a carbonyl functionality on the graphite surface gives rise to an O(1s) binding energy of 531.6 eV. In the case of the fluoroacetate the functionality is closely associated with the CF_3 group and we might have expected a higher binding energy but Holländer's study[39] of TFAA derivatised poly(vinyl alcohol) also assigns the fluoroacetate carbonyl to a peak at 531.6 eV whilst the methoxy type functionality appeared at 534.8 eV. The larger intensity of the 531.6 eV peak compared with the 534.0 peak suggests that some intensity from the original peak before TFAA treatment remains. This implies that either unreacted OH groups remain at the surface after TFAA treatment or a second species is present with a binding energy similar to the OH. Exposure of the *heated* HCl treated surface to TFAA results in negligible reaction; the small amount of fluorine that is present is probably due to hydroxyl groups formed at the HOPG surface whilst the sample was transferred from the oven to the treatment chamber. Clearly heating to 573 K is sufficient to convert all of the surface hydroxyl groups to other functionalities, confirming the observations from Fig. 2.

Identifying carboxylic acid groups at carbon surfaces. Trifluoroethanol (TFE, CF_3CH_2OH in the presence of DTBC) reacts selectively with carboxylic acid groups at carbon surfaces, eliminating water with a 1 : 1 stoichiometry and high selectivity:[16–18]

$$CF_3CH_2OH\ (g) + -CO_2H\ (a) \rightarrow CF_3C(OH)O\ (a) + H_2O\ (g) \tag{2}$$

Clean, HCl treated and heated HOPG surfaces were exposed to the TFE–DTBC mixture for 24 hours and the resulting XP spectra are shown in Fig. 4. No reaction was observed between the TFE and the clean HOPG surface and exposure of the HCl treated surface to the TFE–DTBC mixture for 24 hours also results in only a minimal uptake in fluorine containing molecules. The corresponding O(1s) spectra are also within 0.1 eV of the spectra observed without TFE treatment. We can conclude that mild HCl treatment of the HOPG surface does not lead to carboxylic functionality. Fig. 4(c) shows that there is also no reaction between TFE and either of the states generated by heating the acid treated surface. We can therefore rule out the assignment to carboxylic groups of either of the two species with O(1s) binding energies of 531.5 and 533.3 eV.

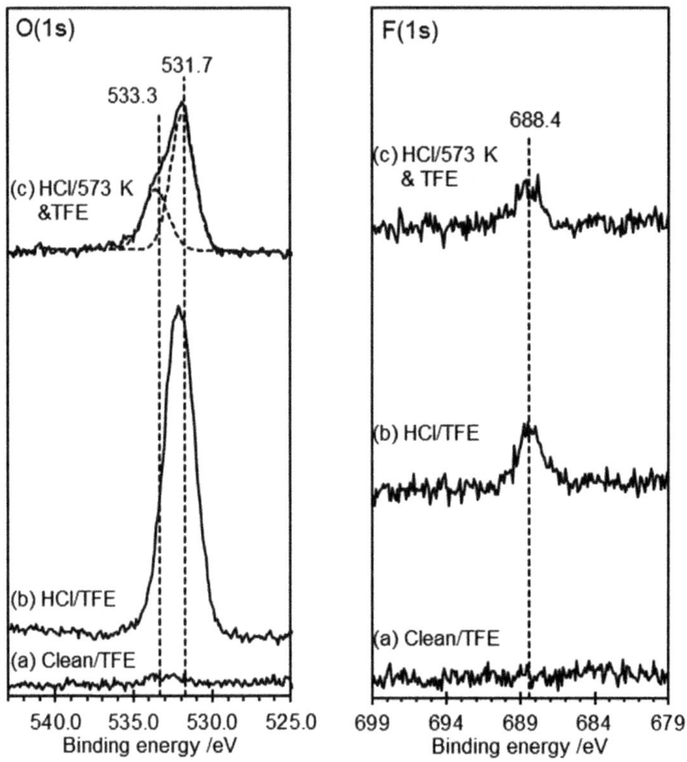

Fig. 4 XP spectra of HCl treated HOPG surface exposed to TFE vapour. (a) Clean HOPG surface exposed to TFE; (b) HOPG surface treated with 0.5 M HCl and subsequently exposed to TFE; (c) HOPG surface treated with 0.5 M HCl and heated for 1 h to 573 K before exposure to TFE.

Identifying ketone groups at carbon surfaces. Trifluoroethyl hydrazine (TFH, $CF_3CH_2NH–NH_2$) reacts selectively with carbonyl groups at carbon surfaces, eliminating water with a 1 : 1 stoichiometry:[16–18]

$$CF_3CH_2NH–NH_2\ (g) + –CO\ (a) \rightarrow CF_3CH_2NH–N{=}\ (a) + H_2O\ (g) \qquad (3)$$

The F(1s) spectra in Fig. 5 demonstrate that there is no reaction between TFH and the clean HOPG surface and there is also only minimal reaction when the HOPG washed with HCl is exposed to TFH. However, exposure of the HOPG surface to TFH after HCl washing and subsequent heating to 573 K results in much more extensive reaction. A strong peak develops in the fluorine region at 688.4 eV accompanied by a strong peak at 400.5 eV in the N(1s) region. Interestingly, although the reaction stoichiometry (eqn (3)) indicates complete removal of the O(1s), two oxygen features remain after TFH treatment; the peak at 533.3 eV is unaffected whereas the peak at 531.6 eV is shifted up to 531.9 eV. A possible explanation is that the water is not completely desorbed but rather reacts with the lattice in a concerted reaction with the hydrazine reaction, furthermore a hydroxyl

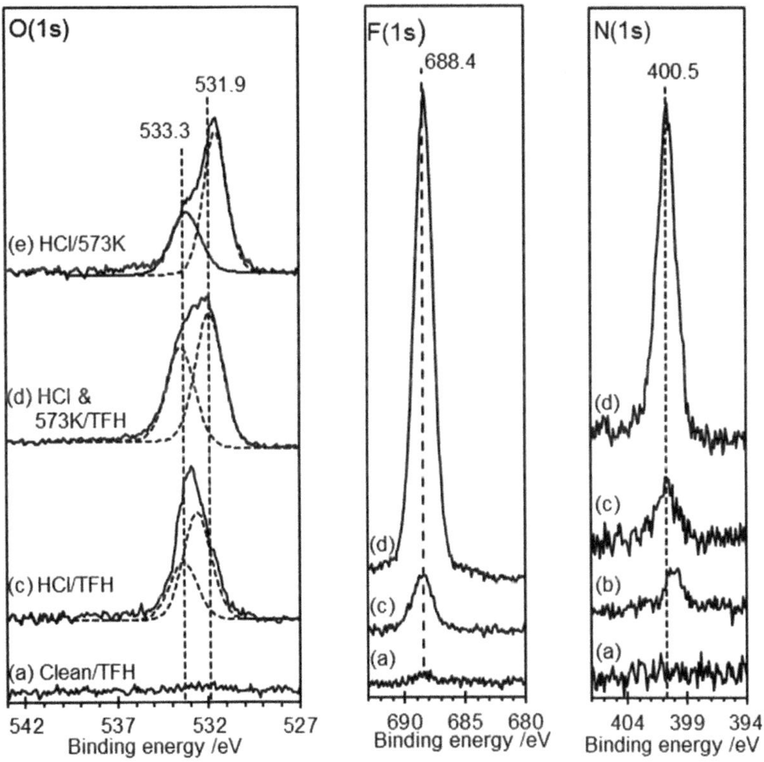

Fig. 5 XP spectra of HCl treated HOPG surface exposed to TFH vapour. (a) Clean HOPG surface exposed to TFH; (b) (N(1s) spectra only), HOPG surface treated with 0.5 M HCl; (c) HOPG surface treated with 0.5 M HCl and subsequently exposed to TFH; (d) HOPG surface treated with 0.5 M HCl and heated for 1 h to 573 K before exposure to TFH. (e) (O(1s) spectra only), HCl treated surface heated to 573 K surface *without* TFH exposure for comparison.

group would be a possible assignment for the peak at 531.9 eV since its close proximity to the hydrazine group could be sufficient to shift the OH binding energy significantly from that of the hydroxyl group observed after HCl treatment.

Gold deposition at acid treated and heated HOPG surfaces

The AFM images of the gold deposited on heated and non-heated HCl treated HOPG (Fig. 6) show similar topography to each other with neither surface exhibiting the raised features originally created by the HCl treatment. This is consistent with the flat surface reported in Fig. 1 for the heated sample but implies that the gold deposition process can also remove these features. The AFM images of the gold deposited on a non-heated surface, Fig. 6(a), show evidence for a discontinuous film, where the gaps between sections of the film indicate a thickness of approximately 2 nm. The films deposited on the heated surface are much smoother, broken only by what appear to be the step edges in the graphite substrate. XP spectra of the surfaces are shown in Fig. 7, and from the attenuation of the C(1s) signal we estimate at least a monolayer of gold or gold chloride is present in each case.

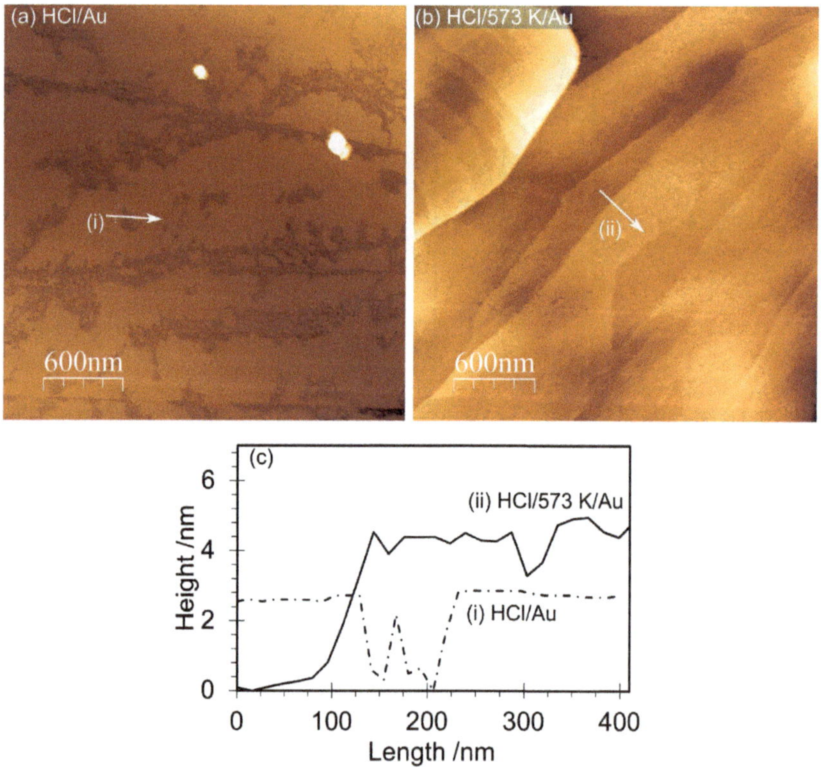

Fig. 6 AFM images of the HOPG surface after gold deposition from a solution of chloroauric acid: (a) HOPG surface treated with 0.5 M HCl, dried and then exposed to a 6 × 10^{-6} M solution of $HAuCl_4$. (b) HOPG surface treated with 0.5 M HCl, dried and heated to 573 K (1 hour) before cooling and being exposed to 2 × 10^{-6} M solution of $HAuCl_4$. (c) Line profiles from the AFM images.

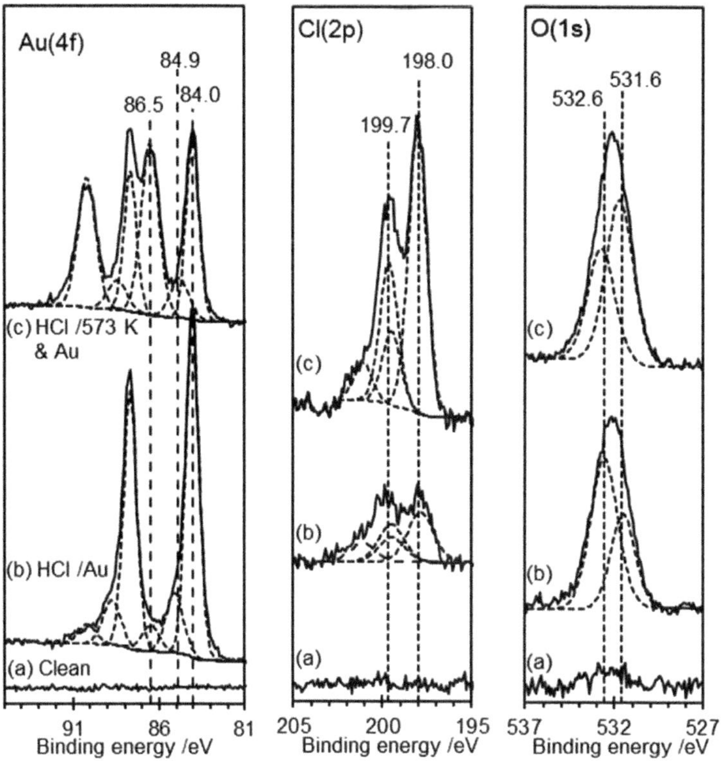

Fig. 7 XP spectra of the HOPG surface after gold deposition from a solution of chloroauric acid: (a) clean surface; (b) HOPG surface treated with 0.5 M HCl, dried and then exposed to a 6×10^{-6} M solution of $HAuCl_4$. (c) HOPG surface treated with 0.5 M HCl, dried and heated to 573 K (1 hour) before cooling and being exposed to 2×10^{-6} M solution of $HAuCl_4$.

In both types of experiment the XP spectra show at least two gold states with characteristic Au($4f_{7/2}$) peaks at 84.0 and 86.5 eV, which can be assigned to Au^0 and Au^{3+}, respectively.[40] Curve fitting of the data unambiguously indicates the presence of a third state in both experiments with a Au($4f_{7/2}$) peak at 84.9 (±0.2) eV. This state has two likely origins, a Au^+ state or a nano-particulate Au^0 (since small gold islands are expected to show a shifted binding energy[41,42]). In view of the relatively smooth films evident in the AFM images, the former explanation is more likely in the present case. Support for these assignments comes from the Cl(2p) spectra, which show two chlorine species with characteristic binding energies of 198.0 and 199.7 eV. Kishi reported similar binding energies for chlorine bound to the Au^{3+} and Au^+ species after exposure of gold metal to chlorine gas.[43]

DFT studies of the stability of functional groups on HOPG surfaces

The functionalization of graphene as a function of chemical potential of reactants has been studied extensively using DFT approaches by Mauri and co-workers covering hydrogenation, and reaction with O_2 and water.[44–46] As part of the present study we have calculated the structure of the most favourable terminations of step edges on graphite using PBE-D2.

Graphite has a hexagonal crystal structure with the graphitic sheets parallel to the (0001) surfaces. A step edge can be represented by introducing higher Miller indices at the expense of increasing the cell size and so the computational cost of the calculations. We choose to use a Miller index of 5 in the *c*-direction so that functional groups at a step edge have a terrace region of at least one aromatic ring separating them from the next edge in the simulation cell.

We initially considered two possible step edge structures: the zigzag (z), in which the step edge is formed with single C atoms from individual 6-rings exposed, and the armchair (a), in which two C atoms from each individual 6-ring appear (see Fig. S1†). The calculated energy required to create a non-functionalised graphite step edge with z-termination is much higher (1.5 eV $Å^{-1}$) than that for the a arrangement (1.1 eV $Å^{-1}$), an effect which can be understood in terms of Clar's rules:[47] on hydrogenation the a-edge leads to C=C double bonds that are effectively isolated from the aromatic system of the graphite sheet whereas the aromatic system is disrupted to a greater extent for the z edge.

The experimental procedure begins with materials with no oxygen present and so we take the a-edge as the best representation of the steps on the unreacted surface. For comparison with the experimental data presented here we considered the reaction of the isolated double bonds of the armchair step edge with water:

$$\mathrm{C}(10\bar{1}5) + N_{\mathrm{H_2O}}\mathrm{H_2O} {=} N_{\mathrm{H_2O}}\mathrm{C}(10\bar{1}5)(-\mathrm{OH})(-\mathrm{H}) \quad (4)$$

The energy required for this process can be calculated using the PBE-D2 energies following:

$$E(\mathrm{edge}) = \frac{1}{2LN_L}\left[E(\mathrm{slab}) - N_{\mathrm{C}}E(\mathrm{bulk}) - N_{\mathrm{H_2O}}E(\mathrm{H_2O})\right] \quad (5)$$

Here the simulation cell is taken to have a length, L, along the step direction, and there are N_L steps within each simulation cell ($N_L = 2$ in this work). N_{C} is the number of carbon atoms in the simulation and $N_{\mathrm{H_2O}}$ the number of water molecules required to fully react the step edges. E(slab), E(bulk) and $E(\mathrm{H_2O})$ are the calculated energies of the simulation cell for the hydroxylated step edge, the energy per C atom of the optimised graphite bulk structure and the energy of an isolated water molecule, respectively. All calculations are performed at the same level of accuracy as described in the Experimental section.

Eight possible terminations of the slab edges were considered (Fig. S2 and S3†) but the calculated edge energies showed that these could be classified into two sets. Fig. 8(a) and (c) show the lowest energy structure obtained with each of the isolated C=C bonds of the armchair edge functionalised with an OH group and H atom such that the OH group is roughly in the plane of the graphite sheet and orientated toward the H on the same C=C bond. This structure has a calculated formation energy, E(edge), of 0.282 eV $Å^{-1}$, a little higher than that reported for a single graphene sheet by Mauri and co-workers[45] (0.247 eV $Å^{-1}$). This difference is likely due to the interaction of the functional groups at the step edge with C atoms in the terrace below.

Rotation of the OH group by roughly 180° results in a member of the higher energy set (Fig. 8(b), $E(\mathrm{edge}) = 0.331$ eV). It can be seen in the figure that there is a

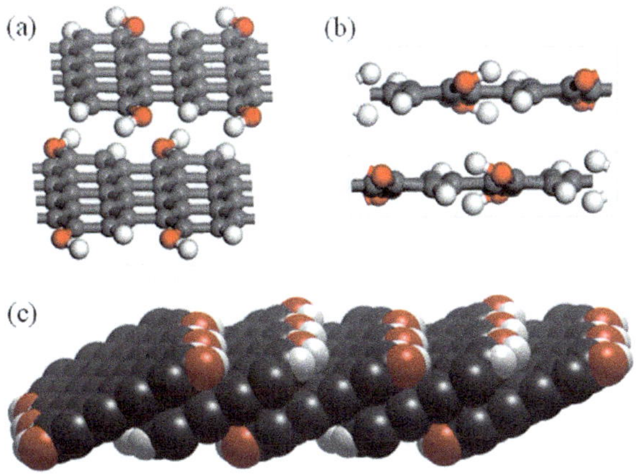

Fig. 8 PBE-D2 calculated structures for the hydroxylated step edge on graphite using the (10$\bar{1}$5) surface. (a) The most stable arrangement and orientation of hydroxyl groups at the step edge found. (b) An example structure with the hydroxyl groups rotated through roughly 180° resulting in a higher energy structure. (c) Space fill representation of the stepped surface from (a) showing an expansion of the simulation cell by 2 into the plane of the figure and 2.5 across the page. Atoms coloured according to C: grey; O: red; H: white.

steric clash between the OH and an H atom on the neighbouring C=C section that leads to the higher energy of this arrangement.

It is interesting that the edge energy is positive for all terminations studied, which indicates that in pure water functionalization will not take place; this may be why acid conditions are required to stabilise the OH functional groups experimentally. In addition, the close proximity of the OH functional groups along the step edge suggests that the addition of the probe molecules used in our experimental work will lead to significant crowding and so can explain the residual population of OH groups observed in the XPS spectrum.

Conclusions

We have shown previously that washing HOPG surfaces with low concentrations of HCl gives rise to areas of local delamination. In the present work, using selective derivatization, we have established that the effect of the acid washing is to generate hydroxyl groups almost exclusively, and we can now identify a characteristic O(1s) binding energy of 532.7 eV for OH(a) on the HOPG surface. Even allowing for the slight difference in calibration used between studies this is slightly lower than that observed[13] with aliphatic polymers and significantly lower than that of hydroxyl groups associated with aromatic polymers, which might have been expected to be a good match to the HOPG surface. It does, however, lie within the ranges (albeit at the lower end) suggested by Yue[15] and Zielke[14] for hydroxyls on carbon fibres. The XP data suggest that either exposure to TFAA did not remove all of the hydroxyl groups present after acid treatment or that a second unreactive species is present. Our DFT calculations, which show that the hydroxyls tend to be closely aggregated, support the former model with steric hindrance preventing complete reaction.

Acid treatment of carbon surfaces has previously been associated with the formation of carboxylic functionality but those studies[48,49] have generally utilised much stronger concentrations of acid than used here. In this work no evidence for carboxylic acid surface groups was obtained, suggesting that the hydroxyl groups are the first stage in the oxidation process and perhaps carboxylic acid functionality develops later. This aspect will be explored in future work.

Heating the hydroxyls leads to the formation of two different oxygen species. The first, with a characteristic O(1s) binding energy of 533.3 eV, is unaffected by treatment with any of the reagents used in this study. We provisionally assign this state to an ether type group (–C–O–C–). The second, which has a characteristic O(1s) binding energy of 531.6 eV, reacts strongly with TFH and can be assigned to a surface ketone.

Conte *et al.*[4] have recently pointed out the importance of the metal/functional group interaction in deposition onto carbon surfaces and the particular paucity of information on the deposition of gold on carbon. Whilst it is clear that we have deposited far more gold in the present experiments than would be desirable in most applications, the striking differences seen here between the oxidation states of gold deposited on the hydroxylated and ketone rich surfaces demonstrate that this is a rich area to be explored. The results provide an interesting glimpse into the anchoring of metal nanoparticles onto carbon surfaces and suggest several future pathways for investigation.

Acknowledgements

This research was supported by EPSRC Grant EP/I038748/1. TJ thanks Cardiff University for provision of a summer studentship under the CUROP scheme. Computing resources were obtained locally from the Cardiff ARCCA division and *via* our membership of the UK's HPC Materials Chemistry Consortium (EPSRC EP/L000202). This work made use of the facilities of HECToR and ARCHER, the UK's national high-performance computing service, which is funded by the Office of Science and Technology through EPSRC's High End Computing Programme.

References

1 P. Serp and J. L. Figueiredo, *Carbon materials for catalysis*, John Wiley & Sons, Hoboken, N.J, 2009.
2 E. Bouleghlimat, P. R. Davies, R. J. Davies, R. Howarth, J. Kulhavy and D. J. Morgan, *Carbon*, 2013, **61**, 124–133.
3 X. Zhang, Y. C. Guo, Z. Cheng Zhang, J. S. Gao and C. M. Xu, *J. Catal.*, 2012, **292**, 213–226.
4 M. Conte, C. J. Davies, D. J. Morgan, T. E. Davies, D. J. Elias, A. F. Carley, P. Johnston and G. J. Hutchings, *J. Catal.*, 2013, **297**, 128–136.
5 M. Conte, A. F. Carley, C. Heirene, D. J. Willock, P. Johnston, A. A. Herzing, C. J. Kiely and G. J. Hutchings, *J. Catal.*, 2007, **250**, 231–239.
6 K. Balasubramanian and M. Burghard, *Small*, 2005, **1**, 180–192.
7 M. N. Tchoul, W. T. Ford, G. Lolli, D. E. Resasco and S. Arepalli, *Chem. Mater.*, 2007, **19**, 5765–5772.
8 S. Banerjee, T. Hemraj-Benny and S. S. Wong, *Adv. Mater.*, 2005, **17**, 17–29.

9 J. Zhang, H. Zou, Q. Qing, Y. Yang, Q. Li, Z. Liu, X. Guo and Z. Du, *J. Phys. Chem. B*, 2003, **107**, 3712–3718.
10 H. P. Boehm, *Carbon*, 2002, **40**, 145–149.
11 F. Rodríguez-reinoso, *Carbon*, 1998, **36**, 159–175.
12 G. Goncalves, P. A. A. P. Marques, C. M. Granadeiro, H. I. S. Nogueira, M. K. Singh and J. Grácio, *Chem. Mater.*, 2009, **21**, 4796–4802.
13 G. Beamson and D. Briggs, *High resolution XPS of organic polymers: the Scienta ESCA300 database*, Wiley, Chichester, England, New York, 1992.
14 U. Zielke, K. J. Hüttinger and W. P. Hoffman, *Carbon*, 1996, **34**, 983–998.
15 Z. R. Yue, W. Jiang, L. Wang, S. D. Gardner and C. U. Pittman Jr, *Carbon*, 1999, **37**, 1785–1796.
16 D. S. Everhart and C. N. Reilley, *Anal. Chem.*, 1981, **53**, 665–676.
17 C. D. Batich, *Appl. Surf. Sci.*, 1988, **32**, 57–73.
18 A. Chilkoti, B. D. Ratner and D. Briggs, *Chem. Mater.*, 1991, **3**, 51–61.
19 L. A. Langley, D. E. Villanueva and D. H. Fairbrother, *Chem. Mater.*, 2006, **18**, 169–178.
20 K. A. Wepasnick, B. A. Smith, K. E. Schrote, H. K. Wilson, S. R. Diegelmann and D. H. Fairbrother, *Carbon*, 2011, **49**, 24–36.
21 M. Bowker, P. R. Davies, R. J. Davies and D. J. Morgan, *Phys. Chem. Chem. Phys.*, 2014, in preparation.
22 J. K. Edwards, B. Solsona, E. N. N., A. F. Carley, A. A. Herzing, C. J. Kiely and G. J. Hutchings, *Science*, 2009, **323**, 1037–1041.
23 N. Fairley, *CasaXPS*.
24 A. F. Carley and M. W. Roberts, *Proc. R. Soc. London, Ser. A*, 1978, **363**, 403–424.
25 A. F. Carley, P. R. Davies, R. V. Jones, K. R. Harikumar, G. U. Kulkarni and M. W. Roberts, *Surf. Sci.*, 2000, **447**, 39–50.
26 I. Horcas, R. Fernandez, J. Gomez-Rodriguez, J. Colchero, J. Gomez-Herrero and A. Baro, *Rev. Sci. Instrum.*, 2007, **78**, 013705.
27 G. Kresse and J. Hafner, *Phys. Rev. B: Condens. Matter Mater. Phys.*, 1993, **47**, 558–561.
28 G. Kresse and J. Hafner, *Phys. Rev. B: Condens. Matter Mater. Phys.*, 1994, **49**, 14251–14269.
29 G. Kresse and J. Furthmüller, *Phys. Rev. B: Condens. Matter Mater. Phys.*, 1996, **54**, 11169–11186.
30 G. Kresse and J. Furthmüller, *Comput. Mater. Sci.*, 1996, **6**, 15–50.
31 H. J. Monkhorst and J. D. Pack, *Phys. Rev. B: Solid State*, 1976, **13**, 5188–5192.
32 J. P. Perdew, K. Burke and M. Ernzerhof, *Phys. Rev. Lett.*, 1996, **77**, 3865–3868.
33 S. Grimme, *J. Comput. Chem.*, 2006, **27**, 1787–1799.
34 G. Kresse and D. Joubert, *Phys. Rev. B: Condens. Matter Mater. Phys.*, 1999, **59**, 1758–1775.
35 P. Blochl, *Phys. Rev. B: Condens. Matter Mater. Phys.*, 1994, **50**, 17953–17979.
36 R. Franklin, *Acta Crystallogr.*, 1951, **4**, 253–261.
37 G. Bacon, *Acta Crystallogr.*, 1951, **4**, 558–561.
38 G. Graziano, J. Klimes, F. Fernandez-Alonso and A. Michaelides, *J. Phys.: Condens. Matter*, 2012, **24**, 424216.
39 F. Pippig, S. Sarghini, A. Holländer, S. Paulussen and H. Terryn, *Surf. Interface Anal.*, 2009, **41**, 421–429.
40 Y.-Y. Fong, B. R. Visser, J. R. Gascooke, B. C. C. Cowie, L. Thomsen, G. F. Metha, M. A. Buntine and H. H. Harris, *Langmuir*, 2011, **27**, 8099–8104.

41 T. K. Sham, A. Hiraya and M. Watanabe, *Phys. Rev. B: Condens. Matter Mater. Phys.*, 1997, **55**, 7585–7592.
42 H. Altass, A. F. Carley, P. R. Davies, and R. J. Davies, 2013, in preparation.
43 K. Kishi and S. Ikeda, *J. Phys. Chem.*, 1974, **78**, 107–112.
44 A. P. Seitsonen, A. M. Saitta, T. Wassmann, M. Lazzeri and F. Mauri, *Phys. Rev. B: Condens. Matter Mater. Phys.*, 2010, **82**, 115425.
45 T. Wassmann, A. P. Seitsonen, A. M. Saitta, M. Lazzeri and F. Mauri, *Phys. Status Solidi B*, 2009, **246**, 2586–2591.
46 T. Wassmann, A. P. Seitsonen, A. M. Saitta, M. Lazzeri and F. Mauri, *Phys. Rev. Lett.*, 2008, **101**, 096402.
47 T. Wassmann, A. P. Seitsonen, A. M. Saitta, M. Lazzeri and F. Mauri, *J. Am. Chem. Soc.*, 2010, **132**, 3440–3451.
48 I. W. Chiang, B. E. Brinson, R. E. Smalley, J. L. Margrave and R. H. Hauge, *J. Phys. Chem. B*, 2001, **105**, 1157–1161.
49 N. Dementev, X. Feng and E. Borguet, *Langmuir*, 2009, **25**, 7573–7577.

Faraday Discussions

PAPER

Aqueous dispersions of oligomer-grafted carbon nanomaterials with controlled surface charge and minimal framework damage†

Sheng Hu,[a] Shu Chen,[b] Robert Menzel,[a] Angela D. Goode,[b] Mary P. Ryan,[b] Alexandra E. Porter[b] and Milo S. P. Shaffer*[a]

Received 22nd May 2014, Accepted 23rd June 2014
DOI: 10.1039/c4fd00116h

Functionalised carbon nanomaterials (CNMs), with an undamaged carbon framework and controlled physiochemical properties, are desirable for a wide range of scientific studies and commercial applications. The use of a thermochemical grafting approach provides a versatile means to functionalise both multi-walled carbon nanotubes (MWCNTs) and carbon black (CB) nanoparticles without altering their inherent structures. The functionalisation process was investigated by employing various types of grafting monomers; to improve water solubility, reagents were chosen that introduced an ionic character either intrinsically or after further chemical reactions. The degree of grafting for both MWCNTs and CB ranged from 3–27 wt%, as established by thermal gravimetric analysis (TGA). Raman spectroscopy confirmed that the structural framework of the MWCNTs was unaffected by the thermochemical treatment. The effectiveness of the surface modification was demonstrated by significantly improved dispersibility and stability in water, and further quantified by zeta-potential analysis. The concentration of stable, individualised and grafted MWCNTs in water ranged from ~30 to 80 μg mL^{-1} after centrifugation at 10 000 g for 15 min, whereas functionalised CB in water showed improved dispersibility up to ~460 μg mL^{-1}. The successful preparation of structurally identical but differently functionalised nanoparticle panels, with high water compatibility and minimal framework damage, is useful for controlled experiments. For example, they can be used to explore the relationship between toxicological effects and specific physiochemical properties, such as surface charge and geometry.

[a]Department of Chemistry and London Centre for Nanotechnology, Imperial College London, Exhibition Road, London SW7 2AZ, UK. E-mail: m.shaffer@imperial.ac.uk
[b]Department of Materials and London Centre for Nanotechnology, Imperial College London, Exhibition Road, London SW7 2AZ, UK

† Electronic supplementary information (ESI) available. See DOI: 10.1039/c4fd00116h

1. Introduction

Carbon nanomaterials (CNMs), *e.g.*, graphenes, carbon nanotubes (CNTs), and carbon black (CB) nanoparticles, have a significant place in nanoscience due to their extraordinary thermal, mechanical and electronic properties;[1] they have been proposed for an extensive range of applications including in biomedical contexts, such as photothermal therapy,[2,3] drug delivery[4,5] and bioimaging.[6,7] However, the hydrophobic nature of CNMs hinders straightforward liquid-phase dispersion, especially in aqueous media, or in a host polymeric matrix. Consequently, chemical functionalisation methods, including both covalent and non-covalent approaches, have been intensively investigated in an attempt to increase their solution stability. Improved processing and compatibility are considered crucial in order to fully exploit their impressive intrinsic properties.[8–10]

One of the most commonly used covalent functionalisation strategies involves aggressive oxidisation of CNMs with strong acids,[11,12] particularly mixtures of HNO_3 and H_2SO_4, to produce highly concentrated aqueous dispersions. The carboxylic acid functional groups generated during such processes allow amidation or esterification reactions to alter the surface properties of the carbon materials further.[13,14] However, the oxidation process inevitably introduces defects into the carbon framework and reduces the dimensions of both CNTs and graphene (oxide),[15,16] leading to diminished properties.[17,18] In addition, the resulting CNMs are typically contaminated with oxidation debris.[19,20] Direct reaction with the carbon framework can also be achieved by using highly reactive intermediates, such as nitrenes, carbenes and ylides.[21] Unfortunately, these reactions typically involve time-consuming filtrations, expensive reagents and the production of large amounts of liquid waste. These functionalisation routes have limited scalability and hence limited applicability to the current scale of CNM production.

Non-covalent interactions between CNTs and organic materials are attractive since the surface of the CNTs can be modified without disturbing the characteristic π system.[22,23] Numerous studies have shown that non-covalent functionalisation using polymers or surfactants can effectively disperse nanotubes by utilizing multiple weak interactions such as π–π interactions, electrostatic interactions, and hydrophobic–hydrophilic competition, *etc.*[21] However, the major drawback of non-covalent functionalisation lies in the lack of stability of the resulting assemblies due to the intrinsically weak interactions, and the usual need to maintain excess free surfactant (or other modifier) in the suspension.

An attractive thermochemical approach to grafting MWCNTs has been reported, which takes advantage of the intrinsic defect groups on the so-called 'pristine' MWCNT surfaces.[24] Heat treatment under vacuum or in an inert atmosphere generates surface radicals, which are suitable for further reaction. Recently, the grafting approach has been applied to the preparation of highly water-compatible MWCNTs, stabilised by cationic, anionic, or non-ionic grafted species; the thermal activation process was followed by 'grafting from' polymerisations, and hydrophilising post-functionalisation reactions on the initially non-ionic monomers.[25] In contrast to traditional acid oxidation protocols, the thermochemical method has the significant advantages of minimising damage to the nanotube framework whilst avoiding production of debris or the use of corrosive solutions.[24] Here, the thermochemical functionalisation method is extended by using a broader range of monomers with intrinsic cationic or anionic functional groups to modify the surface character of MWCNTs. The goal was to increase the surface charge on the MWCNTs using the same set of pre-existing reactive sites, and to simplify the reaction scheme. Additionally, the approach was applied to CB nanoparticles in order to demonstrate its versatility and to provide a panel of CNMs with identical surface chemical properties but different geometries.

2. Experimental methods

2.1. Materials

MWCNTs (diameter 12.1 nm, s.d. 3.7 nm), synthesised by CVD, were obtained from Arkema SA (Lacq-Mourenx, France). Carbon black nanoparticles (Printex 90, primary particle diameter 21.1 nm, s.d. 6.2 nm) were provided by Degussa-Hüls (Frankfurt, Germany). Potassium 3-(methacryloyloxy)propane-1-sulfonate (SPMAK, 98%), (3-acrylamidopropyl)trimethylammonium chloride solution (75% in H_2O), methyl methacrylate (MMA, >98.5%), 4-vinyl pyridine (4-VP, 95%), poly(ethylene glycol) methacrylate (PEGMA, average $M_n = 530$), 1-iodododecane (IDD, 98%), iodomethane (IMe, ≥99%), and lithium hydroxide (LiOH, 98%) were purchased from Sigma-Aldrich for CNM functionalisation. Before use, all liquid monomer chemicals were passed through a chromatographic column consisting of neutral and basic aluminium oxide powders (aluminium oxide 90 (0.063–0.200 mm), activity stage I for column chromatography, Merck Millipore, Germany) and further

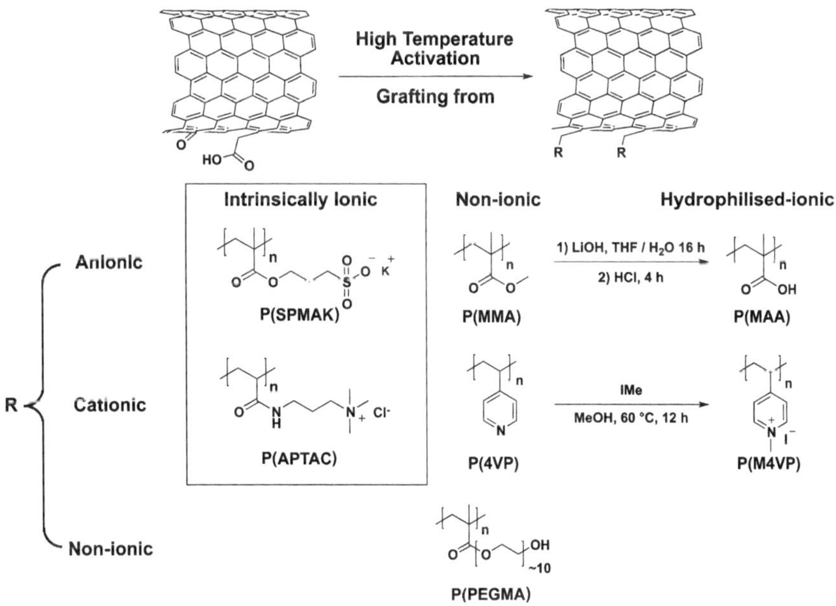

Scheme 1 Reaction scheme for the thermochemical 'grafting from' MWCNTs, comparing intrinsically ionic monomers, non-ionic hydrophilic monomers (PEGMA), and non-ionic monomers (MMA and 4VP) that can be hydrophilised by subsequent reactions to form ionic species (MAA and M4VP respectively).

degassed by bubbling N_2 gas for at least 30 minutes, in order to remove any radical inhibitors and oxygen.

2.2. Functionalisation of CNMs

2.2.1. Thermochemical grafting. The thermal activation process was carried out in a custom-made 30 mm diameter quartz tube attached to a sample flask, and the whole setup was connected to a vacuum system. In a typical experiment, 100 mg of CNM was heated to 1000 °C at a constant ramping rate of 15 °C min^{-1} under vacuum (~5 × 10^{-4} mbar), in a three-zone tube furnace (PTF 12/38/500, Lenton Ltd, UK) and held at the activation temperature for 2 hours. After the activation step, the quartz tube was slowly removed from the heating zone and allowed to cool to room temperature under vacuum. The CNM was then transferred to the connected round bottom flask by gravity. 8 mL of the reactant (for SPMAK, 1 g of compound was dissolved in degassed 1 : 1 v/v ethanol–H_2O co-solvent) was then injected into the flask containing the thermally-activated CNM. The reaction mixture was stirred at room temperature overnight. The unreacted reactant was removed *via* filtration through a 0.10 or 0.45 μm pore size polytetrafluoroethylene (PTFE) membrane (Whatman, UK) under vacuum. The product was thoroughly washed with 3 × 90 mL of washing solvent (details of which solvent was used for each grafting reagent can be found in the ESI†), and then dispersed in 90 mL of solvent and bath sonicated (USC300T, 45 kHz, 80 W, VWR International, USA) for 15 minutes. The filtration–sonication cycle was repeated three times in order to remove any physically absorbed reactants. The functionalised CNMs are named by the abbreviation of the grafted polymer and CNM type: *e.g.* P(SPMAK)-MWCNTs. The other sample codes can be found in Table 1.

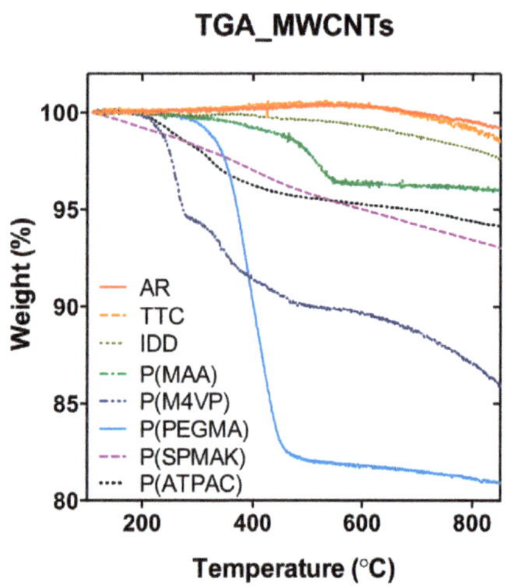

Fig. 1 TGA weight loss profiles of as-received (AR) MWCNTs, thermal treated control (TTC) MWCNTs and various f-MWCNTs, heated in a N_2 atmosphere with a 10 °C min^{-1} ramping rate from 100 °C to 850 °C. Note that the TGA curves and other data within this paper regarding P(MAA), P(M4VP), P(PEGMA), and IDD grafted MWCNTs have been adapted from our previous report.[25]

Table 1 Grafting ratios deduced from weight loss, wt% (weight percentage of grafted oligomers in total grafted sample weight), grafting concentration (moles of monomer per gram of MWCNTs), and estimated degree of polymerisation (number of repeating monomeric units). Repeated measurements give grafting ratios with a standard deviation around 10% of the calculated value. The table summarises the results for the new reagents and those previously reported,[25] calculated based on TGA profiles (Fig. 1)

Sample code	Grafted compound (chemical name)	M_w	Grafting ratio (wt%)	Grafting concentration (μmol g^{-1})	Monomer repeats
P(SPMAK)-MWCNTs	3-Sulfopropyl methacrylate potassium salt	206.7	9.2	490.2	5.8
P(APTAC)-MWCNTs	(3-Acrylamidopropyl) trimethyl ammonium chloride	246.3	5.0	213.7	2.5
P(MAA)-MWCNTs	Methacrylic acid	86.1	3.4	408.8	4.9
P(M4VP)-MWCNTs	*N*-Methyl-4-*vinylpyridine* iodide	247.1	8.7	385.6	4.6
P(PEGMA)-MWCNTs	Poly(ethylene glycol) methacrylate	528.9	18.3	423.5	5.0
IDD-MWCNTs	1-Iodododecane	169.3	1.4	83.9	1.0

2.2.2. Post-functionalisation of f-CB

2.2.2.1. Synthesis of P(M4VP)-CB. P(4-VP)-CB (20 mg) was dispersed in 10 mL of methanol (99.8%, Sigma-Aldrich) by sonication for 5 minutes; IMe (3.12 mL, 50.0 mmol) was added drop-wise, and the reaction mixture was heated to 60 °C overnight under a N_2 atmosphere.[27] Afterwards, the mixture was cooled to room temperature and filtered through a 0.10 μm PTFE membrane. The CB was washed with 3 × 30 mL of ethanol, and then dispersed in 30 mL of ethanol and bath sonicated for 15 minutes. The filtration–sonication cycle was repeated three times in order to remove any physically absorbed reactants.

2.2.2.2. Synthesis of P(MAA)-CB. LiOH (40 mg) was dissolved in 20 mL 10 : 1 v/v THF–water co-solvent before adding 20 mg of P(MMA)-CB. The reaction mixture was bath sonicated for 5 min to obtain a good dispersion, and then stirred at room temperature overnight. Subsequently, 37% hydrochloric acid (HCl, AnalaR grade, BDH) was added drop-wise until the pH value of the solution reached pH 2.[28] The mixture was stirred for another 12 hours, filtered on a 0.10 μm PTFE membrane, and washed with water (3 × 30 mL). The CB residue was dispersed in 30 mL of water by bath sonication for 15 minutes. The filtration–sonication cycle was repeated three times in order to remove any remaining salt and acid.

2.3. Characterisation of the CNMs

2.3.1. Thermogravimetric analysis (TGA). TGA was carried out using a Perkin Elmer Pyris 1, by heating 1.8 ± 0.2 mg CNM samples to 100 °C under a N_2 atmosphere (60 mL min^{-1}), and holding isothermally for 30 minutes to remove residual water and/or solvent; the temperature was then increased from 100 to 850 °C at a constant ramping rate of 10 °C min^{-1} under flowing N_2 (60 mL min^{-1}).

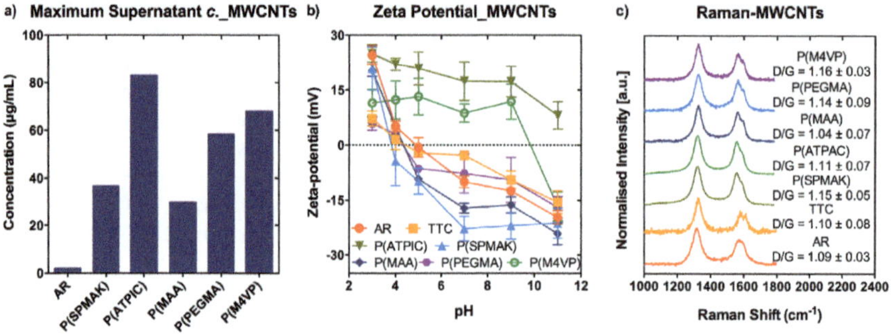

Fig. 2 (a) The supernatant concentrations of various f-MWCNTs were quantitatively determined by their UV absorbances, with a loading concentration of 2 mg mL^{-1}; (b) zeta-potential analysis of AR, TTC and f-MWCNTs as a function of pH. Error bars are mean ± standard deviation (SD) ($n = 5$) and (c) Raman spectroscopy analysis of AR, TTC and f-MWCNTs.

Table 2 Maximum supernatant concentrations (with loading concentrations of 2 mg mL^{-1}), zeta-potentials in pH 7 HPLC water, isoelectric points (IEPs) and D/G ratios of AR and grafted MWCNTs

Sample code	Maximum supernatant concentration in water (µg mL^{-1})	Zeta-potential (mV) in pH 7 HPLC water	IEP	D/G Ratio
AR-MWCNTs	6.0 ± 1.1	−10.0 ± 1.6	4.9	
P(SPMAK)-MWCNTs	36.5 ± 6.5	−22.9 ± 3.4	3.8	1.11 ± 0.07
P(APTIC)-MWCNTs	82.9 ± 2.8	19.2 ± 1.4	N/A	1.15 ± 0.05
P(MAA)-MWCNTs	29.6 ± 7.8	−17.2 ± 1.3	4.3	1.03 ± 0.07
P(PEGMA)-MWCNTs	58.3 ± 4.1	−7.8 ± 5.3	4.6	1.13 ± 0.09
P(M4VP)-MWCNTs	67.8 ± 9.4	8.7 ± 2.4	9.8	1.04 ± 0.03

2.3.2. Zeta-potential measurements. Zeta-potential analysis was performed on a Brookhaven ZetaPALS. f-CNM samples were dispersed in HPLC water with a concentration of 50 µg mL^{-1} by bath sonication for 15 minutes. The pH was adjusted using 0.1 or 0.01 M HCl–NaOH, measured by a digital pH meter (VWR sympHony™ meter, VWR International, USA).

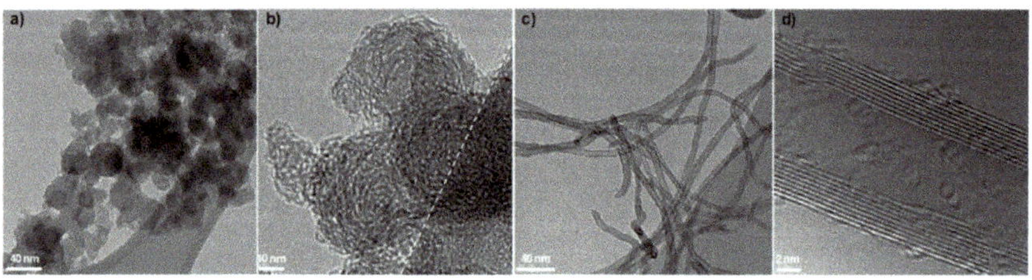

Fig. 3 (a and c) Low resolution and (b and d) high resolution bright field TEM images showing the morphology and crystallinity of CB (a and b) nanoparticles and MWCNTs (c and d). The white dashed line in (b) indicates the edge of the underlying amorphous carbon film of the TEM grid.

2.3.3. Raman spectroscopy. Raman spectra (1000–1800 cm^{-1}) were collected on a LabRam Infinity Raman spectrometer, using a 532 nm laser (scan time 90 s, average 3 scan cycles). The D/G ratio was determined from the ratio of integrated intensity under the Raman bands at around 1350 cm^{-1} (D-band) and 1580 cm^{-1} (G-band) respectively.[29] Average values and standard deviations were obtained from five independent measurements.

2.3.4. UV-vis spectroscopy. The water compatibility of the f-CNMs was quantified by UV-vis spectroscopy (Lambda 950, Perkin Elmer). For water solubility tests, the as-received and f-CNMs were bath sonicated (45 kHz, 80 W, VWR International, USA) in HPLC water for 15 minutes with initial powder loading concentrations of 2 mg mL^{-1}. CNM aggregates were settled by centrifugation for 15 minutes at 10 000*g*. The supernatant was carefully decanted and the concentration of grafted MWCNTs determined by UV absorbance and application of the Beer–Lambert law, $A = \varepsilon cd$, where A is the measured UV absorbance, ε is the extinction coefficient (35.10 mL mg^{-1} cm for Arkema MWCNTs at 800 nm,[30] 19.45 mL mg^{-1} cm for CB nanoparticles at 800 nm) and d is the light path length (1 cm cuvette length in this study).

2.3.5. Specific surface areas. The measurements of adsorption and desorption isotherms of nitrogen at 77 K were carried out on 100 mg CNMs using Micromeritics ASAP 2010 apparatus. Specific surface areas were calculated according to the Brunauer, Emmett and Teller (BET) equation from the adsorption isotherms in the relative pressure range of 0.05–0.20p/p_0.

3. Results and discussion

3.1. MWCNT functionalization

In our previous studies, a thermochemical approach was successfully applied for the preparation of a panel of cationic, anionic and non-ionic functionalized-MWCNTs (f-MWCNTs), *via* modification of the initially uncharged grafted oligomers by hydrolysis or quaternisation. Specifically, poly(methyl methacrylate) (PMMA) was hydrolysed to poly(methacrylic acid) (P(MAA)), *N*-methyl 4-vinylpyridine iodide (M4VP) was obtained from the methylation of 4-*vinylpyridine* (4VP); and the non-ionic poly(ethylene glycol) methacrylate (PEGMA) monomer was grafted as a bulky side chain acrylate (reactions shown in Scheme 1). The f-MWCNTs retained their carbon framework and displayed significantly enhanced water compatibility.[24,25] In order to simplify the synthetic process and further increase the magnitude of surface charge (and hence the colloidal stability of the system), two intrinsically ionic monomers, 3-sulfopropyl methacrylate potassium salt (SPMAK) and (3-acrylamidopropyl) trimethylammonium chloride (APTAC), with explicit anionic and cationic functional groups on their backbones, respectively, were employed to modify the surface characteristics of MWCNTs using the same thermochemical reaction mechanism (refer to Scheme 1).

3.1.1. Characterisation of the f-MWCNTs. The grafting ratios and degrees of polymerisation of the new ionic monomers were characterised based on TGA analysis (Fig. 1 and Table 1). In contrast to the as-received (AR) and non-grafted thermal treated control (TTC) MWCNTs, significant weight losses occurred for the f-MWCNTs, which are attributed to the decomposition of the grafted organic oligomers/polymers. This step indicates that successful oligomer/polymer grafting was achieved. The grafting ratio was calculated based on the difference

Table 3 Grafting ratios of weight loss, wt% (weight percentage of grafted oligomers in the total grafted sample weight), grafting concentrations (moles of monomer per gram of CB), and estimated degrees of polymerisation (number of monomer repeats), calculated based on TGA profiles (Fig. 4b), maximum supernatant concentrations (with a loading concentration of 2 mg mL^{-1}), zeta-potentials in pH 7 HPLC water, isoelectric points (IEPs) and D/G ratios of AR and f-CB

Sample code	Grafting ratio (wt%)	Grafting concentration (μmol g^{-1})	Monomeric repeats	Solubility in water (μg mL^{-1})	Zeta-potential (mV) in pH 7 HPLC water	IEP
AR-CB	N/A	N/A	N/A	149.0 ± 11.7	−3.7 ± 3.2	5.0
TTC-CB	N/A	N/A	N/A	172.1 ± 2.9	−11.0 ± 2.9	5.6
P(MMA)-CB	8.6	940.0	1.9	63.9 ± 3.2	−10.6 ± 1.4	3.8
P(MAA)-CB	5.9	728.2	1.4	205.9 ± 4.6	−14.6 ± 3.2	3.4
P(4-VP)-CB	8.9	929.5	1.8	170.0 ± 5.1	0.7 ± 3.0	7.0
P(M4VP)-CB	13.6	637.0	1.3	230.2 ± 2.7	8.6 ± 2.0	8.1
P(PEGMA)-CB	27.6	720.8	1.4	465.8 ± 9.9	−3.3 ± 1.9	N/A
IDD-CB	7.9	506.7	1.0	N/A	N/A	N/A

Table 4 Comparisons of IDD grafted CB and MWCNTs[25] in terms of grafting ratios of weight loss, wt% (weight percentage of grafted oligomers in total grafted sample weight), grafting concentrations (moles of IDD per gram of CNMs), BET surface areas, and another estimated grafting concentration (moles of IDD per m^2 of carbon surface)

Sample code	Grafting ratio (wt%)	Grafting concentration (μmol g^{-1})	BET surface area (m^2 g^{-1})	Grafting concentration (μmol m^{-2})
IDD-CB	7.9	506.7	270	1.88
IDD-MWCNTs	1.4	84.1	227	0.37

between the weight percentage of the f-MWCNTs and TTC-MWCNTs in order to exclude the effect of minor volatile/absorbed impurities within the AR-MWCNTs. The TTC sample was obtained using the same activation protocol as the functionalisation, but exposing activated MWCNTs to air rather than a monomer at ambient temperatures.[24,25] In order to estimate the number of active sites, the MWCNTs were grafted with the non-polymerisable reagent 1-iodododecane (IDD).[25] A further control of treating TTC-MWCNTs with the monomer MMA, followed by standard workup, indicated that any unreacted monomer was completely removed; thus simple physisorption of monomers on MWCNTs does not contribute to the grafting ratios obtained (Fig. S1†).

The negatively charged P(SPMAK)-MWCNTs have a grafting ratio of 9.2 wt% with a calculated repeating monomeric unit of 5.8, similar to the previously reported anionic P(MAA)-MWCNTs (Table 1). The two types of anionically charged MWCNTs displayed similar decreasing trends in their zeta-potential curves on increasing the solution pH due to the protonation/deprotonation of either sulfonyl or carboxylic acid groups. The P(SPMAK)-MWCNTs displayed a lowered iso-electric point (IEP) of 3.8 compared to P(MAA)-MWCNTs of 4.3, as expected qualitatively from the relative pK_a values of the respective acids. In addition, a

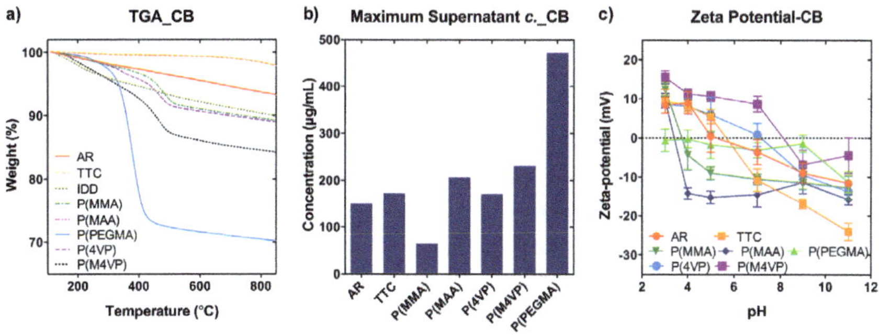

Fig. 4 (a) TGA weight loss profiles of as-received (AR) CB, thermally treated control (TTC) CB and various functionalised CB (f-CB), heated in a N_2 atmosphere with a 10 °C min^{-1} ramp rate from 100 to 850 °C; (b) the supernatant concentrations of various f-CB were quantitatively determined by UV absorbance with a loading concentration of 2 mg mL^{-1}; (c) zeta-potential analysis of AR, TTC and f-CB as a function of pH. Error bars are mean ± standard deviation (SD) ($n = 5$).

more negative zeta-potential in pH 7 aqueous solution was obtained (−22.9 mV) for P(SPMAK)-MWCNTs than P(MAA)-MWCNTs (−17.2 mV), presumably due to the higher grafted concentration. As a result, P(SPMAK)-MWCNTs showed even better water compatibility, with a 23% improvement of water dispersibility compared to P(MAA)-MWCNTs, as measured by UV analysis.

The positively charged P(APTAC)-MWCNTs displayed a 5.0 wt% grafting ratio with a low monomer repeat of only 2.5. Although the grafting concentration and number of monomeric repeats were lower than for the other types of f-MWCNTs, the water dispersibility of P(ATPAC)-MWCNTs was still improved compared to the positively charged P(M4VP)-MWCNTs. The aqueous dispersion of P(ATPAC)-MWCNTs displayed the highest concentration of 82.9 μg mL^{-1} within the f-MWCNTs panel. Again, the surface charge properties were quantified by zeta-potential analysis. The results showed that the P(ATPAC)-MWCNTs sample retained a positive zeta-potential in aqueous solutions even up to pH 11. In pH 7 aqueous solution, the zeta-potential of P(ATPAC)-MWCNTs was +19.2 mV, which is much improved compared to P(M4VP)-MWCNTs (+8.7 mV).

As anticipated, the new ionic monomer grafted MWCNTs appeared undamaged by the process, as semi-quantitatively confirmed by the unchanged D/G ratio and characterized by Raman spectroscopy (Fig. 2c and Table 2).

3.2. Carbon black NP functionalisation

As mentioned earlier, the thermochemical functionalisation approach has great potential to be applied to other types of CNMs, as their surface chemistry is expected to be similar. CNMs produced by gas phase processes, in particular, may be expected to have a similar set of defective surface sites that can be converted to active radicals. As proof of such a concept, CB was selected as a case study to investigate the versatility of the method. Although likely more defective, CB nanoparticles are chemically related, especially to CVD MWCNTs, and hence can provide an interesting control system for the panel of f-MWCNTs, with a different geometry.

As shown in Fig. 3b, the AR-CB displayed layered carbon-sheet structures, but with a much lower crystallinity than the MWCNTs (Fig. 3d). The CB nanoparticles employed in this study had an average primary particle diameter of 21.1 ± 6.2 nm ($N = 100$), as determined by analysis of LR-TEM images (Fig. 3a), which is only slightly larger than that of the MWCNTs used whose average diameter is 12.1 ± 3.7 nm ($N = 100$). Given the similarities in chemical components, diameters and other properties (which will be further discussed below), the two materials can provide interesting comparisons in order to study the geometric effects systematically.

Using an identical preparation process as described above, anionic CB (P(MAA)-CB) was obtained by hydrolysis of hydrophobic methyl methacrylate grafted CB in an alkaline solution. The weak base P(4-VP) was quaternised[26] using iodomethane (IMe) to obtain more stable aqueous dispersions of the protonated polymer (P(M4VP)) grafted CB. TGA confirmed the successful grafting of CB nanoparticles. The TGA profile of the AR-CB displayed continuous weight loss between 100 and 850 °C, due to the lower thermal stability of a less crystalline carbon framework compared to MWCNTs. In contrast to the as-received (AR) and non-grafted thermal treated control (TTC) CB, a second range of significant weight loss occurred at around 200–500 °C for f-CB, which is attributed to the decomposition of the grafted organic oligomers/polymers. This step indicates successful oligomer/polymer grafting.

The grafting ratio of the P(MMA)-CB (8.6 wt%) decreased slightly after the hydrolysis treatment (5.9 wt%) due to the intended loss of methyl groups. The grafting ratio of 4-VP-grafted CB was 8.9 wt%, and the degree of polymerisation was very similar to that of P(MMA)-CB. The TGA profile of the quaternised product retained the features of the P(4-VP)-CB, associated with combustion of the grafted oligomer, with an additional weight loss (4.7%) between 100 and 300 °C, attributed to loss of the methyl group and the associated iodide counter-ion. The higher grafting ratio of P(PEGMA) grafted CB, 27.6 wt%, is partly attributed to the high molecular weight of the side chain of the PEGMA monomer (Table 3). Compared to previously reported f-MWCNTs, the cationic (M4VP), anionic (MAA) and non-ionic (PEGMA) grafted CB samples displayed a similar trend in grafting ratio, namely P(PEGMA) > P(M4VP) > P(MAA) (Table 3).

Based on the grafting ratios of the non-polymerisable reagent IDD, the reactive site concentrations on CB and MWCNTs can be compared. As might be expected from the more defective structure, the number of active sites identified was higher on the CB (approximately 1 per 200 carbons) than the MWCNTs (1 per 1000), both by mass and when normalised to surface area (Table 3). However, the cationic (M4VP), anionic (MAA) and non-ionic (PEGMA) grafted CB did not show a significant increase in the total grafting ratios compared to the MWCNTs grafted with the same monomers. Instead, the apparent number of monomer repeats was consistently reduced. One possible explanation is that the higher density of radicals on the CB surface not only increased the number of polymerisation initiation sites, but also the rate of termination of the polymerization. Termination may either occur by reaction between two short oligomers or the quenching of a growing oligomer by a nearby radical site. It is worth mentioning that in this study, the grafting density and degree of polymerisation are estimated assuming that IDD reacts stoichiometrically with all of the active radical sites and that the polymerisation processes is initiated at a radical site but can terminate anywhere on the CB or MWCNT surface;

in other words each oligomer/polymer chain consumes one radical site.[25] However, if chain–chain or chain–radical site termination dominates, each chain would be associated with two radical sites in total; implementing this assumption would lead to doubled values for estimated chain lengths (monomer repeats) and halved grafting concentrations per unit area. These two alternative estimates represent limits for the true average value. However, a further complication is that the active radical sites on the carbon surface may not be equally accessible to the IDD probe and the monomers used. If steric effects are significant and the monomers are more bulky, then again the real chain length will be longer and the grafting density lower than currently estimated. This effect may be more significant for the high defect density CBs. Further work on the mechanism is needed; current experiments are exploring the use of alkyl iodides with different chain lengths, to explore the influence of steric constraints.

The water solubility of the AR and f-CB samples has been quantified by UV-vis analysis. Due to the high density of oxygen-containing functional groups on CB surfaces, the AR-CB sample already has a moderately high water solubility of 149 μg mL^{-1}. After functionalisation, f-CB, stabilised by cationic, anionic and non-ionic charges, displayed further improved water dispersibility, reaching 230, 206 and 472 μg mL^{-1} after centrifugation respectively. The higher concentration obtained for the P(PEGMA) grafted CB can be attributed to the high steric stabilisation effect of PEG chains and the much higher grafting ratio than cationic and anionic CB.

Zeta potential analysis further confirmed the successful grafting of the CB with anionic, non-ionic and cationic oligomers/polymers (Fig. 4c). The AR-CB and TTC-CB are weakly acidic, with isoelectric points (IEPs) of 5.0 and 5.6, as expected given the surface oxides, including carboxylic acid groups, present on the CB surface. The lowered IEP of 3.4 for P(MAA)-CB compared to P(MMA)-CB (IEP 3.8), confirmed the successful hydrolysis process. Moreover, the improved plateau voltage value of P(MAA)-CB (~−15 mV) compared to P(MMA)-CB (~−10 mV), again, provided proof of a successful reaction. The 4-VP functionalised systems introduce positively-charged groups, shifting the IEP to a higher pH; by introducing a positive charge through quaternisation, the IEP of P(M4VP)-CB was successfully increased to about pH 8.1. The grafted PEGMA is non-ionic and the zeta potentials of aqueous dispersions (up to pH 9) of P(PEGMA)-CB samples remain close to zero, indicating that the original acidic surface oxides are indeed activated and converted to non-ionic PEG-based chains.

4. Conclusion

In this paper, MWCNTs have been successfully grafted with new intrinsically ionic positive and negative monomers using a thermochemical approach, leading to aqueous dispersions of f-MWCNTs with increased surface charge and enhanced water dispersibility compared to previously published results. The degree of functionalisation was quantified by TGA, and a combination of UV-vis and zeta-potential techniques confirmed the altered surface charge and water compatibility of the panel of f-MWCNTs. Further, the thermochemical method was also successfully applied to another important class of CNMs, specifically carbon black (CB). Again, the TGA, UV-vis, and zeta-potential techniques confirmed the successful grafting and water compatibilisation.

Building on previous work, these charged MWCNTs can serve as model materials which can be delivered in aqueous conditions for cytotoxicity and cell uptake investigations; the new monomer systems offer simplified preparation protocols and further improved water solubility. The f-CB samples, with closely matched surface chemical properties, provide excellent controls for the toxicity tests using *in vitro* cell models for exploring geometric effects, and are under investigation. In addition, water dispersible CNMs are desirable in a wide range of applications, from printing conductive inks to assembling filters and electrodes. The versatility of the functionalisation route presented, to address both different classes of carbons and different types of surface grafting, whilst minimizing carbon framework damage, is very attractive. Pure gas phase embodiments of the process avoid time-consuming filtration steps, minimize waste, and are readily scalable. Current investigations are exploring the modification of other types of carbon nanomaterials, including single-walled carbon nanotubes, MWCNT grafted carbon fibres,[31] carbon aerogels, and graphene, as well fundamental aspects of the carbon surface chemistry. The potential applications of such materials will be explored and may be extended by including functional components, for example, pseudocapacitive groups for supercapacitors or metal[24]–oxide[32] catalyst binding groups for heterogeneous reactions.

Acknowledgements

The authors would like to thank the NIEHS (grant # U19ES019536) for the funding of this project. AP acknowledges an individual ERC starting grant for additional funding for SC and ADG (project number: #257182). MSPS/RM acknowledge the Engineering and Physical Sciences Research Council (EPSRC) for additional financial support (grant # EP/G007314/1). The views expressed in this paper are solely of the authors and do not necessarily reflect those of the funding agencies.

References

1 Y. Gogotsi and V. Presser, *Carbon nanomaterials*, CRC Press, 2013.
2 X. Wang, C. Wang, L. Cheng, S.-T. Lee and Z. Liu, *J. Am. Chem. Soc.*, 2012, **134**, 7414–7422.
3 K. Yang, L. Hu, X. Ma, S. Ye, L. Cheng, X. Shi, C. Li, Y. Li and Z. Liu, *Adv. Mater.*, 2012, **24**, 1868–1872.
4 A. A. Bhirde, V. Patel, J. Gavard, G. Zhang, A. A. Sousa, A. Masedunskas, R. D. Leapman, R. Weigert, J. S. Gutkind and J. F. Rusling, *ACS Nano*, 2009, **3**, 307–316.
5 M. Chen, Y. He, X. Chen and J. Wang, *Bioconjugate Chem.*, 2013, **24**, 387–397.
6 A. de la Zerda, Z. Liu, S. Bodapati, R. Teed, S. Vaithilingam, B. T. Khuri-Yakub, X. Chen, H. Dai and S. S. Gambhir, *Nano Lett.*, 2010, **10**, 2168–2172.
7 L. G. Delogu, G. Vidili, E. Venturelli, C. Ménard-Moyon, M. A. Zoroddu, G. Pilo, P. Nicolussi, C. Ligios, D. Bedognetti, F. Sgarrella, R. Manetti and A. Bianco, *Proc. Natl. Acad. Sci. U. S. A.*, 2012, **109**, 16612–16617.
8 N. Karousis, N. Tagmatarchis and D. Tasis, *Chem. Rev.*, 2010, **110**, 5366–5397.
9 V. Georgakilas, M. Otyepka, A. B. Bourlinos, V. Chandra, N. Kim, K. C. Kemp, P. Hobza, R. Zboril and K. S. Kim, *Chem. Rev.*, 2012, **112**, 6156–6214.

10 S. W. Kim, T. Kim, Y. S. Kim, H. S. Choi, H. J. Lim, S. J. Yang and C. R. Park, *Carbon*, 2012, **50**, 3–33.
11 Y. Wang, Z. Iqbal and S. Mitra, *J. Am. Chem. Soc.*, 2006, **128**, 95–99.
12 M. Shaffer, X. Fan and A. H. Windle, *Carbon*, 1998, **36**, 1603–1612.
13 J. M. Gónzalez-Dominguez, M. Gonzalez, A. Anson-Casaos, A. M. Diez-Pascual, M. A. Gomez and M. T. Martinez, *J. Phys. Chem. Lett.*, 2011, **115**, 7238–7248.
14 K. Li, C. Zhang, Z. Du, H. Li and W. Zou, *Synth. Met.*, 2012, **162**, 2010–2015.
15 M. N. Tchoul, W. T. Ford, G. Lolli, D. E. Resasco and S. Arepalli, *Chem. Mater.*, 2007, **19**, 5765–5772.
16 V. Datsyuk, M. Kalyva, K. Papagelis, J. Parthenios, D. Tasis, A. Siokou, I. Kallitsis and C. Galiotis, *Carbon*, 2008, **46**, 833–840.
17 I. Mazov, V. L. Kuznetsov and I. A. Simonova, *Appl. Surf. Sci.*, 2012, **258**, 6272–6280.
18 Y. Zhu, S. Murali, W. Cai, X. Li, J. W. Suk, J. R. Potts and R. S. Ruoff, *Adv. Mater.*, 2010, **22**, 3906–3924.
19 S. Fogden, R. Verdejo, B. Cottam and M. Shaffer, *Chem. Phys. Lett.*, 2008, **460**, 162–167.
20 J. P. Rourke, P. A. Pandey, J. J. Moore, M. Bates, I. A. Kinloch, R. J. Young and N. R. Wilson, *Angew. Chem., Int. Ed.*, 2011, **50**, 3173–3177.
21 S. Banerjee, T. Hemraj Benny and S. S. Wong, *Adv. Mater.*, 2005, **17**, 17–29.
22 M. Tunckol, S. Fantini, F. Malbosc, J. Durand and P. Serp, *Carbon*, 2013, **57**, 209–216.
23 Y. Zhao and J. F. Stoddart, *Acc. Chem. Res.*, 2009, **42**, 1161–1171.
24 R. Menzel, M. Q. Tran, A. Menner, C. Kay, A. Bismarck and M. S. P. Shaffer, *Chem. Sci.*, 2010, **1**, 603–608.
25 S. Chen, S. Hu, E. F. Smith, P. Ruenraroengsak, A. J. Thorley, R. Menzel, A. E. Goode, M. P. Ryan, T. D. Tetley, A. E. Porter and M. S. P. Shaffer, *Biomaterials*, 2014, **35**, 4729–4738.
26 L. Franck-Lacaze, P. Sistat and P. Huguet, *J. Membr. Sci.*, 2009, **326**, 650–658.
27 S. Alunni, V. Laureti, L. Ottavi and R. Ruzziconi, *J. Org. Chem.*, 2003, **68**, 718–725.
28 S. P. Runyon, L. E. Brieaddy, S. W. Mascarella, J. B. Thomas, H. A. Navarro, J. L. Howard, G. T. Pollard and F. I. Carroll, *J. Med. Chem.*, 2010, **53**, 5290–5301.
29 M. S. Dresselhaus, G. Dresselhaus, R. Saito and A. Jorio, *Phys. Rep.*, 2005, **409**, 47–49.
30 M. Q. Tran, C. Tridech, A. Alfrey, A. Bismarck and M. Shaffer, *Carbon*, 2007, **45**, 2341–2350.
31 H. Qian, A. Bismarck, E. S. Greenhalgh and M. S. P. Shaffer, *Compos. Sci. Technol.*, 2010, **70**, 393–399.
32 A. Garcia-Gallastegui, D. Iruretagoyena, M. Mokhtar, A. M. Asiri, S. N. Basahel, S. A. Al-Thabaiti, A. O. Alyoubi, D. Chadwick and M. S. P. Shaffer, *J. Mater. Chem.*, 2012, **22**, 13932–13940.

PAPER

Water-soluble fullerene materials for bioapplications: photoinduced reactive oxygen species generation

Yoko Yamakoshi,*[ab] Safwan Aroua,[ab] Thach-Mien Duong Nguyen,[b] Yuko Iwamoto[a] and Tsuyoshi Ohnishi[c]

Received 23rd April 2014, Accepted 16th May 2014
DOI: 10.1039/c4fd00076e

The photoinduced reactive oxygen species (ROS) generation from several water-soluble fullerenes was examined. Macromolecular or small molecular water-soluble fullerene complexes/derivatives were prepared and their $^1O_2$ and $O_2^{\bullet -}$ generation abilities were evaluated by ESR spin-trapping methods. As a result, efficient $^1O_2$ generation was detected from photoexcited C_{60}, not only in organic solvents, but also in aqueous media and especially from small molecule C_{60}–carboxylic acid derivatives. Whereas efficient $O_2^{\bullet -}$ generation was observed in the aqueous solution of the C_{60}/γ-CD complex under photoirradiation.

1 Introduction

Due to the unique physical and chemical properties of fullerenes – such as photosensitivity and metal encapsulation – biological applications of fullerenes for photodynamic therapy (PDT) and MRI contrast enhancement are considered promising areas of research. The photosensitivity of C_{60} and C_{70}, which was initially reported by Foote and co-workers in the early 1990s,[1,2] is unique and useful due to (1) high quantum yield in $^1O_2$ generation and (2) long excitation wavelength. The quantum yield of $^1O_2$ generation by benzene was estimated to be 1.00. The lifetimes of the singlet excited state and the triplet excited state of C_{60} are 1.2 ns and 250 μs respectively. Since both of these properties are ideal for PDT agents, many research groups became interested in using fullerenes as photobiomaterials. The main obstacle of such applications of fullerenes was their extremely low solubility, especially in water or water-miscible solvents due to their hydrophobic surfaces. Therefore, the derivatization or complexation of fullerenes with water-

[a]Laboratorium für Organische Chemie, ETH-Zürich, Vladimir-Prelog-Weg 3, CH8093 Zürich, Switzerland. E-mail: yamakoshi@org.chem.ethz.ch

[b]Department of Radiology, University of Pennsylvania Perelman School of Medicine, Philadelphia PA19104, USA

[c]Department of Biochemistry and Biophysics, University of Pennsylvania Perelman School of Medicine, Philadelphia PA19104, USA

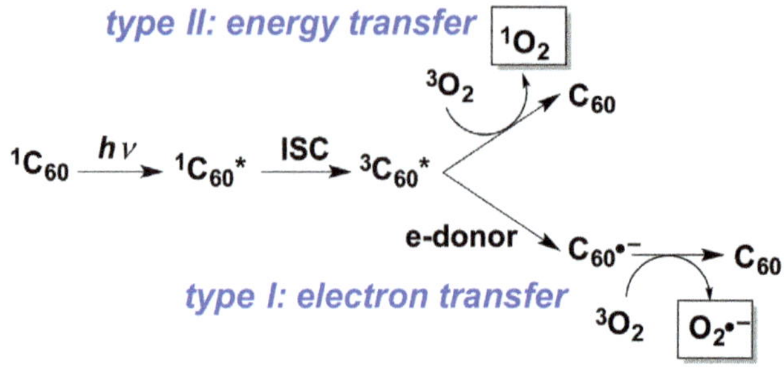

Fig. 1 Two pathways of generating reactive oxygen species (ROS) from photoexcited C_{60}. The *type II* pathway is *via* an energy transfer reaction from the triplet excited state of C_{60} ($^3C_{60}^*$) to generate singlet oxygen ($^1O_2$) and the *type I* pathway is *via* an electron transfer reaction from the C_{60} radical anion ($C_{60}^{\bullet-}$) to generate the superoxide radical anion ($O_2^{\bullet-}$) which will be further converted to the more reactive hydroxyl radical ($^{\bullet}OH$) to cleave DNA molecules.

soluble moieties (ideally biocompatible ones) is essential. The initial studies on bioactive C_{60} derivatives, reported independently in 1993 by the Nakamura group[3] and the Wudl group,[4] showed the preparation of water-soluble C_{60} materials by covalent attachment of water-soluble moieties that enabled *in vitro* biological assays of C_{60}. In their study, the Nakamura-group reported the $^1O_2$ generation from their C_{60} derivatives under photoirradiation, and DNA damage *in vitro*.

According to the reports by Foote and co-workers, photoexcited C_{60} can cause both an energy transfer reaction (*type II* in Fig. 1) to generate a singlet oxygen ($^1O_2$)[1] and an electron transfer reaction (*type I* in Fig. 1) to generate a C_{60} radical anion ($C_{60}^{\bullet-}$).[1,5] The *type I* reaction mechanism is possible due to the redox potential of the triplet excited state of C_{60} ($^3C_{60}^*$) ($^3C_{60}^*$: $E_1 = +1.14$ V *vs.* SCE in PhCN, $^1C_{60}$: $E_1 = -0.42$ V *vs.* SCE in PhCN), and an electron transfer reaction can occur easily in the presence of an electron donor to provide $C_{60}^{\bullet-}$ (the quantum yield of the electron transfer of C_{60} was 1.00 in benzonitrile and 0.52 in benzonitrile–benzene (1 : 1)). In our previous study, we reported the observation of a superoxide radical anion ($O_2^{\bullet-}$) *via* $C_{60}^{\bullet-}$ generated in the presence of physiological concentrations of a biological electron donor (NADH) and efficient DNA cleavage.[6,7] However, the mechanism and conditions that control which reaction pathway (*type I* or *II*) is favoured are still unclear. It is important to clarify the mechanism of ROS generation from photoexcited C_{60} in bio-relevant conditions for future clinical applications and for evaluation of the safety/toxicity of these nanomaterials.

In this study, we have prepared four kinds of water-soluble C_{60} materials by (1) complexation with poly(vinylpyrrolidone) (PVP) (C_{60}/PVP complex, **1** in Fig. 2), (2) copolymerization with PVP (C_{60}–PVP copolymer, **2** in Fig. 2), (3) complexation with γ-cyclodextrin (γ-CD) (C_{60}/γ-CD, **3** in Fig. 2), and (4) chemical functionalization of C_{60} (C_{60}–carboxylic acid, **4** in Fig. 2). We have tested their ability to generate ROS and evaluated their properties as PDT agents.

2 Experimental

2.1 Preparation of water-soluble C_{60} complexes or derivatives

General. C_{60} (99.5%) was purchased from MTR Ltd. (Cleveland, OH, USA). All of the other reagents were purchased from corresponding suppliers and purified

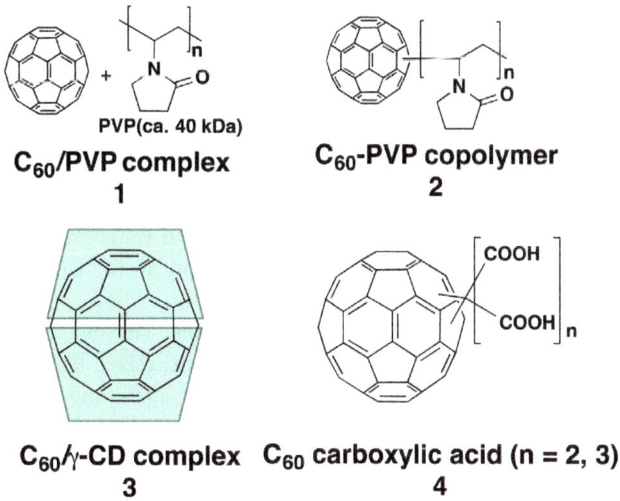

Fig. 2 The water-soluble C_{60} materials used in this study. Two macromolecular materials, C_{60}/PVP complex **1** and C_{60}–PVP copolymer **2**, and two small molecular materials C_{60}/γ-CD complex **3** and C_{60}–carboxylic acid **4**.

as described when needed. All of the solvents were purchased from Acros Organic (Thermo Fisher Scientific, Inc., Geel, Belgium).

C_{60}/PVP complex. The C_{60}/PVP complex was prepared from C_{60} and poly(vinylpyrrolidone) K-30 (Alfa Aesar, Warf Hill, MA, USA) based on the previously reported procedure.[8] A toluene solution of C_{60} and a chloroform solution of PVP were mixed well and slowly concentrated *in vacuo* to give a thin film of the complex. After drying the residue under high vacuum, Milli-Q water was added to provide a transparent brownish solution of the C_{60}/PVP complex.

C_{60}–PVP copolymer. The C_{60}–PVP (1 : 200) copolymer[9] was prepared by the copolymerization of C_{60} and 200 equiv. of *N*-vinylpyrrolidone (NVP, Acros, distilled before use) in the presence of azobisisobutyronitrile (AIBN, Aldrich, 98%, 0.2 equiv. to NVP) as a radical initiator. The reaction was carried out in dichlorobenzene at 60 °C (Scheme 1).[9] A solution of C_{60} (72.0 mg, 0.10 mmol), AIBN (656 mg, 4.0 mmol), and *N*-vinylpyrrolidone (distilled, 2.22 g, 20.0 mmol) in *o*-dichlorobenzene (2.5 mL) was degassed four times and stirred at 60 °C under a N_2 atmosphere until the substance became very viscous (2 days). $CHCl_3$ (2 mL) was added to the reaction mixture to provide a uniform dark brown solution, which was subsequently added to Et_2O (200 mL) to precipitate the C_{60} polymer. The resulting precipitate was separated by centrifugation and the residue was

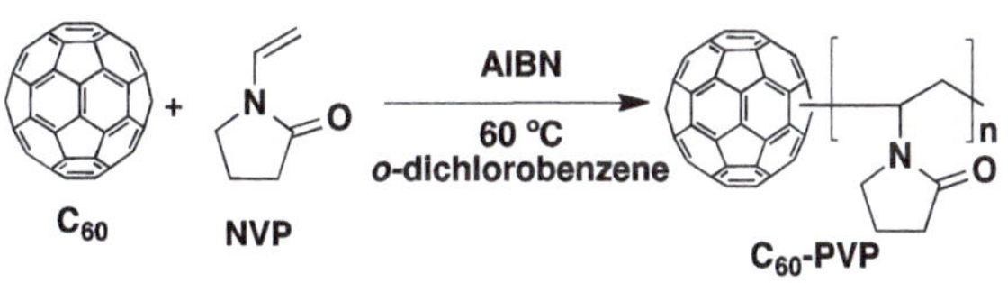

Scheme 1 Preparation of C_{60}–PVP by radical polymerization of NVP in the presence of C_{60} as an end-cap reagent.

again dissolved with $CHCl_3$ and precipitated in Et_2O (2 times). The final powder was collected and dried under vacuum to obtain the C_{60}–PVP copolymer (1 : 200) (2.2 g, isolated yield 96%).

C_{60}/γ-cyclodextrin complex. Solutions of C_{60} (40 mg) in toluene (20 mL) and γ-cyclodextrin (γ-CD, TCI, ≥98%, 134 mg) in water (20 mL) were mixed in a flask and refluxed for 30 hours with vigorous stirring. The reaction mixture was cooled to room temperature and the aqueous layer was separated and centrifuged for 30 min (4000 rpm). After removing the supernatant, cold water was added to the precipitate and centrifuged again (4000 rpm, 40 min at 10 °C). The obtained precipitate was dried under high vacuum for 15 hours.

Bis- and tris-Bingel C_{60}–carboxylic acid derivatives. The bis- and tris-Bingel adducts of C_{60} ($C_{60}(C(COOH)_2)_n$, $n = 2$ or 3) were synthesized based on a previously reported procedure.[10] Briefly, to a solution of C_{60} (493 mg, 0.68 mmol) in toluene (500 mL), NaH (60%, Aldrich, 1.46 g, 34 mmol as NaH) and subsequently 3 equiv. of diethyl bromomalonate (350 μL, 2.05 mmol, prepared by bromination of diethylmalonate) were added and stirred at room temperature under N_2 for 12 hours. Additional NaH (60%, 1.16 g) was added and the reaction mixture was stirred for another 12 hours. The reaction mixture was filtered through Celite, concentrated *in vacuo*, and purified by SiO_2 column chromatography (SiO_2 150 g, solvent: hexane–toluene) to separate the mono adduct (eluted by hexane–toluene (3 : 7)), bis-adducts (eluted by hexane–toluene (1 : 4)) and tris-adducts (eluted by hexane–toluene (1 : 1)). The obtained bis- and tris-adducts were then subjected to deprotection (by NaH and MeOH) without further isolation of each regioisomer (regioisomeric mixture).

2.2 Detection of reactive oxygen species by ESR and chemical reaction

General. ESR spectra were recorded on a Bruker spectrometer (Bruker BioSpin GmbH, Rheinstetten, Germany) equipped with a microwave bridge X-band ER 082C. Light irradiation and ESR measurements were carried out on the sample in a disposable capillary with 50 μL volume (Wiretrol, Drummond, Broomall, PA, USA), which was placed inside an ESR tube with 4 mm thin wall precision quartz, 250 mm in length (Wilmad, Vineland, NJ, USA).

$^1O_2$ generation by the ESR spin-trapping method. $^1O_2$ was detected by an ESR method using 2,2,6,6-tetramethyl-4-piperidone (4-oxo-TEMP, Acros, 98%) as a spin-trapping reagent. To 0.5 mM fullerene (20 μL) in distilled water, 1 M 4-oxo-TEMP (10 μL), 1 M phosphate buffer (5 μL), and distilled water (65 μL) were added and mixed well under aerobic conditions. The mixed solution was introduced into a capillary, placed inside an ESR tube, and irradiated by a 200 W photoreflector lamp at a distance of 10 cm for 5 min, and then immediately subjected to ESR measurements. The generation of singlet oxygen was detected by an ESR signal corresponding to 4-oxo-TEMPO, formed by the reaction of $^1O_2$ with 4-oxo-TEMP. Measurement conditions: temperature 296 K, microwave frequency 9.375 GHz, microwave power 10 mW, receiver gain 2.0×10^3, modulation amplitude 4.00 G, modulation frequency 100 kHz, sweep time 20.48 s.

$^1O_2$ generation by the ADPA method. A mixture of C_{60} material (50 μM) and anthracene dipropionic acid (ADPA, Sigma-Aldrich, 50 μM) in PBS(−) (1 mL) was irradiated in a quartz cell by a 200 W photoreflector lamp at a distance of 15 cm for 1 min. The decrease in the absorbance at 400 nm was measured as a relative amount of $^1O_2$ generation, which converted ADPA (with absorbance at 400 nm) to a cycloadduct (no absorbance at 400 nm).

$O_2^{\bullet-}$ generation by the ESR spin-trapping method. $O_2^{\bullet-}$ was detected by an ESR method using 5-(diethoxyphosphoryl)-5-methyl-1-pyrroline-*N*-oxide (DEPMPO, Enzo Life Sciences, ≥99%) as a spin-trapping reagent. To a 0.5 mM fullerene aqueous solution (20 μL), 565 mM DEPMPO in water (8 μL), NADH (10 μL), 5 mM diethylenetriaminepentaacetic acid (20 μL), DMSO (26 μL), 1 M phosphate buffer (5 μL), and water (11 μL) were added and mixed well under aerobic conditions. The mixed solution was collected in a capillary (50 μL) inside of an ESR tube and irradiated with a 200 W photoreflector lamp at a distance of 10 cm for 1 min. After light irradiation, the sample was immediately subjected to ESR measurements. The generation of $O_2^{\bullet-}$ was detected as signals corresponding to DEPMPO–OOH, formed by the reaction of $O_2^{\bullet-}$ with DEPMPO. Measurement conditions: temperature 296 K, microwave frequency 9.375 GHz, microwave power 10 mW, receiver gain 2.0×10^3, modulation amplitude 4.00 G, modulation frequency 100 kHz, sweep time 20.48 s.

3 Results and discussion

Four kinds of water-soluble C_{60} materials (Fig. 2) were prepared to compare their ROS generation properties. The C_{60}/PVP complex and C_{60}–PVP copolymer are both polymeric materials made by a combination of C_{60} with the water-soluble non-ionic polymer PVP. The advantage of using PVP is its lack of toxicity, and it has been used as a dispersion reagent in medicines and cosmetics for a long time. In addition, such polymeric materials have advantages in their enhanced permeation and retention (EPR) effects on tumors and other inflammatory diseases. Compared to C_{60}/PVP, with a water-solubility of 0.56 mM, the C_{60}–PVP copolymer with covalently bound PVP displayed better solubility (10–15 mM). Both the PVP complex and copolymer can also be obtained using C_{70} as the fullerene core instead of C_{60}. C_{60}/γ-CD was prepared by complexation in a two-solvent system (toluene and water) and displayed sufficient water-solubility. The tris-Bingel adduct of the C_{60}–carboxylic acid was prepared by the Hirsch–Bingel reaction of C_{60} and a bromomalonate derivative, and isolated as a mixture of regioisomers of tris-adducts (separated from the mono- and bis-adducts). Since the tris-adduct was thoroughly soluble in water after deprotection of the ethyl esters compared to the mono- and bis-adducts (the deprotected bis-adduct was also water-soluble but the solubility of the tris-adduct was better), the tris-adduct (as a mixture of regioisomers) was used for the ROS generation study.

First, $^1O_2$ generation through the *type II* energy transfer pathway of each water-soluble C_{60} material was measured under photoirradiation by an ESR method in the presence of 2,2,6,6-tetramethyl-4-piperidone (4-oxo-TEMP) as a spin-trapping reagent, which reacts with $^1O_2$ to provide 4-oxo-TEMPO (see the scheme in Fig. 3). Rose bengal was used as a standard compound for photoinduced $^1O_2$ generation in aqueous solution. As shown in Fig. 3a, strong peaks corresponding to the 4-oxo-TEMPO were observed in the rose bengal solution. Interestingly, although there is no generation of $^1O_2$ in the C_{60}/PVP complex **1** solution (Fig. 3c), a significant generation of $^1O_2$ was observed for the C_{60}–PVP copolymer **2** solution, in good agreement with our previous report (Fig. 3d). This $^1O_2$ generation was observed with even higher efficiency with the C_{70}–PVP copolymer (data not shown). In our

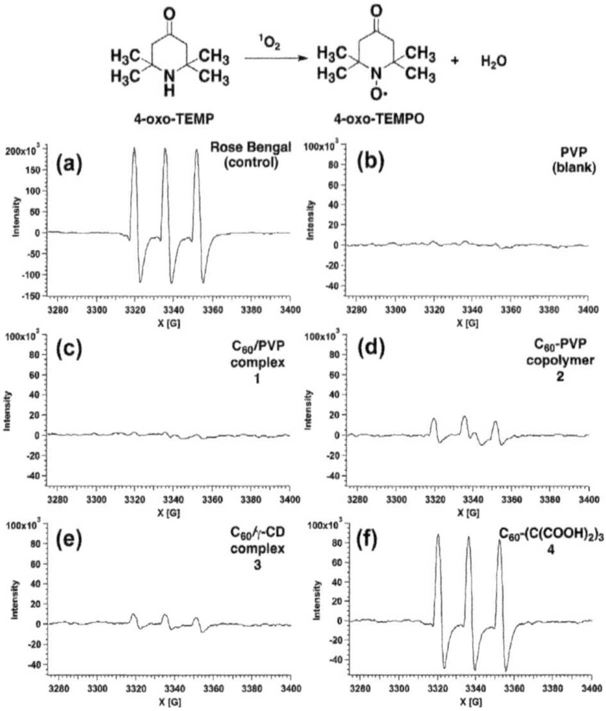

Fig. 3 X-band ESR spectra of the 4-oxo-TEMP adduct with $^1O_2$ generated in rose bengal (a), PVP (b), C_{60}/PVP complex (c), C_{60}–PVP copolymer (d), C_{60}/γ-CD complex (e), and C_{60}–carboxylic acid tris-Bingel derivative (f) in aqueous solution under irradiation of a 200 W photoreflector lamp (1 min). 4-Oxo-TEMP 80 mM in 50 mM phosphate buffer (pH 7). Rose bengal or C_{60} 0.04 mM, PVP 0.4%. Experimental conditions: temperature 296 K, microwave frequency 9.375 GHz, microwave power 10 mW, modulation amplitude 4.00 G, modulation frequency 100 kHz, scan time 20.5 s.

previous paper, we thought that photoexcited C_{60} provides $^1O_2$ in less polar organic solvents such as benzene or benzonitrile, but not in aqueous media.[7] However based on this current result, photoexcited C_{60} can indeed generate $^1O_2$ in polar solvents. The difference in $^1O_2$ generation between the C_{60}/PVP complex and C_{60}–PVP copolymer could be explained by the rate of aggregation. It is known that pristine C_{60} aggregates do not generate ROS due to the self-quenching of $^3C_{60}^*$.[11] Since the solubility of the C_{60}–PVP copolymer is better than the C_{60}/PVP complex, the C_{60}–PVP solution is less aggregated than the C_{60}/PVP solution, which allows better $^1O_2$ generation.

Interestingly, the two small molecule C_{60} materials (C_{60}/γ-CD **3** and C_{60}–carboxylic acid **4**) showed completely different $^1O_2$ generation properties. Whereas the C_{60}/γ-CD complex showed only a little generation of $^1O_2$ (Fig. 3e), the C_{60}–carboxylic acid showed very efficient $^1O_2$ generation (Fig. 3f). The $^1O_2$ generation of the C_{60}–carboxylic acid tris-Bingel adduct **4** was higher than for the bis-Bingel adduct with less substitution (data not shown), presumably due to the solubility differences (generally $^1O_2$ generation from C_{60} derivatives decreases with more substitution[12]). This $^1O_2$ generation from **4** was high and in the same range as from the rose bengal.

$^1O_2$ generation was also measured by an alternative chemical method using anthracene dipropionic acid (ADPA) as a diene for the cycloaddition of $^1O_2$. The relative amount of $^1O_2$ generation was measured by the decrease of OD_{400}, corresponding to the decay of ADPA to a cycloadduct, and was in good agreement with the results detected by the ESR method (Fig. 4).

Next, the generation of $O_2^{\bullet-}$ through the alternative *type I* electron transfer pathway was measured by the ESR method using DEPMPO (5-(diethoxy-phosphoryl)-5-methyl-1-pyrroline-*N*-oxide) as a spin-trapping reagent.[13] Compared to DMPO (5,5-dimethyl-1-pyroline-*N*-oxide) which is commonly used for the detection of $O_2^{\bullet-}$, the use of DEPMPO has advantages in the quantitative detection of $O_2^{\bullet-}$, since the $O_2^{\bullet-}$ adduct of DEPMPO (DEPMPO–OOH) has a sufficiently longer lifetime. Photoirradiation was carried out in the presence of a physiological concentration of NADH as a biological electron donor. The peaks of the DEPMPO adducts observed are shown in Fig. 5.

Consistent with our previous results,[7] $O_2^{\bullet-}$ generation from the C_{60}/PVP complex under photoirradiation was clearly observed (Fig. 5a). Similarly, the C_{60}–PVP copolymer also generated $O_2^{\bullet-}$ in the same range of efficiency but with a slight increase (Fig. 5b). Very interestingly, the generation of $O_2^{\bullet-}$ from the C_{60}/γ-CD complex in aqueous solution was very efficient (Fig. 5c). This may be explained by the lack of self-aggregation of C_{60} in the C_{60}/γ-CD system due to the complete bi-cap encapsulation of C_{60} by γ-CD (leaving no chance that any C_{60} can interact with any other C_{60} molecules) and it does not cause any self-quenching of $^3C_{60}^*$.

$O_2^{\bullet-}$ generation was not observed in the C_{60}–carboxylic acid solutions (Fig. 5d). Instead, peaks corresponding to a $^{\bullet}CH_3$ adduct of DEPMPO (DEPMPO–CH_3, scheme in the caption of Fig. 5) were clearly observed. In general, the generation of $^{\bullet}CH_3$ in such an $O_2^{\bullet-}$ generation detection system, with spin-trapping reagents (DMPO or DEPMPO), is observed as a result of the reaction of $^{\bullet}OH$ and DMSO, which is added to quench $^{\bullet}OH$ whilst simultaneously generating $^{\bullet}CH_3$. (The spin-

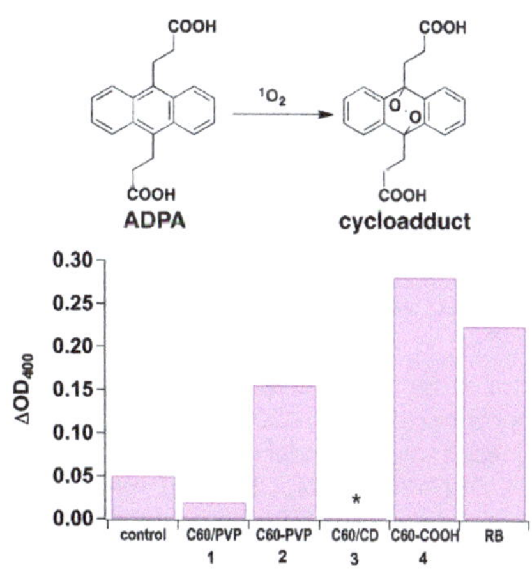

Fig. 4 Detection of $^1O_2$ generation from photoexcited C_{60} materials by the ADPA method. C_{60} material: 50 μM, anthracene dipropionic acid (ADPA): 50 μM, in PBS(−) (1 mL). Irradiation: 200 W photoreflector lamp at a distance of 15 cm for 1 min. The generation of $^1O_2$ was estimated by the decrease in OD_{400}. *In the measurements of the C_{60}/γ-CD complex, insoluble precipitates were observed during the light irradiation and did not provide reliable data.

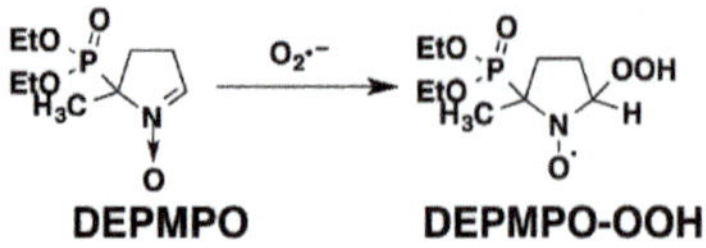

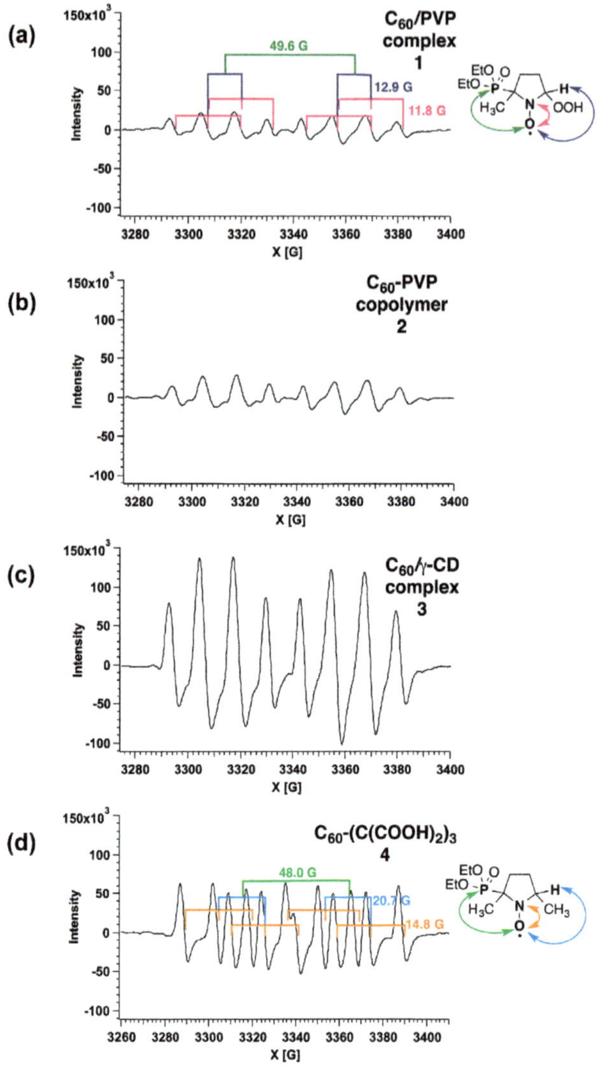

Fig. 5 X-band ESR spectra of the DEPMPO adduct with $O_2^{\bullet -}$ generated by the C_{60}/PVP complex (a), C_{60}–PVP copolymer (b), C_{60}/γ-CD complex (c), and C_{60}–carboxylic acid tris-Bingel derivative (d) in aqueous solution under irradiation of a 200 W photoreflector lamp (1 min). DEPMPO 50 mM, DETAPAC 1 mM, DMSO 3.1 mM in 50 mM phosphate buffer (pH 7), C_{60} 0.2 mM, PVP 2%, NADH 10 mM. Experimental conditions: temperature 296 K, microwave frequency 9.375 GHz, microwave power 10 mW, modulation amplitude 4.00 G, modulation frequency 100 kHz, scan time 20.5 s.

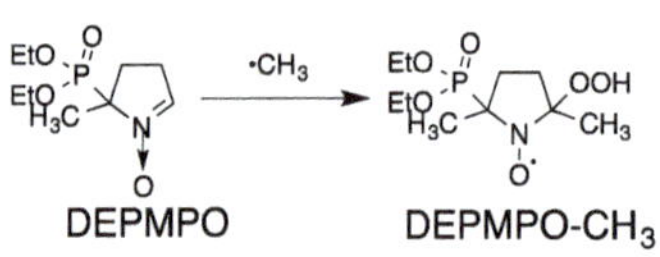

Table 1 Summary of ROS generation from water-soluble C_{60} materials

	C_{60} material	Type of material		Solubility [mM]	$^1O_2$	$O_2^{\bullet -}$
1	C_{60}/PVP	Macromolecule	Complex	0.56 (ref. 8)	—	++
2	C_{60}–PVP	Macromolecule	Derivative	10–15 (ref. 9)	++	++
3	C_{60}/γ-CD	Small molecule	Complex	0.1 (ref. 14)	+	+++
4	C_{60}–$(C(COOH)_2)_3$	Small molecule	Derivative	75 (ref. 15)	+++	($^{\bullet}CH_3$)

trapping reagents can react not only with $O_2^{\bullet -}$ but also $^{\bullet}OH$ with a much faster reaction rate and providing a more stable adduct, which disturbs the detection of the $O_2^{\bullet -}$ adduct's signal.) Therefore, it seemed that $^{\bullet}OH$ was generated much more efficiently in a solution of C_{60}–carboxylic acid. However in a separate experiment of $^{\bullet}OH$ detection using DMPO as a spin-trapping reagent, there is no significant difference in the generation of $^{\bullet}OH$ between the C_{60}–PVP system and the C_{60}–carboxylic acid system (data not shown). Therefore, the mechanism of this $^{\bullet}CH_3$ generation is still not clear.

The results of the ROS generation of four water-soluble C_{60} materials are summarized in Table 1. Two small molecule C_{60} materials, C_{60}/γ-CD complex **3** and C_{60}–carboxylic acid derivative **4** showed completely opposite tendencies in the generation of $^1O_2$ and $O_2^{\bullet -}$. Whereas **4** generated $^1O_2$ (no $O_2^{\bullet -}$), **3** generated $O_2^{\bullet -}$ (little $^1O_2$). It has been reported that the charge separation of a $^3C_{60}$-electron donor is dependent on the polarity of the environment.[16] This phenomenon in the current results may be explained by the polarity around the C_{60} molecules. In C_{60}/γ-CD, the environment around the C_{60} is rather polar (note that CD is a sugar derivative without aromatic moieties), which provides a longer lifetime of charge-separation. Further experiments such as laser-flash-photolysis will be required to prove this speculation. The reason behind the generation of $^{\bullet}CH_3$ in the solution of C_{60}–carboxylic acid is not clear. Further ESR spin-trapping experiments using various scavengers (such as NaN_3, L-His, and 2,5-dimethylfuran for $^1O_2$ and superoxide dismutase for $O_2^{\bullet -}$) in combination with photo DNA cleavage tests are needed to clarify the mechanism.

4 Conclusions

Four different types of water-soluble C_{60} materials were prepared and tested for their photoinduced ROS generation abilities using ESR spin trapping methods. All of the C_{60} materials clearly showed the generation of either $^1O_2$, produced *via* an energy transfer mechanism, or $O_2^{\bullet -}$, *via* an electron transfer mechanism. Although all of the reactions were carried out in aqueous media, the tendency towards the generation of $^1O_2$ or $O_2^{\bullet -}$ was clearly different. We speculate that this is due to the aggregation state and polarity of the environment around the C_{60} itself. Further experiments in combination with various scavengers for $^1O_2$ and $O_2^{\bullet -}$ and *in vitro* photo DNA cleaving tests are required for further evaluation of the mechanism of ROS generation from C_{60} in aqueous media.

Acknowledgements

This research was supported in part by the Swiss National Foundation (200021_140451), ETH Individual Investigator's Research Award (ETH-25 11-1),

PRESTO program from JST, CEET pilot grant from the University of Pennsylvania, and American Heart Association Researcher's Developing Grant (0930140N). Authors thank Prof. Dr. E. Ogiso-Nakamaru and Tomoko Ohnishi (University of Pennsylvania) for their help in ESR measurements.

Notes and references

1 J. W. Arbogast, A. P. Darmanyan, C. S. Foote, Y. Rubin, F. N. Diederich, M. M. Alvarez, S. J. Anz and R. L. Whetten, *J. Phys. Chem.*, 1991, **95**, 11–12.
2 J. W. Arbogast and C. S. Foote, *J. Am. Chem. Soc.*, 1991, **113**, 8886–8889.
3 H. Tokuyama, S. Yamago, E. Nakamura, T. Shiraki and Y. Sugiura, *J. Am. Chem. Soc.*, 1993, **115**, 7918–7919.
4 R. Sijbesma, G. Srdanov, F. Wudl, J. A. Castoro, C. Wilkins, S. H. Friedman, D. L. Decamp and G. L. Kenyon, *J. Am. Chem. Soc.*, 1993, **115**, 6510–6512.
5 J. W. Arbogast, C. S. Foote and M. Kao, *J. Am. Chem. Soc.*, 1992, **114**, 2277–2279.
6 Y. Yamakoshi, S. Sueyoshi, K. Fukuhara and N. Miyata, *J. Am. Chem. Soc.*, 1998, **120**, 12363–12364.
7 Y. Yamakoshi, N. Umezawa, A. Ryu, K. Arakane, N. Miyata, Y. Goda, T. Masumizu and T. Nagano, *J. Am. Chem. Soc.*, 2003, **125**, 12803–12809.
8 Y. Yamakoshi, T. Yagami, K. Fukuhara, S. Sueyoshi and N. Miyata, *J. Chem. Soc., Chem. Commun.*, 1994, 517–518.
9 S. Oriana, S. Aroua, J. O. B. Sollner, X. J. Ma, Y. Iwamoto and Y. Yamakoshi, *Chem. Commun.*, 2013, **49**, 9302–9304.
10 I. Lamparth and A. Hirsch, *J. Chem. Soc., Chem. Commun.*, 1994, 1727–1728.
11 J. Lee, Y. Yamakoshi, J. B. Hughes and J. H. Kim, *Environ. Sci. Technol.*, 2008, **42**, 3459–3464.
12 T. Hamano, K. Okuda, T. Mashino, M. Hirobe, K. Arakane, A. Ryu, S. Mashiko and T. Nagano, *Chem. Commun.*, 1997, 21–22.
13 C. Frejaville, H. Karoui, B. Tuccio, F. Lemoigne, M. Culcasi, S. Pietri, R. Lauricella and P. Tordo, *J. Chem. Soc., Chem. Commun.*, 1994, 1793–1794.
14 K. I. Priyadarsini, H. Mohan, A. K. Tyagi and J. P. Mittal, *J. Phys. Chem.*, 1994, **98**, 4756–4759.
15 L. L. Dugan, D. M. Turetsky, C. Du, D. Lobner, M. Wheeler, C. R. Almli, C. K. F. Shen, T. Y. Luh, D. W. Choi and T. S. Lin, *Proc. Natl. Acad. Sci. U. S. A.*, 1997, **94**, 12241.
16 D. M. Guldi, R. E. Huie, P. Neta, H. Hungerbuhler and K. D. Asmus, *Chem. Phys. Lett.*, 1994, **223**, 511–516.

Faraday Discussions

PAPER

Synthesis and first X-ray structure of a hexa-*peri*-hexabenzocoronene–fullerene-dyad: a model for an inter-carbon-allotrope hybrid†

Andreas Kratzer,[a] Jan M. Englert,‡[a] Dominik Lungerich,‡[a] Frank W. Heinemann,[b] Norbert Jux[a] and Andreas Hirsch*[a]

Received 16th April 2014, Accepted 28th May 2014
DOI: 10.1039/c4fd00069b

The synthesis of a new hexa-*peri*-hexabenzocoronene (HBC)–fullerene dyad **1** was accomplished involving a covalent linkage between both the planar and the spherical conjugated π-system of the two chromophores. We also present the first X-ray single crystal structure of a HBC–fullerene conjugate. A very short HBC–fullerene distance of 3.2 Å is observed. For the synthesis of this molecule, a new versatile applicable template, namely, a mono-functionalized hexa-*peri*-hexabenzocoronene **7** was attached to a fullerene. Absorption and fluorescence spectroscopy, as well as quantum yield measurements of **1**, indicated close electronic communication between the two subunits, which is promising for possible applications in molecular electronics.

Introduction

Hexa-*peri*-hexabenzocoronene (HBC)[1] represents a well defined D_{6h}-symmetrical substructure of graphene[2] and has recently gained increasing attention as an interesting polyaromatic hydrocarbon (PAH). The electronic and self-assembling properties of HBCs and related PAHs open a wide range of possible applications, for example, in molecular electronics. Since wet-chemical manipulations and transformations of graphene itself are still very challenging due to insolubility and difficulties in structural characterization, the investigation of suitable model compounds is highly desirable. Other remarkable and aesthetically pleasing forms of carbon are the spherical fullerenes such as their most prominent

[a]*Department of Chemistry & Pharmacy, Organic Chemistry & Interdisciplinary Center for Molecular Materials (ICMM), Friedrich-Alexander-Universität Erlangen-Nürnberg, Henkestrasse 42, 91054 Erlangen, Germany. E-mail: andreas.hirsch@chemie.uni-erlangen.de; Web: http://www.chemie.uni-erlangen.de/hirsch*

[b]*Department of Chemistry & Pharmacy, Inorganic Chemistry, Friedrich-Alexander-Universität Erlangen-Nürnberg, Egerlandstrasse 1, 91058 Erlangen, Germany*

† CCDC 997740. For crystallographic data in CIF or other electronic format see DOI: 10.1039/c4fd00069b

‡ These authors contributed equally to the work.

representative, namely, the I_h-symmetrical C_{60}. Also the fullerenes offer very promising physical properties and they have been shown to be of technological interest in solar energy conversion, drug delivery and even in cosmetics and other applications.[3]

Owing to the ability of C_{60} to accept reversibly up to six electrons,[4,5] it can be regarded as a superior electron acceptor in hybrid architectures. The systematic chemical functionalization of C_{60} allows for the combination of the outstanding fullerene properties with those of other compound classes. It can pave the way to hybrid molecules and tailor-made high performance materials.[3] For such functionalization approaches cycloaddition reactions to the [6,6]-double bonds of the fullerene core are most suitable and lead to well defined and unambiguously characterized adducts. An example is the fulleroid phenyl-C_{61}-butyric acid methyl ester (PCBM), which is still the best-performing electron acceptor component in organic photovoltaic devices.[6] Higher adducts of C_{60}, such as amphiphilic hexakis-adducts with an octahedral addition pattern exhibit superior supramolecular properties and can form shape-persistent micelles in aqueous solutions.[7] On the other hand the covalent binding of fullerenes to electroactive donor molecules such as porphyrins leads to architectures suitable for photo induced energy and/or electron transfer.[8,9]

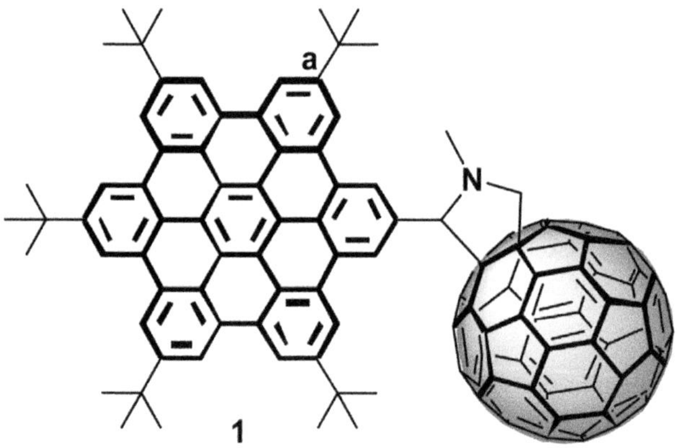

1

In previous work, targeting the PAH–fullerene interactions, assemblies resembling molecular satellite dishes that involved compounds like pyrenes or coronenes attached to heterofullerenes, had been synthesized.[10] Because it is impossible to connect isocyclic C_{60} to just one PAH by a single σ-bond, which in turn would result in radical species, a different synthetic strategy was pursued in this work allowing also for possible diversification. We report here on the synthesis, X-ray structure and physical properties of the new HBC–fullerene dyad **1**, where a planar and a spherical conjugated π-system are linked together by a bridging unit in a covalent fashion. Due to the close contact between the fullerene and the HBC core, a significant amount of mutual influence is expected. Our early investigations already showed that in a direct comparison the HBC will rather act as a donor moiety with respect to the fullerene, which is known to readily accept up to six electrons as a consequence of its triply degenerate LUMO level. These considerations are further corroborated by reports in which hexa-*peri*-hexabenzocoronene derivatives were already used as electron donating species in bulk heterojunction solar cells.[11] Furthermore, after suitable rim functionalization, HBC can potentially act as an efficient columnar templating entity and is anticipated to allow for dynamic supramolecular self-assembly processes, which in turn will provide a new toolset for morphology control. The targeted hybrid molecule **1** represents a suitable model compound to allow direct investigation of the electronic communication between a fullerene and a hexa-*peri*-hexabenzocoronene moiety.

Results and discussion

For the chemical functionalization of fullerenes, usually wet-chemical approaches using homogeneous solutions are applied.[3] Whereas C_{60} exhibits acceptable solubility in polarizable media, the parent HBC is virtually insoluble in such solvents with some exceptions. Initial experiments using soluble hexaphenylbenzene (HPB) derivatives afforded the desired C_{60}–HPB adduct **2**, however, oxidative flattening approaches applying Scholl conditions[12] failed to supply the targeted C_{60}–HBC **3** adduct, presumably due to the very low solubility of the rigid carbon rich structure.

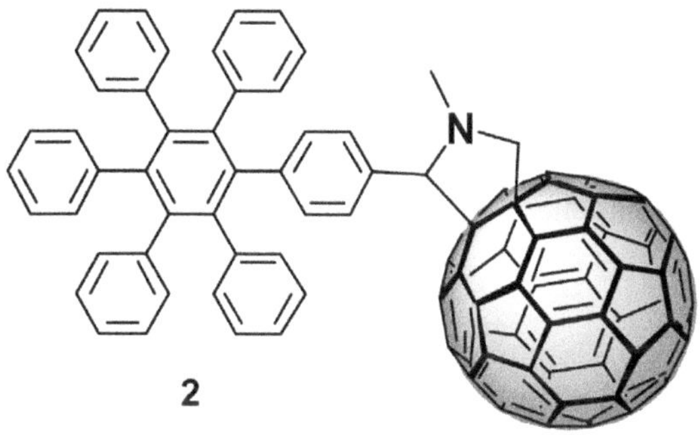

2

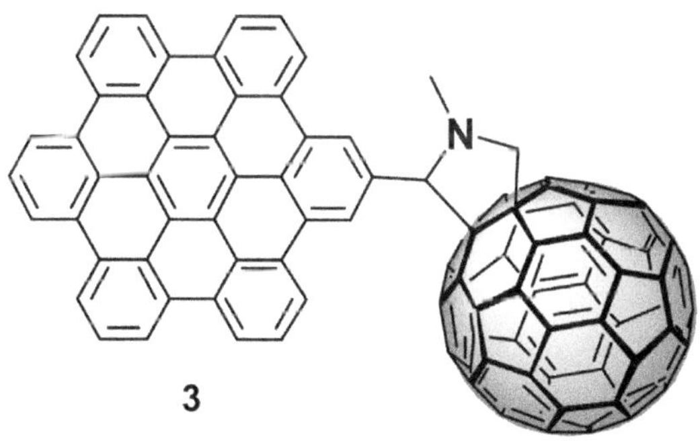

3

In order to overcome this obstacle, we had to introduce solubilizing substituents to the periphery of the HBC part. We decided to introduce solubilizing *t*-butyl groups at position **a** as identified in the representation of molecule **1**. We expected that this would provide sufficient solubility for purification and also for further processing. Accordingly, tetrakis(4-(*tert*-butyl)phenyl)cyclopenta-dienone **5** was synthesized applying a condensation reaction of 1,2-bis(4-*tert*-butylphenyl) ethane-1,2-dione and 1,3-bis(4-*tert*-butylphenyl)acetone. The dione was obtained according to the Mueller-Westerhoff procedure[13] from 1-bromo-4-*tert*-butylbenzene, *n*-butyl lithium and 1,4-dimethylpiperazine-2,3-dione in THF in 65% yield.[14] The second building block was obtained by a two-step one-pot synthesis from 1-(bromomethyl)-4-(*tert*-butyl)benzene. After a intramolecular Claisen condensation in the first step to a β-keto ester, and a subsequent hydrolysis and decarboxylation reaction in the second step, a white powder was obtained in 86% yield.[15] In a Sonogashira reaction in toluene and triethylamine, using 4-*tert*-butyliodobenzene, $[Pd(PPh_3)_2Cl_2]$, CuI and 4-ethynyl-benzaldehyde, we synthesized the required tolane **4**.

[4 + 2] Diels–Alder cycloaddition between an appropriate tetraphenyl-cyclopentadienone derivative **5** and substituted diphenylacetylene **4** in diphenylether afforded the hexaphenylbenzene derivative **6** in 79% yield (Scheme 1). Subsequently, **6** was converted into the corresponding hexa-*peri*-hexabenzocoronene derivative **7** using an oxidative cyclodehydrogenation reaction under Scholl conditions with ferric chloride in nitromethane.[16] For this purpose 2.7 equivalents of $FeCl_3$ per to be formed C–C-bond were pre-dissolved in dry $MeNO_2$ and ice cooled CH_2Cl_2 [12] to obtain **7** in an oxidative flattening progress in quantitative yield. Purification was accomplished by recrystallization from $CHCl_3$ and MeOH or by flash chromatography (DCM–hexanes 1 : 1), if necessary. The confirmation of the successful planarization from **6** to **7** was provided by ^1H NMR spectroscopy. Only the expected ten magnetically inequivalent sets of protons were detected. The exact mass of **7** was recorded by ESI-HR spectrometry, where the corresponding M^+ signal was detected at 830.44. In addition, the appearance of the expanded HBC chromophore was confirmed by UV/vis absorption spectroscopy.

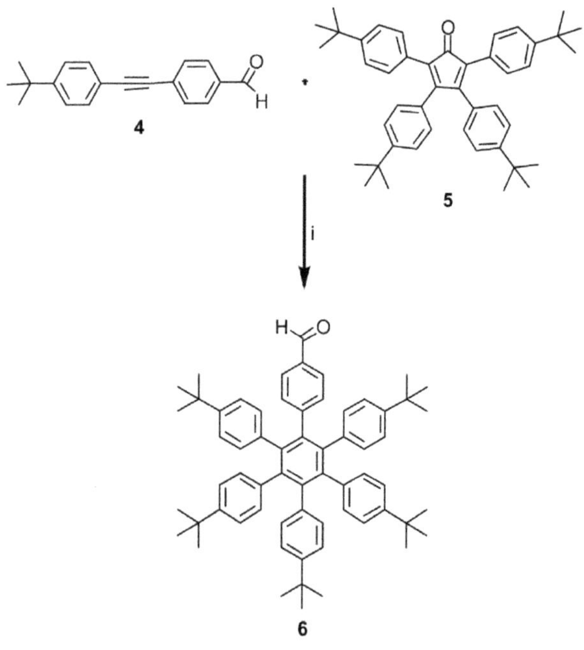

Scheme 1 Synthesis of the formyl-substituted hexaphenylbenzene **6**. (i) Diphenylether, reflux (79%).

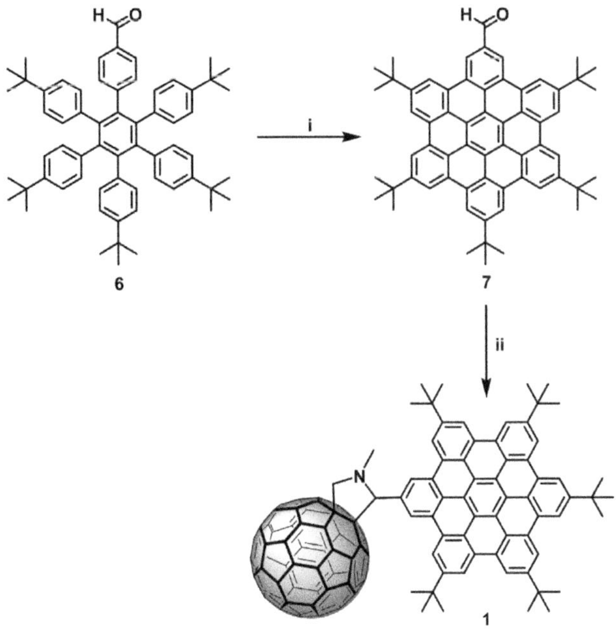

Scheme 2 (i) 16 equiv. dry $FeCl_3$–$MeNO_2$ (300 mg mL^{-1}), CH_2Cl_2, 0 °C, 2.5 h; (ii) 1.2 equiv. C_{60}, 1.2 equiv. sarcosine, toluene, reflux, overnight.

In a final [3 + 2] dipolar cycloaddition, the HBC aldehyde **7**, sarcosine and C_{60} were reacted in refluxing toluene to form the intermediate HBC ylide, which underwent a mono-addition to a [6,6]double bond of the fullerene to give the target molecule **1**. Purification was carried out by flash chromatography (toluene–hexanes 7 : 3) to give **1** as a racemate in a good yield of 79%. The resulting brownish solid was dissolved in CH_2Cl_2 and precipitated with *n*-pentane to minimize solvent inclusions. Characterization by ^{1}H NMR provided evidence for the successful reaction (Scheme 2).

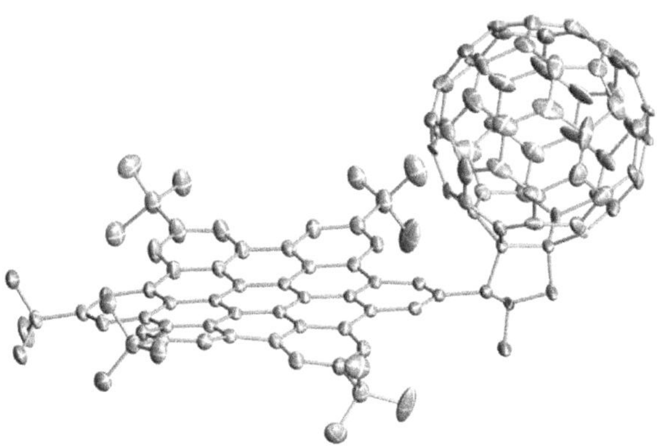

Fig. 1 Thermal ellipsoid representation of the molecular structure of (*R*)-**1** of (*R*/*S*)-**1** in the solid state crystal structure (50% probability ellipsoids; hydrogen atoms, disorder and solvent molecules omitted for clarity).

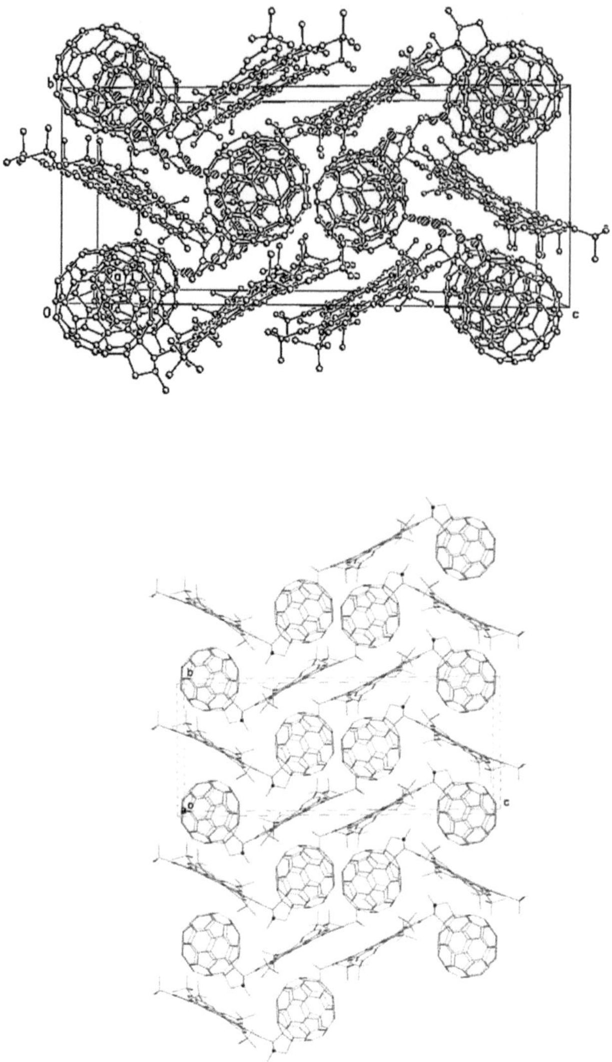

Fig. 2 Crystal packing of the HBC–fullerene hybrid **1**. For reasons of clarity the arrangement of (*R*/*S*)-**1** in the crystal is represented as homochiral strands.

The peaks for the two diastereotopic pyrolidine protons appear as two doublets at 4.24 and 4.94 ppm with a coupling constant of 9.66 Hz. The methyl group at the N-atom is recorded as a singlet at 2.85 ppm. The signal for the aldehyde vanished due to the functionalization with C_{60}. The ^{13}C NMR also confirmed the structure of **1**. The stereogenic carbon center is identified by the signal at 84.2 ppm, the carbon atom of the methyl group at the nitrogen atom likewise resonates at 40.0 ppm. The carbon atom, where the two diasterotopic protons are located, resonates at 68.9 ppm, the signals for the sp^3-hybridized carbon atoms at the fullerene are observed at 68.9 and 69.9 ppm. The typical sp^2-C_{60} signals are detected in the specific region for a monoadduct between 134.7 and 155.9 ppm. The successful formation of **1** was also confirmed by ESI-TOF HR-mass spectrometry and the exact molecular weight was measured as $m/z = 1577.49$. UV/vis absorption spectroscopy indicated the typical HBC bands at 346, 362 and 392, as well as the typical feature for a fullerene monoadduct at 431 nm.[3]

To obtain a better insight into the crystal packing and supramolecular interaction between the two molecular entities, X-ray crystallography was applied.[17] Single crystals of **1** were grown by slow vapor diffusion of pentane into a chloroform solution of compound **1**, which subsequently crystallized as a racemic mixture in the achiral monoclinic space group $P2_1/n$ and contains four molecules of $CHCl_3$ in the asymmetric unit of the unit cell. We want to emphasise in this context, that this is the first reported crystal structure of a hexa-*peri*-hexabenzocoronene–fullerene hybrid (Fig. 1). Due to the statistic arrangement of both (*R*)-**1** and (*S*)-**1** in the crystal, the data set was refined in such a way that (*R*)-**1** and (*S*)-**1** appear as a superposition of each in the asymmetric unit with a distribution of 51.6% and 48.4%. The monomers in the crystal form one-dimensional strands of (*R*/*S*)-**1** with the said distribution, which is inverted in half of the strand due to a center of inversion in the crystal. Some structural features of the crystal structure of **1** are especially noteworthy. There are two interesting π–π-

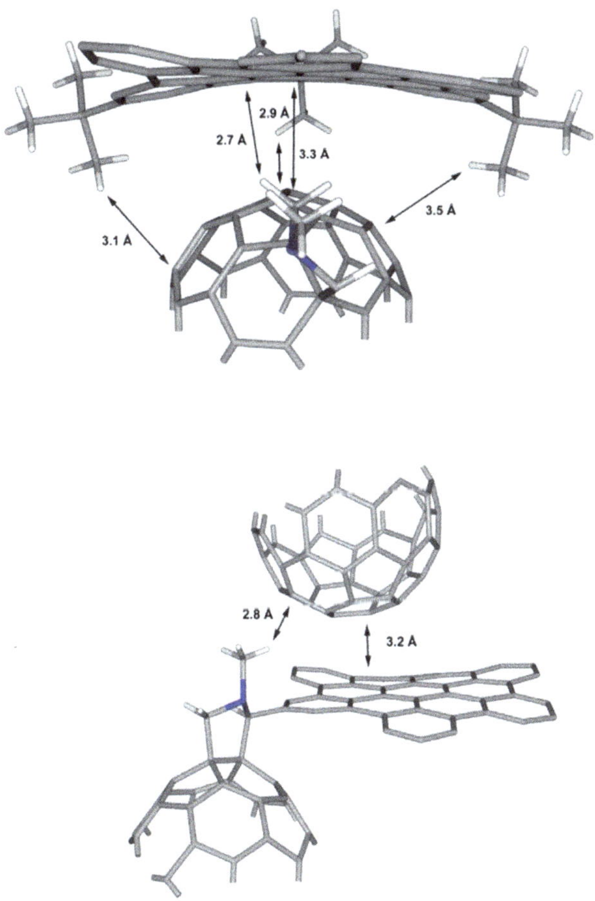

Fig. 3 Distances in the close contact HBC–fullerene hybrid 1.

interactions between the HBC and the fullerene within the crystal packing. The shortest measured sp^2–sp^2-distance between the HBC and the neighbouring fullerene cage is 3.2 Å (Fig. 2) and hence shorter than the distance between two single layers in graphite, which is 3.35 Å.[18] The planar HBC geometry is distorted to accommodate the ball-shaped structure of the fullerene and maximize π–π-interactions between the fullerene and the HBC moiety, as well as the C–H/π-interactions between the *t*-butyl-groups and the fullerene cage.

All these interactions lead to a close contact crystal packing. There are further very short distances within the hybrid material. The methyl group at the nitrogen atom has a distance of only 2.8 Å to the fullerene cage (Fig. 2), the H atoms of the *t*-butyl groups are around 3.3 Å away from the C_{60} moiety. The shortest distance between two fragments, namely one diastereotopic proton at the methylene bridge to the nitrogen atom and the HBC plane, was measured to be around 2.7 Å (Fig. 3).

Furthermore, fundamental photo physical properties of the novel hybrid material were investigated. The presence of HBC was determined by its characteristic absorption/emission spectroscopy of **1** (Fig. 4). The most intense HBC p-band transition of **1** is observed at $\lambda_{max} = 361$ nm.

The extinction coefficient was obtained by fitting the absorbance against the concentration in agreement with Beer's law. Dyad **1** exhibits an extinction coefficient of 200 000 M^{-1} cm^{-1}. Compared to known HBC derivatives the extinction coefficient is relatively high. Functionalized fullerenes, in which chromophores or electroactive moieties are covalently linked to the fullerene cage, offer distinct advantages such as molecular extinction in the visible region of the spectrum, the preference for a population of singlet *versus* triplet excited states and the ultimate fates (*i.e.*, lifetimes and reactivity) of these excited states.[19] This makes hybrid **1** a candidate for applications in bulk heterojunction solar cells. Ongoing studies will clarify if the effect arises from the fullerene moiety. Fluorescence quantum yields of synthesized compounds **1** and **7** were determined by a standard literature procedure using anthracene as a reference standard ($\Phi = 0.33$ in THF).[20] We observed a quantum yield of 12.3% for the HBC–CHO precursor **7**, whereas the

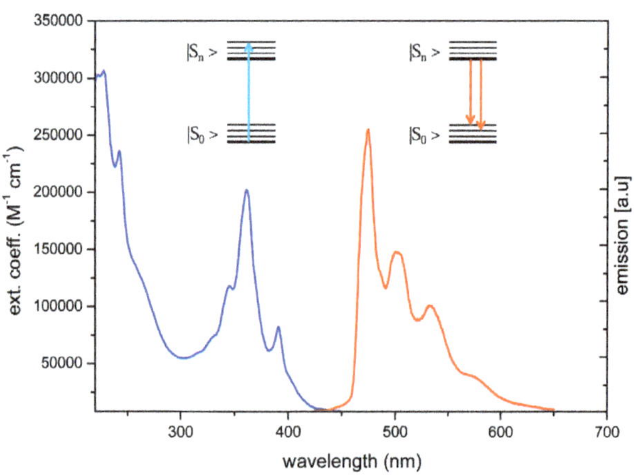

Fig. 4 Absorption/emission spectra of 1.

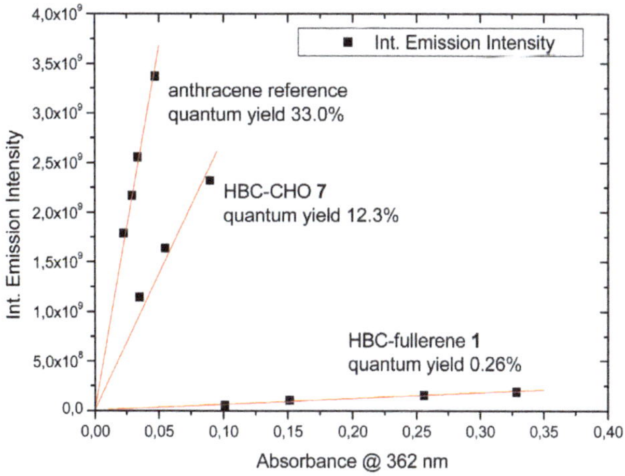

Fig. 5 Quantum yield of anthracene reference, HBC–CHO **7** and HBC–fullerene **1**.

quantum yield of the hybrid material was determined to be as low as 0.26% (Fig. 5). These results indicate strong electronic interaction between the HBC moiety and the fullerene. The quenching of the HBC like emission is explained by

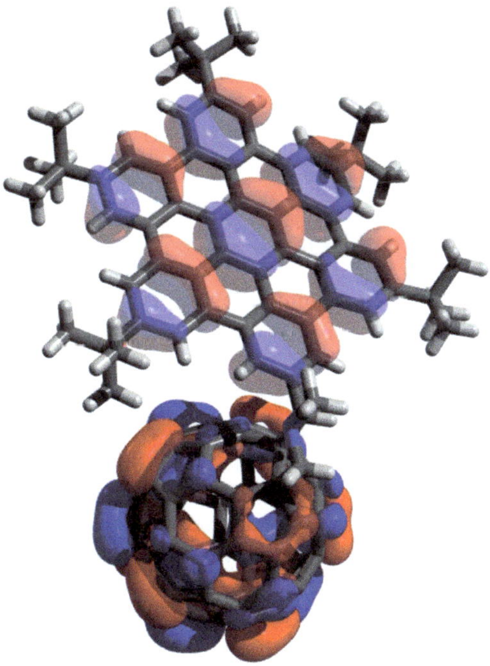

Fig. 6 Molecular orbital (MO) representation of the fullerene–HBC dyad (**1**) on a DFT B3LYP/6-311G quantum mechanical level of theory as calculated using Gaussian '09. The HOMO (translucent) is centered on the HBC unit while the LUMO (solid) is exclusively located on the fullerene cage.

Table 1 Half-wave potentials, HOMO and LUMO energy levels and band gap energy of hybrid 1.[21]

$E_{1/2(\text{red1})}$	$E_{1/2(\text{red2})}$	$E_{1/2(\text{red3})}$	$E_{1/2(\text{red4})}$	$E_{1/2(\text{ox1})}$	E_{HOMO}	E_{LUMO}	E_{gap}
−0.95	−1.35	−1.50	−1.90	1.18	−6.0	−3.9	2.1

strong electronic communication between both – fullerene and HBC – moieties. We have to note though, that an additional reduction of the fluorescence quantum yield is expected due to the presence of the nitrogen lone pair in the pyrolidine ring.

DFT calculations gave good evidence for the location of the LUMO on the fullerene, while the HOMO is centered round the HBC (Fig. 6). We applied cyclic voltammetry to investigate the electrochemical properties of **1** (Table 1). Four reversible reduction processes and one reversible oxidation process were observed. The first half-wave reduction potential is about −0.95 eV, the first half-wave oxidation potential around 1.18 eV, resulting in a HOMO energy level of −6.0 eV and a LUMO energy level of −3.9 eV. For an efficient charge transfer, *e.g.* in polymer solar cells, from the donor to the acceptor fullerene, the relative position of donor LUMO and the acceptor LUMO is crucial.

Compared to PCBM ($E_{\text{HOMO}} = -6.1$ eV, $E_{\text{LUMO}} = -3.7$ eV),[22] the LUMO energy of **1** is around 0.2 eV lower. Also the HOMO–LUMO gap is lower by −0.3 eV, however, these energy differences are rather small. The PM3 calculated differences of the LUMO energies (HyperChem) of PCBM and **1** are $\Delta E_{\text{LUMO}} = 0.07$ eV. This corroborates the experimental results. Both, B3LYP/6-311G and PM3 calculations clearly show that the HOMO of **1** is HBC-centered. The calculated HOMO energies are significantly above the occupied orbital levels of the C_{60} core. As a consequence a comparison between the HOMO–LUMO gap of PCBM and **1** is not suitable.

Conclusion and outlook

In this contribution we have presented the synthesis of a HBC–fullerene hybrid molecule with a covalent linkage between both moieties. For this purpose we have synthesized a new hexa-*peri*-hexabenzocoronene compound, which could serve as a versatile template for optoelectronic applications. Furthermore, we presented the first X-ray crystal structure of a HBC–fullerene hybrid molecule, where a very close contact of 3.17 Å between the two aromatic units was observed. Besides the synthesis and the structure determination, optical and electronic properties were investigated. Steady-state emission spectroscopies and determination of the extinction coefficient as well as quantum yield measurements provide evidence for close electronic communication between the two entities in the dyad. A detailed investigation of possible electron or energy transfer processes in this dyad and related yet to be synthesized systems will be a topic of future research in our laboratories. In this work HBC was implemented to serve as a model compound to assess fundamental structural and photophysical properties of such assemblies. Detailed time resolved spectroscopic investigations, which are expected to allow in depth characterization of the underlying photophysics, will provide a more detailed picture of the roles of the HBC entity as well as its effect on the fullerene acceptor. Concerning the results of previous works,[10] as well as our initial investigations presented in this paper, nearly quantitative transduction of photoexcited states to the rigidly linked buckyball can be anticipated. Also, further functionalization of this hybrid material to yield fullerene hexakis-adducts is currently underway. Such architectures could even provide water solubility. Furthermore, HBC–fullerene hybrids will be tested as self-sensitized acceptor materials in bulk heterojunction organic solar cells.

Experimental section

General remarks and chemicals

Chemicals were purchased from commercial sources and were used without further purification (unless otherwise stated). C_{60} was purchased from IoLiTec Nanomaterials. Solvents were distilled before use; solvents that generated acid had K_2CO_3 in the distillation flask. For reactions with fullerenes, HPLC grade solvents were used. TLC: Merck TLC silica gel 60 F254. UV/vis spectroscopy: Perkin Elmer Lambda 1050, THF as solvent, absorption maxima given in nm, extinction coefficients given in M^{-1} cm^{-1}. IR spectroscopy: attenuated total reflectance (ATR), Bruker FTIR Tensor 27 instrument. NMR spectroscopy: Bruker Avance 400 or Avance 300 spectrometer. Field strengths are given as the resonance frequency of the respective nuclei. The chemical shifts are given in ppm relative to tetramethylsilane (TMS). Abbreviations: s singlet, d doublet, m multiplet, br broad. Spectra were recorded at room temperature. MALDI MS: Shimadzu Axima Confidence instrument (TOF). ESI MS: Bruker Daltronics micrOTOF (TOF) instrument. Matrices: (*E*)-2-(3-(4-(*tert*-butyl)-phenyl)-2-methyl allylidene) malononitrile (dctb), 3,5-dimethoxy-4-hydroxycinnamic acid (sin).

4-((4-(*tert*-Butyl)phenyl)ethynyl)benzaldehyde (4). 4-*tert*-Butyliodobenzene (250 mg, 0.96 mmol, 1.0 equiv.), $[Pd(PPh_3)_2Cl_2]$ (13 mg, 0.02 mmol, 0.2 equiv.), and CuI (7 mg, 0.04 mmol, 0.02 equiv.) were dissolved in dry toluene (5 mL) and triethylamine (5 mL) under an argon atmosphere. Subsequently 4-ethynylbenzaldehyde (150 mg, 1.15 mmol, 1.2 equiv.) was added at once and the mixture was stirred overnight. After filtration the solvent was removed, and the crude mixture was purified by flash chromatography (SiO_2, toluene). The product was obtained as a pale yellow solid (195 mg, 77%).

^1H NMR (400 MHz, $CDCl_3$): δ [ppm] = 1.32 (s, 9H), 7.37–7.39 (m, 2H), 7.46–7.49 (m, 2H), 7.64–7.66 (m, 2H), 7.83–7.85 (m, 2H).

^{13}C NMR (100.5 MHz, $CDCl_3$): δ [ppm] = 31.1, 34.9, 88.0, 93.8, 119.4, 125.5, 129.6, 129.9, 131.5, 132.0, 135.2, 152.4, 191.4.

MS (MALDI, sin): *m*/*z*: 262 [M^+].

IR (ATR, diamond): $\tilde{\nu}$ [cm^{-1}] = 2964, 2867, 2826, 2726, 1696, 1597, 1560, 1362, 1203, 1103, 829, 813.

HPB–CHO (6). Compound **2** (129 mg, 0.49 mmol, 1.0 equiv.) and cyclopentadienone **5** (300 mg, 0.49 mmol, 1.0 equiv.) were dissolved in diphenylether (2.5 mL) and refluxed at 270 °C overnight. After the mixture was allowed to cool to room temperature, methanol was added. The precipitate was filtered and washed several times with cold methanol. The product was obtained as a grey solid (249 mg, 60%).

^1H NMR (300 MHz, $CDCl_3$): δ [ppm] = 1.08 (s, 45H), 6.65 (m, 10H), 6.80 (m, 10H), 7.02 (m, 2H), 7.35 (m, 2H), 9.73 (s, 1H).

^{13}C NMR (100.5 MHz, $CDCl_3$): δ [ppm] = 31.2, 34.0, 3421, 123.1, 123.4, 128.0, 130.9, 131.0, 132.3, 133.1, 137.3, 137.5, 137.6, 138.5, 139.8, 140.8, 141.2, 147.5, 147.6, 148.0, 148.5, 192.4.

MS (MALDI, sin): m/z = 842 [M^+].

HRMS (ESI, MeCN–toluene): calcd for $C_{63}H_{70}O^+$ [M^+]: 842.54266, found: 842.54289.

IR (ATR, diamond): $\tilde{\nu}$ [cm^{-1}] = 2961, 2903, 2866, 1705, 1602, 1392, 1361, 1206, 1018, 861, 832.

HBC–CHO (7). To an ice-cooled, argon-saturated solution of **6** (100 mg, 0.12 mmol, 1.0 equiv.) in 50 mL dry dichloromethane, a freshly prepared solution of anhydrous $FeCl_3$ (352 mg, 2.16 mmol, 18 equiv.) in dry nitromethane (5.0 mL) was added dropwise by syringe through a pressure equilibrated septum while maintaining a low but constant stream of argon passing through the solution. Cooling was maintained for typically one hour and the evaporated solvent was replaced if necessary. Subsequently, the flask was sealed and stirred for 8 hours. Afterwards the reaction was quenched by the addition of 50 mL cold methanol. The resulting mixture was filtered and purified by recrystallization from $CHCl_3$ (10 mL) and methanol (10 mL). Further purification could be performed by column chromatography (silica gel, DCM–hexanes 1 : 1), if necessary. The final product was obtained as a yellow solid in good to quantitative yields.

^{1}H NMR (400 MHz, $CDCl_3$): δ [ppm] = 1.69 (s, 18H), 1.86 (s, 18H), 1.90 (s, 9H), 8.69 (m, 2H), 8.81 (m, 2H), 8.93 (m, 2H), 9.06 (m, 2H), 9.19 (m, 2H), 9.24 (m, 2H), 10.17 (s, 1H).

^{13}C NMR (100.5 MHz, $CDCl_3$): δ [ppm] = 32.0, 32.1, 35.6, 35.7, 35.8, 118.7, 118.8, 118.9, 119.0, 120.0, 120.8, 121.2, 121.5, 123.1, 123.3, 123.4, 128.7, 129.1, 129.8, 129.9, 130.1, 130.3, 130.5, 132.6, 148.6, 148.8, 149.2, 192.5.

MS (MALDI, om): m/z = 831 [M^+].

HRMS (ESI, MeCN–CH_3Cl): calcd for $C_{63}H_{58}O^+$ [M^+]: 830.44876, found 830.44928.

IR (ATR, diamond): $\tilde{\nu}$ [cm^{-1}] = 2954, 2923, 2854, 1697, 1607, 1367, 1184, 867.

UV/vis (THF): λ_{max}: 333, 349, 369, 400, 423, 447, 476 nm.

HBC–fullerene (1). Compound **7** (50 mg, 0.60 mmol, 1.0 equiv.) was dissolved in toluene (50 mL) under inert conditions before C_{60} (52 mg, 0.07 mmol, 1.2 equiv.) was added. The mixture was stirred for 30 minutes under exclusion of light until C_{60} was dissolved completely. Then sarcosine (6.0 mg, 0.07 mmol, 1.2 equiv.) was added and the mixture was refluxed overnight. After cooling down to room temperature and filtration the crude mixture was purified by flash chromatography (SiO_2, toluene–hexanes 7 : 3). The occurring brown solid was dissolved in CH_2Cl_2 and precipitated with n-pentane to exclude solvent inclusions (70 mg, 66%).

^{1}H NMR (400 MHz, $CDCl_3$): δ [ppm] = 1.79 (s, 18H), 1.81 (s, 27H), 2.85 (s, 3H), 4.24 (d, 1H, ^2J = 9.66 Hz), 4.93 (d, 1H, ^2J = 9.66 Hz), 5.46 (s, 1H), 9.26 (s, 2H), 9.29 (s, 2H), 9.30 (s, 6H), 9.57 (s, 2H, br).

^{13}C NMR (100.5 MHz, $CDCl_3$): δ [ppm] = 29.8, 31.7, 31.7, 32.0, 35.1, 35.2, 40.0, 68.9, 69.9, 84.3, 118.8, 118.8, 119.1, 119.2, 120.0, 120.3, 120.5, 120.7, 122.7, 123.6, 123.7, 123.7, 125.9, 129.9, 130.1, 130.2, 130.3, 130.8, 134.7, 135.4, 135.6, 135.9, 136.2, 141.1, 141.3, 141.3, 141.4, 141.5, 141.6, 141.6, 141.7, 141.9, 142.0, 142.0, 142.1, 142.2, 142.3, 142.6, 142.8, 144.0, 144.2, 144.3, 144.7, 144.7, 144.8, 145.1,

145.1, 145.1, 145.2, 145.2, 145.3, 145.3, 145.4, 145.5, 145.7, 145.7, 145.8, 145.9, 146.0, 146.2, 146.8, 146.8, 148.3, 148.4, 148.6, 153.0, 153.4, 153.5, 155.9.

MS (MALDI, dctb): m/z = 1578 [M^+].

HRMS (ESI, MeCN–$CHCl_3$): calcd for $C_{122}H_{65}O_3^+$ [M^+]: 1577.49828, found: 1577.49089.

IR (ATR, diamond): $\tilde{\nu}$ [cm^{-1}] = 2952, 2921, 2852, 1731, 1606, 1579, 1461, 1362, 1259, 868.

UV/vis (THF): λ_{max} (ε, M^{-1} cm^{-1}): 330 (73 000), 345 (117 000), 361 (202 000), 391 (82 000), 431 (10 000) nm.

Acknowledgements

The authors thank the Deutsche Forschungsgemeinschaft (DFG-SFB 953 "Synthetic Carbon Allotropes", Projects A1 and A2) and the Bayrisches Staatsministerium für Bildung und Kultus, Wissenschaft und Kunst ("Solar Technologies Go Hybrid", SolTech; Energiecampus Nürnberg, EnCN) for financial support.

References

1 J. Wu, W. Pisula and K. Müllen, *Chem. Rev.*, 2007, **107**, 718.
2 A. K. Geim and K. S. Novoselov, *Nat. Mater.*, 2007, **6**, 183.
3 A. Hirsch, M. Brettreich and F. Wudl, *Fullerenes: Chemistry and Reactions*, Wiley, 2006.
4 L. Echegoyen and L. E. Echegoyen, *Acc. Chem. Res.*, 1998, **31**, 593.
5 Q. Xie, E. Perez-Cordero and L. Echegoyen, *J. Am. Chem. Soc.*, 1992, **114**, 3978.
6 J. C. Hummelen, B. W. Knight, F. LePeq, F. Wudl, J. Yao and C. L. Wilkins, *J. Org. Chem.*, 1995, **60**, 532.
7 S. Burghardt, A. Hirsch, B. Schade, K. Ludwig and C. Böttcher, *Angew. Chem.*, 2005, **117**, 3036.
8 D. M. Guldi, *Chem. Soc. Rev.*, 2002, **31**, 22.
9 H. Imahori and Y. Sakata, *Adv. Mater.*, 1997, **9**, 537.
10 F. Hauke, S. Atalick, D. M. Guldi, J. Mack, L. T. Scott and A. Hirsch, *Chem. Commun.*, 2004, 766.
11 S. J. Kang, S. Ahn, J. B. Kim, C. Schenck, A. M. Hiszpanski, S. Oh, T. Schiros, Y.-L. Loo and C. Nuckolls, *J. Am. Chem. Soc.*, 2013, **135**, 2207.
12 S. P. Brown, I. Schnell, J. D. Brand, K. Müllen and H. W. Spiess, *J. Am. Chem. Soc.*, 1999, **121**, 6712.
13 U. T. Mueller-Westerhoff and M. Zhou, *J. Org. Chem.*, 1994, **59**, 4988.
14 A. Gourdon, S. K. Sadhukhan and C. Viala, *Synthesis*, 2003, 1521.
15 D. Romer, *Synthesis*, 2011, **2011**, 2721.
16 J. M. Englert, J. Malig, V. A. Zamolo, A. Hirsch and N. Jux, *Chem. Commun.*, 2013, **49**, 4827.
17 ESI.†
18 H. Marsh and F. Rodríguez-Reinoso, *Sciences of carbon materials*, Universidad de Alicante, 2000.
19 D. M. Guldi and N. Martin, *Fullerenes: From Synthesis to Optoelectronic Properties*, Springer, 2003.
20 S. L. Murov, I. Carmichael and G. L. Hug, *Handbook of Photochemistry*, Taylor & Francis, 2nd edn, 1993.

21 T. P. I. Saragi, T. Spehr, A. Siebert, T. Fuhrmann-Lieker and J. Salbeck, *Chem. Rev.*, 2007, **107**, 1011.
22 C. Brabec, U. Scherf and V. Dyakonov, *Organic Photovoltaics: Materials, Device Physics, and Manufacturing Technologies*, Wiley, 2011.

Faraday Discussions

PAPER

Organocatalysis for new chiral fullerene-based materials†

Rosa M. Girón,[a] Silvia Reboredo,[a] Juan Marco-Martínez,[a] Salvatore Filippone*[a] and Nazario Martín*[ab]

Received 11th April 2014, Accepted 22nd April 2014
DOI: 10.1039/c4fd00065j

The cycloaddition reaction of 3-alkynoates to fullerenes has been studied under different experimental conditions and phosphines in the presence of a base which is responsible for the previous isomerization to the corresponding allenoate. The versatility of this reaction allows chiral fullerenes to be obtained at will, thus opening up a new avenue in the use of enantiomerically pure carbon nanostructures in materials science.

Introduction

The advent of new allotropic forms of carbon,[1] namely fullerenes,[2] carbon nanotubes[3] and, more recently, graphene[4] has prompted great excitement in the scientific community inspired by their singular properties and, therefore, by their potential applications in materials and bio-medical sciences.

After the first rush in understanding fullerenes' reactivity and the search for soluble and processable easily to synthesize derivatives, fullerenes chemistry has been focused towards more fine chemical modifications.[5] Thus modern reactions, namely those mediated by transition metals, have been introduced in fullerenes science in order to obtain more sophisticated structures and a better control of their properties.[6] However, despite the great advance in recent synthetic methodologies, to the best of our knowledge, chirality in fullerenes is an important issue that has not been properly addressed so far.[7,8] Actually, besides the traditional fields, chirality is also critical in fields such as nanoscience, where fullerenes are foreseen to have their potential main applications.[9]

The former preparations of chiral fullerene derivatives have been based on the use of chiral starting materials or on the employ of inefficient chiral separations from racemates.[7,8,10] This approach is dependent on the availability of the chiral starting material and the final products are obtained without a good control of the configuration; therefore, this approach is strongly limited.

[a]*Departamento de Química Orgánica I, Facultad de Química, Universidad Complutense, E-28040 Madrid, Spain. E-mail: nazmar@quim.ucm.es; Fax: (+34) 91.394.41.03; Tel: (+34) 91.394.42.27*

[b]*IMDEA –Nanoscience, Campus de Cantoblanco, E-28049 Madrid, Spain. Tel: (+34) 91 299 88 93*

† Electronic Supplementary Information (ESI) available. See DOI: 10.1039/c4fd00065j

On the other hand, and despite the chemical analogy, both metal mediated or organocatalytic methods,[11] developed for the chiral activation of electron poor olefins, cannot be applied to fullerenes. Indeed, the lack of a coordinating group able to interact with a chiral Lewis acid, or of a functional group for a suitable LUMO activation, hampered the use of most modern asymmetric methodologies.

The introduction of chiral metal catalysis to fullerenes chemistry represents a major breakthrough in the preparation of optically active fullerene derivatives. Thus, chiral pyrrolidino[3,4:1,2][60]fullerenes were directly prepared with a total control of the stereochemical outcome by activation of iminoesters in chiral *N*-metallated azomethine ylides.[12] These reactions displayed a high selectivity both in the configuration of the resulting pyrrolidines (*cis* or *trans* adducts were obtained at will with excellent enantiomeric excesses)[13] and in the control of the site- and regio-selectivity when less symmetrical [70]fullerenes[14] or endofullerenes, namely La@$C_{72}(C_6H_3Cl_2)$,[15] were used.

A second important milestone occurred with the application of metal-free catalytic preparation of chiral fullerene derivatives. In this process, easily available chiral phosphines, used as nucleophilic organocatalyst, triggered a HOMO activation of fullerene reagent partners, namely racemic α-allenoates, that lead, through a formal [3 + 2] stereoconvergent cycloaddition, to cyclopenteno[60]-fullerenes with remarkable high enantiomeric excesses.[16]

More recently, the set of chiral catalytic pools available were successfully broadened by the chiral activation of azlactones by the use of both chiral metal catalysis and Brønsted-base organocatalysts. In this process, another class of fullerene derivatives, namely pyrrolino[60]fullerenes, have been directly prepared in good enantiomeric excesses.[17]

The phosphine-catalyzed [3 + 2] cycloaddition methodology afforded a novel family of cyclopentenofullerenes with a wide range of allenoates. However, optimal conditions did not allow us to perform the reaction starting from 2-alkynoates. Thinking about the possible mechanism pathway, we realised that the allenoate intermediate could play a key role in the process. Because of that, we envisioned 3-alkynoates as direct precursors of 2-allenoates. These potential dipoles are readily accessible by coupling of terminal alkynes and diazoacetate derivatives under copper catalysis (Scheme 1).[18]

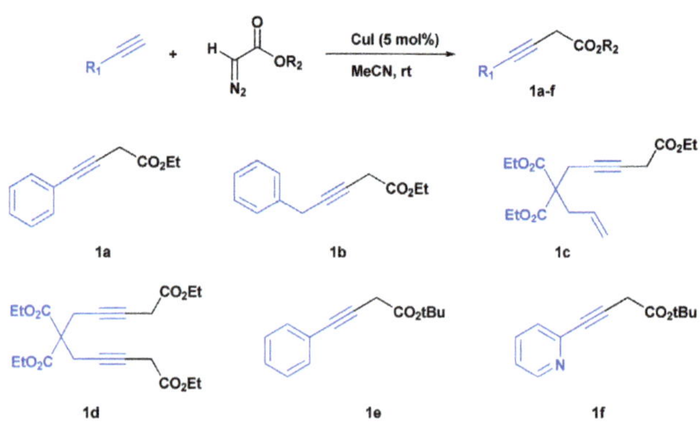

Scheme 1 Synthesis of 3-alkynoates.

Results and discussion

We have carried out the evaluation of the cycloaddition reaction with the previously prepared ethyl-4-phenylbut-3-ynoate **1a**. In order to favour the isomerisation from the 3-alkynoates to the corresponding allenoate, we carried out the reaction under the optimized conditions that we had applied to the allenoates as dipoles but in the presence of an additional base. For instance, when Et_3N was used as base, the corresponding cyclopenteno[60]fullerene **2a** was obtained in 57% conversion and >99% ee. This result encouraged us to evaluate in depth the use of several bases[19] as well as phosphines, both in their chiral and achiral versions (Fig. 1, Table 1).

Using **1a** as a model alkynoate and following the previously optimized reaction conditions for the allenoates, we started evaluating several commercially available chiral phosphines with toluene as the solvent under an argon atmosphere. We also added a base for an easier isomerization of the alkynoate to the corresponding allenoate. As can be seen in Table 1, when Et_3N was used as the base, the reaction did take place with high conversions and variable enantioselectivities depending on the evaluated phosphine (Table 1, entries 1–14). Once again, (*S*,*S*)-*f*-Binaphane (***P*-VII**) was the optimal catalyst when looking for the best combination between conversion and enantioselectivity (entry 7). However, other related

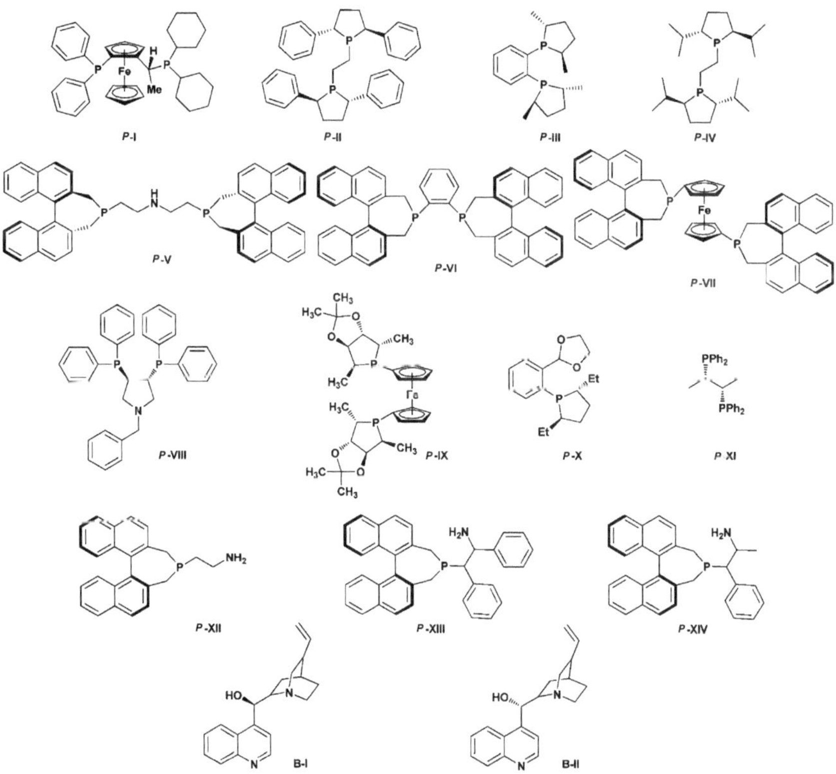

Fig. 1 Commercially available chiral phosphines *P*-I-XIV and bases B-I-II used in this study.

Table 1 Phosphine-base-catalyzed [3 + 2] cycloaddition of alkynoate **1a** to [60]fullerene[a]

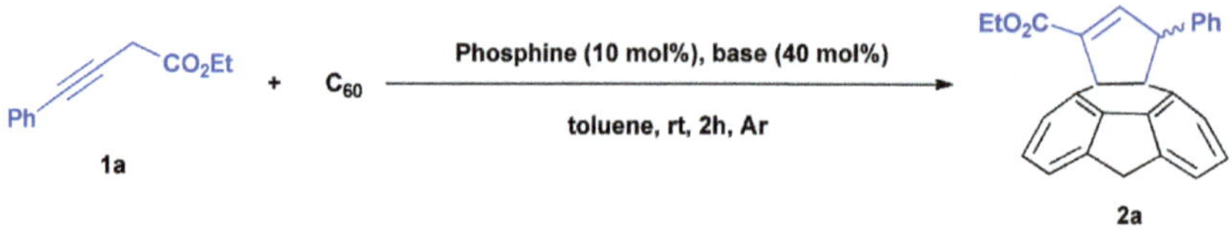

Entry	Base	Phosphine	Conv.[c,d] (%)	ee[c] (%)
1	Et_3N	***P*-I**	49	80 (*R*)
2	Et_3N	***P*-II**	99	4 (*R*)
3	Et_3N	***P*-III**	30	20 (*S*)
4	Et_3N	***P*-IV**	21	6 (*S*)
5	Et_3N	***P*-V**	60	66 (*S*)
6	Et_3N	***P*-VI**	74	54 (*S*)
7	Et_3N	***P*-VII**	57	>99 (*R*)
8	Et_3N	***P*-VIII**	35	24 (*R*)
9	Et_3N	***P*-IX**	12	62 (*S*)
10	Et_3N	***P*-X**	99	72 (*S*)
11	Et_3N	***P*-XI**	35	32 (*S*)
12	Et_3N	***P*-XII**	59	80 (*R*)
13	Et_3N	***P*-XIII**	24	72 (*R*)
14	Et_3N	***P*-XIV**	<5	—
15	**B-I**	—	—	—
16	**B-I**	Bu_3P	77	0
17	**B-I**	***P*-XII**	36	78 (*R*)
18	**B-II**	***P*-XII**	51	82 (*R*)
19[b]	—	***P*-XII**	31	88 (*R*)

[a] A solution of the base (40 mol%) and alkynoate **1a** (0,016 mmol) in dry toluene (1 mL) was stirred for 15 min and then [60]fullerene (0,017 mmol) and the phosphines (10 mol%) were added; the resulting mixture was stirred for 2 h at room temperature under an Ar atmosphere. [b] A solution of the phosphine (10 mol%) and alkynoate **1a** (0,016 mmol) in dry toluene (1 mL) was stirred for 15 min and then [60]fullerene (0,017 mmol) was added; the resulting mixture was stirred for 2 h at room temperature under an Ar atmosphere. [c] Determined by HPLC analysis. Configuration of the chiral center has been determined by analogy with the reported cyclopenteno[60]fullerenes with CD measurements. [d] Based on consumed [60]fullerene.

phosphines also turned out to be efficient in this reaction. Thus, ***P*-I**, with the ferrocene structure being part of the skeleton, afforded similar results to those given by (*S*,*S*)-*f*-Binaphane (***P*-VII**) in the previous work[16] (entry 1). Another dinaphthophosphepine derivative, (***P*-V**), gave access to the opposite enantiomer with a high conversion (entry 5). We then moved to a different type of phosphines, endowed with an amine group as a part of the phosphine. These catalysts (***P*-XII**, ***P*-XIII**, ***P*-XIV**) (entries 12–14), were evaluated, and the cycloadduct was obtained in high enantioselectivities and moderate to good conversions, except for ***P*-XIV**, in which the reaction did not take place (entry 14). The reaction was also carried out with a chiral amine (**B-I**) for the isomerization of the alkynoate and tributylphosphine for the later cycloaddition (entry 16) (the base itself is not capable of performing the cycloaddition process, entry 15). In this instance, the enantioselectivity dramatically dropped down giving rise to a racemate. An additional test using chiral phosphine (***P*-XII**) confirms that the information present in the phosphine is the responsible for the induction of the process (entry 17). Chiral amine (**B-II**), behaved quite similarly to its enantiomer (**B-I**), affording a good conversion and very good enantioselectivity (entry 18). Finally, in order to determine if the presence of a base is needed for the success of the reaction, we run the reaction in the absence of an additional base in the reaction medium. Interestingly, the ee was even better while the conversion slightly decreased (entry 19) (31%, 88% ee). Once we found the best catalyst, we tried to improve the results by modifying other parameters such as temperature, solvent, catalyst loading, *etc.* Nevertheless, the best results were those derived from the conditions that we had previously optimized, including the use of 2-[(11b*S*)-3*H*-binaphtho[2,1-c:1′,2′-e]-phosphepin-4(5H)-yl]ethanamine (***P*-XII**), in 1 mL of anhydrous toluene operating at room temperature.

These optimized conditions were lately applied to the series of synthetized 3-alkynoates **1a–f** and various allenoates **3a–h** (Table 2).

Table 2 Phosphine-base-catalyzed [3 + 2] cycloaddition of alkynoates/allenoates **1a–f**/**3a–h** to [60]fullerene[a]

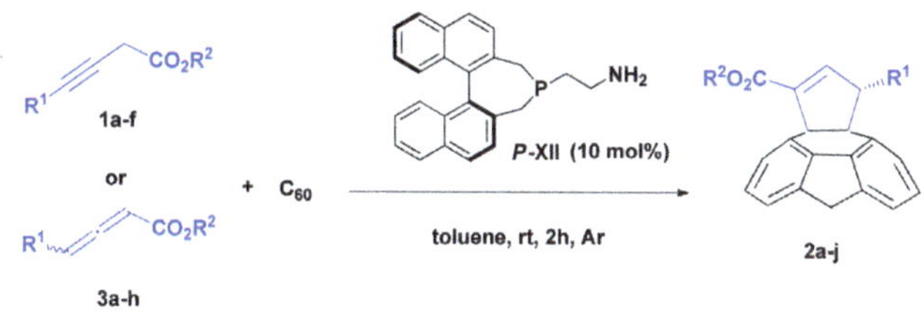

Entry	Dipole	R^1	R^2	Conv.[b,c] (%)	ee[b] (%)	Product
1	**1a**	Ph	Et	31	88	**2a** (*R*)
2	**1b**	Bn	Et	—	—	—
3	**1c**	enyne	Et	19	n.d.	**2b**
4	**1d**	diyne	Et	—	—	—
5	**1e**	Ph	*t*Bu	49	60	**2c** (*R*)
6	**1f**	py	*t*Bu	—	—	—
7	**3a**	Ph	Et	42	86	**2a** (*R*)
8	**3b**	Bn	Et	67	84	**2d** (*S*)
9	**3c**	Me	Et	20	93	**2c** (*S*)
10	**3d**	Et	Et	70	92	**2f** (*S*)
11	**3e**	*i*Pr	Et	42	80	**2g** (*S*)
12	**3f**	4-CHOPh	Et	31	n.d.	**2h**
13	**3g**	Bn	*t*Bu	28	80	**2i** (*S*)
14	**3h**	Ph	Bn	37	99	**2j** (*R*)

[a] A solution of 2-[(11b*S*)-3*H*-binaphtho[2,1-c:1′,2′-e]phosphepin-4(5H)-yl]ethanamine (10 mol%) and alkynoate **1a-f**/allenoate **3a–h** (0,016 mmol) in dry toluene (1 mL) was stirred for 15 min; then [60]fullerene (0,017 mmol) was added and the resulting mixture was stirred for 2 h at rt under Ar atmosphere. Configuration of the chiral center has been determined by analogy with the reported cyclopenteno[60]fullerenes with CD measurements. [b] Determined by HPLC analysis. [c] Based on consumed [60]fullerene. n.d.: not determined.

Taking into account the typical yields reported for fullerene functionalization, for the alkynoate that was precursor of one of the allenoates we had already described, the reaction worked satisfactorily both in terms of conversion and enantioselectivity (Table 2, entries 1, 7). The introduction of a benzyl group at position 4 turned to be unsuccessful. This is probably due to the existence of two available acidic protons in the dipole which trigger the formation of a complex mixture of compounds that we were unable to purify (entry 2). The polyunsaturated substrates, the 1,6-enyne **1c**, gave access to a cyclopentene derivative with an additional double bond in its structure (entry 3). However, we were unable to determine the enantiomeric excess of the derivative. The variation of the enyne to an 1,6-diyne (entry 4), did not afford positive results because a complex mixture was obtained. By varying the ethyl ester group to a *tert*-butyl group, a slight increase in the conversion was observed, albeit with a decrease of the enantioselectivity (88% ee *vs.* 60%) (entry 5). When a piridyl group was present as substituent in position 4, the starting materials were recovered unaltered (entry 6). Those allenoates that had been previously evaluated under (*S*,*S*)-*f*-Binaphane catalysis (entries 7, 9–11, 14), were tested again. Conversions as well as enantioselectivities were similar. The substitution of the phenyl group by a benzyl one (entry 7 *vs.* 8), yielded the cycloadduct in very similar results (conv: 42% *vs.* 67% and 86% *vs.* 84% ee). The modification of the ethyl ester group by a *tert*-butyl ester was also tolerated (entry 13). The additional functionalization of the cyclopentene ring was achieved by the use of an allenoate with a formyl group present in the aromatic ring with a moderate conversion (entry 12). The ee was not determined as a consequence of the impossibility of finding HPLC separation conditions.

With the aim of testing a different type of dipole under the optimised conditions, we used a substrate that had previously been employed in [3 + 2] and [4 + 2] cycloadditions with fullerenes.[20] These unmodified Morita–Baylis–Hillman adducts, DMAP-catalysed and in presence of Ac_2O, gave access to variable penta- and hexacycles. We were interested in checking our reaction conditions using these adducts as dipoles, so we proceeded to prepare one of them (**4**) (Scheme 2).

Then, 1,2-bis(diphenylphosphine)ethane (dppe) was used as the catalyst for the preparation of the racemic derivative. The characterization of the product **5**

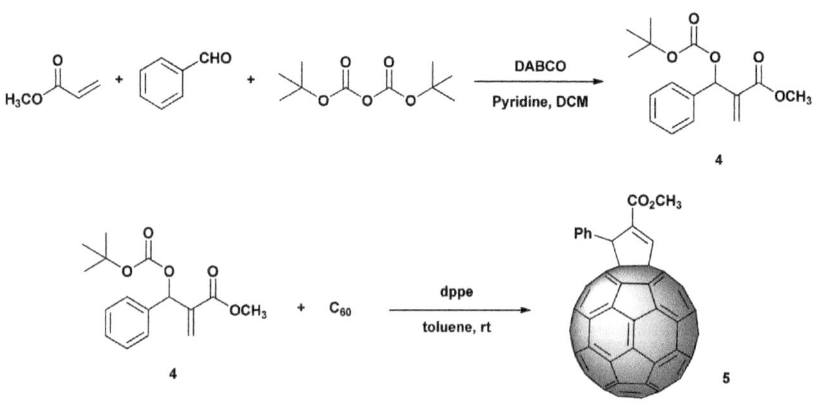

Scheme 2 Synthesis of an unmodified Morita–Baylis–Hillman adduct and phosphine-catalyzed [3 + 2] cycloaddition.

revealed that the obtained cycloadduct had a differently substituted five-membered ring. The chiral version of the reaction with 2-[(11b*S*)-3*H*-binaphtho[2,1-c:1′,2′-e]phosphepin-4(5H)-yl]ethanamine (**P-XII**) acting as the catalyst yielded the product **5** in 36% conversion and 57% ee.

All new [60]fullerene based compounds have been characterized by electrochemical spectroscopy due to their electroactive character when they are subjected to cyclic voltammetry (CV) studies. Thus, we carried out the CV measurements in *o*-DCB/MeCN (4 : 1) containing $Bu_4N^+PF_6^-$ (0.1M) as the supporting electrolyte at room temperature, along with pristine C_{60} as a reference. Moreover, glassy carbon, platinum wire, and Ag/Ag^+ electrodes were used as working, counter and reference electrodes, respectively (see ESI†). As was expected, three reversible reduction waves were observed at similar potential values, with an interesting cathodic shift for the first half-wave (see ESI†), when compared with pristine [60]fullerene. This technique allows us to determine unambiguously the number of saturated double bonds in the fullerene sphere. Therefore, taking into account the reduction potential values obtained for these compounds, we can easily confirm the sole presence of mono-adducts in our samples. Indeed, only the saturation of one of the double bonds of the carbon cage takes place, which produces a rise in the energy of the LUMO.[21]

The aforementioned results and those related to our previous work demonstrate that the most successful phosphines are those bearing at least two alkyl substituents (Fig. 1), which turned to be more nucleophilic than the related aryl-substituted phosphines and, show particularly efficient results for seven-membered phosphacycles. Taking into account the importance of the rigid structure of these phosphorous-containing heterocycles, and since they are not commercially available, we focused on the synthesis of a new four ring derivative. This catalyst would allow a deeper evaluation of the cycloaddition process as well as the possible switching of the enantioselectivity by the use of the opposite chiral starting phosphine. Phosphines were prepared according to a modified previously reported procedure as can be seen in Scheme 3.[22]

The commercial ferrocene phosphine was treated with butyl lithium at −30 °C in THF and then a mesylated diol was added while the temperature was reduced

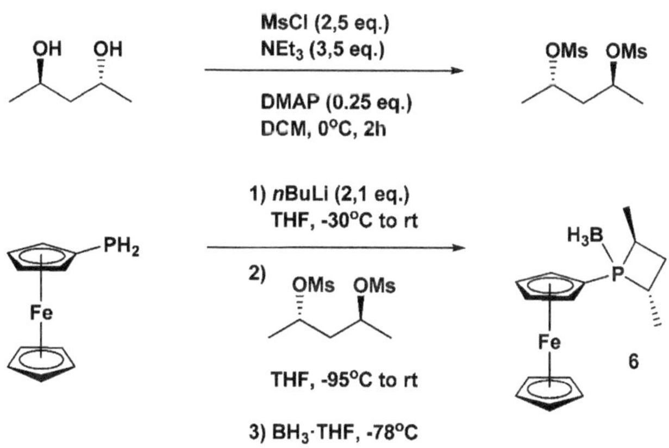

Scheme 3 Synthesis of four-membered phosphacycle.

to −95 °C. The obtained species was stabilized with BH_3 isolating the corresponding final four membered cyclic phosphine **6** in a moderate yield. Despite this type of phosphine being previously employed with no further modifications, when it was evaluated in the [3 + 2] cycloaddition reaction between several allenoates or alkynoates and [60]fullerene, no reaction was observed. Afterwards, we tried the "*in situ*" activation of the phosphine by DABCO addition, observing that high temperatures were needed for a complete deprotection (Scheme 4).

Unfortunately, the obtained results in terms of conversion and enantioselectivity values were found to be significantly lower than those obtained by using the seven-membered ring phosphine. Remarkably, this phosphine gave access to the opposite enantiomer when compared to that obtained from phosphine ***P*-XII**.[23]

Despite the stereoconvergence of cycloaddition reactions already being explained in several previous works related with conventional olefins described in the literature,[24] further investigations were conducted with the target of confirming how [60]fullerene affects the process when it was used as dipolarophile. Thus, we proceeded to separate both enantiomers of the starting allenoate **3a** by using semipreparative HPLC.[25] When they were subjected to reaction under the same optimised reaction conditions, the formation of the corresponding cyclopenteno[60]fullerene with the same enantiomeric form (*R*) and even in the same value (78% ee) took place. Therefore, this experimental finding reveals that the cycloaddition process occurs in a stereoconvergent manner.

Although we do not have real evidence for the mechanism of this cycloaddition reaction, the configuration of the cyclopentene derivatives, and Lu's initial proposal[26] as well as some recent mechanistic studies, allow us to suggest a plausible mechanism (Scheme 5). The process starts with a nucleophilic attack from the phosphine onto the central electrophilic atom of the allenoate derived from the isomerization of the starting alkynoate. Then, the (6,6) double bond of the fullerene would suffer from an attack of the zwitterionic intermediate formed in the first stage (**Ia–Ib**). An intramolecular conjugate addition of the formed intermediate (**II**), followed by a proton transfer of the stabilized ylide (**III–IV**) gives rise to the catalyst elimination and the desired adduct formation.

Once the reaction has been well-established with [60]fullerene as the dipolarophile, we turned our attention to higher fullerenes. In contrast to

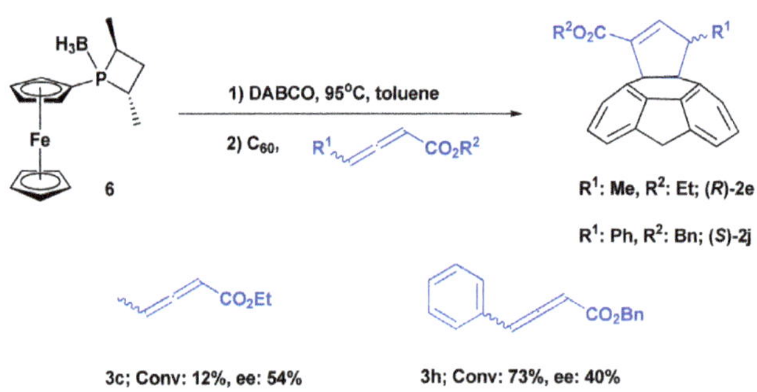

Scheme 4 One-pot deprotection of the four-membered phosphine followed by cycloaddition of allenoate **3c**, **3h** to [60]fullerene.

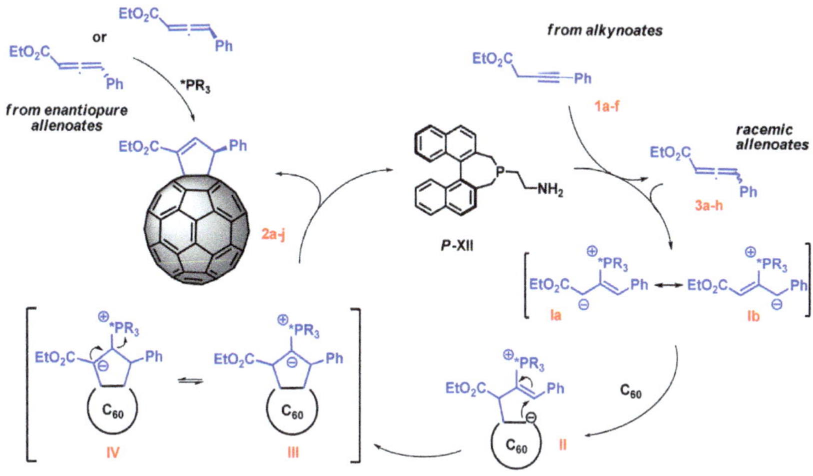

Scheme 5 Mechanism for the phosphine-catalyzed [3 + 2] cycloaddition of alkynoates/allenoates **1a–f**/**3a–h** to [60]fullerene.

[60]fullerene, [70]fullerene has no spherical symmetry and, therefore, four differently reactive double bonds can be found on the cage giving rise to different site-isomers. In this regard, additions occur preferentially on those double bonds located in the polar zone (the most strained fullerene double bonds), in fact, the α site is preferred followed by β and γ sites. On the other hand, the flatter equatorial region is less reactive and the addition only rarely takes place at the δ site (Scheme 6).[27]

Keeping this in mind, [70]fullerene was subjected to the reaction in the presence of a racemic phosphine (dppe, 20 mol%) giving rise to the desired

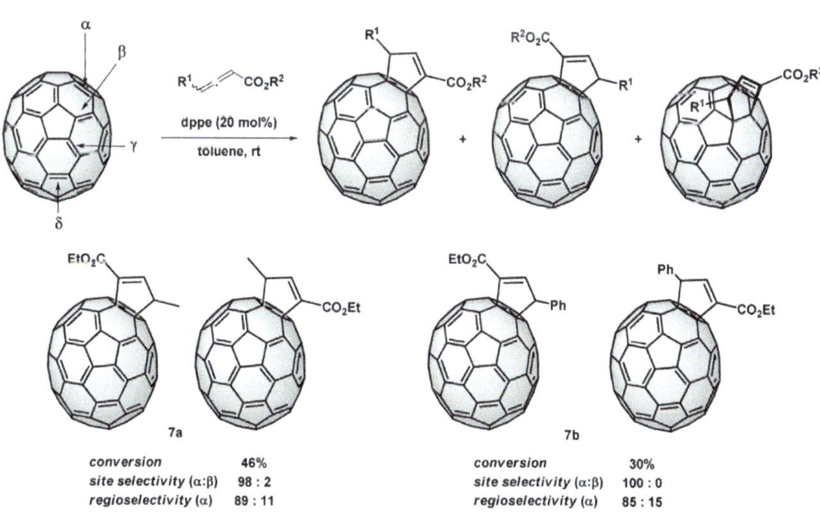

Scheme 6 Phosphine-base-catalyzed [3 + 2] cycloaddition of allenoates **3a**, **3c** to [70]fullerene.

cyclopenteno[70]fullerenes in a mixture of site-isomers α and β (Scheme 6). As expected, when the reaction took place under optimized conditions, excellent site-selectivities were afforded. Thus, in the case of **7a**, the yielded α : β ratio was 98 : 2; and it was even better for **7b**, in which the α site isomer is the unique isomer observed. Moreover, good regioselectivities were also obtained for both cases. Thus, regioisomers with the ester moiety on the polar region and the methyl group (**7a**) or the aromatic ring (**7b**) on the equatorial region were formed preferentially in 89 : 11 and 85 : 15 ratios, respectively. The preference of the ester moiety for the polar region has previously been observed in the [3 + 2] cycloaddition reaction of α-iminoesters onto [70]fullerene.[28] The enantioselective synthesis of the α-regioisomer cyclopenteno[70]fullerene **7b** that is preferentially formed was achieved under phosphine ***P*-XII** catalysis with an 88 : 12 enantiomeric ratio.

Conclusions

The interest in the great variety of carbon nanoforms known so far stems from their mechanical, optical, and electronic properties with a wide number of potential applications. Actually, in addition to the aforementioned nanoforms of carbon, other less-known species such as nanohorns, nano-onions, peepods, nanobuds, nanocups, or carbon dots, just to mention a few, are known. In contrast to fullerenes, most of the chemical and also physical properties of these systems are still scarcely known. Furthermore, the advances in the understanding of the chemical reactivity of fullerenes has been used as a benchmark for the remaining carbon nanoforms and, thus, most of the reactions carried out on fullerenes have also been tested on the following carbon nanoforms, namely nanotubes and graphene.

No doubt, the few examples reported to date on chiral fullerenes reveal that important differences can be found in some physical properties in going from the racemate to the enantiomerically pure species. These former and encouraging results make the synthesis of chiral carbon nanoforms a new and appealing field of research which is almost unexplored. Control on the preparation of chiral monodisperse empty fullerenes and endofullerenes at will represents a first and important step for further studies focused on the more complex polydisperse systems such as carbon nanotubes and graphene.

In this study, we have carried out the cycloaddition reaction of 3-alkynoates to fullerenes under different experimental conditions and with a variety of chiral phosphines under basic conditions. The presence of the base is responsible for the isomerization of the alkynoate to the corresponding allenoate, which reacts with the fullerene (C_{60} and C_{70}).

The controlled introduction of chirality in these systems represents a great synthetic challenge where the imagination of chemists is the main limit. However, the expectations for these new chiral carbon allotropes make this a worthwhile venture.

Acknowledgements

This work was supported by the European Research Council ERC-2012-ADG_20120216 (Chirallcarbon), Ministerio de Economía y Competitividad

(MINECO) of Spain (project CTQ2011-24652; Consolider-Ingenio CSD2007-00010), and the CAM (MADRISOLAR-2 project S2009/PPQ-1533). N. M. thanks the Alexander von Humboldt Foundation.

Notes and references

1 J. L. Delgado, M. A. Herranz and N. Martín, *J. Mater. Chem.*, 2008, **18**, 1417.

2 H. W. Kroto, J. R. Heath, S. C. O'Brien, R. F. Curl and R. E. Smalley, *Nature*, 1985, **318**, 162.

3 S. Iijima, *Nature*, 1991, **354**, 56.

4 K. S. Novoselov, *et al.*, *Science*, 2004, **306**, 666.

5 A. Hirsch and M. Bettreich, *Fullerenes, Chemistry and Reaction*, Wiley-VCH, Weinheim, Germany, 2005.

6 S. Filippone, E. E. Maroto, Á. Martín-Domenech and N. Martín, In *Advances in Organometallic Chemistry and Catalysis*, John Wiley & Sons, Inc., 2013, p. 459.

7 C. Thilgen and F. Diederich, *Chem. Rev.*, 2006, **106**, 5049.

8 C. Thilgen, I. Gosse and F. Diederich, *Top. Stereochem*, John Wiley & Sons, Inc., 2003, 23, p. 1.

9 Y. Wang, J. Xu, Y. Wang and H. Chen, *Chem. Soc. Rev.*, 2013, **42**, 2930; Y. Hizume, K. Tashiro, R. Charvet, Y. Yamamoto, A. Saeki, S. Seki and T. Aida, *J. Am. Chem. Soc.*, 2010, **132**, 6628.

10 B. M. Illescas, N. Martín, J. Poater, M. Solà, G. P. Aguado and R. M. Ortuño, *J. Org. Chem.*, 2005, **70**, 6929; B. Illescas, J. Rifé, R. M. Ortuño and N. Martín, *J. Org. Chem.*, 2000, **65**, 6246; F. Novello, M. Prato, T. Da Ros, M. De Amici, A. Bianco, C. Toniolo and M. Maggini, *Chem. Commun.*, 1996, 903.

11 D. W. C. MacMillan, *Nature*, 2008, **455**, 304.

12 S. Filippone, E. E. Maroto, A. Martín-Domenech, M. Suarez and N. Martín, *Nat. Chem.*, 2009, **1**, 578.

13 E. E. Maroto, S. Filippone, A. Martín-Domenech, M. Suarez and N. Martín, *J. Am. Chem. Soc.*, 2012, **134**, 12936.

14 E. E. Maroto, *et al.*, *Angew. Chem., Int. Ed.*, 2011, **50**, 6060.

15 K. Sawai, *et al.*, *J. Am. Chem. Soc.*, 2011, **133**, 17746.

16 J. Marco-Martínez, V. Marcos, S. Reboredo, S. Filippone and N. Martín, *Angew. Chem., Int. Ed.*, 2013, **52**, 5115.

17 J. Marco-Martinez, S. Reboredo, M. Izquierdo, V. Marcos, J. L. López, S. Filippone and N. Martín, *J. Am. Chem. Soc.*, 2014, **136**, 2897.

18 A. Suárez and G. C. Fu, *Angew. Chem., Int. Ed.*, 2004, **43**, 3580.

19 Some chiral bases allowed us the synthesis of chiral allenoates whose enantiomeric excesses were measured by chiral HPLC (Chiralpak IC). The optimal value (ee: 54%) was obtained by using cinchonidine **B-I** as chiral base.

20 H.-T. Yang, W.-L. Ren, C.-B. Miao, C.-P. Dong, Y. Yang, H.-T. Xi, Q. Meng, Y. Jiang and X.-Q. Sun, *J. Org. Chem.*, 2013, **78**, 1163.

21 L. Echegoyen and L. E. Echegoyen, *Acc. Chem. Res.*, 1998, **31**, 593; N. Martín, L. Sánchez, B. Illescas and I. Pérez, *Chem. Rev.*, 1998, **98**, 2527; J. Heinze, *Angew. Chem., Int. Ed. Engl.*, 1984, **23**, 831.

22 M. Ostermeier, J. Prieß and G. Helmchen, *Angew. Chem., Int. Ed.*, 2002, **41**, 612; A. Marinetti, V. Kruger and F.-X. Buzin, *Tetrahedron Lett.*, 1997, **38**, 2947.

23 The cases in which R^1 = Ph, the priority of substituents changes and leads to an *R* configuration.

24 M. Sampath and T.-P. Loh, *Chem. Sci.*, 2010, **1**, 739.
25 Both enantiomers were separated previously by semipreparative HPLC using Regis WHELK-O1 5/100 25cm × 10.0 mm column, hexane/2-propanol 90 : 10, flow rate 1.50 mL min^{-1}
26 C. Zhang and X. Lu, *J. Org. Chem.*, 1995, **60**, 2906.
27 F. Cozzi, W. H. Powell and C. Thilgen, *Pure Appl. Chem.*, 2005, **77**, 843; W. H. Powell, F. Cozzi, G. P. Moss, C. Thilgen, R. J.-R. Hwu and A. Yerin, *Pure Appl. Chem.*, 2002, **74**, 629; M. S. Meier, G. W. Wang, R. C. Haddon, C. Pratt Brock, M. A. Lloyd and J. P. Selegue, *J. Am. Chem. Soc.*, 1998, **120**, 2337.
28 E. E. Maroto, A. de Cózar, S. Filippone, Á. Martín-Domenech, M. Suárez, F. P. Cossío and N. Martín, *Angew. Chem., Int. Ed.*, 2011, **50**, 6060.

Faraday Discussions

PAPER

Inkjet printing of graphene

Kirill Arapov,*[a] Robert Abbel,[b] Gijsbertus de With[a] and Heiner Friedrich[a]

Received 14th April 2014, Accepted 20th May 2014
DOI: 10.1039/c4fd00067f

The inkjet printing of graphene is a cost-effective, and versatile deposition technique for both transparent and non-transparent conductive films. Printing graphene on paper is aimed at low-end, high-volume applications, *i.e.*, in electromagnetic shielding, photovoltaics or, *e.g.*, as a replacement for the metal in antennas of radio-frequency identification devices, thereby improving their recyclability and biocompatibility. Here, we present a comparison of two graphene inks, one prepared by the solubilization of expanded graphite in the presence of a surface active polymer, and the other by covalent graphene functionalization followed by redispersion in a solvent but without a surfactant. The non-oxidative functionalization of graphite in the form of a donor-type graphite intercalation compound was carried out by a Birch-type alkylation, where graphene can be viewed as a macrocarbanion. To increase the amount of functionalization we employed a graphite precursor with a high edge to bulk carbon ratio, thus, allowing us to achieve up to six weight percent of functional groups. The functionalized graphene can be readily dispersed at concentrations of up to 3 mg ml^{-1} in non-toxic organic solvents, and is colloidally stable for more than 2 months. The two inks are readily inkjet printable with good to satisfactory spreading. Analysis of the sheet resistance of the deposited films demonstrated that the inks based on expanded graphite outperform the functionalized graphene inks, possibly due to the significantly larger graphene sheet size in the former, which minimizes the number of sheet-to-sheet contacts along the conductive path. We found that the sheet resistance of printed large-area films decreased with an increase of the number of printed layers. Conductivity levels reached approximately 1–2 kΩ $\square^{-1}$ for 15 printing passes, which roughly equals a film thickness of 800 nm for expanded graphite based inks, and 2 MΩ $\square^{-1}$ for 15 printing passes of functionalized graphene, having a film thickness of 900 nm. Our results show that ink preparation and inkjet printing of graphene-based inks is simple and efficient, and therefore has a high potential to compete with other conductive ink formulations for large-area printing of conductive films.

[a]*Laboratory of Materials and Interface Chemistry, Dept. of Chemical Engineering and Chemistry, Eindhoven University of Technology, Den Dolech 2, 5612AZ, Eindhoven, The Netherlands. E-mail: h.friedrich@tue.nl*
[b]*Holst Centre – TNO, High Tech Campus 31, 5656AE, Eindhoven, The Netherlands. E-mail: robert.abbel@tno.nl*

1 Introduction

Manufacturing large-area graphene thin films is an important step towards the commercialization of graphene based technologies. The inkjet printing of graphene is one method for the controlled deposition of large-area transparent or non-transparent conductive films.[1–6] In comparison with other approaches such as chemical vapor deposition (CVD),[7,8] inkjet printing is a poor competitor. This holds true especially for transparent graphene thin films,[1,2] where high conductivity and high transparency, together with precise control over the number of layers and defect density is required. Nevertheless, the printing of graphene has a very high potential for applications, where non-transparent but highly conductive patterns are required.[9] Examples of such low-end, high-volume applications could be radio frequency identification tags (RFID tags), electromagnetic shielding, or devices where graphene printing of conductive patterns can be effectively utilized. Graphene inks made from graphite have the potential to revolutionize the field of printed conductors by replacing metallic inks while at the same time reducing biological hazards and production costs.

There are several reports on graphene-based inks encompassing different preparation routes, including the oxidation of graphite to graphite oxide,[1,2,10–12] and extended ultrasonication of graphite with[3,13] or without a surfactant.[6] The majority of studies on non-oxidized graphene report a relatively low graphene concentration,[2,6] poor conductivity[2] and rather low ratio of graphene to surfactant.[9] Recently, the successful non-oxidative functionalization of graphene was shown, leading to slightly improved dispersibility in a number of solvents.[14–16] Functionalization seems to be very promising for graphene inks as it allows the use of surfactants for colloidal stabilization to be avoided, which in turn has been made responsible for decreasing the conductive properties of a printed layer by increasing the sheet-to-sheet contact resistance. Unfortunately, the previously demonstrated amount of graphene functionalization and, thus, product dispersibility, is far from ideal, a point which this paper will address.

In this paper we describe two methods for the preparation of graphene-based inks. First, a simple, efficient and up-scalable method is presented, starting from raw graphite and avoiding any oxidation steps to preserve the conductive properties of the graphene throughout the entire process. Second, a non-oxidative covalent functionalization of graphene is presented resulting in a final product dispersibility of up to 3 mg ml^{-1} in non-toxic solvents. Subsequently we studied the printability of the two inks using two paper substrates, namely FS3 and LumiForte special application papers, which are widely used as substrates for the inkjet and screen printing of conductive patterns. Furthermore, the performance of the two inks was evaluated by means of sheet resistance measurements of large printed areas as a function of the number of printed layers.

2 Experimental

2.1 Preparation of expanded graphite (EG) from Li–THF–GIC

The synthesis of a ternary graphite intercalation compound with Li in THF (Li–THF–GIC) was performed using the procedure of Nomine and Bonnetain.[17] Briefly, 1.28 g of naphthalene $C_{10}H_8$ (0.01 mol, Alfa Aesar, USA) was dissolved in

100 ml of freshly distilled tetrahydrofuran (THF) by vigorous stirring, followed by the addition of 0.12 g of freshly cut metallic lithium (0.017 mol, Alfa Aesar, USA). Then 0.5 g of −10 mesh graphite (0.042 mol, Alfa Aesar, USA or Asbury Graphite, USA) was added to the reaction mixture at once. The reaction flask was then sealed and left stirring for another 72 hours. The Li–THF–GIC was separated from the side products by decantation followed by rinsing with freshly distilled THF. The residue was filtered and dried in ambient conditions for 20 minutes.

The freshly prepared GIC was placed at the bottom of a Vitreosil quartz crucible ensuring that all of the particles were in mutual contact. The crucible was placed in a 2450 MHz 700 W home appliance microwave oven and exposed to irradiation for 1 minute. The as-prepared expanded graphite (EG) was used without any further treatment.

2.2 EG ink preparation with Plasdone S-630

EG (1.4 mg ml^{-1}) was added to a glass vial containing 5 mg ml^{-1} of a 60 : 40 copolymer of *N*-vinyl-2-pyrrolidone and vinyl acetate (Plasdone S-630, Ashland Inc., USA) dissolved in a mixture of isopropanol (IPA) and *n*-butanol (*n*-BuOH) with a volume ratio of 1 : 3. The slurry was ultrasonicated (Branson 1510DTH SB) in ice water for 30 minutes. The as-obtained dispersion was used without any further treatment.

2.3 Functionalization of graphite platelet carbon nanofibers (GCNFs)

The synthesis and functionalization of graphite *via* a potassium intercalation compound (C_8K GIC) were performed in a N_2 glovebox with an O_2 concentration <0.1 ppm and H_2O <0.1 ppm.

Preparation of NaK. A liquid NaK alloy was prepared similarly to ref. 18. Briefly, freshly cut and dry pieces of Na and K with a mass ratio of 1 : 7.4 were pressed together and agitated with a spatula to obtain a liquid state alloy.

Preparation of C_8K GIC. Graphite platelet carbon nanofibers (100 mg, 0.0083 mol, Strem Chemicals, USA) were added to a 50 ml round bottom flask containing 20 ml of absolute THF, followed by the addition of 0.29 ml of freshly prepared NaK alloy. The reaction mass was stirred for 3 days to obtain a dark blue to green dispersion. Next, 0.9 g of *p*-nitrobenzylbromide (0.00416 mol, Merck Chemicals, Germany) was dissolved in 10 ml of absolute THF and added to a GIC dispersion by 0.5 ml every 1.5 hours. After addition the reaction mixture was additionally agitated for 1 day. Afterwards the sample was taken out of the glovebox, followed by quenching with isopropanol (5 ml), ethanol (5 ml), and finally water (5 ml). The reaction mass was then centrifuged at 8000 rpm for 30 minutes and the supernatant was discarded. The product was redispersed in a THF–H_2O (3 : 1) mixture and centrifuged under the same conditions as described above. The washing procedure was repeated with the following solvent mixtures: IPA–H_2O (3 : 1), EtOH–H_2O (3 : 1), EtOH–H_2O (1 : 3), and H_2O. The washed product was finally redispersed in H_2O and filtered through a 0.45 μm pore size PTFE membrane and dried under vacuum at 50 °C to yield a black powder of *p*-nitrobenzyl functionalized graphite platelet carbon nanofibers (PNB–GCNFs). This material was used for the preparation of graphene inks without further treatment.

2.4 Ink preparation with PNB–GCNFs

PNB–GCNF powder (3 mg ml^{-1}) was added to a vial containing propylene glycol diacetate (PGDA, The Dow Chemical Company, USA) and ultrasonicated (Branson 1510DTH SB) in ice water for 2.5 h. The as-prepared dispersion was used for printing without further treatment.

2.5 Printing conditions

Printing tests were performed on a Dimatix DMP 2800 system (Dimatix-Fujifilm Inc., USA), equipped with a 10 pl drop volume cartridge (DMC-11610). The print head contained 16 parallel squared nozzles with a diameter of 30 μm. The dispersion was printed at a voltage of 13 V, using a frequency of 5 kHz and a customized waveform. FS3 paper (Felix Schoeller, Germany) and LumiForte paper (Stora Enso, Finland) were used as substrates. The distance between the printing head and substrate was set to 200 μm. The following patterns were printed: dots (3 lines, 200 μm drop spacing), rectangles of 5 × 35 mm in size, and 3 pixel wide lines with a 20 μm line distance. In the case of the functionalized graphene inks with propylene glycol diacetate as the solvent, the substrate temperature was optimized to 50 °C.

2.6 Thermal gravimetric analysis conditions

TGA measurements were performed using a TGA Q500 (TA Instruments, USA) in the temperature range 35–600 °C under a N_2 gas flow with a heating rate of 10 °C min^{-1}.

2.7 X-ray photoelectron spectroscopy conditions

X-ray photoelectron spectroscopy (XPS) measurements were carried out with a K-Alpha, Thermo Scientific spectrometer using an aluminum anode (Al $K\alpha = 1486.3$ eV) operating at 510 W with a background pressure of 8×10^{-8} mbar. The spectra were recorded using the VGX900 data system collecting 40 scans. The spectra were acquired at a take-off angle of 0° relative to the surface normal, corresponding to a probe depth of around 10 nm.

2.8 X-ray diffraction conditions

X-ray diffractograms were obtained on a Bragg–Brentano Rigaku Geigerflex diffractometer with CuKα irradiation using Lindemann capillaries with an internal diameter of 3 mm.

2.9 Sheet resistance measurement conditions

The sheet resistance was measured with a four-point probe setup (source: Keithley 237 High Voltage Measure Unit, resistance meter: Keithley 6517A High Resistance Meter) with an interprobe distance of 5 mm, in a range of currents from 10 nA to 1 mA.

3 Results and discussion

3.1 Preparation of ink with expanded graphite

The ink preparation workflow starts with a donor-type intercalation of raw graphite as shown in Fig. 1. The donor-type intercalation compound (IC) is formed by donating electrons to the graphite matrix, thus, making it negatively charged, and accompanied by the insertion of positively charged species to sandwich a graphene layer between two layers of intercalants (stage 1 IC).[18,19] In contrast to oxidation, *i.e.*, the withdrawal of electrons from a graphite matrix, such an approach prevents the formation of any functional groups, and hence, preserves the conductive properties of graphene. The formed stage 1 donor-type IC slowly decomposes upon exposure to ambient conditions into mixtures of a higher stage IC.[20] For instance, the IC flake shown in Fig. 1b exhibits charging effects at the edges and planes which can be attributed to lithium salts deintercalating from the graphite matrix. Therefore, the intercalated materials were analyzed or used immediately after their synthesis.

To increase the distance between graphene layers and to promote layer separation, the IC was subjected to high-speed thermal expansion by means of microwave irradiation. Rapid heating of the GIC results in an abrupt conversion of the intercalated species into the gas phase, thus, causing expansion of the graphite planes along the *c*-axis.[21,22] The resulting expanded graphite (EG) has a porous worm-like structure, as shown in Fig. 1c, with a significantly increased specific volume (more than 300 times larger compared to the initial IC). In the next step, EG with concentrations of up to 1.4 mg ml^{-1} was dispersed in a mixture of i-propanol and *n*-butanol to obtain a colloidally stable dispersion (Fig. 1d). For ink formulation we opted for the highest graphene content *vs.* a reasonable colloidal stability. We found experimentally that at a graphene concentration of 1.4 mg ml^{-1} against 5 mg ml^{-1} of surfactant, the ink is colloidally stable for more than 9 months, whilst also providing a maximum graphene to surfactant ratio.

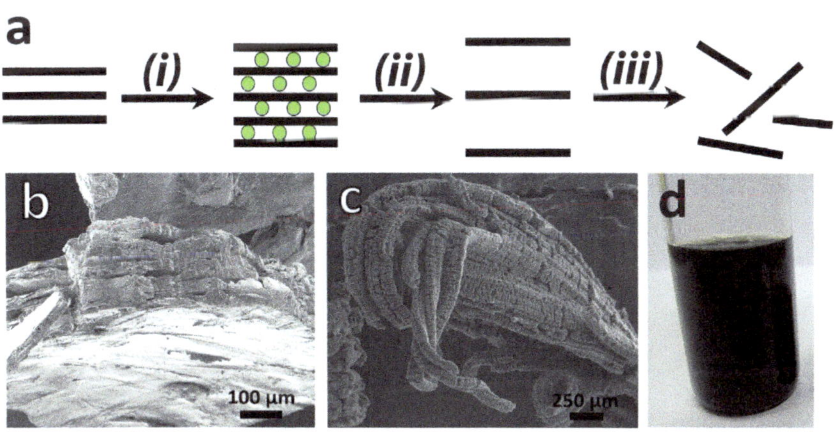

Fig. 1 The preparation of graphene inks from raw graphite *via* intercalation. (a) The overall scheme showing: (i) graphite intercalation, (ii) thermal expansion, and (iii) liquid phase exfoliation; (b) a scanning electron micrograph of a Li(THF) graphite intercalation compound; (c) a scanning electron micrograph of expanded graphite; (d) a graphene dispersion with a concentration of 1.4 mg ml^{-1} prepared from expanded graphite.

3.2 Preparation of ink with functionalized graphene

As an alternative to the surfactant-based graphene ink formulation, we also developed a functionalized graphene ink (Fig. 2a). To date, there have been several studies on graphene functionalization[14,16,23,24] based on the chemistry of donor-type graphite intercalation compounds. Nevertheless, most approaches demonstrate only a very low amount of functionalization (thus poor dispersibility), which is in good agreement with the early works of Bergbreiter and Killough.[25] According to ref. 25, in a donor-type IC the reactivities of edge and bulk carbon atoms are different, with edge atoms undergoing the desired two-electron Birch-type alkylation, while bulk carbon can just give up an electron *via* a single electron transfer, thus acting as a catalyst for the undesired Wurtz-type coupling. This has led to the conclusion that the edge-to-plane ratio of carbon atoms plays a significant role in functionalization efficiency. Therefore, instead of commonly used micrometer-sized graphite, which has a low edge-to-plane carbon ratio, we used platelet-type, *i.e.* graphitic, carbon nanofibers (GCNFs). This type of graphite, with graphene sheet sizes of only 10–60 nm, assures a high edge-to-plane carbon ratio (Fig. 2b).

According to the above concept, a high degree of functionalization can significantly improve the dispersibility of small sheet graphene due to an increased number of interaction sites between functional groups and solvent associates. Thus, we synthesized a potassium graphite intercalation compound with GCNFs as the starting material for further functionalization. Next, we functionalized the as-obtained GCNF IC with *p*-nitrobenzylbromide using an approach similar to the one described by Garst, Barbas, and Barton.[26] In order to minimize the formation of undesired Wurtz-type dimerization side products, the

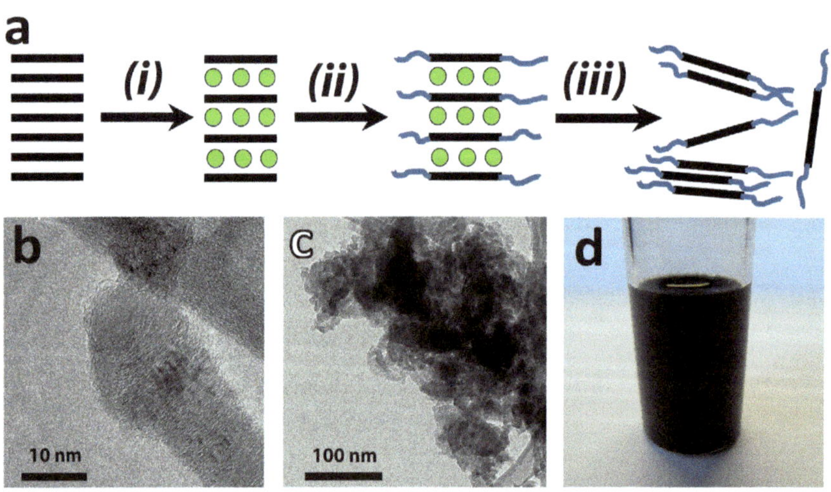

Fig. 2 Preparation of functionalized graphene inks from graphitic carbon nanofibers (GCNFs). (a) Scheme of GCNF functionalization (i) preparation of potassium–GCNF IC, (ii) functionalization of potassium–GCNF IC with *p*-nitrobenzylbromide, and (iii) liquid phase exfoliation; (b) a transmission electron micrograph of the starting GCNF; (c) a transmission electron micrograph of the cast-dried functionalized GCNF; (d) a PNB–GCNF dispersion with a concentration of 3 mg ml^{-1} in propylene glycol diacetate.

concentration of alkyl radicals in solution had to be kept much lower than the concentration of the substrate. Such conditions can be achieved by a slow alkylhalide vapour diffusion into the reaction medium or, as utilized in our procedure, by the stepwise addition of a very dilute (see experimental) alkylhalide solution. Upon the addition of alkylhalide we observed a significant volume expansion of the graphite phase indicating *in situ* exfoliation. The dried reaction product was found to be dispersible in a number of solvents including ethanol, i-propanol, chlorinated solvents and others. Transmission electron microscopy of a cast dried dispersion revealed that most of the GCNFs had been exfoliated into platelets which were prone to agglomeration, as shown in Fig. 2c.

We also performed dispersibility tests in order to find an appropriate solvent for inkjet printing. Our criteria for the final formulation included non-toxicity of the solvent, combined with a high solid content *vs.* colloidal stability, and printability. Experimentally we found that an optimal solvent for *p*-nitrobenzyl functionalized GCNFs (PNB–GCNFs) was propylene glycol diacetate (PGDA), allowing a solid content of 3 mg ml^{-1} without the use of surfactants (Fig. 2d). PGDA is a common solvent used for inkjet printing, with a viscosity of 2.6 mPa s (at 25 °C) and a surface tension of 32.9 mN m^{-1} (at 20 °C). The downside of this solvent is the rather low relative evaporation rate of 0.04 (*n*-butyl acetate = 1.0), which might extend the drying process.

3.3 Analysis of the functionalized graphene

The functionalized GCNF graphene was subjected to thermal gravimetric analysis (TGA). In an inert atmosphere this technique revealed (Fig. 3a) a sharp

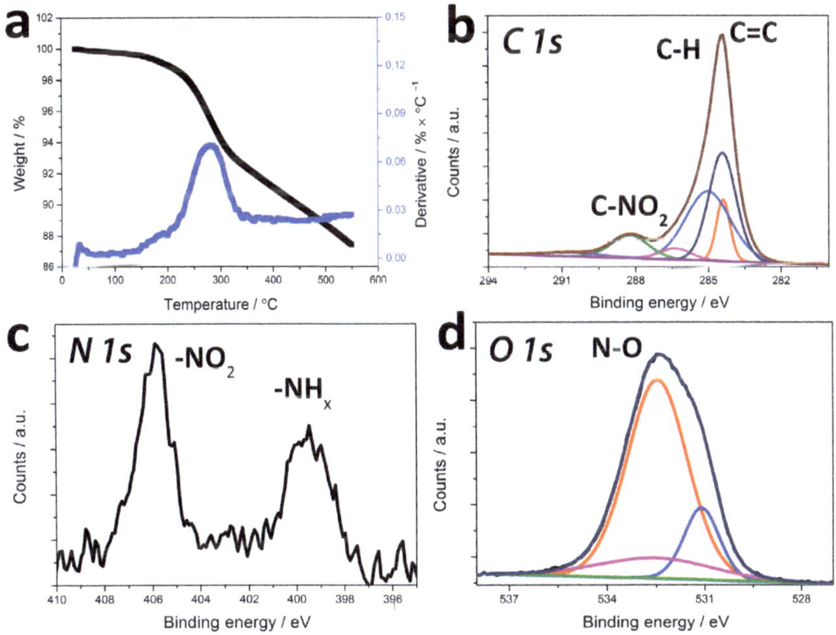

Fig. 3 Analysis of *p*-nitrobenzyl functionalized carbon nanofibers. (a) Thermal gravimetric analysis in dry N_2 with a heating rate of 10 °C min^{-1}; X-ray photoelectron spectra of (b) C 1s, (c) N 1s, and (d) O 1s.

decomposition peak amounting to about 6 weight percent, which was attributed to the presence of functional groups (*p*-nitrobenzyl), relating to approximately one per 166 carbon atoms. This hypothesis was supported by X-ray photoelectron spectroscopy (Fig. 3b–d). Here, XPS analysis confirmed the presence of carbon covalently bonded to a strong electron withdrawing group, *i.e.*, $C–NO_2$ (Fig. 3b).[27,28] Further, XPS showed the presence of nitrogen, presumably in a nitro group ($-NO_2$), as well as in reduced form $-NH_x$ (possibly including the hydroxylamine form), which can be explained by the partial reduction of nitro groups by the NaK alloy (Fig. 3c).[28,29] Also the XPS spectra of oxygen show the presence of a strong electron withdrawing oxygen containing group such as $-NO_2$.[28] Quantification of the XPS data by peak area adjusted with relative sensitivity factors[30] gave a relative carbon to nitrogen ratio of approximately 99 to 1. This corresponds to a functionalization degree of one *p*-nitrobenzyl group per 92 carbon atoms or theoretically 11 wt%. The observed discrepancy between the TGA and XPS data can be attributed to several factors, such as probing depth, and the sample surface roughness of the latter. Nevertheless, the results found by both techniques are comparable when the margins of error are taken into account. Considering all of the above results, combined with the significantly improved dispersibility of the final product, we concluded that functionalization had been successful.

3.4 Inkjet printing of graphene-based inks onto paper substrates

Printing tests of the developed inks were performed on commercially available paper substrates, namely FS3 and Lumi. FS3 is a glossy, polymer-coated paper and Lumi is a matt and rougher paper. The presence of a polymer coating on the FS3 minimizes the surface roughness, making it similar to polymeric foil substrates such as polyethylene terephthalate (PET), and it improves the adhesion of the printed layers. Nevertheless, even with a coating on top, FS3 still maintained its absorbing properties towards solvents, and quickly separates the non-soluble components from soluble ones.

To analyze the spreading behavior of the inks, arrays of dots were printed. In Fig. 4a–d optical micrographs of the corresponding shapes of the printed dots are shown. It can be clearly seen that, independent from the ink formulation, the two substrates have different spreading behavior. Dots printed on FS3 paper have a round shape with sharp edges and no significant signs of any "coffee stain" effects are observed. Contrary to FS3, both of the inks printed on Lumi paper demonstrate extensive spreading with an increase in dot size. In addition, the printed dots have irregular shapes with ill-defined edges, thus significantly lowering the resolution of the printed features. All of the above indicates clearly that independent from the ink formulation, both substrates have different spreading properties, with the FS3 paper suppressing any expansion of the droplet while it is drying and adsorbing, whereas on the Lumi paper the droplet is spreading while it is drying or being absorbed.

An important characteristic of evaluating ink-substrate compatibility is the edge resolution and the ink spreading over large printing areas. To this end, rectangles of 5×35 mm in size were printed with 1, 5, 10 and 15 printing passes on both paper substrates. In Fig. 4e–l optical micrographs of the corners of the printed rectangles are presented. The corners of the rectangles printed on FS3

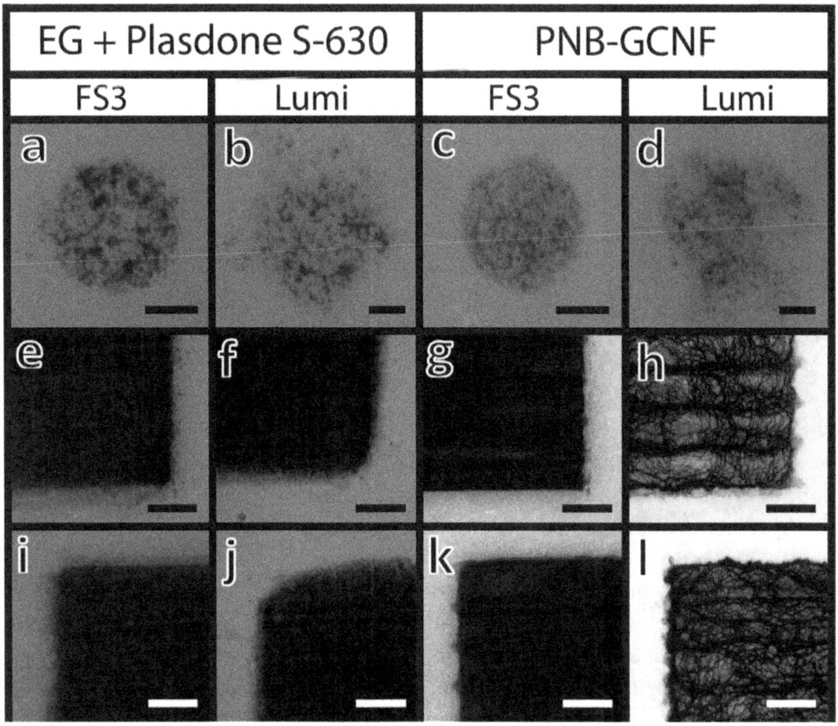

Fig. 4 Graphene based inks printed on FS3 (a, e, i and c, g, k) and Lumi (b, f, j and d, h, l) papers. Expanded graphite with Plasdone S-630 dots (a and b), and large-area rectangles (e, i and f, j), 10 printing passes. *p*-Nitrobenzyl functionalized graphene dots (c and d), and large-area rectangles (g, k and h, l), 10 printing passes. Scale bars: (a–d) – 20 μm, (e–l) – 500 μm.

paper (Fig. 4e, g, i and k) have sharp edges in the printing direction and perpendicular to it, demonstrating a fair printing definition. In the case of EG + Plasdone S-630, it can be seen that the corners are rounded for the ink printed on Lumi paper (Fig. 4f and j), which can be explained by extensive spreading of the ink over the substrate. Overall the ink covered the substrate uniformly. Optical micrographs of the functionalized graphene ink printed on Lumi paper (Fig. 4h and l) also demonstrate a reasonable printing resolution, with a distinct edge roughness, probably caused by the slow solvent evaporation and an increase of the Marangoni flow within a layer.

We found that ink spreading depends not only on the substrate but also on the ink characteristics. The formulation based on expanded graphite with a polymeric surfactant printed on the FS3 substrate (Fig. 4e and i) demonstrated a good spreading and uniform particle distribution. This ink was formulated with a solvent mixture of highly volatile i-propanol and *n*-butanol, that evaporates quickly (evaporation rate of i-propanol = 2.9, *n*-butanol = 0.4, *n*-butyl acetate = 1.0), thus, minimizing any particle segregation and producing a homogeneous layer. For the same ink, a similar behavior was observed in the case of the Lumi paper (Fig. 4f and j), where the ink spreads rather uniformly with minimal drying artefacts. The ink formulation based on functionalized graphene (PNB–GCNF) printed on FS3 paper (Fig. 4g and k) demonstrated a non-uniform though regular particle segregation into lines parallel to the printing direction. This behavior is most likely caused by the slowly evaporating solvent and the ink's instability at locally high concentrations. This creates a pattern where lines of segregated material are poorly connected with each other, potentially causing anisotropy of the conductive properties. The spreading of the PNB–GCNF ink on the Lumi substrate (Fig. 4h and l) was very complex, with particle segregation not only along the printing direction but also perpendicular to it. This demonstrates that extensive spreading in combination with slow drying decreases the printing resolution and overall printing quality.

3.5 Sheet resistance evaluation of inkjet printed graphene films on paper substrates

Along with the printability of inks, another highly important characteristic is the conductivity of the printed layers. We performed four point probe sheet resistance measurements of printed large-area rectangles with different numbers of layers, *i.e.*, 1, 5, 10 and 15 printing passes. We found that in all cases, single-pass printed rectangles were not conductive, indicating that the achieved coverage of conductive particles was below the percolation threshold. However, after subsequent printing passes, a significant increase in the conductivity of the printed layers was observed (Fig. 5).

We observed a significantly lower sheet resistance (R_s) for the ink prepared from expanded graphite and polymeric stabilizer than for the functionalized graphene based ink. As can be seen from Fig. 5, the trend continued with an

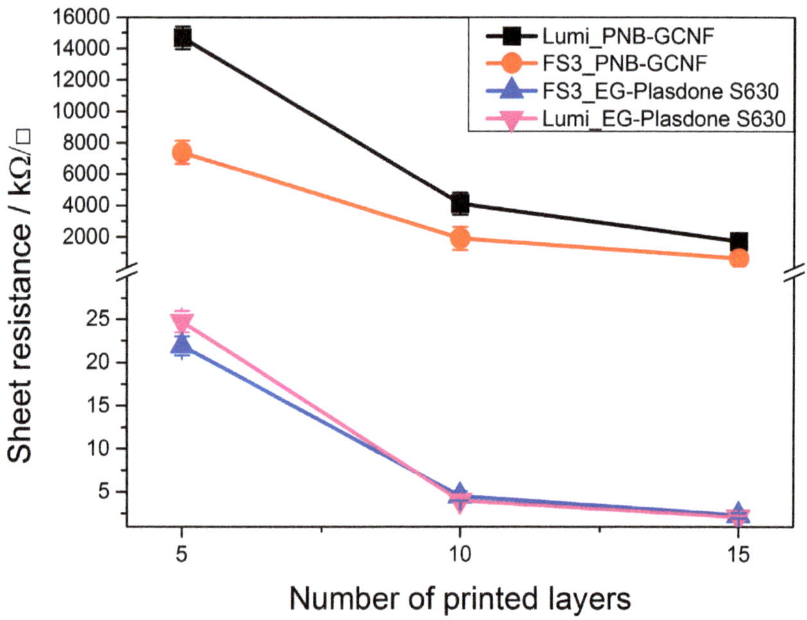

Fig. 5 Sheet resistance of 5 × 35 mm printed rectangles as a function of the number of printing passes.

increase in the number of printing passes for both inks resulting in decreasing sheet resistance. This results, for 15 printing passes, in a film sheet resistance of approximately 1–2 kΩ $\square^{-1}$ for the EG + Plasdone S-630 ink (800 nm thickness) and ~2 MΩ $\square^{-1}$ for the PNB–GCNF ink (900 nm thickness). Remarkably, in the case of EG with Plasdone S-630, we did not notice any significant difference in R_s for the two paper substrates, which is in good agreement with the observed spreading of the ink (Fig. 4e, i and f, j). However, we found a significant (~2 times) difference in R_s for the PNB–GCNF ink, which correlates with the difference in spreading and drying of the ink (Fig. 4g, k and h, l) described earlier. Thus, inkjet printed films with uniform material distribution demonstrate lower and isotropic sheet resistance, whilst printed films with a non-uniform particle distribution show higher and anisotropic sheet resistance.

From the observed sheet resistance data we hypothesize that the size of the graphene sheets plays a role in the conductive properties. If we assume that the R_s of a graphene film scales with the resistance of the contact between two flakes (R(CR)) and the resistance of the graphene flake itself (R(GF)), then the number of contacts would give a major contribution to the overall resistance. Here we assume that R_s for a flake is approximately constant for all sizes as it depends on the defect density,[31–33] or doping,[34–40] *etc.* Since the large (90% of population >1 μm) graphene flakes of EG + Plasdone S-630 assemble into a layer with fewer contacts between each other compared to a layer of the PNB–GCNF graphene (90% of population <60 nm), then the overall sheet resistance of the former film (on both substrates) is much lower than that of the latter.

4 Conclusions

In this paper we demonstrated two approaches for graphene ink formulation: first, based on the solubilization of graphene with a surface active polymer, and, second, based on covalent graphene functionalization. In the first route, we used a donor-type graphite intercalation compound as an intermediate prior to thermal expansion to avoid any oxidation of the graphene. We experimentally found that the presence of a surface active polymer (60 : 40 copolymer of *N*-vinyl-2-pyrrolidone and vinyl acetate, Plasdone S-630) facilitated efficient graphene exfoliation allowing stable colloidal dispersions with concentrations of up to 1.4 mg ml^{-1} at a polymer concentration of 5 mg ml^{-1} in a mixture of non-toxic solvents. The formulated ink was printable on substrates including plastic foils and paper.

In the second route, we synthesized a potassium graphite intercalation compound with graphitic carbon nanofibers as a precursor and utilized it as an intermediate for covalent functionalization. We confirmed the hypothesis that high ratios between edge and bulk carbon atoms would increase the Birch-type alkylation efficiency with *p*-nitrobenzylbromide, allowing up to six weight percent of functional moieties. The functionalized graphene was readily dispersible in a number of solvents at high concentrations including alcohols, ethers, and chlorinated solvents. We found that the optimal solvent was propylene glycol diacetate, which can disperse up to 3 mg ml^{-1} of solid material with a colloidal stability of up to several months.

Both inks were printable on FS3 and Lumi paper substrates with good to satisfactory spreading. From our observations of spreading and drying behavior, we concluded that the optimal combination is an ink of expanded graphite with Plasdone S-630 printed on FS3 paper, as the ink forms a uniform layer with a reasonable print resolution. Further, we analyzed the conductive properties of the inks by measuring the sheet resistance of large-area printed rectangles. The evaluation of sheet resistance as a function of the number of printing passes showed a significant (up to 2000 times) difference between the two inks in favor of the one containing expanded graphite. We hypothesize that this difference could be due to a much smaller number of sheet-to-sheet contacts, and that the overall resistance mainly depends on the resistance of the graphene itself. In contrast to that, for a functionalized graphene printed film, the number of sheet-to-sheet contacts is much larger, therefore, the resistance of the film is limited by inter-layer contact resistance. Such a hypothesis is a task for further investigation. Finally, we found that the sheet resistance decreases with an increase of the number of printing passes, providing the opportunity that further increasing the printed layer thickness could lower the sheet resistance to reach the conduction values achieved by metal based inks. Taking into account the simplicity and cost-effectiveness of our method, we believe that inkjet printing of graphene based inks is a good alternative for the mass production of conductive films and devices on paper.

Acknowledgements

The research leading to these results has received funding from the European Union Seventh Framework Programme (FP7-MC-ITN) under grant agreement no. 264710. The authors would like to thank the Directorate-General for Science, Research and Development of the European Commission for financial support of the research. Authors would like to acknowledge Corne Rentrop (TNO, Eindhoven, The Netherlands) for providing paper for the experiment and valuable comments to the final version of the manuscript.

References

1 L. Huang, Y. Huang, J. Liang, X. Wan and Y. Chen, *Nano Res.*, 2011, **4**, 675–684.
2 J. Li, F. Ye, S. Vaziri, M. Muhammed, M. C. Lemme and M. Ostling, *Adv. Mater.*, 2013, **25**, 3985–3992.
3 S. Lim, B. Kang, D. Kwak, W. H. Lee, J. A. Lim and K. Cho, *J. Phys. Chem. C*, 2012, **116**, 7520–7525.
4 E. B. Secor, P. L. Prabhumirashi, K. Puntambekar, M. L. Geier and M. C. Hersam, *J. Phys. Chem. Lett.*, 2013, **4**, 1347–1351.
5 Y. Su, J. Du, D. Sun, C. Liu and H. Cheng, *Nano Res.*, 2013, **6**, 842–852.
6 F. Torrisi, T. Hasan, W. Wu, Z. Sun, A. Lombardo, T. S. Kulmala, G. W. Hsieh, S. Jung, F. Bonaccorso, P. J. Paul, D. Chu and A. C. Ferrari, *ACS Nano*, 2012, **6**, 2992–3006.
7 X. Li, Y. Zhu, W. Cai, M. Borysiak, B. Han, D. Chen, R. D. Piner, L. Colombo and R. S. Ruoff, *Nano Lett.*, 2009, **9**, 4359–4363.
8 A. Reina, X. Jia, J. Ho, D. Nezich, H. Son, V. Bulovic, M. S. Dresselhaus and J. Kong, *Nano Lett.*, 2009, **9**, 30–35.
9 J. M. Crain, J. S. Lettow, I. A. Askay, S. A. Korkut, K. S. Chiang, C.-H. Chen, R. K. Prud'homme, *USA Pat.*, # 8,278,757, 2012.

10 V. Dua, S. P. Surwade, S. Ammu, S. R. Agnihotra, S. Jain, K. E. Roberts, S. Park, R. S. Ruoff and S. K. Manohar, *Angew. Chem., Int. Ed.*, 2010, **49**, 2154–2157.

11 L. T. Le, M. H. Ervin, H. Qiu, B. E. Fuchs and W. Y. Lee, *Electrochem. Commun.*, 2011, **13**, 355–358.

12 L. T. Le, M. H. Ervin, H. Qiu, B. E. Fuchs, J. Zunino and W. Y. Lee, *IEEE NANO 2011*, Portland, OR, 2011.

13 H. Zhang, A. Xie, Y. Shen, L. Qiu and X. Tian, *Phys. Chem. Chem. Phys.*, 2012, **14**, 12757–12763.

14 J. M. Englert, K. C. Knirsch, C. Dotzer, B. Butz, F. Hauke, E. Spiecker and A. Hirsch, *Chem. Commun.*, 2012, **48**, 5025–5027.

15 L. H. Liu, M. M. Lerner and M. Yan, *Nano Lett.*, 2010, **10**, 3754–3756.

16 B. Genorio, W. Lu, A. M. Dimiev, Y. Zhu, A. R. Raji, B. Novosel, L. B. Alemany and J. M. Tour, *ACS Nano*, 2012, **6**, 4231–4240.

17 M. Nomine and L. Bonnetain, *J. Chim. Phys. Phys.-Chim. Biol.*, 1969, **66**, 1731–1741.

18 L. M. Viculis, J. J. Mack, O. M. Mayer, H. T. Hahn and R. B. Kaner, *J. Mater. Chem.*, 2005, **15**, 974.

19 M. S. Dresselhaus and G. Dresselhaus, *Adv. Phys.*, 2002, **51**, 1–186.

20 M. Inagaki and O. Tanaike, *Carbon*, 2001, **39**, 1083–1090.

21 Y. W. Zhu, S. Murali, M. D. Stoller, A. Velamakanni, R. D. Piner and R. S. Ruoff, *Carbon*, 2010, **48**, 2118–2122.

22 S. Jin, L.-S. Xie, Y.-L. Ma, J.-J. Han, Z. Xia, G.-X. Zhang, S.-M. Dong and Y.-Y. Wang, *J. Phys.: Conf. Ser.*, 2009, **188**, 012040.

23 S. Chakraborty, J. Chattopadhyay, W. Guo and W. E. Billups, *Angew. Chem., Int. Ed.*, 2007, **46**, 4486–4488.

24 J. M. Englert, C. Dotzer, G. Yang, M. Schmid, C. Papp, J. M. Gottfried, H. P. Steinruck, E. Spiecker, F. Hauke and A. Hirsch, *Nat. Chem.*, 2011, **3**, 279–286.

25 D. E. Bergbreiter and J. M. Killough, *J. Am. Chem. Soc.*, 1978, **100**, 2126–2134.

26 J. F. Garst, J. T. Barbas and F. E. Barton, *J. Am. Chem. Soc.*, 1968, **90**, 7159–7160.

27 E. Bekyarova, S. Sarkar, S. Niyogi, M. E. Itkis and R. C. Haddon, *J. Phys. D: Appl. Phys.*, 2012, **45**, 154009.

28 A. K. Haridas, C. S. Sharma, V. Sritharan and T. N. Rao, *RSC Adv.*, 2014, **4**, 12188.

29 G. Yang, H. Hu, Y. Zhou, Y. Hu, H. Huang, F. Nie and W. Shi, *Sci. Rep.*, 2012, **2**, 698.

30 C. D. Wagner, L. E. Davis, M. V. Zeller, J. A. Taylor, R. H. Raymond and L. H. Gale, *Surf. Interface Anal.*, 1981, **3**, 211–225.

31 J.-H. Chen, W. Cullen, C. Jang, M. Fuhrer and E. Williams, *Phys. Rev. Lett.*, 2009, **102**, 236805.

32 K. W. Clark, X. G. Zhang, I. V. Vlassiouk, G. He, R. M. Feenstra and A. P. Li, *ACS Nano*, 2013, **7**, 7956–7966.

33 A. Lherbier, S. M. M. Dubois, X. Declerck, S. Roche, Y.-M. Niquet and J.-C. Charlier, *Phys. Rev. Lett.*, 2011, **106**, 046803.

34 A. Benayad, H. J. Shin, H. K. Park, S. M. Yoon, K. K. Kim, M. H. Jin, H. K. Jeong, J. C. Lee, J. Y. Choi and Y. H. Lee, *Chem. Phys. Lett.*, 2009, **475**, 91–95.

35 A. Kasry, M. A. Kuroda, G. J. Martyna, G. S. Tulevski and A. A. Bol, *ACS Nano*, 2010, **4**, 3839–3844.

36 K. K. Kim, A. Reina, Y. Shi, H. Park, L. J. Li, Y. H. Lee and J. Kong, *Nanotechnology*, 2010, **21**, 285205.
37 T. Kobayashi, M. Bando, N. Kimura, K. Shimizu, K. Kadono, N. Umezu, K. Miyahara, S. Hayazaki, S. Nagai, Y. Mizuguchi, Y. Murakami and D. Hobara, *Appl. Phys. Lett.*, 2013, **102**, 023112.
38 H. Mousavi and R. Moradian, *Solid State Sci.*, 2011, **13**, 1459–1464.
39 Y. J. Ren, S. S. Chen, W. W. Cai, Y. W. Zhu, C. F. Zhu and R. S. Ruoff, *Appl. Phys. Lett.*, 2010, **97**, 053107.
40 S. Tongay, K. Berke, M. Lemaitre, Z. Nasrollahi, D. B. Tanner, A. F. Hebard and B. R. Appleton, *Nanotechnology*, 2011, **22**, 425701.

Faraday Discussions

DISCUSSIONS

Functionalisation, separation and solvation: general discussion

Nazario Martín, Philip Davies, Chris Ewels, Vladimir Falko, Kosmas Prassides, Toshiaki Enoki, Sheng Hu, Robert Mora, David Johnson, Antonio Rodriguez-Fortea, Noreen Hanly, Katsumi Kaneko, Ayse Turak, Rebecca Edwards, Kirill Arapov, Valeri Kovalenko, Santosh Kumar Bikkarolla, David Zitoun, Pulickel Ajayan, Alexander Zöpfl, Andreas Hirsch, Ronan McHale, Malcolm Heggie, Milo Shaffer, Juan Casado, Mary Chan-Park, Pagona Papakonstantinou and Yoko Yamakoshi

DOI: 10.1039/C4FD90045F

Mary Chan-Park opened the discussion of the paper by Philip R. Davies: The O(1) peak is difficult to deconvolute – did you verify the C–O–C group using C(1s) spectra? Or can you use other techniques?

Philip Davies replied: We did look very carefully at the C(1s) spectra but we are working with a single crystal sample rather than a high surface area powder and the surface concentrations we are dealing with (~20–30% monolayer) are too small to make a significant difference against the bulk C(1s) background. However, we are confident of our assignments for a number of reasons: firstly, we only curve fit where there is very good evidence that a peak is present; secondly, the curve fits have been tested against a series of spectra from different samples and the agreement is excellent; finally, we have since repeated this study using nitric acid in place of HCl and have obtained very similar results.[1]

1 C. Buono, R. Burgess, P.R. Davies, R. J. Davies, T. Legge, A. Lai, R. Lewis, D. J. Morgan, D. J. Willock, *J. Catal.*, submitted.

Ayse Turak remarked: The bubbles that you observed in the HOPG surface disappear both with heating and with introduction of gold nanoparticles. With regard to the Au deposition, are the bubbles acting as a template for the nucleation of the nanoparticles? With regard to the heating effect, what do you think is the mechanism? Are the bubbles due to some deformation of the layer, whereby the heating anneals out the defects; or is it due to entrapped gases, therefore the heating releases the gases; or some other effect?

Philip Davies replied: We have some evidence suggesting that the gold deposition occurs at the site of the features observed in the AFM, however we cannot definitively state that this is the case at the moment. We are not certain why the functionalisation of the surface leads to the topographical features we see, but we speculate that it is related to a reduction in bonding between the graphite layers. In this paper we have shown that the effect of the heating is to convert the OH groups into ketones and ethers and we are currently investigating how this might affect the interlayer repulsion using DFT based calculations.

Ayse Turak said: Your experiments are on the surface of HOPG. Have you tried similar experiments on free standing or exfoliated graphene?

Philip Davies answered: We have established a collaboration with Swansea University to look at free standing graphene and hope to try the experiment you suggest in the near future.

David Zitoun commented: You started with catalysis – could you comment on the catalysis itself? Gold is known to diffuse – do you think you need sulfur to strongly graft the particles for catalysis and prevent coalescence?

Philip Davies responded: This is an interesting idea. Whilst we have some evidence that the oxygen functional groups, and particularly the OH, appear to facilitate gold nucleation at the carbon surface, we have not studied the stabilisation effect. Lee *et al.*[1] showed that sulfur could stabilise platinum against sintering on mesoporous carbons but I'm not aware of any work where sulfur has been used to stabilise gold.

1 H. I. Lee, S. H. Joo, J. H. Kim, D. J. You, J. M. Kim, J.-N. Park, H. Chang and C. Pak, *J. Mater. Chem.*, 2009, **19**, 5934–5939

David Zitoun asked: Can you pay the price of inserting thiols in the system to promote grafting, knowing that sulfur usually known as an inhibitor for catalysis?

Philip Davies responded: I'm not aware of any work that has been done on this area with gold. It would be a very interesting idea to explore.

Kirill Arapov asked: What would be a theoretical mechanism for hydroxylation of graphite by means of just hydrochloric acid? What is most probable reactive site: the plane or defects? Can we observe similar hydroxylation reaction with low molecular weight polyaromatic hydrocarbons, such as ovalene or hexa-*peri*-hexabenzocoronene?

Philip Davies replied: The AFM images show features across the terraces as well as at step edges. DFT calculations show that introducing a defect into the plane of the graphite would involve a prohibitively large energy barrier and so we think the hydroxylation must be occurring at existing defective sites. For this reason we think it unlikely that mild acid treatments would have any effect on polyaromatic hydrocarbons. We are currently exploring the question of mechanism with DFT calculations but the reaction probably involves activation of a defect with H_3O^+ in a cluster of water molecules.

Kirill Arapov asked: Do you expect a hydroxylation reaction to occur between polyaromatic hydrocarbons (such as ovalene, hexa-*peri*-hexabenzocoronene) and hydrochloric acid?

Philip Davies responded: We think the hydroxyls form at existing defect sites and so we would not expect any reaction with the polyaromatics.

Vladimir Falko remarked: It is known (from the studies by the Manchester group) that for atoms (*e.g.*, hydrogen) and molecules that require adding some sp^3 to purely sp^2 bonding of carbons in graphene adsorption takes place preferentially on graphene ripples. Do you observe a similar behaviour in your experiment?

Philip Davies responded: The graphene bubbles we observe after acid treatment are dispersed over the terraces and not necessarily located at step sites.[1] We suspect this is due to the presence of existing defects at these sites but we have not been able to detect them before-hand. We also cannot prove that the OH groups are located at these sites. On clean surfaces gold deposition does prefer step edge sites but we now have evidence (using lower gold concentrations than is reported in the present paper)[2] that the acid-treated surface gives a better dispersion of the gold with smaller nanoparticle sizes. We do not know whether the gold is nucleating at the same sites that give rise to the bubbles.

1 E. Bouleghlimat, P. R. Davies, R. J. Davies, R. Howarth, J. Kulhavy, and D. J. Morgan, *Carbon*, 2013, **61**, 124–133.
2 R. J. Burgess, P. R. Davies, A. Lai, T. Legge and D. J. Morgan, in preparation.

Kirill Arapov asked: The hydroxylation of graphitic species will lead to the transformation of sp^2 carbon hybridization state to sp^3. This will be unambiguously detected by Raman scattering as the increase of defects. Did you observe such a behavior during your experiment?

Philip Davies replied: The number of hydroxylated sites formed on the graphite surface is very low (~20% of a monolayer) and the signal is swamped by the bulk crystal signal. We were unable to detect any changes in the Raman spectra after either acid treatment or gold deposition.

Chris Ewels said: Can you infer anything about the chemical environment of the hydroxyl groups from the form of the XPS O(1s) peak (FWHM *etc.*)? *E.g.* whether all hydroxyl groups are in similiar chemical environments?

Philip Davies responded: The FWHM of the O(1s) peak assigned to the hydroxyls is ~ 2 eV, against an expected linewidth of ~1.5 eV suggesting more than a single environment for the OH, but we do not have the resolution to separate out any individual components.

Pagona Papakonstantinou commented: XPS is a surface sensitive technique with a probing depth of up to a few nm. Do you think you are seeing only the top layer?

Philip Davies answered: The mean free path of the photoelectrons in carbon is quite long and it is for this reason the changes made to the carbon atoms in the uppermost layer are completely swamped by the signal from the carbon atoms deeper in the crystal. However, we think its unlikely that the mild acid treatments we are using here will penetrate beyond the uppermost graphene layers and therefore the oxygen signals we see are probably from this region. Some confirmation of this hypothesis comes from the observation that stripping a few monolayers from the HOPG crystals is sufficient to remove all of the oxygen from the XPS spectra.

Rebecca Edwards asked: Given your deposition technique (blow drying an HCl drop onto your surface) and the fact that the disappearance of the features on your surface upon heating coincides with the disappearance of the chlorine signal in the XPS, do you think that the features on your surface could be HCl crystals?

Philip Davies replied: The quantity of chlorine present after treatment with HCl is vanishingly small, according to the XP spectra, so we do not think the HCl is crystallising on the surface. Chlorine is only detected in significant quantities on the deposition of gold from chloroauric acid solution and is probably associated with the gold complex.

Santosh Kumar Bikkarolla commented: I am just wondering which oxidizing method of carbon nanostructures by acids can provide good electron transport in the hybrid materials (gold/carbon nanostructures).

Philip Davies answered: This is not something we have explored but we hope to look at such effects in the future.

Pagona Papakonstantinou asked: Are there any specific guidelines for fitting the C(1s) and O(1s) peaks?

Philip Davies replied: In our experiments the changes in the C(1s) signal were far too small to attempt a curve fit. We did investigate subtraction of the spectra from each other which is often a better technique for looking for small changes, but we were unable to detect any changes in the spectra. Fitting of the O(1s) signals needs to be carried out with care; we have only fitted the spectra with peaks that could be clearly identified in the raw spectra. The parameters we have used are described in the paper.

Ayse Turak opened the discussion of the paper by Sheng Hu: You mentioned that the scalability of other approaches is limited. Can you please comment on the scalability of your approach?

Sheng Hu answered: The lab scale using a 30 mm quartz tube set-up, with a static bed, is up to a few grams. We have systematically studied PMMA grafting on MWCNTs at scales increasing from 0.1 g to 5 g. At the upper end, the grafting ratio begins to fall, we believe that it is simply due to the opacity and low thermal conductivity of the fluffy MWNT powder bed preventing uniform activation. Much larger scale (>100 g) processing has been successfully achieved by our industrial partner.

Philip Davies asked: In your earlier paper[1] on the radical formation process it is stated that only a limited percentage of the functional groups are converted into the radical sites that you make use of. Are you able to say anything about the type of functional groups that give rise to these sites and those that do not?

1 R. Menzel, *et al.*, *Chem. Sci.*, 2010, **1**, 603–608.

Sheng Hu answered: We are currently trying to categorise the different types of surface oxides that are present as a function of annealing temperature and hence determine which type(s) can generate radicals during our thermo-activation process. We are conducting a series of experiments under different activation temperatures whilst studying the progress using XPS, TGA-MS-IR, and the resulting grafting efficiency.

Andreas Hirsch asked: You used TGA - isn't there a problem here seeing differences between mixtures and real functionalisation? Did you try using Raman to confirm your results? Did you get an independent confirmation of functionalisation?

Sheng Hu replied: It is true that TGA cannot always directly distinguish between physically absorbed reagents and covalently functionalised compounds, since the changes in decomposition temperatures/profiles are often modest. However, here we completed a range of control experiments to confirm that the mass losses were related to covalent functionalization of the nanocarbons. First, as shown in the ESI, a control of treating TTC-MWCNTs with various types of monomers, followed by standard workup, indicated that unreacted monomer was completely removed, demonstrating no physical absorption of monomers on MWCNTs.

Further, unreacted monomers in fact have a much lower TGA mass loss temperatures than the grafted oligomers. We have observed this mass loss for samples examined after simple drying rather than the full workup procedures. This observation confirms that remaining monomer would tend to evaporate rather than polymerise.

Lastly, our method takes advantage of existing defective groups on the nanocarbon surface; it only activates defects remaining from the synthesis process, instead of introducing new defects as do most conventional methods. Consequently, we anticipated and observed no change in the D/G ratio determined by Raman analysis, indicating an undamaged carbon nanomaterial framework.

David Johnson asked: Have you tried any radical trapping (spin trapping) experiments? This would confirm the mechanism is definitely radical. Also if you were to add your spin trapped material to the monomer mixture and you see no onset of polymerization this would go a long way to proving that the oligomers you observe are covalently bound to the CNTs rather than merely physically adsorbed.

Sheng Hu replied: Indeed, radical trapping can help to prove the involvement of radicals in our method. We have previously reported the use of the free radical inhibitor galvinoxyl to quench our activated MWCNTs and hence allow the determination of radical concentrations spectroscopically. We have also reported EPR analysis to determine the lifetime of our radicals generated on MWCNTs surfaces. We have also completed a number of control experiments, including the exposure of HTT nanocarbons (that is activated material, exposed to air to quench the radicals) to both monomer and homopolymers, to rule out physisorption. Please see our previous paper for further details of these aspects.[1]

1 R. Menzel *et al.*, *Chem. Sci.*, 2010, **1**, 603–608.

Mary Chan-Park enquired: Does the polymer wrap around the whole bundle or around individual tubes?

Sheng Hu replied: In this paper we discussed the functionalization of CVD-grown multi-walled nanotubes (MWNTs) and carbon black nanoparticles (CB NPs). Neither of these materials form packed bundles, and their accessible BET surface area is similar to the geometric expectations based on their structure. In these cases, there should be no limitation to the functionalization of the whole surface due to the accessibility of monomers during process chain propagation. Local steric effects due to the distribution of active sites, the size of the growing chains, or the presence of slit pores may still play a role. In the case of carbon black, of course, it is not possible to separate the primary particles that are fused at high temperature to form the aggregated structure. However, broadly the 'individual'" structures are functionalised, as imaged by the tagging experiments in our original paper.[1]

1 R. Menzel *et al.*, *Chem. Sci.*, 2010, **1**, 603–608.

Ronan McHale asked: Can you control the polymerisation at all by trapping?

Sheng Hu replied: We are currently conducting experiments to use TEMPO (or other reversible radical terminators) to control the polymerisation to pursue higher molecular weight

Ronan McHale asked: Have you tried to measure the molecular weight of the polymer?

Sheng Hu answered: It is quite difficult to directly measure the molecular weight of the grafted polymer, however, in the paper, we described an approach to determine the polymerisation initiation sites; in combination with the total grafting ratio, we can therefore estimate the average molecular weight of the grafted polymers/oligomers.

In order to determine the number of active sites, the MWNTs were grafted with the non-polymerisable reagent 1-iodododecane (IDD). The grafting density and degree of polymerisation were then calculated, assuming that IDD reacts stoichiometrically with all active radical sites, and the polymerisation processes is initiated at a radical site but can terminate anywhere on the MWNT surface. It is worth noting that the above calculation method assumes one of the two possible termination mechanisms. The other possibility is traditional radical-radical coupling between adjacent chains which would imply oligomer loops of twice the indicated molecular weight. It is not immediately clear which mechanism is dominant.[1]

1 S. Chen *et al.*, *Biomaterials*, 2014, **35**, 4729–4738.

David Johnson commented: Termination rates will always be quite high without introducing controlled polymerization techniques. Your method requires you to generate all your initiating radical prior to the onset of polymerization. As a result termination will be fast with respect to propagation. In a normal free radical polymerization you'd choose and initiator with a long half life to minimize the instantaneous radical concentration. Reversible termination is probably the only way to get higher molecular weights.

Sheng Hu responded: Thank you very much for your comment. Termination rates in our polymerisation/oligomerisation reactions are high, as indicated by the low grafted molecular weights. As discussed in the paper, there are two possible termination mechanisms, one is the traditional radical-radical coupling, the other is termination elsewhere on the nanocarbon surface. It is not immediately clear which mechanism is dominant. However, we are currently trying to use TEMPO to control the polymerisation to pursue higher molecular weight, as you suggest.

Katsumi Kaneko asked: As there are so many studies on improvement of dispersion of carbon nanotube in water, could you show how stable the carbon nanotubes dispersed in water are? In particular, multi-wall carbon nanotubes (MWCNTs) tend to precipitate quickly. How long do MWCNTs stably disperse in water?

Sheng Hu replied: Since our method can covalently modify the surface of MWCNTs, the grafted MWCNTs tend to be quite stable in solution. For MWCNTs, less than 20% of MWCNTs precipitated out after 2 weeks, as determined by UV-vis analysis. It's worth noting that we have no excess surfactant or polymer in solution, only the grafted-MWNTs, and that even then the grafting ratio is low. We also impose a relatively stringent centrifugation regime (10 000 g), to avoid counting agglomerates amongst the 'dispersed' material.

Katsumi Kaneko asked: I want to know how long the single wall carbon nanotubes (SWCNTs) disperse in water stably? Also, is there any depression of the excellent physical properties of the SWCNTs through your surface coating?

Sheng Hu answered: The stabilities of thermally-grafted SWCNT dispersions are currently being investigated, but again, similar to the MWCNTs, we would expect a reasonably good stability of oligomer grafted SWCNTs due to the covalent modification.

Our paper on the functionalization of SWCNTs including a range of physical property characterisation is in preparation. We do not expect the intrinsic

SWCNTs properties to be significantly depressed after the thermochemical functionalization since the functionalization method does not involve harsh oxidation or intense sonication processes which can damage the CNT framework. In addition, the number of covalently-grafted SWNT framework carbons is low, since the overall grafting ratio is relatively low and the grafted species have relatively high molecular weight. One of the most significant advantages of our method is to minimize framework damage, hence maintaining the excellent properties of CNTs.

Sheng Hu turned the discussion to the paper by Yoko Yamakoshi: In terms of the aggregation do you have any evidence for size distribution differences (like DLS characterisation)? What about the effects of the surface charges on the mechanism?

Yoko Yamakoshi responded: There is no particular charges in these molecules. Aggregation of the molecules can be caused by hydrophobic interaction of the buckyball core.

Malcolm Heggie opened a more general discussion: I recall my late friend and colleague, Prof. Roger Taylor, was very keen on nomenclature of the fullerenes. Were he alive I am sure he would contribute strenuously to this discussion. I do recall his insistence that 'incar', short for incarcerated, I believe, should be used instead of the @ notation for endohedral fullerenes. Is this no longer the position?

David Johnson added: When we are talking of modification nomenclature for carbon nanomaterials, perhaps we should consider them as surfaces. Is there anything in the IUPAC definitions for surface modification which we could apply or learn from?

Philip Davies replied: There are surface science conventions for describing ordered arrays of adsorbates at surfaces but these depend on long range order in the adsorbate lattice. It is not clear to me that such long range order is present in any of the systems we are talking about here.

Rebecca Edwards commented: Thermogravimetric analysis is often used to prove successful functionalisation of carbon nanomaterials. This data rarely includes error analysis despite the fact that two thermal profiles of the same material can appear different in the TGA. What are people's thoughts on the use of TGA for identifying the type of functionalisation (covalent, non-covalent) and reliably quantifying it?

Milo Shaffer responded: TGA of grafted nanocarbons intrinsically involves complex kinetics of competing processes. It is possible to improve the repeatability by using a consistent and appropriately chosen sample size. We tend to use around 1mg but the value will depend on the instrument; broadly, there is a balance between instrument sensitivity (given the expected grafting ratio) and the need to maintain uniform conditions within the sample powder bed. When using an inert atmosphere, it often seems to take longer than expected to fully flush out atmospheric oxygen and water; again it will depend on the instrument, but more than an hour's conditioning does not seem to be atypical. Designing the program to have sufficient drying time to remove residual water or other solvent is also important. It is possible to obtain repeatable data with an acceptable error, although not always practical or necessary to repeat every sample.

Estimates of the grafting ratio should be made relative to the appropriate reference, *i.e.* material that has been subjected to the same conditions/work-up, but without the key reagent/activation, rather the simple untreated material. This reference sample should provide a relatively featureless baseline. Other secondary controls may also required to rule out the possibility of physisorption. TGA/MS or TGA/FTIR can be very helpful to confirm that the mass loss corresponds to the intended functionalization rather than solvent, combustion, or other contamination.

Ayse Turak opened the discussion of the paper by Andreas Hirsch: Donor–acceptor dyads of the type you describe here are typically used as donor materials in bulk heterojunction (BHJ) photovoltaic devices or morphology stabilizers in ternary BHJ. Given the location and energy of the LUMO, do you see this material as acting as a good acceptor?

Andreas Hirsch replied: It is a good acceptor, just as are most of the C_{60} monodadducts. However, at the same time it can act also as a donor component because of the hexabenzocoronene added. Whether will be finally a good building block of organic photovoltaic devices has to be investigated.

Ayse Turak asked: It is possible to produce a thin film of this material by solution processing (*i.e.* is it soluble) or by a vacuum evaporation technique (thermal evaporation or e-beam deposition)?

Andreas Hirsch responded: We haven't tried so far. But I expect it should be possible.

Valeri Kovalenko enquired: In your single crystal X-ray experiment did you find any redistribution of the electron density of C_{60} moiety in its close proximity to the hexa-*peri*-hexabenzocoronene part of the adjacent dyad?

Andreas Hirsch replied: So far we didn't look into that.

Toshiaki Enoki commented: The distance between the two components is very short (3.1 Å). You explained that the charge transfer between the two components gives strong interaction, resulting in the short distance. Accordingly it is important to estimate the charge transfer rate.

Nazario Martín added: I guess that the photoinduced electron transfer in your dyad in solution is intra-molecular. If your material is not covalently connected, could you get photoinduced intermolecular electron transfer?

Andreas Hirsch replied: Yes, the photoinduced electron transfer should be intra-molecular. In the solid state photoinduced electron transfer could also be possible if no covalent attachment was involved but if the two chromophors adopted a similar relative orientation. In solution however, such an orientation will not be stable.

Antonio Rodriguez-Fortea commented: The energy difference between the LUMO of acceptor and the HOMO of donor is somewhat related to the open-circuit voltage in dye-sensitized solar cells. It is stated in your paper (page 10) that the LUMO energy of HBC–C_{60} is a little higher (200 mV) than that of PCBM. However, according to the LUMO energies in Table 1 and page 10, these are −3.9 eV for HBC–C_{60} and −3.7 eV for PCBM; it should be the other way around.

Andreas Hirsch answered: This is a good remark and it is indeed correct that the LUMO energy of compound **1** is lower than that of PCBM, which is counter-intuitive. But all our measurements and calculations are correct. We checked them again. We have now updated our manuscript appropriately.

Juan Casado asked: How could you change the molecular structure of your dye in order to have a suitable π-stacking arrangement for photovoltaic applications?

Andreas Hirsch replied: We could even introduce more bulky substituents in order to prevent dimerization.

Juan Casado further asked: What about by promoting H–π interactions?

Andreas Hirsch responed: This is also a possibility that we will investigate in the future.

Kosmas Prassides opened the discussion of the paper by Nazario Martín: Why there is an improvement in the conductivity in the two enantiomers relative to the racemic mixture? What is the origin of this effect? Perhaps it is related to the removal of the disorder associated with the racemic mixture?

Nazario Martín answered: The origin of the observed effect, although not totally unveiled, is likely to be associated to the better crystalline order found in the enantiomer. As Prof. Aida has described in his work,[1] physical properties (in this case charge-carrier mobility) change depending on the morphology of the sample, being a multilamellar sphere for the racemate whereas a rod-like structure is formed for the enantiomer. As a result, the obtained photocurrent is one order of magnitude higher in the *enantio*-enriched compound.

1 Y. Hizume, K. Tashiro, R. Charvet, Y. Yamamoto, A. Saeki, S. Seki and T. Aida, *J. Am. Chem. Soc.*, 2010, **132**, 6628.

Mary Chan-Park asked: Do you know why the racemic mobility wasn't the average of the enantiomers?

Nazario Martín responded: We still have to prove that enantiomers with the same optical purity show exactly the same mobility value. The data available so far seem to indicate this. However, I do not see why the racemic mixture should have the average value of the enantiomers. Actually, the enantiomers should show the same value and, therefore, the racemic mixture being formed by a mixture of compounds should exhibit a lower molecular order and, therefore, influence the morphology and mobility value.

Milo Shaffer asked: Why isn't the electron mobility of the *R* and *S* enantiomers identical?

Nazario Martín replied: Great question. Of course, they should be the same. However, due to the different enantiomeric purity of both enantiomers (ee 98% and ee 94%) the experimental values for the charge mobility is slightly different. However, when having both enantiomers the same ee purity the observed value should be the same.

Juan Casado enquired: Can you comment on the mechanism of the transfer of the chirality to the produced material in order to justify the crystal shape responsible of the electron mobility?

Nazario Martín answered: The produced material from different enantiomers should show the same electron mobility. Of course, this is assuming that the ee values for both enantiomers is the same. According to our experimental data, the racemic mixture affords lower electron mobility than the pure enantiomers. This experimental finding reveals the connection between crystal order, morphology and charge mobility.

Chris Ewels commented: The discussion of functionalised fullerene enantiomers is very interesting, but I don't understand some of your comments concerning crystallisation. Notably you say that for some racemic mixtures there is segregation during crystallisation. This seems understandable for crystal packing reasons, that separate enantiomer crystals may form. But you suggest that in some cases only one of the enantiomers forms crystals while the other remains in solution – how can this be "squared" with the idea that apart from their chiral symmetry flipping, the two species are identical?

Nazario Martín replied: In a racemic mixture where the two enantiomers are present in the same proportion, it is not possible to selectively crystalize only one of two enantiomers without the assistance of an external chiral agent. Depending on the structure of the target compound of crystallization, different compounds are used. Once the agent is added, the enantiomers become diastereoisomers giving raise to different properties and so allowing their selective isolation (racemic resolution).

In our particular case, the X-ray analyzed crystal has been obtained through slow evaporation of a solution of a mixture of CS_2–hexane of the enantiopure cyclopenteno fullerene derivative which has previously been obtained through an enantioselective cycloaddition reaction.

Pulickel Ajayan asked: I can understand the chirality in fullerenes and nanotubes but how do you expect this to manifest in an infinite sheet like graphene?

Nazario Martín responded: This is a very good and intriguing question, difficult to address. I guess that, as typically occurs, it would be nice to explore first other smaller and simpler models such as graphene quantum dots or nanoribbons. In principle, the presence of "external" chiral atoms either in the periphery or basal plane should allow the introduction of chirality in these systems. How to control this chirality is a further, not simple, question...

Noreen Hanly commented: In the results summary table you presented, it is usual to quote the %ee value of the enantiomer you are referring to, if that enantiomer is not pure (as you have actually done in the paper submitted), otherwise results could be a bit misleading for the reader, when attempting to infer their properties.

It is better to represent *R* and *S* enantiomers as mirror images, and not as presented on your summary slide (although these were correct). This also facilitates quick interpretation by a reader in such a mixed audience who may not be an expert in stereochemistry. This was a very nice substantial paper though!

Nazario Martín replied: We agree with the comment. Several ways for referring to the enantiopurity of compounds have been described in the literature. We have chosen that in which the major enantiomer is represented in the scheme and all the enantiomeric values are referred to this enantiomer. In any case, again we agree with this statement.

Robert Mora opened the discussion of the paper by Kirill Arapov: Would you comment on the choice of solvent please? Also has the author tried other solvents to disperse his solute?

Kirill Arapov responded: The choice of solvent is mainly dictated by the tight–loose ion pair theory. You may find more information in the paper by Bank and Bockrath.[1] Briefly, the tighter the ion pair, the more delocalized the carbanion-radical leading to a higher rate of electron transfer from GIC to a substrate. We tried other solvents for discussed reaction (*e.g.* DME) and found that THF is better option.

1 S. Bank, B. Bockrath, Reactions of aromatic radical anions. VI. Kinetic study of the reaction of sodium naphthalene with water, *J. Am. Chem. Soc.*, 1971, **93(2)**, 430–437.

Mary Chan-Park asked: Does the size of the grapheme have an effect on solubility?

Kirill Arapov replied: Yes, the size of graphene has an effect on solubility. The smaller the size, the better the solubility.

Pagona Papakonstantinou enquired: When use DMF/NMP solvent do you get any functional groups (*e.g.* nitrogen) on graphene?

Kirill Arapov answered: In our work we did not use NMP or DMF, therefore we cannot provide you any specific data on that. However, from a chemical point of view one should not expect the formation of any groups when dispersing graphite/graphene in a solvent.

Alexander Zöpfl commented: Stabilization of graphene in suspension is the key step for reliable printing. The usage of surfactants can be a solution, but this has negative impact on the sheet resistance of the final printed graphene layer. Are there any post treatments or techniques known to get rid of the surfactants in the final layer (something like washing or evaporation)?

Also, what are the resolution limits for inkjet printed structures in your setup and other state-of-the-art works?

Kirill Arapov responded: The typical inkjet inks consist of not only conductive filler and solvent but also a binder, usually a polymer to improve the colloidal stability of the inks and provide adhesion to the substrate. The presence of the binder indeed worsens sheet resistance, however, the removal of surfactant/binder results in the ink's poor adhesion, loss of flexibility, agglomeration and other issues. There is a possibility to remove the excess of surfactant (see the work of the group of A. Ferrari), however, the technologically does not seem to be promising.

In our work the major resolution limiting factor is the interaction of the inks with a substrate. The substrate causes significant spreading of the ink, while the ink solvent does not evaporate as quickly as needed. This combination leads to lower resolution and lower specific density of the deposited material (*e.g.* grams per area). In the best case scenario we can achieve resolution in the micrometer range.

Faraday Discussions

PAPER

Gel electrophoresis using a selective radical for the separation of single-walled carbon nanotubes

Sara Mesgari, Ashok Kumar Sundramoorthy, Leslie S. Loo* and Mary B. Chan-Park*

Received 5th May 2014, Accepted 9th June 2014
DOI: 10.1039/c4fd00092g

We have applied agarose gel electrophoresis (AGE) to single-walled carbon nanotubes (SWNTs) that have been pre-reacted with metallic-selective ionic radicals and then re-suspended with sodium cholate (SC) surfactant to obtain highly purified (up to 98%) semiconducting single-walled carbon nanotubes (s-SWNTs). The proposed combination method exploits the preferential reactivity with the metallic nanotube of the radicals generated from an azo naphthalene compound (Direct Blue 71 (I)) to preferentially increase the surface charge, and therefore the electrophoretic mobilities, of the metallic nanotube population under the influence of the electric field in AGE. The excellent separation achieved was verified by UV-vis-NIR and Raman spectroscopy as well as by the performance of field effect transistors fabricated with semiconducting-enriched SWNTs. FETs fabricated with I-assisted AGE-separated semiconducting nanotubes exhibited mobilities of ~3.6 to 11.7 cm^2 V^{-1} s^{-1} and on/off ratios from 10^2 to 10^6.

1. Introduction

Current state-of-the-art single-walled carbon nanotube (SWNT) synthesis methods invariably produce mixtures of metallic (m-SWNTs) and semiconducting (s-SWNTs) nanotubes. However, most electronic applications of SWNTs, such as in nanoelectronics and thin film transistors, require electronically homogeneous SWNTs, *i.e.* either metallic or semiconducting species. The electronic heterogeneity of SWNTs is a major obstacle to the widespread use of SWNTs in electronic applications. If not electronically purified, the parasitic effects of the unwanted type of SWNT significantly degrade the properties of the devices made with these nanotubes. Various metallicity-based separation techniques have been proposed to separate metallic from semiconducting nanotubes, including covalent or non-covalent selective chemistry, chromatography, and density-gradient ultracentrifugation.[1–4] However, ultra-high purity (>98%) and high process yield/throughput

School of Chemical & Biomedical Engineering, Nanyang Technological University, 62 Nanyang Drive, Singapore 637459, Singapore. E-mail: mbechan@ntu.edu.sg; Fax: +65 6794 7553; Tel: +65 6790 6064

techniques which can achieve highly homogeneous nanotubes at an affordable cost, which is necessary for industrial scaling and widespread use, as well as methods for the characterization[5] of the ultra-high purity nanotubes produced with such techniques, are still the subject of intense research.

Various covalent and non-covalent chemistries for the metallicity-based separation of nanotubes have been reported[4–11] and they appear to offer simplicity. Non-covalent chemistries include the selection of metallic nanotubes by amines[6,7] such as propylamine and isopropylamine; on the other hand, the selection of semiconducting SWNTs can be achieved using aromatic molecules[8,9] and porphyrin.[10] For selective covalent functionalization, others have reported the preferential functionalization of metallic nanotubes using diazonium salts,[11–13] dichlorocarbene[14] and nitronium ions;[15–17] on the other hand, semiconducting SWNTs are selectively reacted by hydrogen peroxide[18,19] and azomethine ylides.[20] The degree of selectivity of these chemistries and their compatibility with the subsequent solution processing required to achieve high purity separation vary widely.[4–20] Among the reported chemistries, radicals have been shown to result in considerable separation and thus have been the focus of a number of studies.[21–27] Radicals are known to attack both metallic and semiconducting SWNTs, but they have been shown to have preferential reactivity with m-SWNTs due to the availability of electrons near the Fermi level in the metallic nanotubes.[24,26,28]

Recently, we reported a novel dual selective naphthalene-based azo dispersing agent (Direct Blue 71(**I**)) which decomposes under sonication to produce radicals that preferentially attack metallic species.[29] The preferential attack on metallic nanotubes by the resulting radicals may be combined with other separation techniques to improve the yield and/or purity. Agarose gel electrophoresis (AGE) has recently been shown to be a high yielding technique for the metallicity-based separation of SWNTs; relatively high s-SWNTs purities (85–95%) and process yields (up to 25%) have been reported.[30,31] The separation effectiveness of AGE appears to be affected by the surfactants used; sodium dodecyl sulfate (SDS)[30] and chondroitin sulfate (CS-A)[31] are the only surfactants reported to result in considerable separation with AGE. We hypothesize that the combination of selective chemistry with AGE may result in further improvements in AGE to achieve even higher purity nanotubes.

In this paper, we preferentially attack metallic nanotubes with decomposed ionic **I**-derived radicals and then subject the reacted mixture to AGE. A functionalization process involving a fairly long sonication time (3 h) and a higher **I** concentration than previously reported[29] is used to achieve the preferential attack of m-SWNTs by the decomposed **I** radicals and to prepare the mixture for subsequent use in AGE. The reaction mixture is then suspended in sodium cholate (SC) prior to AGE. The proposed method (hereafter referred to as **I**-AGE) relies on the combination of the higher reactivity of the radicals generated from Direct Blue 71(**I**) with metallic nanotubes, the relatively high charge density of the generated ionic radicals of **I** and also the effective debundling ability of SC to achieve excellent separation using the potentially high yield AGE method.

Selective enrichment of the s-SWNTs is achieved with a simple six step process (shown in Fig. 1a): (i) dispersion of purified arc-discharge P2-SWNTs with direct blue 71 (**I**) with heating and sonication, (ii) storage of the mixture in the dark for 48 h, (iii) filtration to remove any unreacted and excess **I**, (iv) re-dispersion of the radical-attacked SWNTs in an aqueous sodium cholate (SC) surfactant solution,

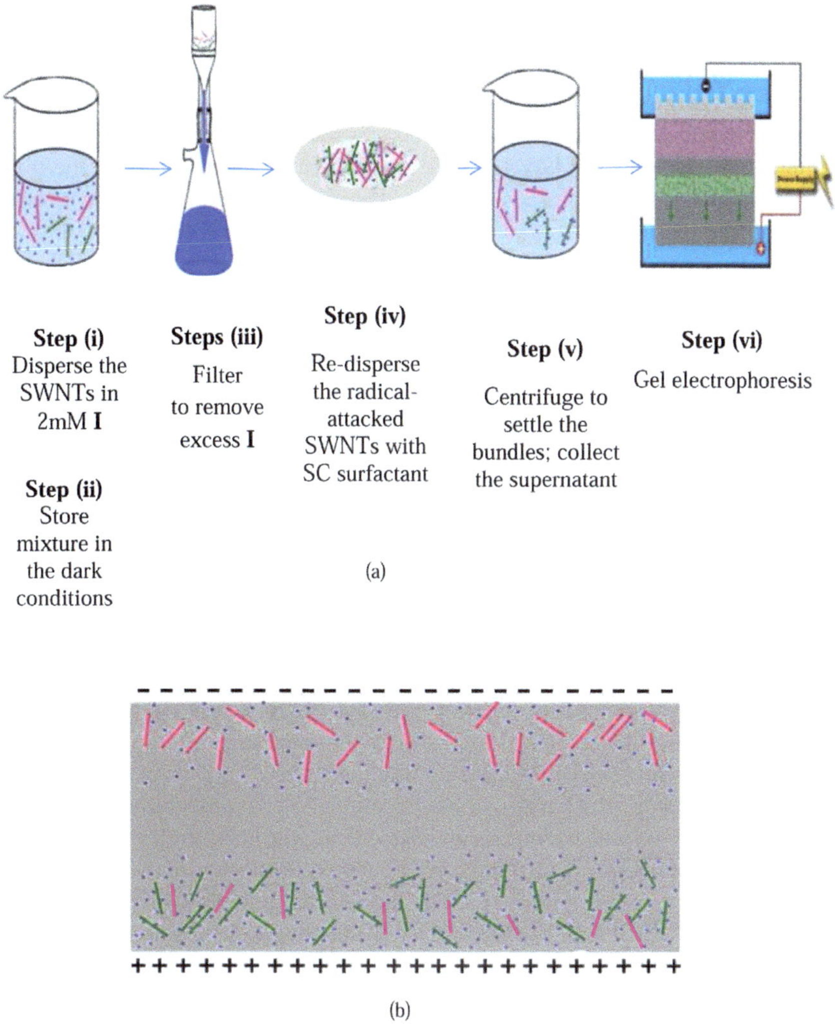

Fig. 1 (a) Schematic representation of the separation procedure in the I-AGE method. (b) Schematic illustration of the separation mechanism in I-AGE.

(v) centrifugation of the solution to remove nanotube bundles and (vi) agarose gel electrophoresis of the supernatant to obtain highly enriched semiconducting SWNTs. Hybrids formed by the reaction of the ionic degradation products of **I** with SWNTs have a significantly higher charge density and, consequently, mobility under an applied electric field than unreacted SWNTs do. The significant preference of these degraded ionic radicals for metallic SWNTs results in the preferential charging and mobility of nearly all of the m-SWNTs, leading to the movement of m-SWNTs towards the bottom gel fraction, with significant amounts of unreacted s-SWNTs lagging and remaining in the top gel fraction (Fig. 1b). Sodium cholate (SC) is added to the reacted nanotube mixture to promote the dispersion of the suspended nanotubes into individuals.

To the best of our knowledge, the present work is the first report of selective chemistry-assisted gel electrophoresis which achieves higher purity than previously proposed AGE techniques using SDS or CS-A. The enrichment was confirmed by UV-visible-Near Infrared spectroscopy (UV-vis-NIR) and Raman spectroscopy. The increased ratio of the D peak to the G peak (D/G) indicated by the Raman spectroscopy of the metallic-enriched eluted SWNTs obtained by AGE, confirms the functionalization and corroborates that the naphthyl-based radicals preferentially react with m-SWNTs. The enrichment of the s-SWNTs in the top fraction of the agarose gel was also confirmed by the high performance of network field effect transistors (FETs) fabricated using enriched nanotubes.

2. Results

During gel electrophoresis of the **I**-functionalized SC-dispersed P2 SWNTs (hereafter denoted "P2/SC-**I**"), some SWNTs migrated down the agarose gel while other SWNTs remained immobile in the top part of the gel. The nanotubes collected from the top gel fraction after AGE (hereafter denoted "P2/SC-**I**/AGE/Top") and the nanotubes obtained from the bottom gel fraction ("P2/SC-**I**/AGE/Bottom") were analyzed using UV-vis-NIR and Raman spectroscopy.

Fig. 2a shows the UV-vis-NIR spectra of the semiconducting- and metallic-enriched SWNT fractions purified using the **I**-AGE technique. For P2 SWNTs, the first van Hove electronic transition of the metallic nanotubes (M_{11}) and the second van Hove electronic transition of the semiconducting nanotubes (S_{22}) occur in the 600–800 nm and 920–1050 nm ranges, respectively.[32] For P2/SC-**I**/AGE/Top, which are semiconducting-enriched SWNTs, the M_{11} peak was decreased significantly compared to that of the un-separated pristine P2 samples (hereafter denoted "P2/Pristine"), confirming the reduced content of the metallic nanotubes after AGE. The purity of the semiconducting nanotubes obtained from the top gel fraction (pink curve) was estimated using the areas under the metallic M_{11} and semiconducting S_{22} bands to be about 98%, while that of the pristine P2 samples was estimated to be 78%. The significant reduction in the metallic content of the P2/SC-**I**/AGE/Top samples confirms the excellent selectivity of our combined functionalization cum AGE technique. The more reacted metallic nanotubes are postulated to move agilely towards the bottom electrode under the electric field. The bottom fraction (green curve, P2/SC-**I**/AGE/Bottom) is still dominated by semiconducting nanotubes, but is enriched (to about 30%) in metallic nanotube species compared with the input P2 SWNTs. To confirm the contribution of **I** functionalization in achieving good separation using AGE, a control AGE separation was also performed with non-**I** functionalized P2 SWNTs dispersed in 2% SC (denoted "P2/SC/AGE/Top"). As seen in Fig. 2a, this sample has only about 85% semiconducting purity in the top fraction, confirming the important role that **I** functionalization plays in achieving high purity semiconducting nanotubes through AGE.

The separation efficiency of the proposed **I**-AGE technique was also examined using Raman spectroscopy. The radial breathing modes (RBM) of the two separated fractions ("top" and "bottom") were measured with 633 nm Raman laser excitation, which excites both the metallic and semiconducting species of P2 SWNTs.[3,33,34]

Fig. 2b shows the Raman spectra with 633 nm excitation. The pristine P2 shows two major peaks at 155 cm^{-1} and 170 cm^{-1} due to the semiconducting species and one peak at 200 cm^{-1} due to the metallic species. The metallic peak is

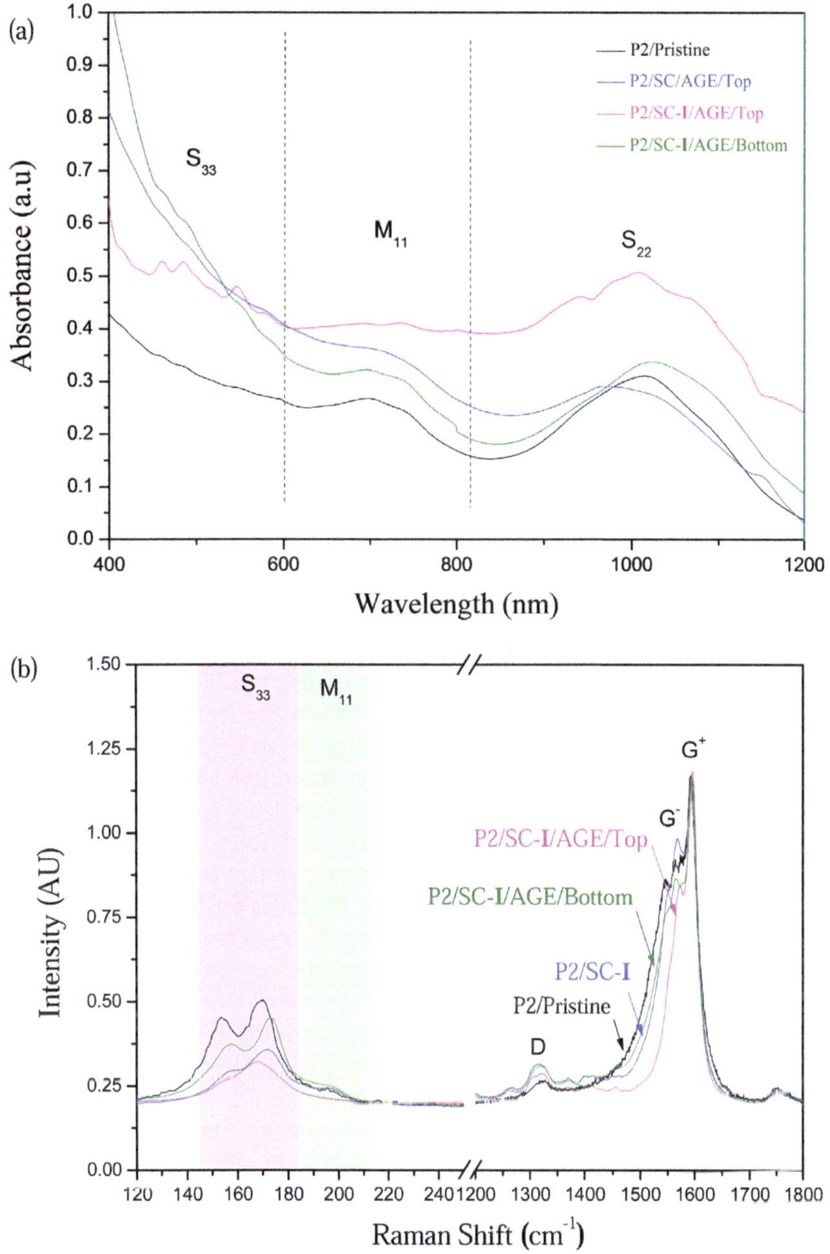

Fig. 2 (a) UV-vis-NIR spectra of P2/Pristine, P2/SC/AGE and P2/SC-I after gel electrophoresis. (b) Raman spectra of P2/Pristine and P2/SC-I before and after gel electrophoresis, collected using a 633 nm laser.

plainly present in the P2/Pristine sample (black curve) but is nearly absent from the P2/SC-**I**/AGE/Top curve (pink). The metallic peak of the P2/SC-**I**/AGE/Bottom samples (green curve) is enhanced relative to the semiconducting peaks. This corroborates the excellent enrichment of the semiconducting species in the top gel after **I**-AGE.

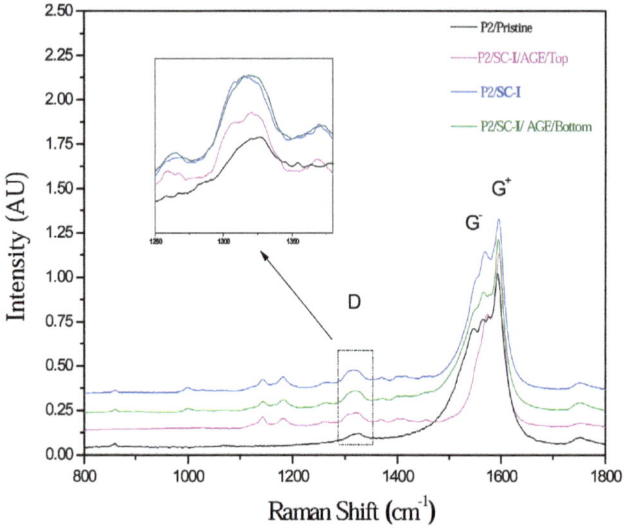

Fig. 3 Raman spectra (633 nm) of the pristine and separated P2 SWNTs; inset is the enlarged view of the D-band evolution of the samples.

The G$^-$ band also confirms the reduction of the metallic nanotube species from the top fraction. For metallic nanotubes, due to the strong coupling in the density of states, the G$^-$ feature is broad and asymmetric while the G$^-$ band for the semiconducting nanotubes exhibits a Lorentzian line shape.[3,35] As seen in Fig. 2b, the G$^-$ feature of the P2/SC-**I**/AGE/Top sample (at approximately 1550–1580 cm^{-1}) is narrowed compared to those of the P2/SC-**I**/AGE/Bottom and pristine samples, corroborating the decreased metallic content in this fraction.

Raman spectroscopy was also used to verify the preferential covalent reaction of the **I** fragments with the metallic nanotubes. The ratio of the D peak to the G peak (D/G) is related to the extent of nanotube sidewall functionalization.[12] As shown in Fig. 3 and Table 1, the D/G ratio was considerably higher for P2/SC-**I**/AGE/Bottom SWNTs than for the pristine sample or the P2/SC-**I**/AGE/Top fraction. This result confirms the higher extent of functionalization of the SWNTs in the bottom gel fraction, which is also the m-SWNT-enriched fraction. This corroborates with our previous finding that **I**-derived radicals preferentially attack metallic species, thereby enhancing their mobility under AGE compared with the less-attacked semiconducting species.

To demonstrate the suitability of the nanotubes separated by this method for use in electronic devices, the purified semiconducting SWNTs (P2/SC-**I**/AGE/Top) were applied in bottom-gated FETs using the dip coating technique (described in Section 4.4). For comparison purposes, a control sample of FETs (with a similar nanotube density) was fabricated using a P2/Pristine solution. A total of 20 FETs were fabricated using each of the above solutions.

Fig. 4a shows the variation of the drain current (I_d) with gate voltage (V_{gs}) at a drain voltage (V_{ds}) of 1.5 V, for a representative device fabricated using a P2/SC-**I**/AGE/Top solution. This FET shows typical p-type transistor characteristics and a carrier mobility of 6.7 cm^2 V^{-1} s^{-1}. Fig. 4b shows the variation in the output characteristic drain current (I_d) with the drain voltage (V_d), measured at several

Table 1 The ratio of the disorder mode (D peak) and the tangential mode (G peak) of pristine and separated P2 SWNTs

Sample	D/G ratio
P2/Pristine	0.082 ± 0.021
P2/SC-**I**	0.149 ± 0.009
P2/SC-**I**/AGE/Top	0.098 ± 0.003
P2/SC-**I**/AGE/Bottom	0.146 ± 0.005

gate voltages over the range −150 to 150 V. Fig. 4c shows the histogram of the on/off ratios of the 20 FETs fabricated using purified semiconducting SWNTs as well as that of 20 FETs made using the un-separated pristine P2 solution (with similar nanotube densities). The FETs fabricated using the P2/SC-**I**/AGE/Top solution exhibit significantly higher on/off ratios (predominantly ≥ 10^4) than the FETs made with the P2/Pristine solution (on/off ratios mostly < 10^2). Fig. 4d shows the trend of the on/off ratio with mobility for the devices fabricated using the purified semiconducting SWNTs (P2/SC-**I**/AGE/Top). The FETs fabricated using our

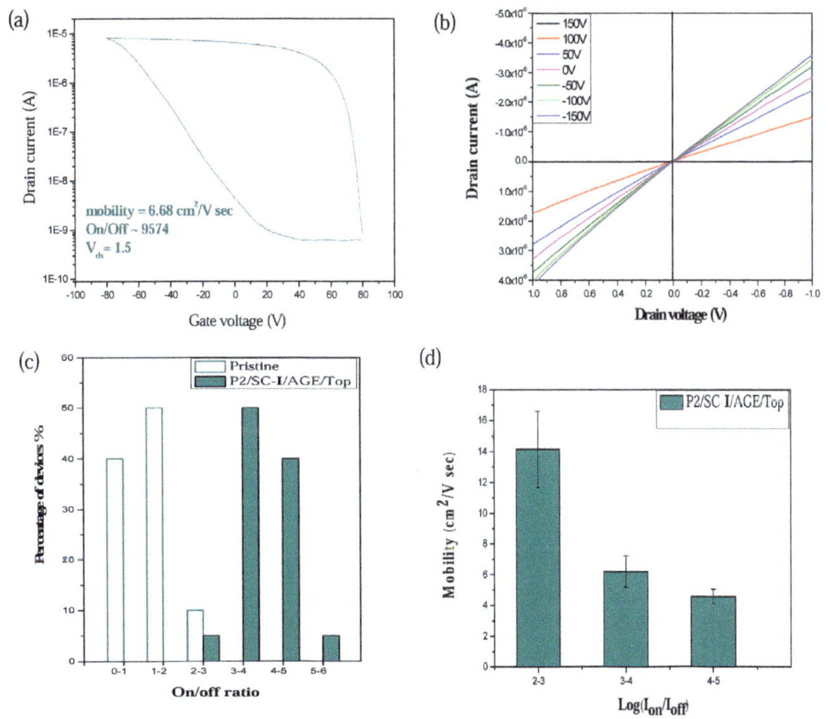

Fig. 4 (a) Transfer characteristics (V_{ds} = 1.5 V) of I_{ds} *vs.* V_{gs} for bottom-gated FETs fabricated using the P2/SC-I/AGE/Top solution. (b) Output characteristics, I_{ds} *vs.* V_{ds}, measured at gate voltages of 150 (top) to −150 V (bottom) with 50 V intervals. (c) Histogram of on/off ratios of 20 devices made with P2/SC-I/AGE/Top SWNTs (blue) and 20 devices made with pristine SWNTs (white). (d) The trend of the on/off ratio with device mobility for the devices fabricated using the P2/SC-I/AGE/Top solution.

enriched semiconducting SWNTs show promising performance metrics: on/off ratios ranging from 10^2 to 10^6 and mobilities ranging from 3.6 to 11.7 $cm^2 V^{-1} s^{-1}$. These metrics are significantly higher than those of the devices fabricated using un-separated nanotubes.

Fig. 5a shows the morphology of the nanotube network as measured by AFM. The device channel area is observed to be covered by a film-like layer of SWNTs with an estimated density of ~20 tubes per μm^2. The length distribution of nanotubes after **I**-AGE is shown in Fig. 5b. The average length of nanotubes after enrichment with **I**-AGE was estimated to be about 0.5 μm, indicating some shortening of the nanotubes after enrichment (length of pristine SWNTs = ~1 μm). The shortening of the nanotubes in **I**-AGE is mainly due to the long sonication duration (3 h) required to prepare samples for **I**-AGE.

The results of the UV-vis-NIR and Raman spectroscopy, and the performance of the FETs presented above confirm the excellent separation of the metallic from the semiconducting nanotubes using our proposed **I**-AGE technique. The process yield of our **I**-AGE technique based on the original nanotube input was measured to be about 18%, which is significantly higher than the yield of other highly selective techniques such as **I**-DGU (<5%).[29] The simultaneous high selectivity with high yield of the proposed technique makes it appealing for industrial scale production of high purity semiconducting nanotubes.

The promising separation achieved using this **I**-AGE technique originates from the unique characteristics of **I**, including its capability to generate radicals which preferentially attack unwanted metallic species and its relatively high charge density.[29] Due to the relatively high negative charge of **I** and the preferential radical reactivity with metallic nanotubes, a significant differential charge density develops between the reacted nanotubes, which contain most of the metallic species as well as many of the semiconducting species, and the unreacted nanotubes, which are almost entirely semiconducting. In the presence of an applied electric field, non-covalently bound **I** molecules are likely to be stripped from the nanotube surfaces so that the s-SWNTs, which are non-covalently functionalized with undecomposed **I** molecules, will be relatively immobile. The significantly higher charge of the reacted nanotubes compared to the unreacted ones, results in considerable differential electrophoretic mobility between the two populations under a uniform electric field during AGE. The more mobile

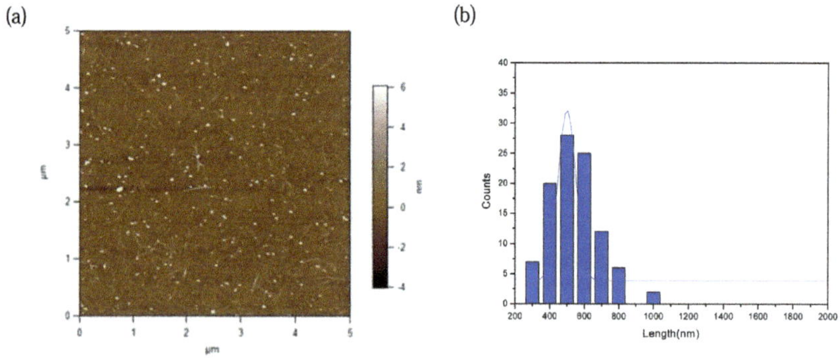

Fig. 5 (a) Atomic force microscopy (AFM) image of a representative FET fabricated using the P2/SC-I/AGE/Top solution. (b) Length distribution of the P2/SC-I/AGE/Top SWNTs.

negatively charged fraction, which is enriched in metallic species, moves towards the anode electrode. The less charged fraction, which is highly depleted in metallic species, remains in the upper portion of the gel. Based on the results presented here and the results of previous studies using CS-A and SDS,[30,31] we hypothesize that achieving good separation with AGE requires (i) the use of highly charged dispersants/functionalization agents to increase the mobility of the SWNTs under an electric field and (ii) the selectivity of the charged dispersing/functionalization agents for metallic or semiconducting nanotubes, leading to the development of differential mobility between the two species. The higher purity achieved using **I** rather than CS-A, despite both having relatively similar charge densities,[29,31] indicates that while a high charge density of the dispersing/functionalization agent is a requirement of separation, the key to achieving higher purities with AGE is to increase the selectivity of the charged dispersing/functionalization agents for metallic or semiconducting nanotubes.

Due to its relatively high selectivity, process yield and throughput, the method developed in this study has great potential for scaling and commercialization. However, industrial scale up of this method requires further investigation into the potential effects that increasing the throughput may have on purity and yield of the semiconducting SWNTs produced. In addition, the commercialization potential of this method may be increased by shortening the process duration of the various steps involved including AGE, acid treatment and selective functionalization.

3. Conclusions

We have shown that sonication of a novel naphthalene-based azo compound prior to dispersing the mixture with sodium cholate and then subjecting the mix to AGE can separate most of the metallic nanotubes from the immobile nanotubes, considerably increasing the semiconducting purity of these SWNTs. The higher s-SWNT purity achievable using our proposed technique compared to previously proposed AGE techniques using SDS and CS-A, confirms the effect of the **I**-radicals on separation. The high process yield (~18%) and high purity (~98%) achieved when using our proposed technique, together with its affordability and scalability, render it highly appealing for industrial scaling. Network FET devices made with the semi-enriched SWNTs show promising performance metrics (mobility in the range of ~3.6 to 11.7 $cm^2\ V^{-1}\ s^{-1}$ and on/off ratios from 10^2 to 10^6). The efficiency of our proposed **I**-AGE technique originates mainly from the preferential reactivity of the radicals of **I** with metallic nanotubes as well as the relatively high charge density of these radicals.

4. Experimental section

4.1. Materials

Direct Blue 71 (**I**), sodium cholate (SC), sodium dodecyl sulphate (SDS), boric acid, tris hydroxymethyl aminomethane, dimethylformamide (DMF) and nitric acid were purchased from Sigma-Aldrich (Singapore) and used as received. Purified arc discharge SWNTs (P2) were purchased from Carbon Solutions, Inc. (Riverside, CA, USA). Agarose was purchased from NacalaiTesque (Japan).

4.2. Solution preparation and selective functionalization

15 mg of purified arc discharge (P2) nanotubes were mixed with 2 mM of Direct Blue 71 in 50 ml deionized water. The solution was stirred for 45 min at 60 °C and was probe-sonicated (SONICS, VCX-130) in an ice bath for 3 h (70% amplitude, 5 s on, 2 s off). The solution was stored in the dark for 48 h and was subsequently filtered and then washed with water two to three times to remove any unreacted **I**. The reacted SWNT powder was re-dispersed in an aqueous solution containing 2 wt% of sodium cholate and was centrifuged (Sartorius, SIGMA@3K30) at 50 000*g* for 1 h to settle the SWNT bundles. The supernatant was collected for further purification using agarose gel electrophoresis.

4.3. Gel electrophoresis

1 wt% agarose was added to 0.2 wt% surfactant in a 2× transport buffer (TB) solution (pH = 8.5; 48.5 mM boric acid and 50 mM tris hydroxymethyl aminomethane). The agarose gel solution was then melted in a microwave oven and used to fill the gel container (0.3 cm × 20 cm × 20 cm plate container) of the gel electrophoresis apparatus (PROTEAN II XI, Bio Rad Laboratories, Singapore) up to about 2/3 of its height. Subsequently, 20 ml of the functionalized SWNT solution, prepared as described in Section 4.2, was mixed with a similar volume (20 ml) of 2 wt% agarose solution and used to fill the remaining 1/3 space of the plate container, after being melted in a microwave oven. The gel electrophoresis apparatus was then run at 200 V for 4 h. After AGE, the gel was cut into three fractions distinguished by differences in color.[31,36] The top and bottom gel fractions containing the purified semiconducting and metallic-enriched SWNTs, respectively, were degraded in 10% nitric acid at 80 °C for 18 h followed by a basic wash (1 M NaOH) to remove the agarose gel. The purified SWNTs were washed twice with distilled water and sonicated in DMF solvent three times to remove **I**. They were then redispersed in a (1 : 4) mixture of SDS and SC solution using probe sonication (Sonics, 130 W, 20 kHz). After probe sonication, the SWNT solution was centrifuged at 50 000*g* for 1 h and the supernatant was then collected and used for characterization and, in the case of the s-SWNTs, for fabrication of field effect transistors. The s-SWNTs were annealed after acid and base treatments. Annealing was performed in argon for 1 h at 400 °C.

4.4. Characterization

The clean SWNTs obtained from the top and bottom gel fractions were examined by UV-vis-NIR and Raman spectroscopy. UV-vis-NIR spectroscopy was performed using a Varian Cary 5000 UV-vis-NIR spectrophotometer. Raman spectroscopy was conducted using a Renishaw inVia Raman microscope equipped with a 633 nm (1.96 eV) laser. The purity of the semiconducting nanotubes obtained through **I**-AGE was estimated using the $A_S/(A_M+A_S)$ ratio, where A_S and A_M are respectively, the areas under the semiconducting (S_{22}) and metallic (M_{11}) absorption spectral bands in the UV-vis-NIR spectra, after baseline subtraction.

The semiconducting SWNTs obtained from the top gel fraction were used to fabricate field effect transistors (FETs) using the dip coating method. All devices were fabricated on silicon substrates covered with a 300 nm thick SiO_2 layer which served as the gate dielectric. The surface of the substrate was cleaned with piranha solution and, prior to nanotube deposition, was functionalized with aminopropyltriethoxysilane (APTES) to improve the uniformity of the nanotube network.[37] To create networks of nanotubes, the wafer was repeatedly immersed into the purified s-SWNT solution until sufficient nanotube density was deposited on the wafer. The wafer was then rinsed with deionized water to completely remove the surfactants and was dried with nitrogen gas. The wafers were then annealed in argon for 1 h at 400 °C and later in vacuum at 250 °C for 2 h. After annealing, source and drain gold electrodes were deposited on the substrate by photolithography and the active channel regions were then covered with a protective passivation layer. After depositing the protective layer, the remaining unprotected SWNTs outside the FET channel were etched away by oxygen plasma (100 W, 2 min). The FET channel width and length were 50 and 100 μm, respectively. The nanotube network morphology was measured by atomic force microscopy (AFM) with a MFP 3D microscope (Asylum Research) in AC mode.

A Keithley semiconductor parameter analyzer Model 4200-SCS was used to characterize the FET performance. The mobility was estimated using the following well-known equation:[38]

$$\mu = \frac{L}{C \times W \times V_\mathrm{d}} \times \frac{\mathrm{d}I_\mathrm{d}}{\mathrm{d}V_\mathrm{g}} \tag{1}$$

where L is the channel length, W is the channel width, V_d is the drain voltage, I_d is the drain current, V_g is the gate voltage and C is the capacitance per unit area. C was computed by:

$$C = \frac{D}{\left(C_\mathrm{Q}{}^{-1} + (1/2\pi\varepsilon_0\varepsilon_{0_x})\ln\left(\sinh(2\pi L_{0_x} D)/R\pi D\right)\right)} \tag{2}$$

where C_Q (4×10^{-10} [F m^{-1}]) is the quantum capacitance of carbon nanotubes, D is the density of the nanotubes, L_ox is the SiO_2 thickness, ε_ox is the dielectric constant of SiO_2, ε_0 is the permittivity of free space ($\varepsilon_0 = 8.85 \times 10^{-12}$ F m^{-1}) and R is the nanotube radius.[39]

Acknowledgements

This work was supported by a Competitive Research Program grant from the Singapore National Research Foundation (NRF-CRP2-2007-02) and an ASTAR grant (102 170 0141). Sara Mesgari acknowledges the support of Nanyang Technological University through a Research Scholarship.

References

1 J. Liu and M. C. Hersam, Recent developments in carbon nanotube sorting and selective growth, *MRS Bull.*, 2010, **35**(4), 315–321.

2 M. C. Hersam, Progress towards monodisperse single-walled carbon nanotubes, *Nat. Nanotechnol.*, 2008, **3**(7), 387–394.

3 C. H. Liu and H. L. Zhang, Chemical approaches towards single-species single-walled carbon nanotubes, *Nanoscale*, 2010, **2**(10), 1901–1918.

4 S. A. Hodge, *et al.*, Unweaving the rainbow: A review of the relationship between single-walled carbon nanotube molecular structures and their chemical reactivity, *Chem. Soc. Rev.*, 2012, **41**(12), 4409–4429.

5 H. Park, *et al.*, High-density integration of carbon nanotubes via chemical self-assembly, *Nat. Nanotechnol.*, 2012, **7**(12), 787–791.

6 Y. Maeda, *et al.*, Large-scale separation of metallic and semiconducting single-walled carbon nanotubes, *J. Am. Chem. Soc.*, 2005, **127**(29), 10287–10290.

7 Y. Maeda, *et al.*, Dispersion and separation of small-diameter single-walled carbon nanotubes, *J. Am. Chem. Soc.*, 2006, **128**(37), 12239–12242.

8 R. Voggu, *et al.*, A simple method of separating metallic and semiconducting single-walled carbon nanotubes based on molecular charge transfer, *J. Am. Chem. Soc.*, 2010, **132**(16), 5560–5561.

9 P. Anilkumar, *et al.*, Non-covalent interactions of derivatized pyrenes with metallic and semiconducting single-walled carbon nanotubes, *J. Phys. Chem. C*, 2011, **115**(22), 11010–11015.

10 H. P. Li, *et al.*, Selective interactions of porphyrins with semiconducting single-walled carbon nanotubes, *J. Am. Chem. Soc.*, 2004, **126**(4), 1014–1015.

11 W. J. Kim, M. L. Usrey and M. S. Strano, Selective functionalization and free solution electrophoresis of single-walled carbon nanotubes: Separate enrichment of metallic and semiconducting SWNT., *Chem. Mater.*, 2007, **19**(7), 1571–1576.

12 M. S. Strano, *et al.*, Electronic structure control of single-walled carbon nanotube functionalization, *Science*, 2003, **301**(5639), 1519–1522.

13 W. J. Kim, *et al.*, Connecting Single Molecule Electrical Measurements to Ensemble Spectroscopic Properties for Quantification of Single-Walled Carbon Nanotube Separation, *J. Am. Chem. Soc.*, 2009, **131**(9), 3128–3129.

14 H. Hu, *et al.*, Sidewall Functionalization of Single-Walled Carbon Nanotubes by Addition of Dichlorocarbene, *J. Am. Chem. Soc.*, 2003, **125**(48), 14893–14900.

15 K. H. An, *et al.*, A diameter-selective attack of metallic carbon nanotubes by nitronium ions, *J. Am. Chem. Soc.*, 2005, **127**(14), 5196–5203.

16 C. M. Yang, *et al.*, Selective removal of metallic single-walled carbon nanotubes with small diameters by using nitric and sulfuric acids, *J. Phys. Chem. B*, 2005, **109**(41), 19242–19248.

17 K. H. An, *et al.*, A diameter-selective chiral separation of single-wall carbon nanotubes using nitronium ions, *J. Electron. Mater.*, 2006, **35**(2), 235–242.

18 M. Yudasaka, M. Zhang and S. Iijima, Diameter-selective removal of single-wall carbon nanotubes through light-assisted oxidation, *Chem. Phys. Lett.*, 2003, **374**(1–2), 132–136.

19 Y. Miyata, Y. Maniwa and H. Kataura, Selective oxidation of semiconducting single-wall carbon nanotubes by hydrogen peroxide, *J. Phys. Chem. B*, 2006, **110**(1), 25–29.

20 C. Ménard-Moyon, *et al.*, Separation of semiconducting from metallic carbon nanotubes by selective functionalization with azomethine ylides, *J. Am. Chem. Soc.*, 2006, **128**(20), 6552–6553.

21 G. Schmidt, *et al.*, Labile diazo chemistry for efficient silencing of metallic carbon nanotubes, *Chem. – Eur. J.*, 2011, **17**(5), 1415–1418.

22 G. Schmidt, *et al.*, Mechanism of the Coupling of Diazonium to Single-Walled Carbon Nanotubes and Its Consequences, *Chem. – Eur. J.*, 2009, **15**(9), 2101–2110.

23 M. L. Usrey, E. S. Lippmann and M. S. Strano, Evidence for a two-step mechanism in electronically selective single-walled carbon nanotube reactions, *J. Am. Chem. Soc.*, 2005, **127**(46), 16129–16135.
24 Z. Dai, *et al.*, Selective Small-Diameter Metallic Single-Walled Carbon Nanotube Removal by Mere Standing with Anthraquinone and Application to a Field-Effect Transistor, *J. Phys. Chem. C*, 2010, **114**(49), 21035–21041.
25 J. B. Li, *et al.*, On-chip diameter-dependent conversion of metallic to semiconducting single-walled carbon nanotubes by immersion in 2-ethylanthraquinone, *RSC Adv.*, 2012, **2**(4), 1275–1281.
26 J. W. Zhao, *et al.*, Solution-processable semiconducting thin-film transistors using single-walled carbon nanotubes chemically modified by organic radical initiators, *Chem. Commun.*, 2009, 7182–7184.
27 C. W. Lee, *et al.*, Solution-processable carbon nanotubes for semiconducting thin-film transistor devices, *Adv. Mater.*, 2010, **22**(11), 1278–1282.
28 J. Lu, *et al.*, Selective Interaction of Large or Charge-Transfer Aromatic Molecules with Metallic Single-Wall Carbon Nanotubes: Critical Role of the Molecular Size and Orientation, *J. Am. Chem. Soc.*, 2006, **128**(15), 5114–5118.
29 A. K. Sundramoorthy, *et al.*, Scalable and effective enrichment of semiconducting single-walled carbon nanotubes by a dual selective naphthalene-based AzO dispersant, *J. Am. Chem. Soc.*, 2013, **135**(15), 5569–5581.
30 T. Tanaka, *et al.*, High-yield separation of metallic and semiconducting single-wall carbon nanotubes by agarose gel electrophoresis, *Appl. Phys. Express*, 2008, **1**(11), 1140011–1140013.
31 S. Mesgari, *et al.*, High Selectivity cum Yield Gel Electrophoresis Separation of Single-Walled Carbon Nanotubes Using a Chemically Selective Polymer Dispersant, *J. Phys. Chem. C*, 2012, **116**(18), 10266–10273.
32 M. J. O'Connell, *et al.*, Band gap fluorescence from individual single-walled carbon nanotubes, *Science*, 2002, **297**(5581), 593–596.
33 M. C. LeMieux, *et al.*, Self-sorted, aligned nanotube networks for thin-film transistors, *Science*, 2008, **321**(5885), 101–104.
34 N. Anderson, A. Hartschuh and L. Novotny, Chirality changes in carbon nanotubes studied with near-field Raman spectroscopy, *Nano Lett.*, 2007, **7**(3), 577–582.
35 S. M. Keogh, *et al.*, Spectroscopic analysis of single-walled carbon nanotubes and semiconjugated polymer composites, *J. Phys. Chem. B*, 2004, **108**(20), 6233–6241.
36 S. Mesgari, *et al.*, Polymer removal from electronic grade single-walled carbon nanotubes after gel electrophoresis, *J. Mater. Chem. C*, 2013, **1**(41), 6813–6823.
37 D. Chattopadhyay, I. Galeska and F. Papadimitrakopoulos, A route for bulk separation of semiconducting from metallic single-wall carbon nanotubes, *J. Am. Chem. Soc.*, 2003, **125**(11), 3370–3375.
38 C. Wang, *et al.*, Wafer-scale fabrication of separated carbon nanotube thin-film transistors for display applications, *Nano Lett.*, 2009, **9**(12), 4285–4291.
39 Q. Cao, *et al.*, Gate capacitance coupling of singled-walled carbon nanotube thin-film transistors, *Appl. Phys. Lett.*, 2007, **90**(2), 023516-1–023516-3.

Faraday Discussions

PAPER

Single-walled carbon nanotube networks in conductive composite materials

Oana A. Bârsan,*[ab] Günter G. Hoffmann,[ab] Leo G. J. van der Ven[ac] and G. (Bert) de With[a]

Received 1st May 2014, Accepted 28th May 2014
DOI: 10.1039/c4fd00087k

Electrically conductive composite materials can be used for a wide range of applications because they combine the advantages of a specific polymeric material (*e.g.*, thermal and mechanical properties) with the electrical properties of conductive filler particles. However, the overall electrical behaviour of these composite materials is usually much below the potential of the conductive fillers, mainly because by mixing two different components, new interfaces and interphases are created, changing the properties and behaviours of both. Our goal is to characterize and understand the nature and influence of these interfaces on the electrical properties of composite materials. We have improved a technique based on the use of sodium carboxymethyl cellulose (CMC) to disperse single-walled carbon nanotubes (SWCNTs) in water, followed by coating glass substrates, and drying and removing the CMC with a nitric acid treatment. We used electron microscopy and atomic force microscopy techniques to characterize the SWCNT films, and developed an *in situ* resistance measurement technique to analyse the influence of both the individual components and the mixture of an epoxy/amine system on the electrical behaviour of the SWCNTs. The results showed that impregnating a SWCNT network with a polymer is not the only factor that affects the film resistance; air exposure, temperature, physical and chemical properties of the individual polymer components, and also the formation of a polymeric network, can all have an influence on the macroscopic electrical properties of the initial SWCNT network. These results emphasize the importance of understanding the effects that each of the components can have on each other before trying to prepare an efficient polymer composite material.

Introduction

The potential of carbon nanotubes (CNTs) in many different fields has led to an exponential growth of interest in these materials over the past twenty years.[1] Not

[a]*Laboratory of Materials and Interface Chemistry, Eindhoven University of Technology, Den Dolech 2 5612 AZ, Eindhoven, the Netherlands. E-mail: O.A.Barsan@tue.nl*
[b]*Dutch Polymer Institute (DPI), P.O. Box 902, 5600 AX, Eindhoven, the Netherlands. E-mail: info@polymers.nl; Fax: +31 40 247 24 62; Tel: +31 40 247 56 29*
[c]*Akzo Nobel Automotive & Aerospace Coatings, Sassenheim, the Netherlands*

only their structure and size, but also their excellent mechanical, thermal, optical and electrical properties have widened the range of applications and scientific fields in which there has been an increased interest in understanding and exploiting them.[2–6] Single-walled carbon nanotubes (SWCNTs) in particular are gaining popularity due to their large aspect ratios, narrow diameter distribution, semiconductor properties and also their increasing availability on the market. These materials have been used to prepare thin conductive and transparent films,[7–9] and also electrically conductive,[10] mechanically reinforced[11] or thermally conductive[12] polymer composites. Because SWCNTs exert strong van der Waals attractive forces, they are very difficult to disperse without the use of surfactants or other dispersion agents.[13–16] Functionalization has proved to be efficient in increasing SWCNT dispersibility in different solvents or polymers.[17,18] However, the process involves damaging the surface of the SWCNTs, which in turn affects their mechanical and electrical properties. Despite their potential and the numerous studies on SWCNTs, there are many unresolved issues when practical applications, such as electrically conductive composites, are concerned. These materials have not yet surpassed the efficiency of existing composite materials such as graphite or metal fiber/flake reinforced polymers.[19,20]

Electrically conductive composites are typically prepared by dispersing SWCNTs, directly or with the aid of solvents, into a polymer.[10,21] This approach has led to the preparation of inhomogeneous composites with poorly dispersed SWCNTs and low conductivity values, even though percolation thresholds as low as 0.001 wt% have been achieved.[10,21–23] On the one hand, in order to obtain a good and stable dispersion of SWCNTs in a polymer, good compatibility between the two components is necessary. On the other hand, surrounding the conductive particles with an insulating polymer layer leads to individually insulated SWCNTs, dispersed throughout the polymer matrix. Such composites would either be insulators or have very low conductivities. This means that the contacts between the conductive particles inside a polymer matrix play a crucial role in the electrical properties of composite materials. Having a uniform SWCNT network, with good physical and electrical contacts within a polymer matrix, is a factor required for achieving good conductivities in composite materials.

Recently, a different approach was successfully used to obtain SWCNT/polymer composite thin films[24] with sheet resistance (R_S) values in the range of 100–1000 ohm sq^{-1}. This technique involves two main steps: (1) preparing a uniform SWCNT thin film on a glass substrate; (2) coating the SWCNT film with a specific polymer followed by drying/curing the polymer and peeling the film off the glass substrate. A free-standing composite film is obtained in this way. This approach brings the advantages of first having a SWCNT network with good physical and electrical properties, and then filling the gaps with the desired polymer. The idea is particularly appealing since it ensures good physical and electrical contacts between the SWCNTs, and allows a proper characterization of the network's properties before actually adding the polymer matrix. However, the influence of the newly formed interfaces on the electrical properties of the polymer impregnated SWCNT network has not been studied in detail. We have adapted and improved the technique and designed a method for directly measuring and characterizing the influence of the polymer matrix on the SWCNT network. The purpose of our study is to analyze how the polymeric network affects the electrical behaviour of the already existing SWCNT network.

Experimental

SWCNT film preparation

All chemicals were used as received, without any further purification. Super high purity SWCNTs manufactured by the CVD method were purchased from US Research Nanomaterials, Inc. 0.1% solutions of SWCNTs were prepared by adding 10 mg SWCNTs as purchased and 300 mg of sodium carboxymethyl cellulose (CMC), 90 000 MW ($C_{28}H_{30}Na_8O_{27}$), from Sigma Aldrich to 10 mL of distilled water. The solutions were kept cool in an ice bath while being sonicated for 15 minutes using a SONICS Vibra-Cell sonicator with a 6.4 mm tip diameter and 60% amplitude (as described by Li *et al.*[24]). The solution was then centrifuged at 8500 rpm for 1 hour in a 1.5 cm diameter vessel, using a SIGMA 3-30K centrifuge and only the upper 80% of the supernatant was kept. The glass substrates were cleaned by immersion into a RCA cleaning solution (5 parts H_2O, 1 part NH_4OH, 30%, 1 part H_2O_2, 30%) without heating, left overnight and then rinsed with distilled water. Thin SWCNTs + CMC films were deposited on 7 cm × 7 cm glass substrates with a 30 μm gap doctor blade at a speed of 10 mm s^{-1}, using a COATMASTER 509 MC. The films were then dried for 20 h in an oven at 45 °C. The as-prepared samples were immersed into an acid bath (HNO_3, 9 M) for 4 hours at 40 °C, and thereafter taken out and dried at 45 °C for 2 hours under vacuum. The samples were subsequently kept for 20 h at 150 °C under vacuum for acid residue removal, and for two hours at 400 °C in a nitrogen atmosphere for CMC residue removal. Thermogravimetric analysis (TGA) on pure CMC was performed using Q500 equipment, with a heating rate of 10 °C per minute between 25–600 °C both in air and nitrogen atmospheres.

SWCNT film characterization

Sheet resistance measurements were performed using standard four point probe measurements with a 4 mm probe spacing. A Keithley electrometer model 6517A and source-meter model 237 were used to apply current and measure the resulting voltage drop. The results were converted to sheet resistance using correction factors for rectangular thin films as shown by Smits *et al.*[25] Scanning electron microscopy (SEM) images were taken with a Quanta 3D FEG microscope and scanning transmission electron microscopy (STEM) images were acquired using a Titan electron microscope operated at 300 kV with a High Angle Annular Dark Field (HAADF) detector. Atomic force microscopy (AFM) was performed using a SOLVER NEXT in tapping mode, and a Veeco 150 Dektak profilometer (Bruker) was used for thickness measurements. Raman spectra were taken with a LabRAM confocal spectrometer (Horiba Scientific) with an excitation wavelength of 632 nm. Transmittance measurements were performed using a SHIMADZU UV-3102 PC spectrophotometer in the range of 350–700 nm.

Polymer impregnation

An epoxy/amine system was used as the polymer matrix. A bisphenol A based epoxy resin (Epikote 828, Resolution Nederland BV) with an equivalent weight (eqw) per epoxide group of 187 g mol^{-1}, and Jeffamine D-230 (Huntsman Holland

BV) with an NH-eqw of 60 g mol^{-1} were chosen as the model system. The polymer was prepared by mixing the two components at a 1 : 1 epoxy/amine ratio (based on an epoxy group/NH equivalent), using a RCT basic IKA magnetic stirrer for 5 minutes at 700 rpm for 5 minutes, followed by 3 minutes degassing in a BRANSON 1510 ultrasonic cleaner. The mixture was then coated onto the SWCNT film using a 60 μm gap doctor blade, and cured in an oven at 100 °C for 4 hours. Contact angle measurements on the SWCNT films were performed at room temperature using an OCA 30 System.

In situ resistance measurements

The previously prepared 7 cm × 7 cm SWCNT films on glass substrates were cut to sample sizes of 2 cm × 3.5 cm. An EMITECH K550 sputter coater at a current of 20 mA was used to deposit two 1 cm wide gold electrodes on each sample, leaving a 1.5 cm uncovered strip across the width of the samples (example in Fig. 6). A VOLTCRAFT VC840 multimeter was connected to the gold electrodes and a computer to enable *in situ* resistance measurements on each sample to be performed. The polymer was manually applied on the uncovered area of the SWCNT film while the resistance was being measured.

Results and discussion

SWCNT film preparation and characterization

In the first part of our research, SWCNT films were prepared using a technique adapted from Li *et al.*,[24] from a water solution of SWCNTs dispersed with CMC. The water was then removed by drying in an oven while the CMC was removed by nitric acid treatment, followed by several temperature treatments (45 °C, 150 °C and 400 °C) for the subsequent removal of acid and CMC residues. The prepared films (see the experimental section) were characterized by SEM and TEM after each step of the process to ensure the complete removal of CMC residues, and the nanometer and micrometer scale homogeneity of the SWCNT films. The majority of the CMC from the films was removed after the acid treatment. However, islands of CMC residue and also structures similar to polymer wrapping of SWCNTs[26–28] were visible under SEM and TEM measurements, respectively (Fig. 1b, c and e). Because the TGA results showed that CMC degrades between 250–300 °C, the samples were also subjected to a 400 °C final thermal treatment, during which the remaining CMC residue was fully removed (Fig. 1d and f).

The samples were also subjected to four-point probe measurements after each of the thermal treatment steps to monitor the evolution of the films' sheet resistance. After drying at 45 °C and 150 °C all of the films had a sheet resistance (R_S) of R_S = 200–250 ohm sq^{-1} and 300–500 ohm sq^{-1}, respectively, suggesting that the water and nitric acid were gradually removed from the films. The final 400 °C treatment of the films led to a sheet resistance of R_S = 1000–1500 ohm sq^{-1}. This last step had the double purpose of removing the remaining CMC residue from the network but also of removing the p-type doping effect[29] of the remaining nitric acid adsorbed into the SWCNT network.

Raman spectra taken of the pristine SWCNTs and the films after the final thermal treatment show an increase in the intensity of the D-band, generally

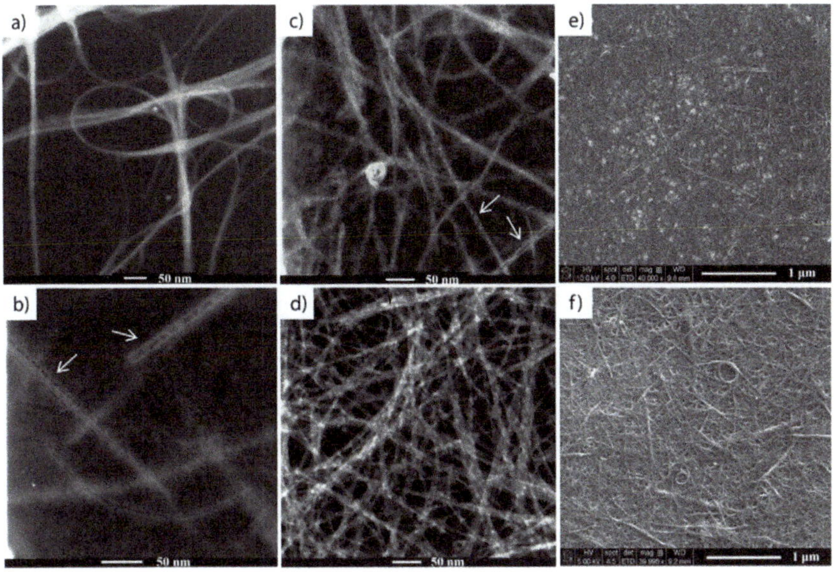

Fig. 1 TEM images of (a) as-purchased SWCNTs, (b) SWCNTs + CMC film, and TEM and SEM images of the same SWCNT film after nitric acid treatment (c and e) and after the final 400 °C thermal treatment (d and f), respectively.

associated with the amount of defects in SWCNTs[30–32] (Fig. 2), when compared to the intensity of the G-band. The I_G/I_D ratio drops from 27 for the pristine material to 16 for the treated SWCNTs, which could be due to a shortening of the tubes during ultrasonication, or the introduction of some defects during the nitric acid treatment. Even if some damage occurred to the SWCNTs during the preparation procedures, the quality of the materials remained high.[33,34]

The thickness of the films was determined by scratching the films and measuring the height profile across the scratch with a profilometer. Even if the initial SWCNTs + CMC films had thicknesses of approximately 500 nm, after removing the CMC they were reduced to 20–40 nm (Fig. 3a and b). AFM

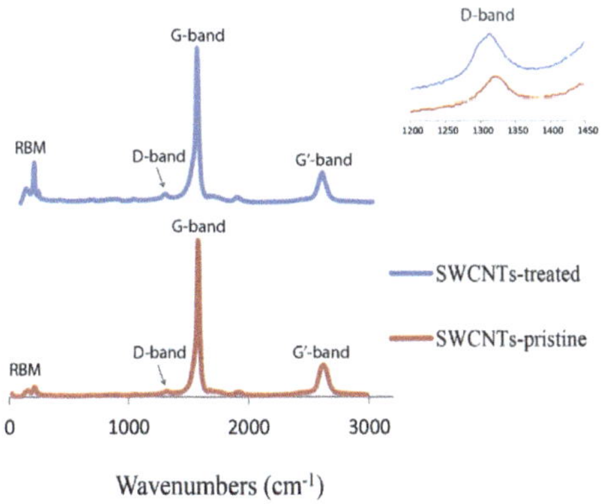

Fig. 2 Raman spectra of the SWCNTs as purchased and after film preparation.

measurements were consistent with the profilometer measurements and showed a cross-section roughness of approximately 40 nm (Fig. 3c–e). AFM, SEM and profilometer measurements all revealed a uniformly dispersed SWCNT network with several large ropes of entangled SWCNTs at the surface. These ropes at as high as 100 nm formed at the surface of the films, most likely during the acid treatment where the tubes at the surface had the freedom to rearrange themselves into a more stable configuration.

Polymer impregnation

The second part of our research focused on impregnating the SWCNTs with a polymer. Li *et al.*[24] used the SWCNT films to deposit a layer of polymer on top of the films, and after drying or curing, peeling-off the films from the glass substrates. Assuming that the polymer fully penetrated the SWCNT network, the free standing film obtained contained a SWCNT network embedded on one side of the polymer film. For most of the polymers they tested, the free-standing films had an up to 10 times lower sheet resistance on the bottom side of the film.

We focused on impregnating the SWCNTs with an epoxy/amine polymer system (Epikote 828 and Jeffamine D-230 (ref. 35)). Droplets of the epoxy/amine uncured mixture and the individual components were added onto the SWCNT film to determine whether the polymers had sufficient compatibility with the SWCNTs to fully impregnate the film. Contact angle measurements showed that even if the epoxy resin had a contact angle of 97.4°, the polymer mixture (75.7 wt% epoxy resin) had a contact angle of 30.2°, which is much closer to the value of the amine alone of 14.7° (Fig. 4). The fact that the contact angle of the uncured mixture dropped from 30.2° to 13.3° within the first 2 minutes, indicated that the polymer had already started to infiltrate into the SWCNT network at room temperature. A free standing composite film was prepared by coating a SWCNT film (R_S = 698 ohm sq^{-1}) with a 60 µm thick film of the polymer mixture. After curing the polymer at 100 °C for 4 hours, the film was peeled off the glass substrate and a free standing composite film with a sheet resistance of the bottom

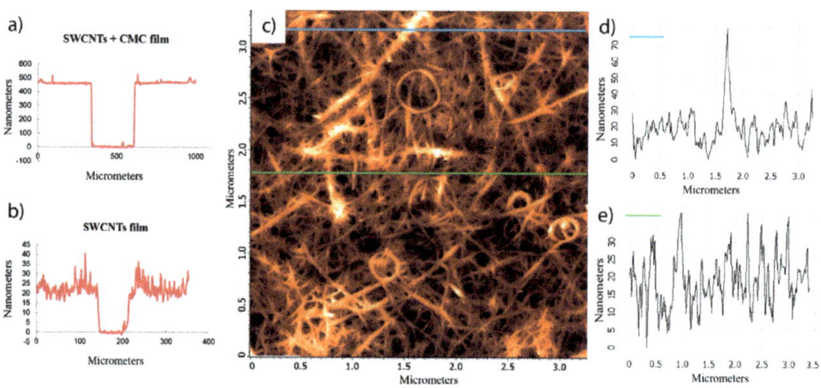

Fig. 3 Profilometer height profile across a scratch in (a) a SWCNTs + CMC film and (b) a SWCNT film. AFM topography image of (c) a SWCNT film and (d and e) cross-section roughness profiles.

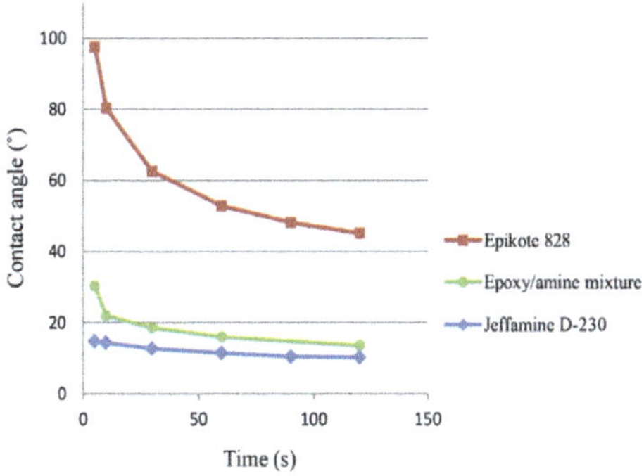

Fig. 4 Contact angle measurements of the epoxy/amine polymer mixture and the individual components on a SWCNT film.

side of the film of $R_S = 4200\text{–}4800$ ohm sq^{-1} was obtained, which is consistent with previous results.[24]

SEM images taken of the bottom side of the peeled-off film show domains of plain polymer in areas with SWCNTs sticking out of the polymer matrix (Fig. 5). The inhomogeneity of the film surface may be a major cause of the decrease in the film resistance measured on the bottom side. However, previous experience with similar CNT–polymer composites[36–38] prove that no matter what the polymer is, the CNTs are far from reaching their maximum potential (that of the pure CNTs) once they are inside a polymer matrix. This may also be due to the different natures of the interfaces and interphases formed when mixing two different components. However, measuring the resistance of the bottom side of a peeled-

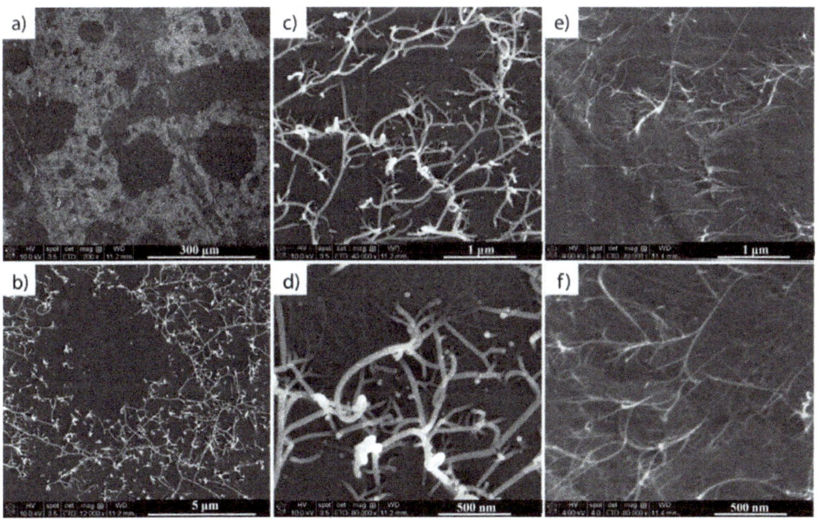

Fig. 5 SEM images taken of the bottom of a peeled off SWCNT/polymer composite film (a–d) with a Au coating for emphasized contrast and (e and f) without a Au coating for emphasized aspect ratio.

off film makes it virtually impossible to know the resistance of the SWCNT network alone after embedding it into the polymer.

In situ resistance measurements during polymer impregnation

In order to have direct access to the SWCNTs after embedding them into a polymer matrix, and to be able to measure the resistance of the network during polymer impregnation, and even during the curing process, we developed a system that allowed us to perform *in situ* resistance measurements. For this purpose, two gold electrodes were applied *via* sputtering onto two opposite edges of the SWCNT film. These electrodes were connected to a multimeter to allow direct measurement of the film's resistance while the polymer was subsequently added (Fig. 6). Because the electrodes were in direct contact with the SWCNT network before polymer impregnation, and the entire film area between them was fully covered by the polymer, the resistance measured between the two electrodes was the resistance of the SWCNT network before, during and after polymer impregnation. The same system can be used to analyze the influence of the individual components of the epoxy/amine polymer mixture on the SWCNT network by applying only one component at a time and comparing the results with the overall effect of the polymer mixture.

SWCNTs are known to be highly sensitive to different gas exposures[39,40] which can cause significant changes in their electronic and transport properties due to chemical doping.[41] The adsorption of different gas molecules on the surface of the nanotubes can modulate the carrier concentrations by generating a surplus or a deficiency of valence electrons and implicitly modifying the CNTs' electrical

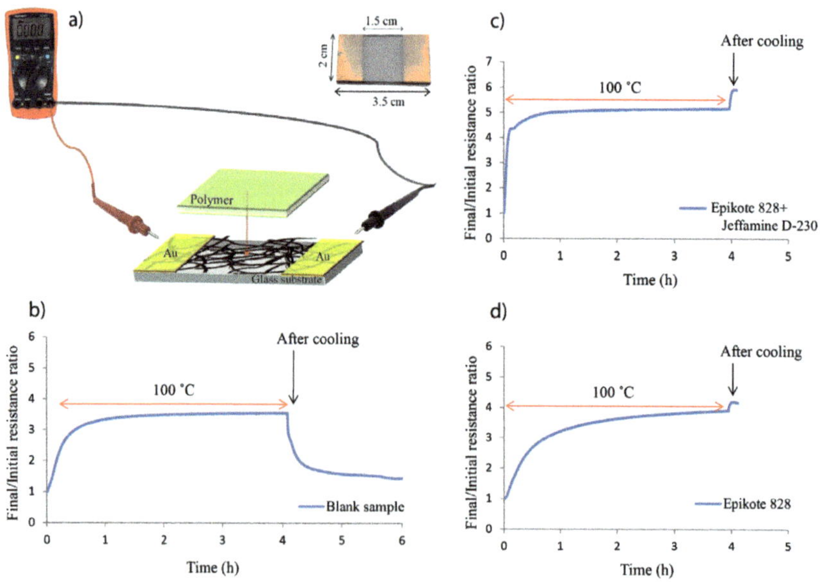

Fig. 6 (a) Schematic of our *in situ* resistance measurement setup and examples of the final/initial resistance ratio curves obtained during (b) a blank measurement, (c) a polymer mixture impregnation and (d) an epoxy resin impregnation.

behaviour. The dopant component can act as an electron donor (n-type doping) or electron acceptor (p-type doping). Studies have shown that a SWCNT film held under vacuum at room temperature can have a 15% decrease in resistance upon air exposure,[42] while exposing unpurified SWCNTs to humid air after heating them up to 350 °C can lead to an up to 4 fold decrease in resistance.[43] Other researchers support the hypothesis that the adsorption of gas molecules changes the value of the Schottky barrier build-up between the SWCNTs and the metal electrode.[44] They support the idea that the gas molecules in the junction region can change the metal work function and the Fermi level alignment. Therefore resistance fluctuations upon heating or air exposure can be due to changes in the Schottky barriers of the metal–nanotube junctions, and not necessarily due to the doping/dedoping of the SWCNTs.

In order to avoid resistance fluctuations due to gas and water adsorption from air, after the final 400 °C thermal treatment, all of the samples were exposed to air overnight to reach saturation and a constant resistance value. Blank *in situ* resistance measurements were performed on a series of samples in an oven at 100 °C for 4 hours, the conditions needed for curing the polymer mixture. The results showed an average resistance increase of 3.1 fold before cooling (example in Fig. 6b), most likely due to desorption of water and gasses from the films upon heating. This value rapidly decreased as soon as the samples were taken out of the oven for cooling due to the readsorption of gasses and water at room temperature.

The same type of measurements were performed while adding only one component of the polymer mixture to separate any possible individual effects on the resistance of the SWCNT network. Measurements performed after impregnating the film with Epikote 828 alone for 4 hours at 100 °C show an average resistance increase of 3.6 fold before cooling (example in Fig. 6d), slightly higher than for the blank samples. Because Jeffamine D-230 is a volatile amine, measurements for the amine component alone were performed at room temperature, showing an average resistance increase of 4.9 fold. If we take into consideration the ratios of the blank samples, we can estimate a resistance increase of up to 15 fold after amine impregnation upon heating at 100 °C for 4 hours. Amines have been shown to have a n-type doping effect on SWCNTs[45,46] and are expected to influence the resistance of the SWCNT network to a certain degree, especially if the carbon nanotubes are p-type doped by the oxygen in air[42,47] or other gas molecules.

When mixing the two components together and impregnating the SWCNT film, the final/initial resistance ratio after curing the polymer for 4 hours at 100 °C was 4.3 before cooling (example in Fig. 6c). In order to obtain a 1 : 1 epoxy/amine ratio (based on an epoxy group/NH equivalent), the amine represents only 24.3 wt % of the epoxy/amine mixture. Therefore the resistance increase may be due not only to a cumulated effect of the two components, but also to the formation of a polymeric network during the curing process.

All of the samples coated with a polymer (epoxy resin alone or the epoxy/amine mixture) and heated at 100 °C had a resistance increase of 1.15 fold after cooling, due to the semiconductor behaviour of the SWCNTs. When the SWCNTs are covered in a polymer, they can no longer adsorb water and gasses when cooled in air.

Each SWCNT film prepared is unique and can have a slightly different thickness compared to others, so the question of whether or not the results of different experiments can be compared is important. In order to ensure the reproducibility of our experiments, we prepared SWCNT films with thicknesses varying from 10 to 140 nm. Thickness measurements involve damaging the samples to measure the height profiles across scratches in the film, therefore SWCNT films were prepared on large glass substrates which were subsequently cut into 2 cm × 3.5 cm samples. Half of these were used for correlating the thickness with resistance and transmittance measurements, while the other half were used for polymer impregnation. The absorbance as a function of thickness curve shows a pseudo-linear dependency within the 10–140 nm thickness range (Fig. 7a). This curve, along with the resistance *vs.* thickness (Fig. 7b) curve, can be used to approximate the thicknesses of the SWCNT films used for polymer impregnation.

In situ resistance measurements performed on films of different thicknesses after impregnating with the epoxy/amine mixture and curing the polymer for 4 hours at 100 °C, show that there is no major difference in the final/initial resistance ratios between 10 nm films and 140 nm films (Fig. 7c). The average final/initial resistance ratios (after cooling) increase from 5 to 6 within this thickness range. This means that experiments performed on SWCNT films of 20–40 nm thickness are comparable despite minor thickness variations. These results are similar to those of Noh *et al.*[48] who impregnated 12 μm thick multi-walled carbon nanotube (MWCNT) mats with cyclic butylene terephthalate (CBT), by compressing them with a heating press. The resistance of the MWCNT mats increased 2 to 9 fold (depending on the compression pressure) after polymer addition.

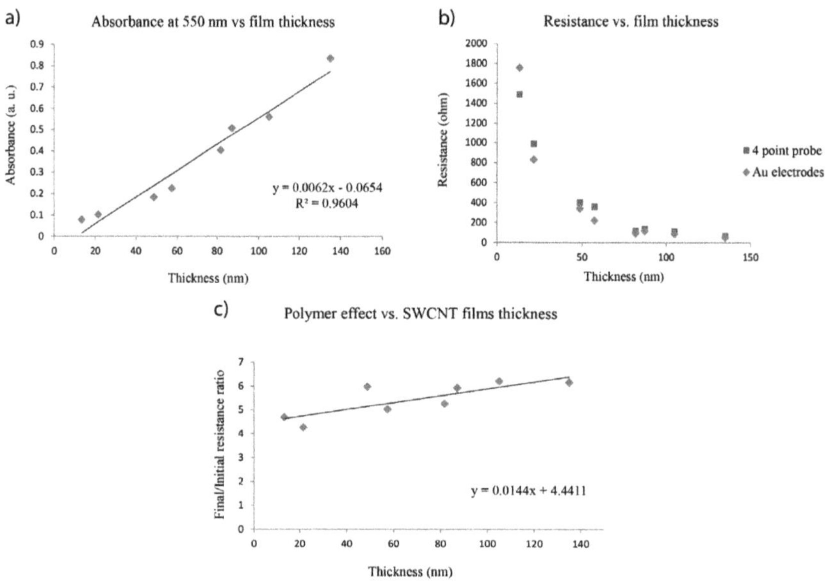

Fig. 7 (a) Absorbance, (b) resistance and (c) polymer impregnation effect depending on the SWCNT film thickness.

Conclusions

We have adapted and improved a preparation technique for SWCNT thin films which represent our reference networks with good electrical properties. These films were used to study the polymer matrix effect on an already existing SWCNT network during polymer impregnation. We developed an *in situ* resistance measurement technique that allowed us to have direct access to the conductive network during polymer impregnation, and also during the curing process of the polymer matrix. These measurements showed that polymer impregnation is not the only factor that can influence the behaviour of the carbon nanotubes; they showed that different components of a polymer mixture can have different effects on the electrical properties of a SWCNT network aside from the formation of a polymeric network during the curing process. Blank measurements showed that air and temperature exposure can also affect the electrical behaviour of the carbon nanotubes, even without the addition of a polymer component. These results prove that in order to prepare an efficient composite material, it is important first to understand the effects that any factor introduced in the preparation process can have on each of the components. Further studies will be performed in order to understand how the polymer impregnation process and the newly formed interfaces affect the electrical contacts between the carbon nanotubes. These studies will contribute to the improvement of future polymer composites efficiencies.

Acknowledgements

This research forms part of the research programme of the Dutch Polymer Institute (DPI), project #756 (CoCoCo). The authors would also like to thank Dr J. G. P. Goossens (TU/e) for his support.

References

1 W. Marx and A. Barth, *Phys. Status Solidi B*, 2008, **245**, 2347–2351.
2 J. N. Coleman, U. Khan, W. J. Blau and Y. K. Gun'ko, *Carbon*, 2006, **44**, 1624–1652.
3 J. Bernholc, D. Brenner, M. B. Nardelli, V. Meunier and C. Roland, *Annu. Rev. Mater. Res.*, 2002, **32**, 347–375.
4 J. Hone, M. C. Llaguno, M. J. Biercuk, A. T. Johnson, B. Batlogg, Z. Benes and J. E. Fischer, *Appl. Phys. A: Mater. Sci. Process.*, 2002, **74**, 339–343.
5 M. S. Dresselhaus, G. Dresselhaus, J. C. Charlier and E. Hernandez, *Philos. Trans. R. Soc., A*, 2004, **362**, 2065–2098.
6 H. Kataura, Y. Kumazawa, Y. Maniwa, I. Umezu, S. Suzuki, Y. Ohtsuka and Y. Achiba, *Synth. Met.*, 1999, **103**, 2555–2558.
7 Z. C. Wu, Z. H. Chen, X. Du, J. M. Logan, J. Sippel, M. Nikolou, K. Kamaras, J. R. Reynolds, D. B. Tanner, A. F. Hebard and A. G. Rinzler, *Science*, 2004, **305**, 1273–1276.
8 F. Mirri, A. W. K. Ma, T. T. Hsu, N. Behabtu, S. L. Eichmann, C. C. Young, D. E. Tsentalovich and M. Pasquali, *ACS Nano*, 2012, **6**, 9737–9744.
9 D. S. Hecht, A. M. Heintz, R. Lee, L. B. Hu, B. Moore, C. Cucksey and S. Risser, *Nanotechnology*, 2011, **22**, 075201.

10 M. B. Bryning, M. F. Islam, J. M. Kikkawa and A. G. Yodh, *Adv. Mater.*, 2005, **17**, 1186–1191.

11 Y. Zhao, H. P. Hong and J. S. Welsh, *Proc. SPIE7404, Nanostructured Thin Films II*, 2009, 74040H.

12 M. J. Biercuk, M. C. Llaguno, M. Radosavljevic, J. K. Hyun, A. T. Johnson and J. E. Fischer, *Appl. Phys. Lett.*, 2002, **80**, 2767–2769.

13 K. D. Ausman, R. Piner, O. Lourie, R. S. Ruoff and M. Korobov, *J. Phys. Chem. B*, 2000, **104**, 8911–8915.

14 Y. Y. Huang and E. M. Terentjev, *Polymers*, 2012, **4**, 275–295.

15 N. Minami, Y. J. Kim, K. Miyashita, S. Kazaoui and B. Nalini, *Appl. Phys. Lett.*, 2006, **88**, 093123.

16 T. Takahashi, K. Tsunoda, H. Yajima and T. Ishii, *Jpn. J. Appl. Phys., Part 1*, 2004, **43**, 3636–3639.

17 U. Dettlaff-Weglikowska, J. M. Benoit, P. W. Chiu, R. Graupner, S. Lebedkin and S. Roth, *Curr. Appl. Phys.*, 2002, **2**, 497–501.

18 Y. Zhao, B. Mannhalter, H. P. Hong and J. S. Welsh, *J. Nanosci. Nanotechnol.*, 2010, **10**, 5776–5782.

19 D. M. Bigg, *Polym. Eng. Sci.*, 1979, **19**, 1188–1192.

20 G. H. Zheng, J. S. Wu, W. P. Wang and C. Y. Pan, *Carbon*, 2004, **42**, 2839–2847.

21 J. Z. Kovacs, B. S. Velagala, K. Schulte and W. Bauhofer, *Compos. Sci. Technol.*, 2007, **67**, 922–928.

22 C. A. Martin, J. K. W. Sandler, M. S. P. Shaffer, M. K. Schwarz, W. Bauhofer, K. Schulte and A. H. Windle, *Compos. Sci. Technol.*, 2004, **64**, 2309–2316.

23 J. K. W. Sandler, J. E. Kirk, I. A. Kinloch, M. S. P. Shaffer and A. H. Windle, *Polymer*, 2003, **44**, 5893–5899.

24 X. K. Li, F. Gittleson, M. Carmo, R. C. Sekol and A. D. Taylor, *ACS Nano*, 2012, **6**, 1347–1356.

25 F. M. Smits, *Bell Syst. Tech. J.*, 1958, **37**, 711–718.

26 M. J. O'Connell, P. Boul, L. M. Ericson, C. Huffman, Y. H. Wang, E. Haroz, C. Kuper, J. Tour, K. D. Ausman and R. E. Smalley, *Chem. Phys. Lett.*, 2001, **342**, 265–271.

27 W. Gomulya, G. D. Costanzo, E. J. F. de Carvalho, S. Z. Bisri, V. Derenskyi, M. Fritsch, N. Frohlich, S. Allard, P. Gordiichuk, A. Herrmann, S. J. Marrink, M. C. dos Santos, U. Scherf and M. A. Loi, *Adv. Mater.*, 2013, **25**, 2948–2956.

28 W. Choi, S. Ohtani, K. Oyaizu, H. Nishide and K. E. Geckeler, *Adv. Mater.*, 2011, **23**, 4440–4443.

29 W. Zhou, J. Vavro, N. M. Nemes, J. E. Fischer, F. Borondics, K. Kamaras and D. B. Tanner, *Phys. Rev. B: Condens. Matter Mater. Phys.*, 2005, **71**, 205423.

30 R. Saito, M. Hofmann, G. Dresselhaus, A. Jorio and M. S. Dresselhaus, *Adv. Phys.*, 2011, **60**, 413–550.

31 S. Costa, E. Borowiak-Palen, M. Kruszynska, A. Bachmatiuk and R. J. Kalenczuk, *Mater. Sci.-Pol.*, 2008, **26**, 433–441.

32 A. Jorio, M. A. Pimenta, A. G. Souza, R. Saito, G. Dresselhaus and M. S. Dresselhaus, *New J. Phys.*, 2003, **5**, 139.

33 Y. Miyata, K. Mizuno and H. Kataura, *J. Nanomater.*, 2011, 786763.

34 R. Schonfelder, F. Aviles, A. Bachmatiuk, J. V. Cauich-Rodriguez, M. Knupfer, B. Buchner and M. H. Rummeli, *Appl. Phys. A: Mater. Sci. Process.*, 2012, **106**, 843–852.

35 N. N. A. H. Meis, L. G. J. van der Ven, R. A. T. M. van Benthem and G. de With, *Prog. Org. Coat.*, 2014, **77**, 176–183.
36 J. Sandler, M. S. P. Shaffer, T. Prasse, W. Bauhofer, K. Schulte and A. H. Windle, *Polymer*, 1999, **40**, 5967–5971.
37 A. Allaoui, S. Bai, H. M. Cheng and J. B. Bai, *Compos. Sci. Technol.*, 2002, **62**, 1993–1998.
38 N. Grossiord, P. J. J. Kivit, J. Loos, J. Meuldijk, A. V. Kyrylyuk, P. van der Schoot and C. E. Koning, *Polymer*, 2008, **49**, 2866–2872.
39 J. J. Zhao, A. Buldum, J. Han and J. P. Lu, *Nanotechnology*, 2002, **13**, 195–200.
40 Y. Wang and J. T. W. Yeow, *J. Sens.*, 2009, **2009**, 24.
41 J. E. Fischer, *Acc. Chem. Res.*, 2002, **35**, 1079–1086.
42 P. G. Collins, K. Bradley, M. Ishigami and A. Zettl, *Science*, 2000, **287**, 1801–1804.
43 Y. Maniwa, Y. Kumazawa, Y. Saito, H. Tou, H. Kataura, H. Ishii, S. Suzuki, Y. Achiba, A. Fujiwara and H. Suematsu, *Mol. Cryst. Liq. Cryst.*, 2000, **340**, 671–676.
44 P. Bondavalli, P. Legagneux and D. Pribat, *Sens. Actuators, B*, 2009, **140**, 304–318.
45 R. B. Koizhaiganova, D. H. Hwang, C. J. Lee, S. Roth and U. Dettlaff-Weglikowska, *Phys. Status Solidi B*, 2010, **247**, 2793–2796.
46 D. R. Samarajeewa, G. R. Dieckmann, S. O. Nielsen and I. H. Musselman, *Carbon*, 2013, **57**, 88–98.
47 A. Tchernatinsky, S. Desai, G. U. Sumanasekera, C. S. Jayanthi, S. Y. Wu, B. Nagabhirava and B. Alphenaar, *J. Appl. Phys.*, 2006, **99**, 034306.
48 Y. J. Noh, H. S. Kim and S. Y. Kim, *Carbon Letters*, 2012, **13**, 243–247.

Faraday Discussions

PAPER

Few layer graphene–polypropylene nanocomposites: the role of flake diameter

Cristina Vallés,[a] Amr M. Abdelkader,[ab] Robert J. Young[a] and Ian A. Kinloch[a]

Received 16th May 2014, Accepted 1st July 2014
DOI: 10.1039/c4fd00112e

Graphene shows excellent potential as a structural reinforcement in polymer nanocomposites due to its exceptional mechanical properties. We have shown previously that graphene composites can be analysed using conventional composite theory with the graphene flakes acting as short fillers which have a critical length of ~3 μm which is required for good reinforcement. Herein, polypropylene (PP) nanocomposites were prepared using electrochemically-exfoliated few layer graphene (FLG) with two different flake diameters (5 μm and 20 μm). The crystallization temperature and degree of crystallinity of the PP were found to increase with the loading of FLG, which suggests that the flakes acted as crystallisation nucleation sites. Mechanical testing showed that the 5 μm flakes behaved as short fillers and reinforced the PP matrix poorly. The modulus of the 20 μm flake composites, however, increased linearly with loading up to 20 wt%, without any of the detrimental aggregation effects seen in other graphene systems. The mechanical data were compared with our previous work on other graphene composite systems and the apparent need to balance the degree of functionalization to improve matrix compatibility whilst not encouraging aggregation is discussed.

1 Introduction

Graphene is an atomically-thin, two-dimensional honeycomb material with extraordinary electronic, thermal and mechanical properties,[1–3] which makes it an ideal candidate for a wide variety of applications including sensors, batteries, supercapacitors and hydrogen storage.[4–7] A promising application is the incorporation of graphene into polymer matrices in order to improve the electrical, thermal and/or mechanical properties.[8–11] The micromechanics of such graphene composites have been studied extensively using Raman spectroscopy to probe the behaviour of individual flakes of graphene embedded in polymer beams under strain.[4,12,13] Examination of the shift rate of the 2D Raman band position with

[a]*School of Materials, University of Manchester, Oxford Road, Manchester, M13 9PL, UK. E-mail: ian.kinloch@manchester.ac.uk; Tel: +44 (0)161 3063615*
[b]*School of Chemistry, University of Manchester, Oxford Road, Manchester, M13 9PL, UK*

strain has shown that monolayer graphene can achieve an effective modulus of 1 TPa within such a polymer composite system.[12] However, this modulus drops as the flake becomes thicker due to the internal shear between the graphene layers within the flake, so that a 5 layer flake has a theoretical modulus of 0.6 TPa.[14] This decrease in modulus with thickness would initially suggest that one would use monolayer graphene to reinforce a composite as it has the highest modulus. However, the degree of reinforcement is a function of the loading fraction and modulus. The maximum achievable loading fraction is strongly influenced by the graphene being thinner than the polymer molecules surrounding it; monolayer graphene has a theoretical maximum volume fraction of ~1/7 assuming a perfect dispersion compared to ~1/3 for trilayer graphene. Thus 4–5 layer graphene is found to be the optimum thickness for reinforcement, taking into account the balance between the increase in volume fraction and drop in modulus with flake thickness. This result is fortuitous as the majority of scalable graphene production routes produce few layer graphene (FLG).

Micromapping of the Raman spectra within individual flakes has demonstrated that the strain distribution within the flakes can be described by conventional shear lag theory where the strain tends to zero at the edges of the flakes. This phenomenon is known as the "short fibre effect" in fibre composites and states that there is a critical length beneath which the end effects dominate and leads to no reinforcement from the filler. Gong *et al.*[12] found that this critical length was ~3 μm for unfunctionalised graphene, due to the relatively weak interfacial strength of 1 MPa compared to the 30 MPa found for carbon fibre–epoxy composites. (Graphene has none of the chemical or mechanical interactions of a sized carbon fibre.) Thus in summary, micromechanics predict that ideally a nanocomposite should use FLG typically 4–5 layers thick with a length of >3 μm, and more preferably >30 μm.

There is a wide range of literature on the processing and properties of bulk graphene nanocomposites using elastomers, thermosets and thermoplastics.[5,6,8,15] However, most of this literature has used either graphene oxide (GO) or graphite nanoplatelets due to the limited availability of graphene on a sufficient scale. Typically these composites show good improvement in the modulus of the composites at very low loadings of graphene (~1 wt%), but the degree of reinforcement then falls at higher loadings, possibly due to the formation of agglomerates in the polymer matrix.[4,16] Such aggregation is common in nanomaterial systems due to their high surface areas and have been seen in other nanocomposites such as carbon nanotubes. One key difference though between graphene and nanotube systems is the relatively low viscosity of the platelet systems; carbon nanotubes were ultimately limited in their application due to the viscosity of a nanotube-polymer dispersion increasing to the extent that the dispersion could not be processed by ~5 wt%, whereas herein graphene-related composites have been compounded and injection moulded at loadings of 20 wt%.

Graphene and graphite nanoplatelets have been found previously to increase the crystallization temperature, crystallization rate, and degree of crystallinity of a PP matrix. It was suggested that the flakes act as a nucleating agent to reduce the barrier to nucleation and modify the crystallite structures of the PP.[17–19] Kalaitzidou *et al.*[18,19] found good enhancement of both the mechanical and electrical properties of PP with the addition of exfoliated graphite nanoplatelets at medium loadings (up to 5 vol%). Functionalization of graphene oxide has also been investigated to improve the dispersion of the flakes in the PP matrix and increase the interfacial adhesion between the filler and the non-polar PP. For example, the reinforcing effects of alkylated graphene oxide (AGO) on a nonpolar PP matrix[20] or the covalent modification of graphene oxide with *p*-phenylenediamine and cyanuric chloride and then grafted with maleic anhydride polypropylene (MAPP)[21] have both been studied. Long chain alkylamines with varying chain lengths have also been chemically grafted to the GO surface in a PP matrix in the presence of a maleic anhydride-*g*-polypropylene (MA-PP) compatibilizer through a melt processing technique.[22,23] An alternative, more ecologically friendly strategy for nanocomposites has been proposed of first coating graphene with polypropylene (PP) latex and then melt-blending the coated graphene with PP.[24] Significant improvements in the mechanical and thermal properties of PP were also found by adding only few percent loading of chemically-reduced graphene oxide,[25] or polypropylene-*graft*-reduced graphene oxide (PP-*g*-rGO) as a novel compatibilizer for PP/polystyrene (PP/PS).[26] However, even after chemically modifying the surfaces of the flakes using such approaches, the challenge of achieving good dispersions of the fillers at high loadings limited the properties of the final composites.

Herein, we investigate the effect of flake diameter on the mechanical and thermal properties of bulk PP nanocomposites at loadings relevant to structural applications. Gram quantities of few layer graphene (FLG) were prepared by reductive electrochemical exfoliation, with the diameter of the flakes being controlled by the graphite grain size in the initial electrode.[27] FLG–PP composites were then prepared from 0.5 to 20 wt% of two different diameter FLG samples (5 μm and 20 μm) by melt-compounding followed by injection moulding. The crystallization temperature and degree of crystallization of the PP and the graphene–PP nanocomposites were determined using DSC. The mechanical properties of the composites were evaluated by tensile testing and related to the thermal properties. The effect of exfoliated graphene with different flake sizes as reinforcement in bulk PP composites was then compared with the analogous PMMA composites.

2 Experimental

2.1 Preparation of the graphene and the graphene–PP nanocomposites

Graphene powders with two mean flake sizes, 5 μm (5-FLG) and 20 μm (20-FLG), and similar thickness (<5 nm) were prepared through a continuous electrochemical exfoliation of graphite, following a procedure described elsewhere by Abdelkader *et al.*[27] All the flakes were under 5 nm in thickness, with ~5% of the sheets being monolayer. 5-FLG–PP and 20-FLG–PP composites were prepared with loadings from 0.5 to 20 wt% by melt mixing using a twin screw extruder. Graphene flakes and PP pellets (Ineos, 100-CA50) were fed into a twin-screw extruder (Thermo Scientific HAAKE MiniLab micro compounder) and the material was cycled for 15 minutes at 190 °C at a screw speed of 100 rpm before being extruded. The extruded samples were further processed into bone shaped specimens by injection moulding (HAAKE MiniJet Piston Injection Moulding System, $T_{cylinder} = 190$ °C, $T_{mould} = 65$ °C, pressure = 740 bar kept for 10 s). Neat PP control samples were processed using the same steps.

2.2 Characterization of the graphene–PP nanocomposites

Scanning electron microscopy (SEM) was used to characterize the morphology and the average lateral diameter of the graphene 5-FLG and 20-FLG flakes using a Philips XL30 FEG SEM, operating at an accelerating voltage of 5 kV.

Differential Scanning Calorimetry (DSC) measurements of the polymer and nanocomposites were carried out under nitrogen gas from 20 °C to 200 °C using a DSC-Q100 from TA Instruments. The samples were first heated to 200 °C at a rate of 10 °C min^{-1} to remove their thermal history, and after that they were cooled to −40 °C. Both cooling and heating were performed at the same rate of 10 °C min^{-1}. To determine the crystallization and melting properties the second heating and first cooling cycles were considered. The associated thermal parameters of crystallization (T_c) and melting (T_m) temperatures, crystallization enthalpy (ΔH_c), heat of melting (ΔH_m), and the percentage of crystallinity (X_c) were extracted. The relative X_c was determined from the melting curves using the following expression:

$$X_c = (\Delta H_m/(1 - x)\Delta H_0) \times 100$$

where ΔH_m is the heat of melting of the samples, ΔH_0 is the theoretical value of the melting heat for 100% crystalline PP which is 165 J g^{-1},[28] and x is the weight fraction of FLG in the sample.

Stress–strain curves were obtained using dogbone shaped specimens in an Instron 4301 machine, using a tensile rate of 0.5 mm min^{-1} with a load cell of 5 kN. The weight fractions were transformed to volume fractions taking the density of the graphene powders as 2.0 g cm^{-3} and that of PP with 51% crystallinity equal to 0.901 g cm^{-3} (the density of amorphous PP, $\rho_{amorphous} = 0.855$ g cm^{-3} and that of crystalline PP, $\rho_{crystalline} = 0.946$ g cm^{-3}).

3 Results and discussion

3.1 Characterization of the graphene flakes

Fig. 1 shows micrographs of typical flakes in the 5-FLG and 20-FLG samples. The 5-FLG graphene powder was composed of mainly FLG flakes with diameters of ~5 μm, whereas most of the flakes were ~20 μm in diameter in the 20-FLG powder. All of the flakes were under 5 nm in thickness, with an average thickness of 2–3 nm and ~5% of the sheets being monolayer.[27] We have found previously with a PMMA system that melt processing of nanocomposites does not significantly reduce these flakes sizes.

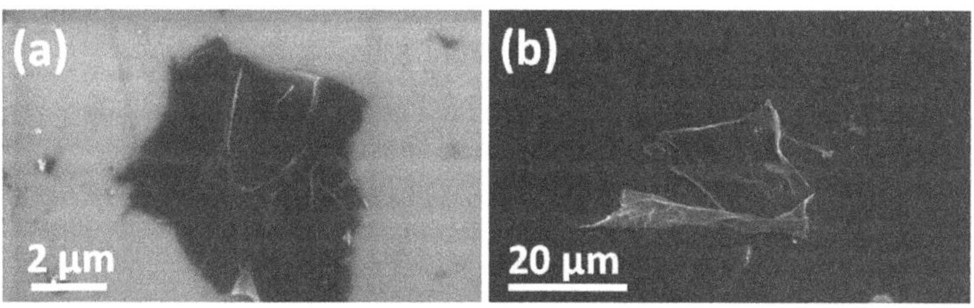

Fig. 1 SEM micrographs of the (a) 5-FLG and (b) 20-FLG graphene powders.

3.2 Crystallization and melting properties

The properties, particularly the mechanical properties, of semicrystalline thermoplastic materials are strongly related to their internal microstructure and crystallinity. The crystallization and melting behaviours of the PP matrix were studied under non-isothermal conditions as a function of graphene loading. Fig. 2 shows the crystallization and melting curves from the second heating and the first cooling cycles (from DSC) for neat PP and the graphene–PP nanocomposites. The thermal parameters extracted from these curves are summarized in Table 1.

The crystallisation temperature was found to increase with the addition of the FLG, going from 119.6 °C for pure PP to maximum temperatures of 126.8 °C and 129.3 °C for 20 wt% 5-FLG–PP and 20-FLG-PP, respectively. The degree of crystallinity values also increased with the addition of FLG but the degree of increase was independent of the flake diameter, with a crystallinity increase of ~10% observed at 20 wt% FLG for both flake diameters. The increases observed in the T_c values indicate that the graphene acted as heterogeneous nucleation sites which facilitated the crystallization of PP during cooling. The results from the heating curves (Fig. 2) found similar values of T_m for both composite systems and the neat PP. For the neat PP, the melting peak at 164.8 °C is attributed to the melting of α-crystals, which are the most common monoclinic crystal structure of PP. (The melting of β-crystals gives a peak at 147.5 °C). The results from the melting curves revealed that no β-crystalline structure was present in the neat PP nor in the graphene–PP nanocomposites. This will be investigated further by X-ray spectroscopy in future work.

3.3 Mechanical properties

All the samples were found to behave as glassy polymers (Fig. 3) with tensile fracture strain decreasing with increasing FLG loading. This embrittlement with the addition of FLG is similar to that reported for other nanotubes and graphene composites.[29] The tensile strengths of the 20-FLG remained constant with loading, whereas the tensile strength of the 5-FLG dropped for the medium loaded samples (Fig. 4).

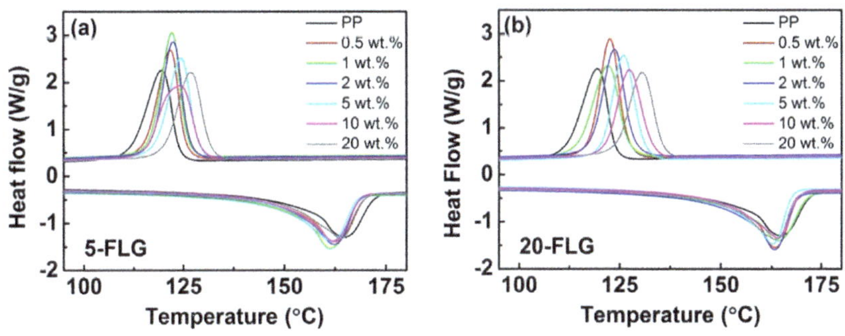

Fig. 2 Crystallization thermograms (first cooling cycle) and melting thermograms (second heating cycle) of neat PP and the 5-FLG–PP (a) and 20-FLG–PP (b) nanocomposites.

Table 1 Summarized results of DSC analysis of neat PP and the 5-FLG–PP and 20-FLG–PP nanocomposites with different loadings. (Crystallization, T_c, and melting, T_m, temperatures; crystallization enthalpy, ΔH_c; heat of melting, ΔH_m and the percentage of crystallinity, X_c.) The standard deviations of the temperature values were typically less than ±1 °C for the three samples and the standard deviations of the percentage of crystallinity values were typically less than 1.5

Sample	wt%	vol%	T_c (°C)	ΔH_c (J g^{-1})	T_m (°C)	ΔH_m (J g^{-1})	X_c
5-FLG–PP	0.0	0.0	119.6	97.1	164.8	91.6	55.5
	0.5	0.2	121.8	96.4	162.7	95.4	58.1
	1.0	0.5	122.1	97.6	162.2	99.2	60.7
	2.0	0.9	122.5	94.6	162.3	95.4	59.0
	5.0	2.3	124.3	92.6	161.9	94.4	60.3
	10.0	4.5	124.3	90.5	162.2	91.7	61.2
	20.0	9.0	126.8	86.0	163.9	86.9	65.8
20-FLG–PP	0.0	0.0	119.6	97.1	164.8	91.6	55.5
	0.5	0.2	122.7	94.5	163.4	94.9	57.8
	1.0	0.5	122.4	98.2	163.7	95.1	58.2
	2.0	0.9	123.1	97.1	163.7	96.3	59.5
	5.0	2.3	125.0	94.0	163.1	96.0	61.2
	10.0	4.5	127.8	86.8	164.0	90.3	60.8
	20.0	9.0	129.3	82.8	163.9	86.0	65.1

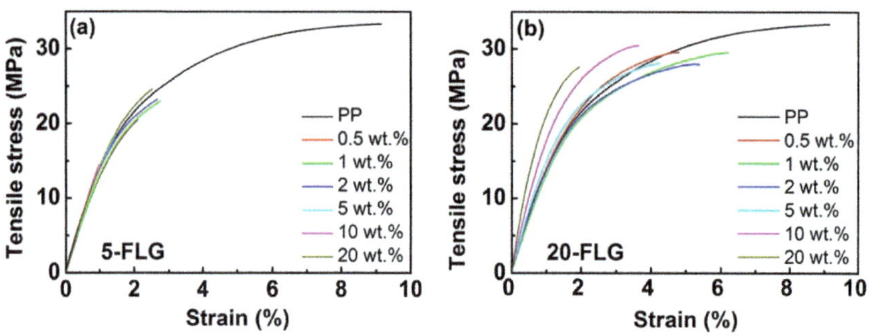

Fig. 3 Stress–strain curves of 5-FLG–PP (a) and 20-FLG–PP (b). Fig. 4b shows the values of the Young's modulus (E) determined by tensile testing plotted as a function of the filler content.

There was no significant change within the experimental error of the Young's modulus, E, of the 5-FLG–PP composites with loading (Table 2). However, for the 20-FLG–PP composites a linear increase of E with loading was observed with the modulus doubling at a 20 wt% loading (2.95 GPa).

3.4 Discussion of the mechanical properties of graphene reinforced thermoplastics

3.4.1 Discussion of the PP system. The mechanical properties of the composites are strongly related to the morphology of the polymer and any reinforcement from the filler. If, as seen herein, the addition of filler induces crystallization, then reinforcement from both the change in polymer morphology and

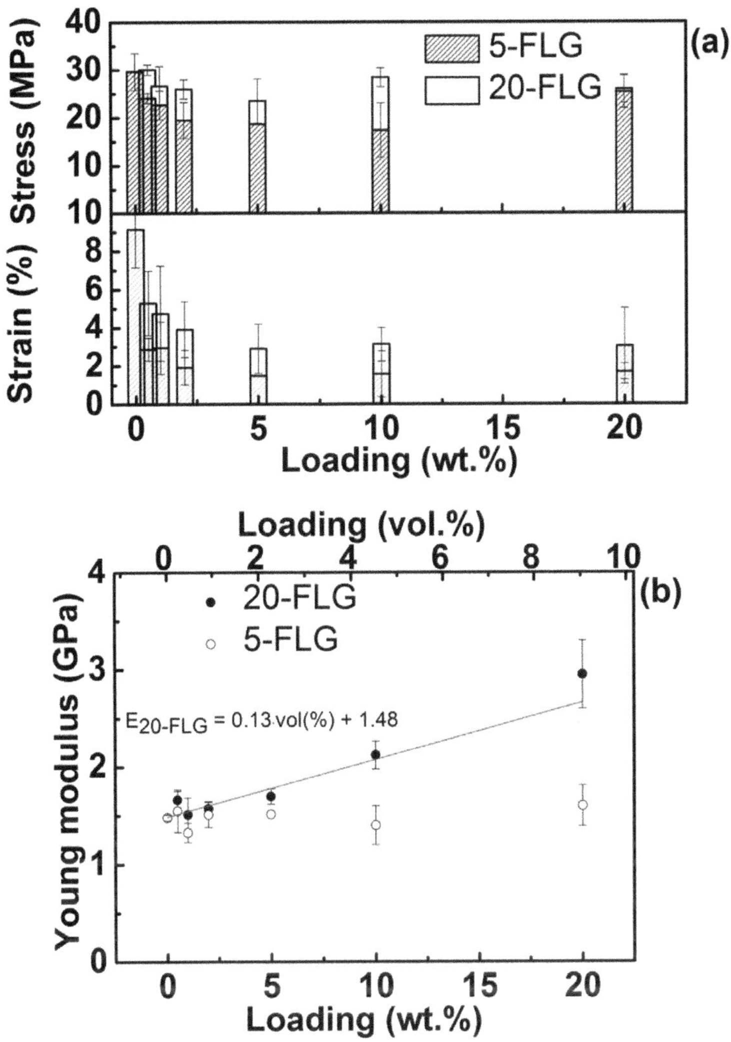

Fig. 4 (a) Variation of the strain and stress at break of the composites with loading. (b) Variation of Young's modulus, *E*, with loading for the two series of nanocomposites. The error bars represent the standard deviation calculated from between 4 and 7 samples.

the FLG need to be considered.[30] In general the nucleation of PP crystals and an increase in crystallinity will lead to an increase in the Young's modulus and yield stress of the PP matrix and possibly to a reduction of the elongation at failure.[31,32] Fortunately, both the 5-FLG and 20-FLG composite systems had a similar change in crystallinity (10%), allowing the effect of crystallinity to be separated from the effect of flake diameter. Given that the 5-FLG composites had a constant modulus with loading, it could be assumed that neither the small increase in crystallinity nor the flakes significantly reinforced the polymer. Thus the increase in modulus for the 20 μm flakes can be approximated to be entirely from the contribution of the larger flakes.

Table 2 The Young's modulus values of the composites as a function of loading and flake diameter

		E of the composites with:	
Loading (wt%)	Loading (vol%)	5 μm flakes	20 μm flakes
0.0	0.0	1.5 ± 0.0	1.5 ± 0.0
0.5	0.2	1.5 ± 0.2	1.7 ± 0.1
1.0	0.5	1.3 ± 0.1	1.5 ± 0.2
2.0	0.9	1.5 ± 0.1	1.6 ± 0.1
5.0	2.3	1.5 ± 0.0	1.7 ± 0.1
10.0	4.5	1.4 ± 0.2	2.1 ± 0.1
20.0	9.0	1.6 ± 0.2	2.9 ± 0.4

The Young's modulus, E_c, of a particulate composite, in the case of uniform strain (and assuming good dispersions of graphene at loadings below the optimal), is given by the modified rule of mixtures as:

$$E_c = E_p \cdot V_p \eta_0 \eta_l + E_m \cdot V_m$$

where E_p is the Young's modulus of the particles, E_m is the Young's modulus of the polymer matrix, and V_p and V_m are the volume fractions of particles and matrix, respectively, within the composites. η_0 is the Krenchel orientation factor and depends on the average orientation of the particles to the stress. η_l is between 0 and 1 and adjusts for the poor stress transfer arising from poor interfaces and/or short fibers, with 1 being a continuous fibre. Frequently the effective modulus, E_{eff}, of a particle is considered which is given by $\eta_l \times E_p$ as it is frequently not possible to separate these two values in a bulk system.

For the 5 μm flakes (5-FLG) there was no reinforcement as the flakes were behaving as short fibres, hence η_l and E_{eff} are equal to 0. For the 20 μm flakes (20-FLG) $E_{eff} \cdot \eta_0$ can be estimated from the gradient of the E *vs.* volume fraction, which gives a value of 13 GPa. We have previously used polarised Raman spectroscopy to assess the degree of orientation in PMMA–FLG injection moulded composites and found that it was random due to the relatively low shear in the small scale injection moulding machine. We therefore assume that the PP composites also have randomly aligned flakes, thus we can take η_0 as ~3/8 (the Krenchel factor for a random 2D orientation of graphene flakes), leading to a value of 35 GPa for the 20 μm flakes, compared to the predicted theoretical maximum of ~600 GPa.[14]

3.4.2 Comparison with previously studied systems. Fig. 5 summarises our studies to-date on thermoplastics (semi-crystalline PP and amorphous poly methyl methacrylate, PMMA) using FLG and graphene oxide (GO) as reinforcements with the modulus normalised to that of the matrix being used. The steeper the slope of the graph the higher the effective modulus of the graphene, as more reinforcement is achieved per flake in the system. However, the ultimate performance of the composite also matters at useful loadings; there is little application value in a composite system which has outstanding reinforcement at low loadings (*e.g.* <5 wt%) and then does not improve its absolute properties at higher loadings (*e.g.* 20–30 wt%). Thus the ideal composite system in Fig. 5 would have a high gradient and linearly increase across all volume fractions, without levelling off.

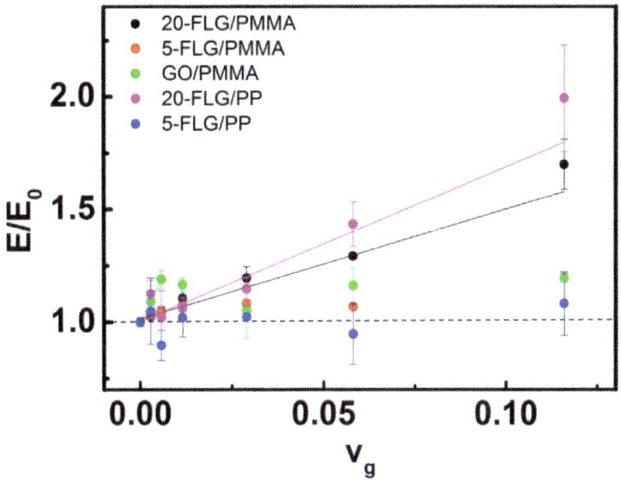

Fig. 5 Variation of E/E_0 with loading for different graphene–polymer systems. The error bars represent the standard deviation calculated from between 4 and 7 samples.

The importance of having a sufficiently large flake diameter is demonstrated for the electrochemically exfoliated FLG in both the PP and PMMA systems, with the 5 µm diameter flakes showing little or no improvement in modulus up to 20 wt% loading. However, 20 µm diameter flakes did show linear improvements in both systems, with the highest degree of reinforcement in PP, possibly due to the increase in crystallinity. So far the literature has tended to show that good reinforcements are only found at relatively low levels of graphene loading due to the difficulty of obtaining good dispersions at higher loadings.[4] Obtaining a good distribution of nano-reinforcement is one of the greatest challenges in the preparation of polymer-based nanocomposites[33] since the properties of nanocomposites can be compromised by a poor dispersion, as will be demonstrated. Herein we show the achievement of good mechanical properties for high volume fractions of unfunctionalized graphene.

GO is frequently used for reinforcement in polymers because it offers two important advantages with respect to "clean" graphene: (i) production in high yield and scale and (ii) the presence of a high number of oxygen-containing functionalities providing a good dispersion of flakes in the matrix at low concentrations through electrostatic dispersion and enhanced graphene–polymer interactions. In particular, the surface chemistry allows shorter flakes to be used as the critical length is dependent on the interfacial shear strength, which depends on the matrix–polymer interactions. (For example, the GO flakes used here are <10 µm in diameter.) As shown in Fig. 5, the surface chemistry does lead to a higher degree of reinforcement than the FLG at low loadings due to the homogeneous distribution of flakes in the polymer. The effective modulus values at these loadings are in the range of 200–300 GPa, and are in total agreement with values determined theoretically that can be found in the literature. However, at higher loadings of graphene oxide the mechanical properties were found to deteriorate due to the formation of agglomerates of graphene oxide, possibly as the GO flakes get close enough to enter the deep energy minimum for electrostatic dispersions where agglomerates are formed.

This comparison of GO and FLG highlights the key challenges now facing the field; functionalization definitely improves the interface of graphene, improving the interface sufficiently to compensate for the drop in modulus which the functionalization causes. (The modulus is inversely related to the cross-sectional area of the flake and the functional groups significantly increase the cross-sectional area of an atomically thin material, which is the predominant reason why the modulus of monolayer GO is less than that of monolayer graphene.) However in most cases to date, the functionalization of graphene appears to increase the rate of aggregation compared to unfunctionalized graphene. Thus a balance of the degree and type of functionalization is required, possibly using steric groups to increase the matrix interface and preventing further aggregation.

4 Conclusions

Electrochemically exfoliated few layer graphene with two different diameters (5 μm and 20 μm) and similar thicknesses were incorporated in a PP matrix by melt mixing using a twin screw extruder at loadings of 0 to 20 wt% followed by injection moulding. Increases in the crystallization temperature and degree of crystallinity with loading were found for the two sizes of graphene, which suggested that the flakes are acting as nucleation sites for the polymer. Mechanical testing showed that the 5 μm flakes behaved as short fibres and poorly reinforced the PP matrix compared to the 20 μm flakes, in line with previous predictions on the critical length of graphene. Significantly, these larger flakes gave a linear increase in the modulus even at high loadings, without any of the detrimental aggregation effects seen in most other graphene systems. This short fibre effect found in the graphene–PP system was in agreement with a previously studied graphene–PMMA system. The FLG-based systems were also compared with previously reported graphene oxide-based systems. The graphene oxide was found to provide a stronger interface with the polymer relative to the unfunctionalized or "clean" graphene, giving high levels of reinforcement with relatively low optimal loadings. At higher loadings the formation of agglomerates of graphene oxide in the matrix causes the mechanical properties of the material to deteriorate.

It is unlikely that graphene will compete with carbon fibre reinforced plastic (CFRP) composites in the short term due to the high degree of orientation and loading of carbon in CRFP composites. However, the modulus of the unoptimised 9 vol% 20-FLG–PP composite is the same as that of the corresponding randomly orientated E-glass fibre composite, as flakes are more efficient at forming isotropic systems than fibres. However, the FLG system weighs 6% less due to the low density of carbon. Therefore it is plausible that with future work on sizing the FLG, graphene could be a viable competitor to injection moulded glass-fibre composites for mass produced products.

Acknowledgements

The authors are grateful to the EPSRC (grant no: EP/I023879/1 and EP/K016946/1) and the European Union Seventh Framework Programme (grant no:604391 Graphene Flagship).

References

1 S. Park and R. S. Ruoff, *Nat. Nanotechnol.*, 2009, **4**, 217–224.
2 A. K. Geim, *Science*, 2009, **324**, 1530–1534.
3 A. K. Geim and K. S. Novoselov, *Nat. Mater.*, 2007, **6**, 183–191.
4 R. J. Young, I. A. Kinloch, L. Gong and K. S. Novoselov, *Compos. Sci. Technol.*, 2012, **72**, 1459–1476.
5 R. Verdejo, M. M. Bernal, L. J. Romasanta and M. A. Lopez-Manchado, *J. Mater. Chem.*, 2011, **21**, 3301–3310.
6 V. Singh, D. Joung, L. Zhai, S. Das, S. I. Khondaker and S. Seal, *Prog. Mater. Sci.*, 2011, **56**, 1178–1271.
7 S. Stankovich, D. A. Dikin, G. H. B. Dommett, K. M. Kohlhaas, E. J. Zimney, E. A. Stach, R. D. Piner, S. T. Nguyen and R. S. Ruoff, *Nature*, 2006, **442**, 282–286.
8 H. Kim, A. A. Abdala and C. W. MacOsko, *Macromolecules*, 2010, **43**, 6515–6530.
9 X. Huang, X. Qi, F. Boey and H. Zhang, *Chem. Soc. Rev.*, 2012, **41**, 666–686.
10 J. R. Potts, D. R. Dreyer, C. W. Bielawski and R. S. Ruoff, *Polymer*, 2011, **52**, 5–25.
11 R. J. Young and I. A. Kinloch, in *Nanoscience: Volume 1: Nanostructures through Chemistry*, The Royal Society of Chemistry, 2013, pp. 145–179.
12 L. Gong, I. A. Kinloch, R. J. Young, I. Riaz, R. Jalil and K. S. Novoselov, *Adv. Mater.*, 2010, **22**, 2694–2697.
13 T. M. G. Mohiuddin, A. Lombardo, R. R. Nair, A. Bonetti, G. Savini, R. Jalil, N. Bonini, D. M. Basko, C. Galiotis, N. Marzari, K. S. Novoselov, A. K. Geim and A. C. Ferrari, *Phys. Rev. B: Condens. Matter Mater. Phys.*, 2009, **79**, 205433.
14 L. Gong, R. J. Young, I. A. Kinloch, I. Riaz, R. Jalil and K. S. Novoselov, *ACS Nano*, 2012, **6**, 2086–2095.
15 T. K. Das and S. Prusty, *Polym.–Plast. Technol. Eng.*, 2013, **52**, 319–331.
16 Z. Li, R. J. Young and I. A. Kinloch, *ACS Appl. Mater. Interfaces*, 2013, **5**, 456–463.
17 C. I. Ferreira, C. Dal Castel, M. A. S. Oviedo and R. S. Mauler, *Thermochim. Acta*, 2013, **553**, 40–48.
18 K. Kalaitzidou, H. Fukushima, P. Askeland and L. Drzal, *J. Mater. Sci.*, 2008, **43**, 2895–2907.
19 K. Kalaitzidou, H. Fukushima and L. T. Drzal, *Composites, Part A*, 2007, **38**, 1675–1682.
20 Y. S. Yun, Y. H. Bae, D. H. Kim, J. Y. Lee, I.-J. Chin and H.-J. Jin, *Carbon*, 2011, **49**, 3553–3559.
21 B. Yuan, C. Bao, L. Song, N. Hong, K. M. Liew and Y. Hu, *Chem. Eng. J.*, 2014, **237**, 411–420.
22 S. H. Ryu and A. M. Shanmugharaj, *Mater. Chem. Phys.*, 2014, **146**, 478–486.
23 S. H. Ryu and A. M. Shanmugharaj, *Chem. Eng. J.*, 2014, **244**, 552–560.
24 P. Song, Z. Cao, Y. Cai, L. Zhao, Z. Fang and S. Fu, *Polymer*, 2011, **52**, 4001–4010.
25 M. El Achaby, F. E. Arrakhiz, S. Vaudreuil, A. El Kacem Qaiss, M. Bousmina and O. Fassi-Fehri, *Polym. Compos.*, 2012, **33**, 733–744.
26 F. You, D. Wang, X. Li, M. Liu, Z. M. Dang and G. H. Hu, *J. Appl. Polym. Sci.*, 2014, 131.

27 A. M. Abdelkader, I. A. Kinloch and R. A. W. Dryfe, *ACS Appl. Mater. Interfaces*, 2014, **6**, 1632–1639.
28 J. E. Mark, *Polymer Data Handbook*, Oxford University Press, New York, 1999.
29 J. Yang, X. Yan, M. Wu, F. Chen, Z. Fei and M. Zhong, *J. Nanopart. Res.*, 2012, 14.
30 J. N. Coleman, M. Cadek, R. Blake, V. Nicolosi, K. P. Ryan, C. Belton, A. Fonseca, J. B. Nagy, Y. K. Gun'ko and W. J. Blau, *Adv. Funct. Mater.*, 2004, **14**, 791–798.
31 Y. Mubarak, P. J. Martin and E. Harkin-Jones, *Plast., Rubber Compos. Process. Appl.*, 2000, **29**, 307–315.
32 F. Stan, L. I. Sandu and C. Fetecau, *Composites, Part B*, 2014, **59**, 109–122.
33 M. Moniruzzaman and K. I. Winey, *Macromolecules*, 2006, **39**, 5194–5205.

Faraday Discussions

PAPER

Do we need covalent bonding of Si nanoparticles on graphene oxide for Li-ion batteries?

Yana Miroshnikov, Gal Grinbom, Gregory Gershinsky, Gilbert D. Nessim and David Zitoun*

Received 2nd May 2014, Accepted 12th June 2014
DOI: 10.1039/c4fd00089g

In this manuscript, we report our investigation of anode materials for Li-ion batteries based on silicon–graphene oxide composites. Previous reports in the literature on silicon–graphene oxide (GO) composites as anodes have shown a large discrepancy between the electrochemical properties, mainly capacity and coulombic efficiency. In our research, the surface chemistry of Si nanoparticles has been functionalized to yield a chemical bond between the Si and GO, a further annealing step yields a Si–reduced GO (Si–rGO) composite while controlled experiments have been carried on mechanical mixing of GO and Si. For all samples, including a simple mixing of Si nanoparticles and GO, a high specific capacity of 2000 mA h g_{Si}^{-1} can be achieved for 50 cycles. The main difference between the samples can be observed in terms of coulombic efficiency, which will determine the future of these composites in full Li-ion cells. The Si–rGO composite shows a very low capacity fading and a coulombic efficiency above 99%. Furthermore, the Si–rGO composite can be cycled at very high rate to 20 C (charge in 3 minutes).

1. Introduction

Silicon is extensively investigated as an anode material for lithium-ion batteries (LIBs), due to its high theoretical capacity (4200 mA h g^{-1}) and its natural abundance.[1] However, silicon undergoes a large volume change of 300–400% upon cycling, which leads to electrode cracking and failure of the battery at some point.[2] Many strategies to overcome these challenges have been proposed, such as the use of nanoparticles dispersed in a binder, the preparation of composites and the use of amorphous Si.[3–5] Yet, all these strategies have shown their limitations and a practical solution for overcoming the problems associated with silicon anodes is far from being found.

Department of Chemistry and Bar Ilan Institute of Nanotechnology and Advanced Materials (BINA), Bar Ilan University, Ramat Gan, Israel. E-mail: David.zitoun@biu.ac.il

Graphene, a one atom thick, sp^2-bonded carbon layer, has been widely studied, due to its properties in terms of conductivity, flexibility and large surface area. Graphene can be produced through many different routes, such as CVD growth, epitaxial growth on SiC and graphene oxide (GO) reduction.[6–9] Reduced GO (rGO) remains flexible, displays chemical reactivity, can be dispersed in solvent and is conductive enough for electrochemical applications.[10–13]

Therefore, the elaboration of Si–rGO composite materials, where Si nanoparticles (NPs) are anchored on the rGO layers, might be a way to ease the effect of the large volume change. A few routes have been already explored to synthesize these nanocomposites using mixing procedures of either rGO or GO with Si NPs.[14–17] Several principles govern the elaboration of composites; Si NPs should be homogeneously spread on GO and the bonding in the composite has to be strong but flexible to accommodate for the volume changes of Si.[18] Research groups reported on composites of GO and Si NPs by mechanical mixing and obtained Si–rGO nanocomposite paper.[19] Others report on the chemical grafting of Si NPs on GO.[20–22] The nanocomposite shows better performance than the pristine Si NPs but the specific capacity rapidly fades after a few tens of cycles which draws serious concerns on the benefits of rGO in composite anodes. Moreover, the significantly large distributions of values found in the literature for the same composite commands a detailed study of the impact of each synthetic step on the electrochemical properties.

In this research, a Si NPs–GO composite is prepared by modification of the surface of Si NPs, covalent bonding to GO and further annealing to yield Si–rGO. The influence of each step is monitored by a full spectroscopic and microscopic characterization of composites. Each composite (before grafting Si/GO, after grafting Si–GO and after annealing Si–rGO) is tested as an anode material in Li-ion batteries to understand the significance of each chemical treatment. For all the samples, a high specific capacity has been achieved but significant differences in coulombic efficiency can be attributed to the chemical nature of the Si–GO bonds.

2. Materials and methods

Silicon nanoparticles of sub-50 nm size (nanopowder 98%, Aldrich), 3-aminopropyltriethoxysilane (APTES) (ACROS, 99%), graphite flakes (ASBURY carbons), mesitylene (FLUKA ≥99%), H_3PO_4 (ERBA 85%), H_2SO_4 (96%, FRUTAROM), $KMnO_4$ (ALDRICH ≥99.0%), and H_2O_2 (FISHER >30%) were used as received.

Synthesis of graphene oxide (GO)

GO was obtained by using an improved Hummers' method.[23] Briefly a 9 : 1 mixture of concentrated H_2SO_4–H_3PO_4 (120 : 13.3 mL) was added to a mixture of graphite flakes (1.0 g) and $KMnO_4$ (6.0 g) producing a slight exotherm to 35–40 °C. The reaction mixture was then heated to 50 °C and stirred for 12 hours. The mixture was cooled to room temperature and poured into ice (~133 mL) with 30% H_2O_2 (1.0 mL). The filtrate was centrifuged (4000 rpm for 4 hours) and the supernatant was decanted away. The remaining solid material was then washed in succession with DI water and ethanol. Then the obtained suspension was centrifuged (4000 rpm 30 min), the supernatant was decanted away and the filtrate was vaccum-dried overnight. A purified graphite oxide was dispersed in water and bath sonicated for 30 min to obtain graphene oxide (GO).

Grafting of APTES (aminopropyltriethoxysilane) on the silicon nanoparticles surface

0.3 g of Si nanopowder is dispersed by magnetic stirring in 20 mL of mesitylene degassed under N_2. 0.318 mL of APTES is added dropwise at room temperature and stirred at 170 °C under N_2 for 4 hours. The dispersion is centrifuged at 6000 rpm for 10 min to precipitate the brown colloids and redispersed in ethanol; the procedure is repeated twice to remove the excess of silane. The sample is then dried under vacuum for 12 hours.

Synthesis of GO and reduced GO (rGO)–silicon composites

A few drops of HNO_3 (0.078 mol L^{-1}) are added to APTES-terminated silicon nanoparticles dispersed in water, and the pH is brought to ~3. A GO dispersion in water is added to the silicon dispersion. The dispersion is heated to 80 °C and stirred for 72 hours. The mixture is cooled to room temperature and freeze-dried at −105 °C overnight to yield the Si–GO compound. The eventual reduction step is achieved by placing the dry sample in a tubular furnace under forming gas (Ar/H_2 95/5, 99.999%). The sample is heated to 1000 °C using a one hour ramp heating procedure and maintaining the temperature for 5 min at 1000 °C, to yield a reduced GO–silicon (Si–rGO) composite. The as prepared samples are transferred and stored in a glove box filled with argon.

Alternatively, a mixture of pristine Si NPs is used in place of APTES-terminated Si NPs following the same procedure with GO (heating in mesitylene under N_2 at 170 °C for 4 hours). The resulting brown nanopowder is named Si/GO assuming a non-covalent bonding between Si and GO.

Characterization

The structure of silicon NPs is investigated by transmission electron microscopy (TEM) on a FEI Tecnai-12 operated at 120 kV or a JEOL JEM-1400 (LaB_6) at 120 kV. The sample is prepared by drop-casting a sonicated dispersion in ethanol on a Lacey carbon grid.

Raman spectra of the samples are collected on a Jobin-Yvon Horiba Raman spectrometer equipped with an Olympus BX41 microscope (×100 lens) and the Raman spectrometer uses a red laser source.

FT-IR spectra of the samples are collected on a Thermo Scientific Nicolet IS20 in ATR mode on a diamond window in the 4000–400 cm^{-1} region.

Electrodes of nanocomposites

All the work has been achieved in an Ar-filled glovebox. The dry composite powder has been laminated onto a roughened copper foil (Oxygenfree, SE-Cu58, Schlenk Metallfolien GmbH & Co. KG). The powder has been equally spread on the surface and adhesion to the current collector has been enhanced by applying a manual pressure with a Kimwipes paper. For the electrochemical half-cell studies of the composite, we have used an electrode with surface area of 1 cm^2.

Coin cells

Anodes are tested in coin-type cells (2523, NRC, Canada) *vs.* lithium metal (Chemetall Foote Corporation, USA). The electrolyte solution is a mixture of fluoro-ethylene carbonate (FEC, Aldrich, <20 ppm water) and dimethyl carbonate (DMC, Aldrich, <20 ppm water) (1 : 4 ratio) with 1 mol L^{-1} $LiPF_6$ (Aldrich). The cells are assembled in an argon-filled glove box, with a purifying system (MBraun GmbH, Germany), oxygen and water contents are below 1 ppm.

Coin-cells are cycled at 30 °C using BT2000 battery cycler (Arbin Instruments, USA) (three cycles at C/10, therefore C rate: all capacity in 1 hour). The active mass of the silicon anodes is measured by dissolving the Si and dosing with ICP-AES three times.

3. Results and discussion

The synthesis of three samples of Si nanoparticles–graphene oxide (Si NPs–GO) composites is achieved following the experimental procedure described in the experimental section. The first sample, named Si/GO, is obtained by simple mixing of the two nanomaterials, and should not present any covalent bonding between the GO sheets and the Si surface. During the synthesis and centrifugation, this sample clearly shows two different phases, according to their difference in color: the dark upper phase of graphene oxide and a brown lower phase of the Si NPs. Another sample, named Si–GO, is synthesized by covalently bonding between GO sheets and Si surface. This sample is prepared by reacting APTES (aminopropyltriethoxysilane)-grafted Si nanoparticles with GO sheets under reflux in mesitylene, yielding a nucleophilic attack of the amine group on the epoxy and carboxylic groups of the GO. During the synthesis, this sample forms a single continuous brownish phase. The as-prepared Si–GO sample can be annealed in a reducing atmosphere to produce a black powder of silicon reduced graphene oxide, a sample named Si–rGO.

APTES, a widely used grafting agent,[24] is chosen for the functionalization of the Si surface. This compound yields a propylamine chain, short enough for electron hopping between GO and Si. Moreover, the siloxane group can easily bind to the native oxide surrounding the Si NPs through a sol–gel process, while the amine group can react on the epoxy and carboxyl groups present in GO to yield a covalent bond. Scheme 1 describes the detailed synthesis route for the preparation of the covalently bonded Si–GO composite. Only this sample has been annealed to reduce the GO and the SiO_2 shell since the Si/GO composite without covalent bonds does not show a good homogeneity.

Fig. 1 presents the TEM images of Si NPs after APTES grafting (A and B), graphene oxide (C) and an AFM image of graphene oxide (D). Si NPs are dispersed on the TEM grid, and tend to agglomerate into large aggregates. Electron diffraction analysis exhibits the typical diffraction of the cubic crystal structure of silicon ($Fd\bar{3}m$, a = 5.4309 Å, JPCDS 27-1402). The GO sheets exhibit a typical wrinkled morphology, with sheets larger than 3 μm (from TEM observation) and average thickness of 1.5 nm (from AFM images). This suggests that the as prepared GO are three layers thick. Fig. 2 displays the TEM images of the as prepared composites, Si–GO (A–C), Si–rGO (D and E) and Si/GO mix (F). The Si–GO sample (A and B) reveals the presence of GO sheets wrapped up on Si

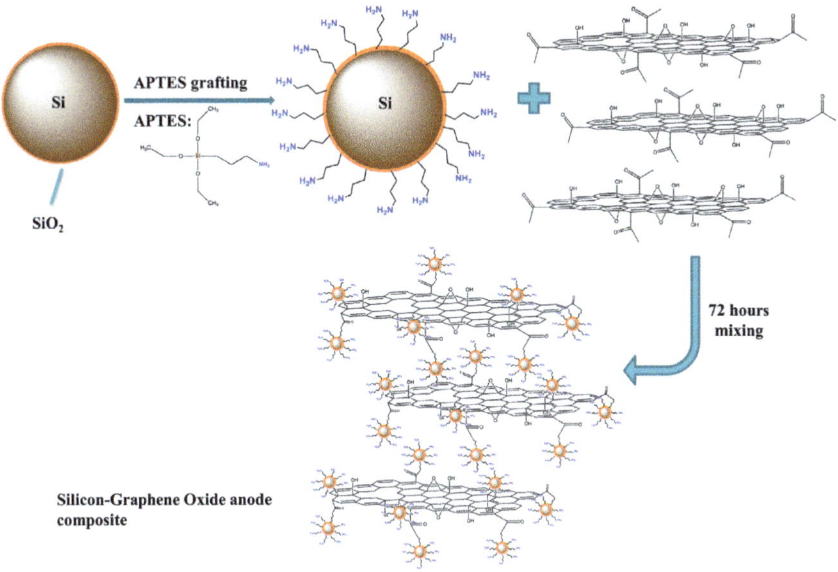

Scheme 1 The detailed synthesis for the preparation of the Si–GO anode composite.

nanoparticles and the electron diffraction shows only the cubic crystalline phase of Si nanoparticles; the Si NPs still form agglomerates which are attached to the GO sheets and dispersed on their surface. Dark field TEM (Fig. 2C) reveals crystalline Si spread on amorphous GO sheets. The Si–GO sample slightly differs from

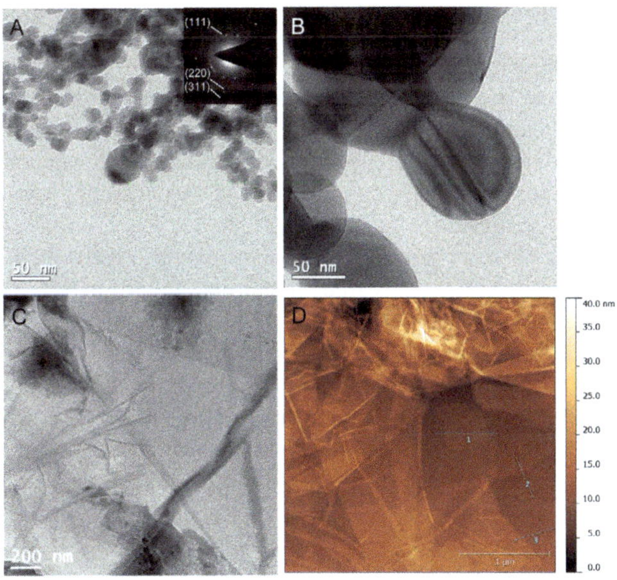

Fig. 1 TEM images of Si after APTES grafting (A and B), graphene oxide (C); AFM image of graphene oxide (D).

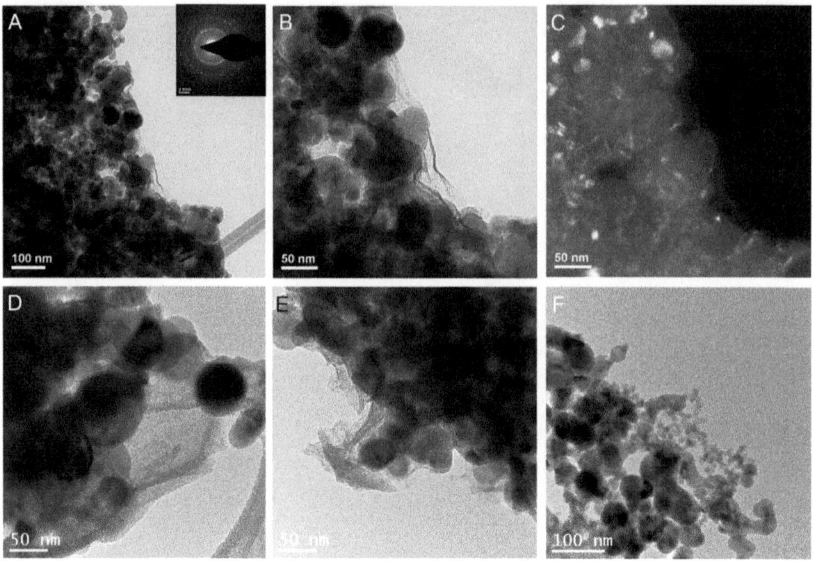

Fig. 2 TEM images of Si–GO composite (A and B), dark field (C), Si–rGO composite (D and E), Si/GO mix (F).

the scheme presented on Fig. 1 since the GO sheets follow the morphology of the Si NPs and wrap up the particles. Si–rGO sample also shows rGO sheets wrapped up on Si (Fig. 2C and D). The morphology of the two samples can hardly be distinguished, allowing us to compare the two samples assuming a simple chemical reduction of the GO and the Si surface, while the overall morphology remains. In Si/GO mixed sample, graphene oxide sheets are not seen on the entire TEM grid and only Si nanoparticles are spread on the TEM grid (Fig. 2F).

Raman spectroscopy of carbon materials is an excellent technique to measure the quality of the carbonaceous material synthesized; it can be used for the characterization of the graphitic carbon (G-band) and disordered carbon (D-band). The G-band corresponds to sp^2 bonds of graphitic carbon, and the D-band is related to defects. Fig. 3 summarizes the Raman spectra of the Si–GO and Si–rGO composite, while pristine GO and surface-modified Si NPs are displayed as control samples.

The peaks at 1330 cm^{-1} and 1590 cm^{-1} correspond to the D band and the G band, respectively, indicating the carbon fingerprint and the defects arising from the harsh oxidation, a typical GO spectrum. However, we can see that Si–GO and Si–rGO also exhibit a peak at 520 cm^{-1}, which corresponds to the presence of crystalline Si in the produced composite. It is important to mention, that after thermal annealing of the Si–GO sample, the intensity ratio of I_D/I_G has increased from 1.29 for Si–GO to 1.39 for Si–rGO, indicating the reduction of GO to rGO.[25–28]

Fig. 4 presents the FTIR spectra of APTES-treated Si nanoparticles (Si–NH_2), Si and GO without covalent bonding (Si/GO), Si graphene oxide composite (Si–GO) and GO. The large peaks at 1093 cm^{-1} that are present in Si–NH_2, Si–GO and Si/GO samples are assigned to Si–O–Si stretching vibrations, which indicates the SiO_x presence on the surface of the Si nanoparticles. The GO sample spectrum

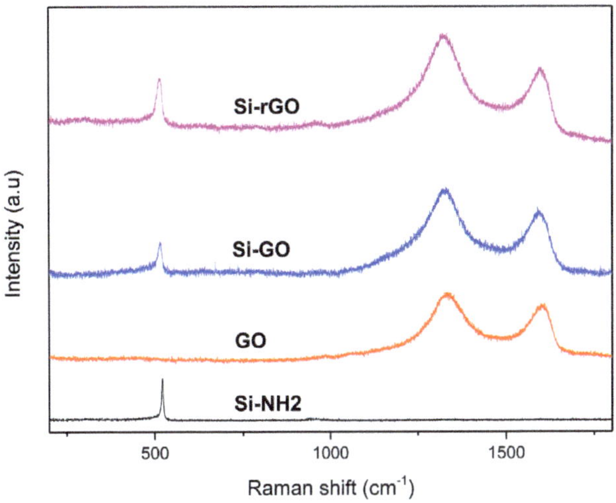

Fig. 3 Raman spectra of Si treated with APTES (Si–NH_2), graphene oxide (GO), silicon–graphene oxide composite (Si–GO) and silicon–reduced graphene oxide composite (Si–rGO).

displays a peak at 1757 cm^{-1} attributed to carboxylic groups (-COOH). A broad peak at ~1200 cm^{-1} can be attributed to the epoxy species present in the GO and a peak at 1511 cm^{-1} indicates the presence of alcohols. The Si–GO sample shows two characteristic peaks at 1550 cm^{-1} and 1690 cm^{-1} for CNH stretch and C=O amide group. These peaks indicate the presence of the covalent bonds formed during the grafting. The FTIR spectrum of Si–rGO shows a significant decrease of the Si–O–Si band which can be attributed to the reduction of the native oxide. t is worth noting that a new band at ~1400 cm^{-1}, consistent with a Si–C bond, could explain the passivation of the sample by the formation of a silicon carbide layer, as demonstrated on the nanocomposite of SiO_2/C submitted to a reducing atmosphere.[29] The peak at ~1635 cm^{-1} can be attributed to the remaining chemical groups in reduced GO.[30]

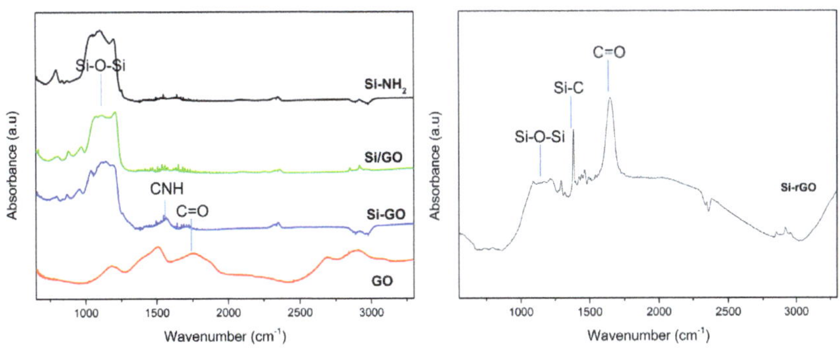

Fig. 4 FTIR spectra of graphene oxide (GO), silicon–graphene oxide composite (Si–GO), Si mixed with GO (Si/GO), Si treated with APTES (Si–NH_2).

All the materials are tested as anodes in Li-ion half cells *vs.* Li metal. The electrode is prepared following the procedure reported in the experimental part. In brief, the composite is spread on a roughened Cu foil as a thin film and inserted in a coin cell *vs.* Li metal with an electrolyte solution of $LiPF_6$ 1 mol L^{-1} in fluoroethylene carbonate–dimethylcarbonate (FEC–DMC), 1 : 4 ratio. Cycling is performed at 1 C (charge in 1 hour and discharge in 1 hour) and at a controlled temperature of 30 °C.

Fig. 5 presents the specific capacities measured for the Si–GO composite, Si–rGO composite, Si/GO mix and the coulombic efficiency of each sample. All capacities are normalized to the mass of Si since our controlled experiment on pure GO do not show any reversible electrochemical activity of GO in the window of cycling potential (0–0.8 V). All composites exhibit high specific capacity above 1400 mA h g^{-1}. The composite prepared by mixing demonstrates the best capacity after 50 cycles (almost 2000 mA h g^{-1}) but rapidly fades to show poorer performances than the Si–GO covalently bonded sample after 300 cycles. The same trend can be observed with Si–GO compared with Si–rGO. Obviously, the GO stands as a good substrate for the Si and the direct bonding mitigates the capacity fading. The coulombic efficiency of the Si–rGO composite allows a quantification of this observed trend. All coulombic efficiencies are higher than 97%. Nevertheless, the coulombic efficiency of Si–rGO rapidly reaches 99% after 50 cycles while Si–GO reaches a slightly lower value after 150 cycles. Si/GO shows a better efficiency for 30 cycles and poorer performances than the other composites after 100 cycles.

Fig. 6 presents the specific capacity measured for Si–rGO composites for more than 1000 cycles. As observed previously, the Si–rGO composites demonstrate a very stable specific capacity after the first 3 cycles, the value stabilizes at 1400 mA h g^{-1} after 100 cycles until 1000 cycles. The batteries continue their cycle life further without any noticeable capacity fading. The batteries have maintained their capacity for more than 1000 cycles.

The Si–rGO composites have been tested at different speeds of charge–discharge cycling. This capacity rate has been tested after 1000 cycles. Fig. 7 displays the charge capacities and voltage profile of the Si–rGO composite at different rates from C/10 (10 hours) up to 60 C (1 min). The composite exhibits a stable capacity at all the given rates. At 20 C, full charge and discharge taking 3 minutes each, though the capacity has dropped to 600 mA h g^{-1}, the composite

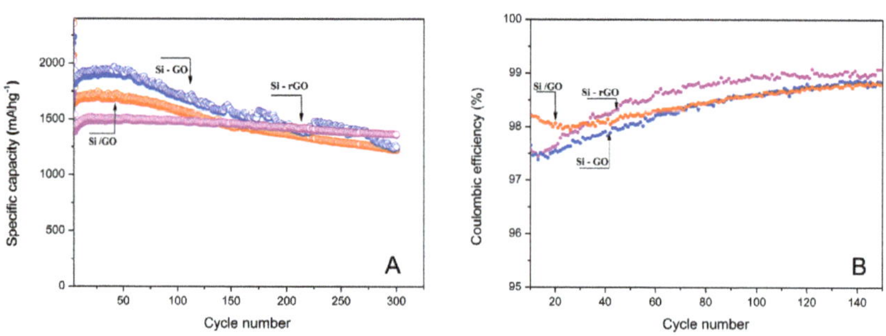

Fig. 5 Cycling performance of Si–GO and Si–rGO composites, compared to the Si/GO mix (a); coulombic efficiency of Si–GO, Si–rGO and Si/GO (b).

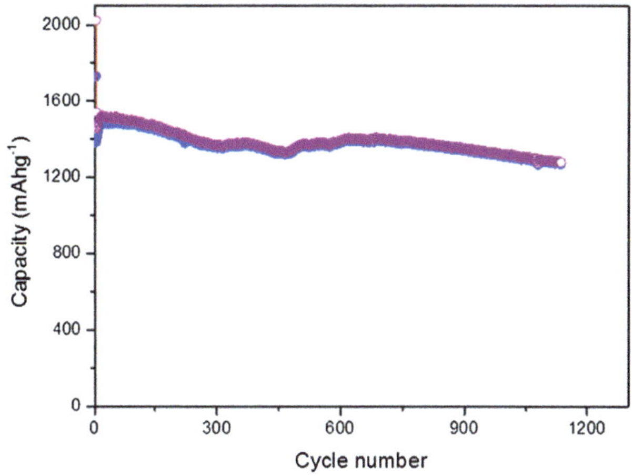

Fig. 6 Gravimetric capacity of rGO–Si composite at 1 C rate.

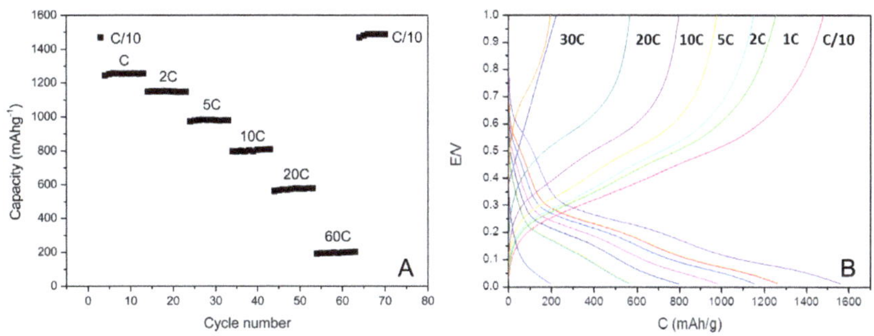

Fig. 7 Rate capability of Si–rGO composite (a) at different rates: C/10, 1 C, 2 C, 5 C, 10 C, 20 C, 60 C; voltage profile of the composite (b).

still outperforms graphite by a factor of two and the anode material demonstrates reversible performance. In a logarithmic scale, the capacity decreases linearly with the rate until 20 C. Beyond this rate (60 C), the capacity significantly drops to less than 200 mA h g^{-1}. Interestingly, when the rate decreases to C/10, full charge and discharge taking 10 hours each, the battery capacity is fully restored. The voltage profile of rGO–silicon exhibits the typical characteristics of Si undergoing lithiation and delithiation, showing two plateaux of potential. At voltages between 0.7 V and 0.3 V another peak, which can be attributed to the rGO, is present.

The first significant result from our investigation is the TEM observation of complete segregation of GO and Si on the TEM grid following a simple Si/GO mix even when performed under the same conditions of temperature and solvent as the covalent grafting (Fig. 2F). However, the grafting of amine function on the surface of Si and the further formation of amides on GO yields a composite of nicely dispersed Si NPs on the GO substrate.

All the samples studied have shown very good performances as Li-ion anodic materials for 50 cycles, with very high specific capacity compared to the values reported in the literature. From the capacity measurements (Fig. 5), Si–GO exhibits the highest gravimetric capacity, around 2000 mA h g^{-1} after 50 cycles. The best composite reported in the literature reaches an initial discharge capacity of about 1750 mA h g^{-1} and 1000 mA h g^{-1} after 200 cycles.[21,22]

Beyond this number of cycles, the performace of Si–GO is getting closer to the Si/GO without covalent bonding, showing a capacity fading. This higher value obtained for the Si–GO composite might be attributed to the fact that Si nanoparticles are evenly distributed on GO which even wrap up the Si NPs. The Si NPs are enclosed in the GO, as shown I the TEM images in Fig. 2A–C, while a simple mixing does not provide a nanoscale mixing of Si NPs and GO. The spectroscopic evidences (Raman and FTIR) for the formation of a composite underlines the formation of an amide bond between the amine-grafted particles and the reactive species of GO (epoxy and carboxylic groups). This amide bond is strong enough to graft the Si NPs on top of GO, or reciprocally GO on top of Si NPs. Nevertheless, this amide bond is also sensitive to the chemical potential. While the anode is cycled between 0 and 1.0 V *versus* Li, amide may react at such a negative potential (−3.0 to −2.0 V *vs.* ENH). One interpretation of the capacity fading could be the reactivity of the amide bond upon cycling, partially releasing the Si from its GO substrate. This bond breaking would explain the behavior of Si–GO which resembles Si/GO after 100 cycles in terms of specific capacity and coulombic efficiency.

Therefore, covalent bonding appears to be an essential synthetic step to graft the electrochemically active Si on its GO substrate. While this step provides an effective composite to 50 cycles, stronger chemical bonds are needed to endure the harsh conditions of potential in the battery.

Therefore, we use the covalent grafting followed by an annealing step to yield Si–rGO composites. As shown by TEM on Fig. 2D and E, the Si NPs are now completely wrapped up by the carbonaceous material which provides a robust matrix. After the annealing of the Si–GO composite in a reducing atmosphere, the aromatic network of GO is partially restored and the defects' density is diminished, as evidenced by Raman spectroscopy (Fig. 3). The annealing step therefore benefits to the 2D order of GO (now rGO) and should improve its electronic properties as reported for similar annealing treatments.[31]

Si–rGO composites retain their capacity as shown from the high values of coulombic efficiency reported in Fig. 5 (>99%). Si particles are scattered on the flexible, partially conductive rGO matrix with a large amount of void spaces, which prevents the cracking of the electrode after the volume expansion of Si during lithiation. Moreover, the concentration of Si nanoparticles on each sheet is quite high (Fig. 2D and E), which might also produce higher conductivity and prevent contact loss.

The Si–rGO composite shows very stable and reversible performance for more than 1200 cycles at a 1C rate (Fig. 6). The composite also shows stable and reversible capacities at different C rates (Fig. 7) and its capacity is fully restored at C/10 rate. On a logarithmic scale, the capacity decreases linearly with the rate to 20 C. At this rate, which corresponds to a charge of 3 minutes, the weight capacity per silicon is still twice that of graphite, which is the standard anode used in commercial Li-ion batteries. Moreover, the coulombic efficiency remains stable and higher than 99% during the rate capability measurements.

4. Conclusions

In this work, we demonstrate the synthesis of composites of graphene oxide (GO) and Si nanoparticles (NPs) as anodic materials for Li-ion batteries. The synthetic pathway makes use of the chemical functions of GO and the amine grafted on Si NPs by a sol–gel process to yield a covalently bond Si–GO. In the absence of this treatment, the two materials show segregation on the microscale. The Si–GO composite shows high specific capacities of 2000 mA h g^{-1} after 50 cycles compared to 1600 mA g^{-1} for the mixed sample. The better capacity is attributed to the homogeneous distribution of Si NPS on GO. A further annealing step is needed to improve the capacity beyond 50 cycles. The Si–rGO composite shows state-of-the-art capacity of 1200 mA h g^{-1} after 1000 cycles with a very low capacity fading and a coulombic efficiency above 99%. Furthermore, the Si–rGO composite can be cycled at a very high rate and shows high specific capacity even at 20 C (charge in 3 minutes).

Acknowledgements

The authors thank Luba Burlaka for her assistance with TEM and Dr Andras Paszternak for the AFM measurements. This work was supported by the I-CORE program of the planning and budgeting committee and the Israel Science Foundation (2797/11).

References

1 D. Larcher, S. Beattie, M. Morcrette, K. Edström, J.-C. Jumas and J.-M. Tarascon, *J. Mater. Chem.*, 2007, **17**, 3759.

2 M.-S. Zheng, Q.-F. Dong, H.-Q. Cai, M.-G. Jin, Z.-G. Lin and S.-G. Sun, *J. Electrochem. Soc.*, 2005, **152**, A2207–A2210.

3 H. Wu, G. Yu, L. Pan, N. Liu, M. T. McDowell, Z. Bao and Y. Cui, *Nat. Commun.*, 2013, **4**, 1943.

4 J. Luo, X. Zhao, J. Wu, H. D. Jang, H. H. Kung and J. Huang, *J. Phys. Chem. Lett.*, 2012, **3**, 1824–1829.

5 E. Markevich, K. Fridman, R. Sharabi, R. Elazari, G. Salitra, H. E. Gottlieb, G. Gershinsky, A. Garsuch, G. Semrau, M. A. Schmidt and D. Aurbach, *J. Electrochem. Soc.*, 2013, **160**, A1824–A1833.

6 S. Mao, H. Pu and J. Chen, *RSC Adv.*, 2012, **2**, 2643.

7 D. C. Marcano, D. V. Kosynkin, J. M. Berlin, A. Sinitskii, Z. Sun, A. Slesarev, L. B. Alemany, W. Lu and J. M. Tour, *ACS Nano*, 2010, **4**, 4806–4814.

8 Y. Zhu, S. Murali, W. Cai, X. Li, J. W. Suk, J. R. Potts and R. S. Ruoff, *Adv. Mater.*, 2010, **22**, 3906–3924.

9 D. R. Dreyer, S. Park, C. W. Bielawski and R. S. Ruoff, *Chem. Soc. Rev.*, 2010, **39**, 228–240.

10 B. Wang, X. Li, X. Zhang, B. Luo, M. Jin, M. Liang, S. A. Dayeh, S. T. Picraux and L. Zhi, *ACS Nano*, 2013, **7**, 1437–1445.

11 T. Kuila, A. K. Mishra, P. Khanra, N. H. Kim and J. H. Lee, *Nanoscale*, 2013, **5**, 52–71.

12 C. Fu, G. Zhao, H. Zhang and S. Li, *Int. J. Electrochem. Sci.*, 2013, **8**, 6269–6280.

13 D. Pan, S. Wang, B. Zhao, M. Wu, H. Zhang, Y. Wang and Z. Jiao, *Chem. Mater.*, 2009, **21**, 3136–3142.
14 K. Eom, T. Joshi, A. Bordes, I. Do and T. F. Fuller, *J. Power Sources*, 2014, **249**, 118–124.
15 H.-C. Tao, L.-Z. Fan, Y. Mei and X. Qu, *Electrochem. Commun.*, 2011, **13**, 1332–1335.
16 X. Zhou, A.-M. Cao, L.-J. Wan and Y.-G. Guo, *Nano Res.*, 2012, **5**, 845–853.
17 R. C. Guzman, J. Yang, M. M.-C. Cheng, S. O. Salley and K. Y. Simon Ng, *J. Mater. Sci.*, 2013, **48**, 4823–4833.
18 H. Xiang, K. Zhang, G. Ji, J. Y. Lee, C. Zou, X. Chen and J. Wu, *Carbon*, 2011, **49**, 1787–1796.
19 J. K. Lee, K. B. Smith, C. M. Hayner and H. H. Kung, *Chem. Commun.*, 2010, **46**, 2025–2027.
20 M. Zhou, F. Pu, Z. Wang, T. Cai, H. Chen, H. Zhang and S. Guan, *Phys. Chem. Chem. Phys.*, 2013, **15**, 11394–11401.
21 G. Zhao, L. Zhang, Y. Meng, N. Zhang and K. Sun, *J. Power Sources*, 2013, **240**, 212–218.
22 Y.-S. Ye, X.-L. Xie, J. Rick, F.-C. Chang and B.-J. Hwang, *J. Power Sources*, 2014, **247**, 991–998.
23 D. C. Marcano, D. V. Kosynkin, J. M. Berlin, A. Sinitskii, Z. Sun, A. Slesarev, L. B. Alemany, W. Lu and J. M. Tour, *ACS Nano*, 2010, **4**, 4806–4814.
24 A. Simon, T. Cohen-Bouhacina, M. C. Porté, J. P. Aimé and C. Baquey, *J. Colloid Interface Sci.*, 2002, **251**, 278–283.
25 K. N. Kudin, B. Ozbas, H. C. Schniepp, R. K. Prud'homme, I. a. Aksay and R. Car, *Nano Lett.*, 2008, **8**, 36–41.
26 X. Díez-Betriu, S. Álvarez-García, C. Botas, P. Álvarez, J. Sánchez-Marcos, C. Prieto, R. Menéndez and A. de Andrés, *J. Mater. Chem. C*, 2013, **1**, 6905.
27 A. C. Ferrari and D. M. Basko, *Nat. Nanotechnol.*, 2013, **8**, 235–246.
28 S. Piscanec, A. C. Ferrari, M. Cantoro, S. Hofmann, J. a. Zapien, Y. Lifshitz, S. T. Lee and J. Robertson, *Mater. Sci. Eng., C*, 2003, **23**, 931–934.
29 P.-C. Gao, Y. Lei, A. F. Cardozo Pérez, K. Rajoua, D. Zitoun and F. Favier, *J. Mater. Chem.*, 2011, 15798–15805.
30 H. Zhang, D. Hines and D. L. Akins, *Dalton Trans.*, 2014, **43**, 2670–2675.
31 H. F. Xiang, Z. D. Li, K. Xie, J. Z. Jiang, J. J. Chen, P. C. Lian, J. S. Wu, Y. Yu and H. H. Wang, *RSC Adv.*, 2012, **2**, 6792.

Faraday Discussions

PAPER

Reduced graphene oxide and graphene composite materials for improved gas sensing at low temperature†

Alexander Zöpfl,[a] Michael-Maximilian Lemberger,[a] Matthias König,[b] Guenther Ruhl,[b] Frank-Michael Matysik[a] and Thomas Hirsch*[a]

Received 30th April 2014, Accepted 27th June 2014
DOI: 10.1039/c4fd00086b

Reduced graphene oxide (rGO) was investigated as a material for use in chemiresistive gas sensors. The carbon nanomaterial was transferred onto a silicon wafer with interdigital gold electrodes. Spin coating turned out to be the most reliable transfer technique, resulting in consistent rGO layers of reproducible quality. Fast changes in the electrical resistance at a low operating temperature of 85 °C could be detected for the gases NO_2, CH_4 and H_2. Especially upon adsorption of NO_2 the high signal changes allowed a minimum detection of 0.3 ppm ($S/N = 3$). To overcome the poor selectivity, rGO was chemically functionalized with octadecylamine, or modified by doping with metal nanoparticles such as Pd and Pt, and also metal oxides such as MnO_2, and TiO_2. The different response patterns for six different materials allowed the discrimination of all of the test gases by pattern recognition based on principal component analysis.

1 Introduction

The simple and reliable monitoring of gas concentrations is important in everyday-life. In industrial processes hazardous gases need to be controlled to guarantee safety. Controlling air quality can save energy in automated air conditioning. Also the detection of environmental pollution (like NO_x) is of great interest.[1] Up to now, solid-state gas sensors based on metal oxide chemiresistors are well established and widely used in detecting gases.[2,3] They are operated at high temperatures,[4] which consumes excessive energy and limits their long-term stability, thus leading to the development of new gas sensor concepts which overcome these drawbacks. In the last decade carbon nanomaterials, like carbon nanotubes[5,6] or graphene,[7–10] were shown to be capable of gas detection due to

[a]Institute of Analytical Chemistry, Chemo- and Biosensors, University of Regensburg, Universitaetsstrasse 31, 93053 Regensburg, Germany. E-mail: alexander.zoepfl@ur.de; michael.lemberger@ur.de; frank-michael.matysik@ur.de; thomas.hirsch@ur.de

[b]Infineon Technologies AG, 93049 Regensburg, Germany. E-mail: Guenther.Ruhl@infineon.com; Matthias.Koenig@infineon.com

† Electronic supplementary information (ESI) available. See DOI: 10.1039/c4fd00086b

their high sensitivity to various gases even at low operating temperatures. Especially graphene has recently attracted intense attention because of its high charge carrier mobility (15 000 cm^2 V^{-1} s^{-1} under ambient conditions) and large surface area as a consequence of having a thickness of only one atom and the absence of any bulk phase.[11,12]

Many different preparation methods for graphene are known so far, but only a few of them are applicable in terms of sensor preparation. Chemically derived graphene obtained by the reduction of graphene oxide (GO)[13,14] is inexpensive and the synthesis is easily scalable. Due to a more defective structure its electrical properties are not as outstanding as those of pristine graphene,[15,16] but are still suitable for sensitive gas detection.[17] It has also been described that graphene with a more defective structure shows an improved adsorption of gas molecules.[18] Furthermore, reduced graphene oxide (rGO) can be dispersed in solutions, simplifying the transfer to a sensor set-up by *e.g.* spraying, printing and casting methods.[19,20]

Similar to solid state gas sensors, gas adsorption on graphene leads to a change in its electrical resistance (Fig. 1). Therefore, any kind of interaction between graphene sheets and adsorbates influences the electronic structure of the graphene, leading to an altered charge carrier concentration or electrical conductance of the material, respectively.[17,21] Nevertheless, up to now most of such sensor concepts have been demonstrated only in inert gas mixtures or at elevated temperatures. For any practical application it is necessary to overcome the limitations of poor selectivity.[7] Many works address this issue by chemically modifying the carbon material with functional groups or doping with metals and metal oxides. Graphene composite materials have been used in many different sensing applications and can be easily produced either by simply mixing two different materials together or by *in situ* nanoparticle growth on graphene in a suspension.[22–24]

Here we report on rGO as a sensitive material for gas detection and the possibilities of functionalization to introduce selectivity. Reduced graphene oxide

Fig. 1 Scheme of the principle of chemiresistor gas sensors based on graphene.

was prepared by the chemical reduction of GO, which was obtained from the oxidation of graphite. The resulting product can be dispersed in water, enabling an easy transfer to pre-structured microelectrodes comprising an interdigital structure. Application of rGO *via* spin coating was optimized and resulted in consistent layers of reproducible quality in terms of the electrical properties. The conductance of such modified electrodes was measured in the presence of various gases diluted in synthetic air (NO_2, CH_4, and H_2) at moderate temperatures (85 °C). Chemical modifications were applied by the insertion of functional groups and by doping with metals and metal oxides. The resulting materials were characterized and tested for their gas sensing behaviour. It was demonstrated that a combination of simple modifications of rGO can lead to sensors with different characteristics, allowing a pattern detection of various gases.

2 Experimental

2.1 Materials and instrumentation

Unless otherwise stated, all chemicals were purchased from Merck (Darmstadt, Germany) or Sigma-Aldrich (Steinheim, Germany) and used without further purification. Ultra-pure water (0.055 μS cm^{-1}) was used in all experiments. All gases and mixtures were purchased from Linde AG.

Raman spectra were recorded on a Thermo Scientific DXR Raman microscope with a 532 nm excitation laser (10 mW). Scanning electron microscopy (SEM) was performed with a Zeiss Ultra 55, EDX on a JOEL JSM-6700F and TGA-FTIR with a Netzsch Iris TG209 connected to a Bruker Equinox 55, all at Infineon Technologies AG Regensburg. Transmission electron microscopy (TEM) was performed using a 120 kV Philips CM12 microscope.

2.2 Synthesis of reduced graphene oxide and modifications

GO was synthesized using a slightly modified Hummers method.[13] Briefly, 100 mg graphite (China flake graphite, K. W. Thielmann & Cie KG) was mixed with 75 mg $NaNO_3$, 7.5 mL conc. H_2SO_4 and 450 mg $KMnO_4$. The mixture was sonicated for 3 h and stirred for 3 days at room temperature. Subsequently, 7.5 mL of 5% H_2SO_4 was added and the mixture was stirred under reflux for 2 h at 100 °C, followed by the addition of 1.5 mL of 30% H_2O_2 under constant stirring for 1 h at room temperature. For purification the obtained GO was washed with the following solutions: four times in 3% H_2SO_4 with 0.5% H_2O_2; two times in 3% HCl; and three times in water. The product was dialyzed against water (14 kDa cut-off) for 10 days.

To obtain rGO,[14] 7 mL of a GO suspension (0.5 mg mL^{-1}) was mixed with 31 μL NH_3 (32% in water). After adding 5 μL of 98% hydrazine hydrate the reaction mixture was refluxed for 1 h at 100 °C. Finally the resulting black suspension was washed with water.

Modification with octadecylamine

To modify rGO with octadecylamine (ODA), the synthesis of surface functionalized graphene by Wang *et al.* was adapted.[25] 5 mg GO was suspended in 5 mL dichlorobenzene. After adding 50 mg ODA, the reaction mixture was sonicated for 24 h at 80 °C. To get rid of excess ODA and dichlorobenzene, the reaction mixture was treated with 40 mL ethanol resulting in a black precipitate. The mixture was centrifuged and the resulting solid was washed twice with ethanol and toluene. This washing step was repeated once more and the product (rGO–ODA) was dried in air.

Modification with MnO_2

A synthesis for GO–MnO_2 nano composites was adapted.[26] 5.1 mg GO and 21 mg $MnCl_2 \cdot 4H_2O$ were dispersed in 5 mL isopropanol and sonicated for 30 min. The reaction mixture was refluxed at 83 °C under vigorous stirring. Afterwards a solution of 12 mg $KMnO_4$ in 0.4 mL water was added and the slurry was refluxed at 83 °C for another 30 min before it was cooled to room temperature. The nano composite was then centrifuged and washed twice with water. The resulting brown product (rGO–MnO_2) was dried at 60 °C overnight.

Modification with TiO_2

For the functionalization of carbon nanomaterials Li *et al.* have reported the modification of single-walled carbon nanotubes with TiO_2.[27] In an analogous way rGO was modified with TiO_2 (rGO–TiO_2): 1 mg of TiO_2 with an average particle size of 1 μm was ground thoroughly. Subsequently the powder was dispersed in 2 mL water, sonicated for 1 h and 1 mL of an aqueous rGO suspension (0.5 mg mL^{-1}) was added. After another sonication for 1 h a grey suspension was formed where the modified rGO flakes tended to aggregate.

Modification with Pd and Pt nanoparticles

Both approaches were adapted from the preparation of graphene–metal nanocomposites by Xu *et al.* and modified slightly.[28] 5 mL of a GO suspension (1 mg mL^{-1}) was mixed with 10 mL ethylene glycol. Afterwards 0.5 mL of a 0.01 M solution of a metal precursor (K_2PtCl_4 or $PdCl_2$) in water was added and the mixture was first stirred for 30 min at room temperature and then for 6 h at 100 °C. During the heating in both cases the colour became black indicating that the GO had been reduced by ethylene glycol. The resulting suspensions containing the metal doped carbon nanomaterials (rGO–Pd and rGO–Pt) were centrifuged for 5 min at 3000 rpm and washed three times with water.

2.3 Electrode preparation

Suspensions of rGO and the modifications were deposited on gold microelectrodes (Fig. 2) *via* spin coating. The respective suspensions in a 1 : 1-mixture of water and isopropanol (0.25 mg mL^{-1}) were sonicated for 10 min before 2 μL of each was spin coated onto the interdigital electrode structure with a Laurell WS-400BZ-6NPP/LITE spin coater (5 s at 500 rpm and 20 s at 3000 rpm). Afterwards the electrodes were heated to approximately 230 °C for 30 s on a hot plate.

2.4 Gas measurements

Gas measurements were performed on a homemade gas mixing device, consisting of mass flow controllers (UFC-8160A and UFC-1660 from Unit Instruments Inc.), a flow cell with temperature control by two heating elements, and a Pt-100 temperature sensor (Fig. SI 1†).

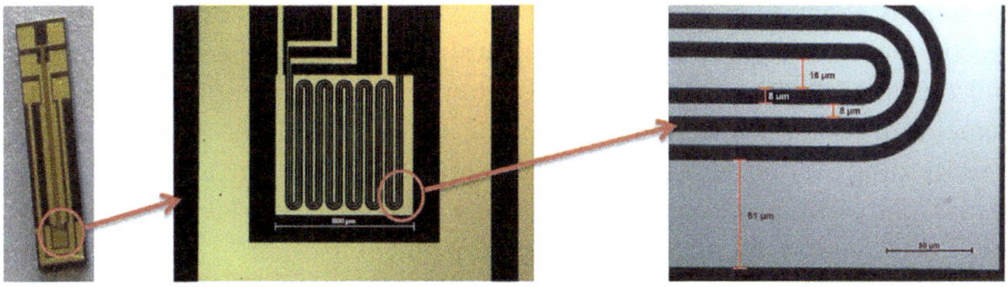

Fig. 2 Microelectrodes with an interdigital structure consisting of gold conducting paths on a Si/SiO_2 wafer substrate.

Measurement of the electrical resistance was accomplished by contacting the two gold conducting paths in the mid-position of the interdigital electrode structure (Fig. 2) using a Keithley 2400 sourcemeter at a constant bias voltage of 50 mV. Synthetic air (N_2: 80%, O_2: 20%) was used as the carrier gas and was mixed with 300 ppm NO_2, 1% H_2 or 1% CH_4. The test gases were all diluted by the synthetic air. Furthermore the gas adsorption tests were accomplished at a constant gas flow of 100 sccm (300 sccm for the concentration dependency of NO_2) and a constant temperature of 85 °C. It was also necessary to heat the electrodes for 20 s at 230 °C before each measurement, ensuring that no analyte gas had been previously adsorbed by reaching the initial resistance. For measuring the influence of air humidity, the gas flow was piped through the head space of a flask containing water.

3 Results and discussion

3.1 Choice of materials and sensor fabrication

The transfer of graphene materials is still the most critical step in sensor preparation using this type of carbon nanomaterial. Graphene prepared by *e.g.* the CVD technique[29] requires a sophisticated transfer method of the carbon nanomaterial from the metallic substrate to the substrate of choice. This can be done *e.g.* by a transfer with stamping methods,[30] which leads to contamination of the graphene and therefore such processes lack reproducibility. In contrast, rGO suspensions can be easily processed. Reduced graphene is usually transferred to electrodes by drop casting, which results in layers of inhomogeneous thickness and low reproducibility. The sensitivity of chemoresistors strongly depends on the thickness of the conductive layer. Therefore we used an interdigital electrode structure which was covered with a thin layer of rGO *via* spin coating. Parameters affecting the spin coating, such as the concentration of rGO, the rotation speed and time, as well as the usage of certain additives like isopropanol for improving the spreading of the suspension were optimized to obtain consistent layers with reproducible total resistance.

3.2 Characterization of the microelectrodes modified with reduced graphene oxide and composite materials

Graphene oxide was prepared by a modified Hummers method and reduced with hydrazine as reported previously.[14] The size of the graphene flakes ranged from

100 nm to about 1 μm as indicated by SEM studies (Fig. SI 2a†). The Raman spectrum of rGO (Fig. 3a) exhibits the typical broad D and G bands at 1342 cm^{-1} and 1602 cm^{-1}.[31] The D/G ratio and position of the G band indicates crystalline domains of several nanometers in size.[32]

Regarding the thermogravimetric analysis (TGA) of GO and rGO (Fig. 3b), a loss of mass can be observed at temperatures of about 200 °C. In FTIR-TGA (Fig. SI 3†) this effect can be ascribed to a thermal reduction, where CO and CO_2 are released.[33] Graphene oxide loses about 25% of its mass, while rGO loses around 15%, which shows that the reduction by hydrazine is incomplete.

Further, the rGO modified electrodes were heated to 230 °C to guarantee a thermal reduction. During this step the electrical resistance dropped from about 8.0 ± 3.8 MΩ to 180 ± 78 kΩ. The oxidation process of graphite introduces many defects into the sp^2 structure which act like a barrier for the electron flux, resulting in better conductance at higher levels of reduction.[34]

Functionalization of rGO was performed by wet chemical methods before the composite material was also transferred to the microelectrodes. Generally all dopants primarily bind/coordinate with the oxygen functionalities of rGO. Whereas ODA reacts most probably with the epoxy groups of GO,[25] doping with metals and metal oxides usually includes the interaction of a precursor metal ion with the carboxyl groups of rGO.[28,35] The successful modification of rGO was revealed by Raman studies. Peaks at 1125 and 1489 cm^{-1} (ODA), 140, 392, 510 and 644 cm^{-1} (TiO_2), and 579 and 633 cm^{-1} (MnO_2) were found (Fig. SI 4†). Graphene composites with conductive Pt- and Pd-nanoparticles could be observed on distinct regions of rGO in SEM and TEM images. The average particle size was below 20 nm (Fig. SI 5 and SI 6†). The elemental composition of the different modified materials was confirmed using energy-dispersive X-ray spectroscopy (EDS) (Fig. SI 7†).

Covalent attachment of functional groups[36] or doping with metals and metal oxides[37] may change the electronic structure and level of n-/p-doping within the material, but may also increase the sensor surface or have a catalytic effect on gas adsorption.

3.3 Gas sensor response

To assess the effect of gas adsorption on the conductance of rGO, the electrical resistance of the coated microelectrodes was measured for different gases of

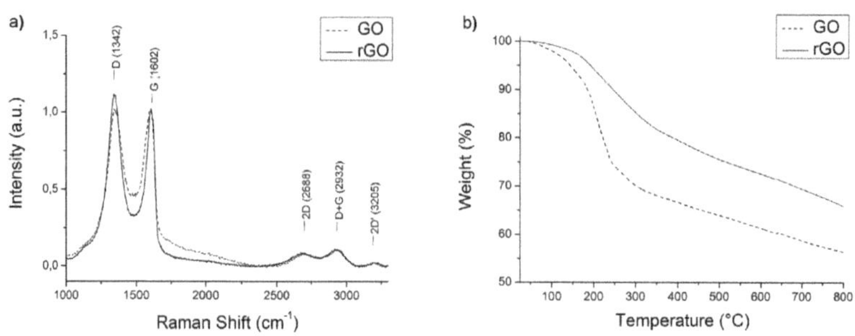

Fig. 3 (a) Raman spectra and (b) thermogravimetric analysis of GO and rGO.

various concentrations. To ensure realistic conditions, a test gas was mixed with a constant flow of synthetic air. An operating temperature of 85 °C was chosen to exclude the influence of humidity on the response of the sensor (Fig. SI 8†). Furthermore, heating to 85 °C improved the response time (<1 min), and an increase in the signal change and recovery rate was found. Fig. 4 displays the change of the relative resistance (R/R_0) in the presence of different concentrations of NO_2. R_0 is hereby the initial resistance right before the addition of a test gas.

The reversibility and concentration dependence of the sensor was evaluated by continuous adsorption and desorption of the different gases. The results for NO_2 are shown in Fig. 4, and the changes in conductivity for other gases are listed in Table 1. A linear behaviour within the applied concentration range (25–150 ppm), a sensitivity of 0.56 ppm^{-1} and a detection limit of 0.3 ppm ($S/N = 3$) were observed for NO_2. In contrast to the signal drop for NO_2, the adsorption of H_2 and CH_4 led to an increase in electrical conductance. The electron withdrawing effect of the adsorbed NO_2 leads to more positive charge carriers since rGO is described to show a p-type semiconducting behaviour.[16,38] Gases such as H_2 or CH_4 cannot act as electron donors or acceptors. It can be assumed that previously physisorbed molecules from synthetic air (most probably oxygen) at the surface of the graphene are replaced by H_2 or CH_4, leading to an increase in electrical resistance.[39,40]

These results clearly demonstrate that rGO on the one hand is an excellent sensor material for detecting gases. In contrast to metal oxide sensors, which are the most common gas sensors so far, sensors based on graphene can be operated at low temperatures with fast response times. On the other hand it is demonstrated that graphene based sensors lack selectivity. Nearly all adsorbates will result in signal changes, therefore they only can be used in detecting analytes in an atmosphere of inert gases.[41–48] To overcome this drawback graphene itself may be doped by nitrogen, as was theoretically calculated by Ma *et al.*[49] Another possibility is to functionalize graphene or to create graphene nanocomposites.[50,51]

Chemical insertion of ODA led to slightly increased signal changes towards H_2 and CH_4. For NO_2 detection almost no influence of the modification could be

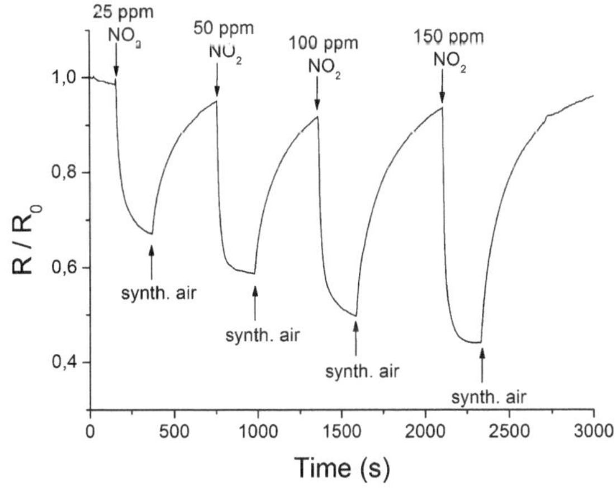

Fig. 4 Changes in the electrical resistance of a rGO layer at 85 °C in the presence of different concentrations of NO_2.

Table 1 Comparison of the response of unmodified rGO and modifications of rGO in the presence of different concentrations of NO_2, H_2 and CH_4 at 85 °C

	Signal [R/R_0]					
Gases	rGO	rGO–ODA	rGO–TiO_2	rGO–MnO_2	rGO–Pd	rGO–Pt
150 ppm NO_2	0.419	0.436	0.718	0.929	0.358	0.758
125 ppm NO_2	0.430	0.437	0.733	0.958	0.387	0.810
100 ppm NO_2	0.445	0.442	0.751	0.959	0.441	0.840
75 ppm NO_2	0.494	0.456	0.768	0.969	0.461	0.849
50 ppm NO_2	0.545	0.481	0.773	0.971	0.489	0.861
25 ppm NO_2	0.578	0.559	0.801	0.974	0.607	0.870
5000 ppm H_2	1.035	1.037	1.026	1.017	1.615	1.135
3000 ppm H_2	1.025	1.027	1.011	1.012	1.508	1.095
1000 ppm H_2	1.012	1.004	1.004	1.010	1.188	1.049
500 ppm H_2	1.005	1.000	1.001	1.004	1.114	1.011
10 000 ppm CH_4	1.033	1.069	1.021	1.022	1.029	1.023
5000 ppm CH_4	1.023	1.048	1.012	1.019	1.019	1.016
3000 ppm CH_4	1.015	1.021	1.008	1.010	1.010	1.010
1000 ppm CH_4	1.006	1.004	1.000	1.006	1.005	1.002

found. The composite material with TiO_2 showed unaffected behaviour towards H_2 and CH_4, but a lower response towards NO_2. This indicates that the adsorption sites for NO_2 may be blocked. Electrodes coated with rGO which was doped with MnO_2 showed almost no signal change for all of the utilized gases. Therefore, this modification could only be used as a reference signal to compensate for small fluctuations in temperature. The most significant change in response was achieved by chemical doping with Pd- and Pt-nanoparticles. The resulting composite materials showed an increased sensitivity towards H_2 (Fig. SI 9b†). The same observations with comparable results have been made with carbon nanotubes decorated with Pt and multilayered graphene nanoribbons doped with Pd.[24,52] Though the composite material with rGO is easier to produce and process.

In summary, all of the composites of rGO presented here are simple to prepare and show different behaviour in their electrical conductance in the presence of various gases. The sensor responses of different functionalized rGO materials towards NO_2, H_2 and CH_4 (Fig. 5) are summarized in Table 1.

3.4 Principal component analysis for pattern recognition of different gases

The altered sensitivity of every sensor towards different analyte gases enables pattern recognition using multivariate analysis based on principal component analysis (PCA). With this chemometric technique it is possible to simplify multidimensional datasets without crucial loss of information. Here the data matrix (Table 1) contains the normalized sensor response of each sensor material to a certain gas and concentration. This multidimensional matrix can be reduced to two principle components (PC1, PC2). The variance of PC1 (95.75%) and PC2 (2.94%) is above 98% and therefore these components already contain significant information to represent the data in two dimensions (Fig. 6). Clear separation between the clusters representing individual gases with no overlap and a

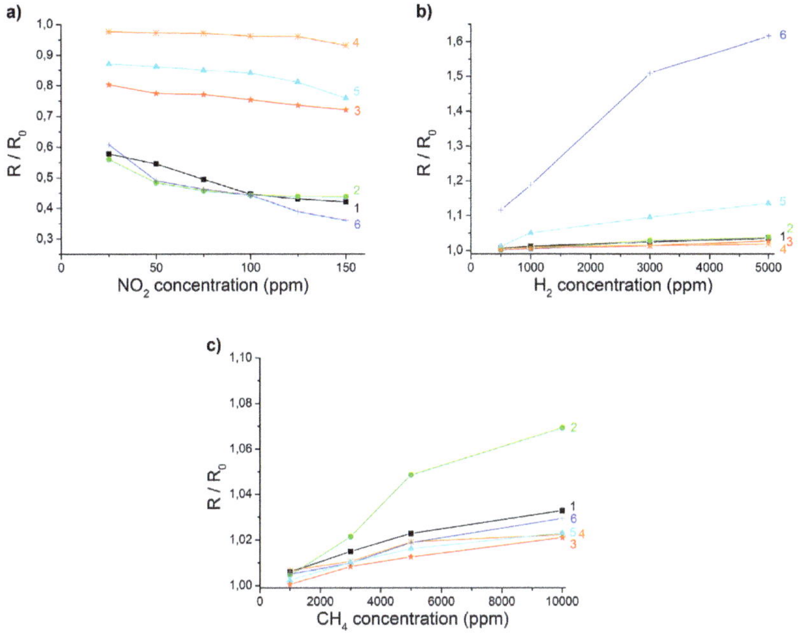

Fig. 5 Changes in electrical resistance of rGO (1) and graphene composites (2 to 6) for different concentrations of NO_2 (a), H_2 (b) and CH_4 (c). The composites are rGO–ODA (2), rGO–TiO_2 (3), rGO–MnO_2 (4), rGO–Pt (5), and rGO–Pd (6).

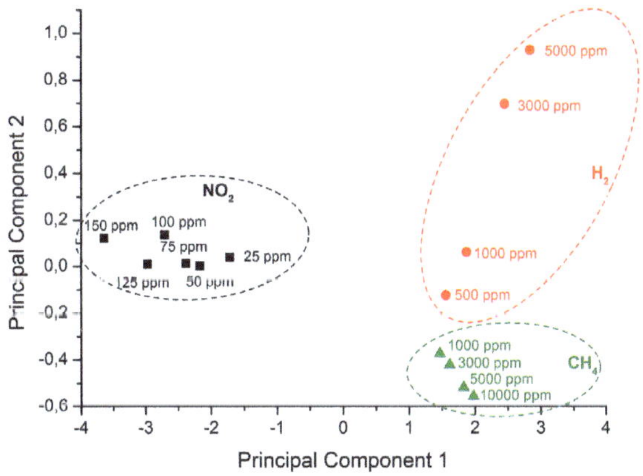

Fig. 6 Pattern analysis based on PCA using six different sensors (modified with rGO, rGO–ODA, rGO–TiO_2, rGO–MnO_2, rGO–Pd, and rGO–Pt) for various concentrations of individual gases (NO_2, H_2, and CH_4).

recognizable trend of concentration allows the identification of all analytes with a set of six different electrodes.

Regeneration of the sensor material still has to be further investigated. In this study the regeneration was mostly performed by a short term heat treatment at

230 °C between measurements. Furthermore this step ensures the complete desorption of any physisorbed impurities as demonstrated by the TGA measurements. In an application this could be realized by different operation temperatures with short regeneration cycles.

4 Conclusion

Reduced graphene oxide is an ideal sensor material for chemiresistive gas sensors due to its simple preparation and functionalization enabling altered sensitivities. Compared to commercial solid state gas sensors, rGO sensors have the advantage of operating at the low temperature of 85 °C to exclude the strong influences of humidity. It was demonstrated that the spin coating of rGO composites results in reproducible sensor behaviour suitable for well established fabrication technologies. In a chemiresistor setup, unmodified rGO showed high sensitivity towards NO_2 on the ppm level at ambient conditions. The sensor was rather unselective and also showed responses towards H_2 and CH_4. Upon different functionalizations it was possible to achieve different sensor behaviors for different gases. This can be used to apply PCA to discriminate between each individual gas. It is expected that this approach could be extended to build up sensor arrays for detecting the concentration of many individual gases in a complex matrix at low temperatures.

Acknowledgements

This work has been supported by Infineon Technologies AG, Regensburg. The authors thank Stefan Wilhelm and Christoph Fenzl for TEM studies and Gerhard Poeppel for the help with PCA.

References

1 J. M. Samet, *Inhalation Toxicol.*, 2007, **19**, 1021–1027.
2 P. T. Moseley, *Meas. Sci. Technol.*, 1997, **8**, 223–237.
3 G. Korotcenkov, *Mater. Sci. Eng., B*, 2007, **139**, 1–23.
4 A. Tricoli, M. Righettoni and A. Teleki, *Angew. Chem., Int. Ed.*, 2010, **49**, 7632–7659.
5 P. Bondavalli, P. Legagneux and D. Pribat, *Sens. Actuators, B*, 2009, **140**, 304–318.
6 S. Mao, G. Lu and J. Chen, *J. Mater. Chem. A*, 2014, **2**, 5573–5579.
7 F. Schedin, A. K. Geim, S. V. Morozov, E. W. Hill, P. Blake, M. I. Katsnelson and K. S. Novoselov, *Nat. Mater.*, 2007, **6**, 652–655.
8 W. Yuan and G. Shi, *J. Mater. Chem. A*, 2013, **1**, 10078–10091.
9 S. Basu and P. Bhattacharyya, *Sens. Actuators, B*, 2012, **173**, 1–21.
10 Q. He, S. Wu, Z. Yin and H. Zhang, *Chem. Sci.*, 2012, **3**, 1764–1772.
11 A. K. Geim and K. S. Novoselov, *Nat. Mater.*, 2007, **6**, 183–191.
12 C. Soldano, A. Mahmood and E. Dujardin, *Carbon*, 2010, **48**, 2127–2150.
13 W. S. Hummers Jr and R. E. Offeman, *J. Am. Chem. Soc.*, 1958, **80**, 1339–1339.
14 D. Li, M. B. Müller, S. Gilje, R. B. Kaner and G. G. Wallace, *Nat. Nanotechnol.*, 2008, **3**, 101–105.

15 C. Gómez-Navarro, R. T. Weitz, A. M. Bittner, M. Scolari, A. Mews, M. Burghard and K. Kern, *Nano Lett.*, 2007, **7**, 3499–3503.
16 S. Gilje, S. Han, M. Wang, K. L. Wang and R. B. Kaner, *Nano Lett.*, 2007, **7**, 3394–3398.
17 J. T. Robinson, F. K. Perkins, E. S. Snow, Z. Wei and P. E. Sheehan, *Nano Lett.*, 2008, **8**, 3137–3140.
18 Y.-H. Zhang, Y.-B. Chen, K.-G. Zhou, C.-H. Liu, J. Zeng, H.-L. Zhang and Y. Peng, *Nanotechnology*, 2009, **20**, 185504.
19 G. Eda, G. Fanchini and M. Chhowalla, *Nat. Nanotechnol.*, 2008, **3**, 270–274.
20 Y. Xu and G. Shi, *J. Mater. Chem.*, 2011, **21**, 3311–3323.
21 O. Leenaerts, B. Partoens and F. M. Peeters, *Phys. Rev. B: Condens. Matter Mater. Phys.*, 2008, **77**, 125416.
22 S. Mao, S. Cui, G. Lu, K. Yu, Z. Wen and J. Chen, *J. Mater. Chem.*, 2012, **22**, 11009–11013.
23 Q. Huang, D. Zeng, H. Li and C. Xie, *Nanoscale*, 2012, **4**, 5651–5658.
24 J. L. Johnson, A. Behnam, S. J. Pearton and A. Ural, *Adv. Mater.*, 2010, **22**, 4877–4880.
25 S. Wang, P.-J. Chia, L.-L. Chua, L.-H. Zhao, R.-Q. Png, S. Sivaramakrishnan, M. Zhou, R. G.-S. Goh, R. H. Friend, A. T.-S. Wee and P. K.-H. Ho, *Adv. Mater.*, 2008, **20**, 3440–3446.
26 S. Chen, J. Zhu, X. Wu, Q. Han and X. Wang, *ACS Nano*, 2010, **4**, 2822–2830.
27 X. Li, J. Niu, J. Zhang, H. Li and Z. Liu, *J. Phys. Chem. B*, 2003, **107**, 2453–2458.
28 C. Xu, X. Wang and J. Zhu, *J. Phys. Chem. C*, 2008, **112**, 19841–19845.
29 S. Bae, H. Kim, Y. Lee, X. Xu, J.-S. Park, Y. Zheng, J. Balakrishnan, T. Lei, H. Ri Kim, Y. I. Song, Y.-J. Kim, K. S. Kim, B. Özyilmaz, J.-H. Ahn, B. H. Hong and S. Iijima, *Nat. Nanotechnol.*, 2010, **5**, 574–578.
30 K. S. Kim, Y. Zhao, H. Jang, S. Y. Lee, J. M. Kim, K. S. Kim, J.-H. Ahn, P. Kim, J.-Y. Choi and B. H. Hong, *Nature*, 2009, **457**, 706–710.
31 S. Stankovich, D. A. Dikin, R. D. Piner, K. A. Kohlhaas, A. Kleinhammes, Y. Jia, Y. Wu, S. T. Nguyen and R. S. Ruoff, *Carbon*, 2007, **45**, 1558–1565.
32 M. A. Pimenta, G. Dresselhaus, M. S. Dresselhaus, L. G. Cançado, A. Jorio and R. Saito, *Phys. Chem. Chem. Phys.*, 2007, **9**, 1276.
33 I. Jung, D. A. Field, N. J. Clark, Y. Zhu, D. Yang, R. D. Piner, S. Stankovich, D. A. Dikin, H. Geisler, C. A. Ventrice and R. S. Ruoff, *J. Phys. Chem. C*, 2009, **113**, 18480–18486.
34 I. Jung, D. A. Dikin, R. D. Piner and R. S. Ruoff, *Nano Lett.*, 2008, **8**, 4283–4287.
35 J. Wöllenstein, M. Burgmair, G. Plescher, T. Sulima, J. Hildenbrand, H. Böttner and I. Eisele, *Sens. Actuators, B*, 2003, **93**, 442–448.
36 K. P. Loh, Q. Bao, P. K. Ang and J. Yang, *J. Mater. Chem.*, 2010, **20**, 2277–2289.
37 S. Bai and X. Shen, *RSC Adv.*, 2012, **2**, 64–98.
38 J. D. Fowler, M. J. Allen, V. C. Tung, Y. Yang, R. B. Kaner and B. H. Weiller, *ACS Nano*, 2009, **3**, 301–306.
39 M. Gautam and A. H. Jayatissa, *Mater. Sci. Eng., C*, 2011, **31**, 1405–1411.
40 Y. Dan, Y. Lu, N. J. Kybert, Z. Luo and A. T. C. Johnson, *Nano Lett.*, 2009, **9**, 1472–1475.
41 H. Y. Jeong, D.-S. Lee, H. K. Choi, D. H. Lee, J.-E. Kim, J. Y. Lee, W. J. Lee, S. O. Kim and S.-Y. Choi, *Appl. Phys. Lett.*, 2010, **96**, 213105.
42 J. L. Johnson, A. Behnam, S. J. Pearton and A. Ural, *Adv. Mater.*, 2010, **22**, 4877–4880.

43 J. Zou, L. J. Hubble, K. S. Iyer and C. L. Raston, *Sens. Actuators, B*, 2010, **150**, 291–295.
44 B. H. Chu, C. F. Lo, J. Nicolosi, C. Y. Chang, V. Chen, W. Strupinski, S. J. Pearton and F. Ren, *Sens. Actuators, B*, 2011, **157**, 500–503.
45 W. Li, X. Geng, Y. Guo, J. Rong, Y. Gong, L. Wu, X. Zhang, P. Li, J. Xu, G. Cheng, M. Sun and L. Liu, *ACS Nano*, 2011, **5**, 6955–6961.
46 V. Tjoa, W. Jun, V. Dravid, S. Mhaisalkar and N. Mathews, *J. Mater. Chem.*, 2011, **21**, 15593–15599.
47 W. Yuan, A. Liu, L. Huang, C. Li and G. Shi, *Adv. Mater.*, 2013, **25**, 766–771.
48 F. Zhang, F. M. Yasin, X. Chen, J. Mo, C. L. Raston and H. Zhang, *RSC Adv.*, 2013, **3**, 25166–25174.
49 C. Ma, X. Shao and D. Cao, *Sci. China: Chem.*, 2014, **57**, 911–917.
50 Y. Yang, C. Tian, J. Wang, L. Sun, K. Shi, W. Zhou and H. Fu, *Nanoscale*, 2014, **6**, 7369–7378.
51 L. Huang, Z. Wang, J. Zhang, J. Pu, Y. Lin, S. Xu, L. Shen, Q. Chen and W. Shi, *ACS Appl. Mater. Interfaces*, 2014, **6**, 7426–7433.
52 A. Kaniyoor, R. I. Jafri, T. Arockiadoss and S. Ramaprabhu, *Nanoscale*, 2009, **1**, 382–386.

Faraday Discussions

PAPER

Oxygen reduction reaction by electrochemically reduced graphene oxide†

Santosh Kumar Bikkarolla,[a] Peter Cumpson,[b] Paul Joseph[c] and Pagona Papakonstantinou*[a]

Received 1st May 2014, Accepted 2nd June 2014
DOI: 10.1039/c4fd00088a

We show that a partially reduced graphene oxide electrocatalyst, synthesized by electrochemical reduction of graphene oxide (GO), displays significantly enhanced catalytic activity towards the oxygen reduction reaction (ORR) in alkaline solutions compared to the starting GO. The electrochemical partial reduction of GO was confirmed by X-ray diffraction and X-ray photoelectron spectroscopy. Electrochemical impedance spectroscopy (EIS) verified the enhanced electron transfer ability of the electrochemically reduced graphene oxide (ErGO) compared to GO. The resultant ErGO electrode showed enhanced capacitance and an ORR onset potential of −0.11 V *vs.* Ag/AgCl, similar to that of a nitrogen doped reduced graphene oxide (NrGO) electrode produced by a hydrothermal process. However the ErGO exhibited considerably lower electron transfer numbers (2.0–3.3 at a potential range of −0.4 V to −1.0 V) indicating that although both catalysts operate under combined $4e^-$ and $2e^-$ ORR processes, ErGO follows a more predominant $2e^-$ pathway. The ORR process in ErGO has been linked to the presence of quinone functional groups, which favour the $2e^-$ ORR pathway.

Introduction

The oxygen reduction reaction (ORR), which generates electricity through the electrochemical reduction of oxygen, underpins many clean-energy technologies such as fuel cells and metal–air batteries. However, progress is hampered by the large intrinsic overpotentials of the ORR on the cathode side of fuel cells, which require the relatively high loading of platinum electrocatalysts, which are both expensive and scarce.[1] As a result, the development of inexpensive,

[a]*School of Engineering, Engineering Research Institute, University of Ulster, Newtownabbey, BT37 0QB, UK. E-mail: p.papakonstantinou@ulster.ac.uk*

[b]*School of Mechanical and Systems Engineering, University of Newcastle, Stephenson Building, Newcastle upon Tyne NE1 7RU, UK*

[c]*School of the Built Environment, University of Ulster, Newtownabbey, BT37 0QB, UK*

† Electronic supplementary information (ESI) available. See DOI: 10.1039/c4fd00088a

stable and catalytically active materials for the ORR that can approach and ideally surpass the performance of Pt based catalysts is of great technological significance.

Recently there has been an explosion of studies on introducing various heteroatoms, *e.g.* N, B, P, S, and I,[2–12] or a combination of these[13–15] into carbon nanomaterials, which demonstrate a competitive ORR activity in alkaline electrolytes compared to the benchmark Pt/C catalyst. This phenomenon has been elucidated using density functional calculations, where it was found that the heteroatoms induce an uneven charge distribution in adjacent sites, which alters the local spin or charge density thus promoting O_2 adsorption and facilitating molecular oxygen activation and subsequent O=O bond breaking.[16,17]

Since the discovery of graphene, immense efforts have been focused on GO because it is the most promising precursor for obtaining large quantities of this unique material. Graphene oxide can be visualized as individual sheets of graphene decorated with oxygen functional groups on both basal planes and edges, which has been prepared by oxidative exfoliation of graphite. GO needs to be reduced to graphene by thermal or chemical reduction to recover its conducting properties. Thermal treatment at 1000 °C,[18] or chemical reduction with hydrazine,[19] sodium borohydride,[20] aluminium hydride,[21] and ammonium hydroxide, or hydrothermal treatment with hydrazine[22] and ammonium thiocynate[23] can partially remove the oxygen functionalities on the basal planes and edges of the graphene oxide sheets. In the reduction process the sp^2 carbon network is partially restored, which improves the electrical conductivity of the material. These methods are suitable for the production of large scale quantities of reduced graphene oxide, however the majority of them necessitate the use of high temperatures and non-environmentally friendly reducing agents. In contrast, electrochemical reduction of GO doesn't require the use of any hazardous chemical reducing agents or high temperatures and the process is fast and simple.[24–34]

ErGO has shown promising characteristics for applications in supercapacitors[30,31,35–37] and biosensors[26,34,38–40] due to its high surface area, presence of functional groups and increased conductivity. However the ORR performance of ErGO has not been investigated systematically. The work by D. Huang *et al.*[29] was centred on the ORR study of electrochemically reduced layer-by-layer (LBL) self-assembly of negatively charged GO and positively charged PDDA rather than the ORR study of electrochemically reduced GO itself. The ORR performance of ErGO in alkaline (0.1 M KOH) solutions is only mentioned peripherally.[29] Y. Shao *et al.*[33] reported the ORR performance of ErGO in acidic media (0.5 M H_2SO_4) in the form of a cyclic voltammogram, with no detailed kinetic or mechanistic studies.

In this work ErGO is synthesised by reducing graphene oxide through repetitive cathodic cycling in the potential range of 0 to −1.5 V in a N_2 saturated 0.1 M Na_2SO_4 solution.[41,42] The material is further explored as a metal-free catalyst for ORR applications in alkaline solutions by carrying out kinetic studies, and evaluating its electrochemically active surface area and methanol tolerance. In addition the ORR performance of ErGO is compared to that of nitrogen doped reduced graphene oxide (NrGO) prepared by a hydrothermal route.

Experimental section

Synthesis of nitrogen-doped reduced graphene oxide (NrGO)

Graphene oxide was prepared by the modified Hummer's method.[18] The GO was then exfoliated in water by ultrasonication for 2 h. The GO dispersion was centrifuged at 1000 rpm for 5 min to remove any unexfoliated graphene oxide and the supernatant was centrifuged at 3000 rpm for 30 min. The obtained graphene oxide pellet was dried in an oven at 60 °C for 48 h. NrGO was synthesized through the hydrothermal reaction of graphene oxide with ammonia and hydrazine. In a typical experiment the as synthesized graphene oxide was suspended in water to give a concentration of 1 mg ml^{-1} followed by sonication of 70 ml of the above solution for 2 h, the pH was adjusted to 10 with ammonia hydroxide and 1 ml of hydrazine (35 wt% in H_2O) was added to the solution and stirred for 15 min. The solution was transferred to a 100 ml Teflon lined vessel to carry out the hydrothermal reaction at 160 °C for three hours. The resulting product was cleaned with plenty of DI (deionized) water and collected by centrifugation at 3000 rpm for 45 min. The pellet was dried under vacuum at 60 °C for 48 h.

Electrochemical reduction of graphene oxide (GO)

A total of 5 mg of graphene oxide was ultrasonicated in 1 ml of DI water for 60 min. To the homogenously dispersed GO solution 50 μl of Nafion was added and further sonicated for 30 min. The GO solution (4 μl) was applied onto a 3 mm glassy carbon electrode (GCE) and dried under an infrared lamp. The electrochemical reduction was carried out by cyclic voltammetry in the potential range of 0 to −1.5 V *vs.* Ag/AgCl at a scan rate of 50 mV s^{-1} in a N_2 saturated 0.1 M Na_2SO_4 solution in a standard three electrode cell with Ag/AgCl as the reference electrode and platinum as the counter electrode. Cyclic voltammograms (CVs) in N_2 saturated 0.1 M Na_2SO_4 are shown in Fig. S3.†

Electrochemical measurements

The oxygen reduction reactions were carried out in a 0.1 M KOH solution. The electrolyte was saturated with O_2 by bubbling O_2 for 15 min. A flow of O_2 was maintained over the electrolyte during the measurements. Cyclic voltammograms were collected at a scan rate 0.1 V s^{-1} after at least 25 cycles of the electrode at 0.1 V s^{-1}. RDE measurements were collected at a scan rate of 10 mV s^{-1}. Kinetic analysis was performed using the Koutecky–Levich equation given below.

$$\frac{1}{J}=\frac{1}{J_L}+\frac{1}{J_K}=\frac{1}{nFkC_0}+\frac{1}{0.62nFD^{\frac{2}{3}}\nu^{-\frac{1}{6}}C_0\omega^{\frac{1}{2}}} \quad (1)$$

where J is the measured current density, J_K and J_L are the kinetic and limiting diffusion current densities, n is the electron transfer number in the ORR process, F is the Faraday constant (96 500 C mol^{-1}), C_0 is the bulk concentration of O_2 (1.2×10^{-6} mol cm^{-3}), D is the diffusion constant of O_2 in 0.1 M KOH (1.9×10^{-5} cm^2 s^{-1}), ν is the kinematic viscosity of the electrolyte (0.01 cm^2 s^{-1}), ω is the angular frequency of the disk ($2\pi N$, where N is the linear frequency) and k is the electron transfer rate constant.

Structural characterization

X-ray photoelectron spectroscopy was carried on a Kratos Axis Nova XPS equipped with an A1 Kα ($h\nu$ = 1486.6 eV) X-ray source at the NEXUS laboratory. X-ray diffraction (XRD) analysis was performed on a Bruker D8-discover diffractometer fitted with a global mirror (Cu Kα radiation source $\lambda = 1.540$ Å). Scanning electron microscopy images were obtained on an FEI quanta 200 3D at a 20 kV accelerating voltage. High resolution transmission electron microscopy (HRTEM) images were obtained on a Jeol JEM 2011. Raman spectroscopy was performed with an ISA LabRAM equipped with a 633 nm laser.

For the XPS measurements GO without Nafion was drop dried on a gold coated substrate and reduced in the potential range 0 to −1.2 V *vs.* Ag/AgCl for 22 CV cycles. For all of the remaining structural characterizations Nafion was present in the GO solution and the measurements were taken on a GCE. The TEM samples were prepared by carefully removing the material from the GCE, followed by ultrasonication in isopropanol and coating the TEM grids with a few microliters of the solution. For the XRD measurements the ErGO was scraped off from the GCE, followed by sonication in 70% ethanol and then drop dried onto a silicon substrate. SEM imaging was carried out on the GCE electrode.

Results and discussion

XPS was used to determine the elemental composition and bonding states of the GO, ErGO (GO reduced for 22 CV cycles) and NrGO. Fig. 1(a)–(c) show the C 1s spectra of GO, ErGO and NrGO. The high resolution C 1s spectra of GO show two well defined peaks which is a signature of extreme oxidation of GO. The C 1s core level spectra for the GO and ErGO samples contain four major peaks at 284.4 eV, 285.1 eV, 286.46 eV and 287.9 eV which are attributed to the sp^2 carbon, hydroxyl (C–OH), and epoxide (C–O–C) groups, and the combined contribution from carbonyl (>C=O) and carboxyl (HO–C=O) groups.[18,43,44] The C 1s of NrGO can be

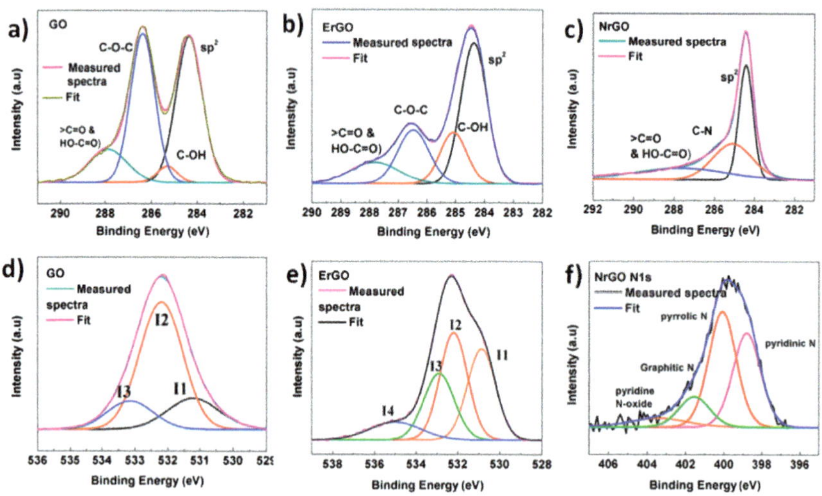

Fig. 1 (a–c) High resolution C 1s XPS spectra of GO, ErGO and NrGO. (d and e) High resolution O 1s XPS spectra of GO and ErGO. (f) High resolution N 1s XPS spectra of NrGO.

fitted to three components at 284.4 eV, 285.1 eV and 287.7 eV which correspond to the sp^2 carbon, C–N and a contribution from carbonyl and carboxylic acid groups, respectively.[45]

Fig. 1(d) and (e) show the O 1s spectra of GO and ErGO. Deconvolution of the O 1s spectra produces 3 main peaks at around 531.20, 532.2 and 533.3 eV assigned to C=O (oxygen doubly bonded to an aromatic carbon – denoted as I1), C–O (oxygen singly bonded to an aliphatic carbon – denoted as I2), and phenolic (oxygen singly bonded to an aromatic carbon – denoted as I3) groups respectively. The formation of quinone groups on ErGO is clearly evidenced from the broadening of the O 1s peak in the low energy region. Fig. 1(f) shows the N 1s spectra of NrGO. The N 1s in NrGO can be divided into four components, the pyridinic N (398.8 eV), pyrrolic N (400.0 eV) graphitic N (401.54 eV) and pyridinic N-oxide (403.4 eV). These assignments are in agreement with the literature.[22]

The C/O ratios in the case of GO, ErGO and NrGO are 2.5, 4.0 and 9.6 respectively. The high oxygen content in the reduced sample could be due to the limited number of cycles employed here (22 cycles) for preparing the sample for XPS measurements, as reduction may take place only on the first few surface layers, whereas the XPS is able to probe underneath the highly oxidised layers (XPS probing depth can be as deep as 10 nm).[25,33]

It is worth mentioning that wide XPS spectra of ErGO revealed the presence of Na and S elements, which originated from the Na_2SO_4 electrolyte used in the reduction process (see Fig S1†). Similar impurities were also present in previously reported work on the electrochemical reduction of GO over an extended number of cycles.[33] In our case reducing the GO for a larger number of cycles didn't improve the ORR performance, as discussed later. Although the presence of sulphur in carbon related materials has been claimed to act as an active site for the ORR, here its presence is in minute amounts compared to that of oxygen.

Fig. 2(a) shows the powder XRD spectra of GO, ErGO/Nafion and NrGO. The GO spectrum shows a characteristic peak at 11.9°. After the electrochemical reduction of GO there are five characteristic peaks 11.9°, 17.5°, 24°, 38.5° and 44°. The peaks at 24° and 44° correspond to the (002) and (001) planes of graphitic carbon, respectively. The peak at 38.5° corresponds to Na_2SO_4 (Fig. S6†). The low intensity of the (002) peak reveals that only the top surface of the GO is reduced, which confirms that the electrochemical reduction is active only for the first few surface

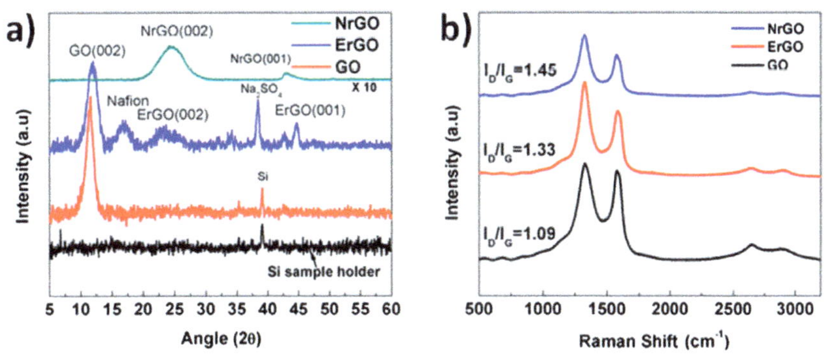

Fig. 2 (a) XRD spectra of GO, ErGO, NrGO and sample holder. (b) Raman spectra of GO, ErGO and NrGO.

layers. The peak at 17.5° corresponds to Nafion, which helps in binding the graphene oxide to the GCE. The XRD pattern of NrGO shows two distinctive peaks at 24.5° and 43.0° corresponding to the (002) and (100) reflections of graphitic carbon. The average interlayer distance was calculated as 3.62 Å, which is greater than the graphene interlayer distance of 3.33 Å, which indicates a moderate oxidation level present in the NrGO.

Raman spectroscopy is an important tool in characterising carbon materials because it is able to provide information on the crystalline size, degree of disorder and can distinguish between single layer graphene and multilayered graphene. The G and D bands of carbon materials are assigned to the E_{2g} phonon mode of sp^2 carbon atoms and the breathing mode of κ-point phonons of A_{1g} symmetry. Fig. 2(b) shows the Raman spectra of GO, ErGO and NrGO. The Raman spectrum of GO shows two bands at 1324 cm^{-1} and 1579 cm^{-1} with an I_D/I_G ratio of 1.09. After the electrochemical and hydrothermal reduction of GO the I_D/I_G ratios are increased to 1.33 and 1.45 respectively, indicating a decrease in the size of the sp^2 domain and a restoration of the sp^2 network upon the electrochemical reduction of GO.[18,43,46]

Fig. 3(a) and (b) show SEM images of ErGO. These images indicate that the ErGO has a 3D porous morphology similar to that of reduced GO sheets derived from a hydrothermal reduction. Further information on the morphology of the ErGO is revealed by TEM. As can be seen in Fig. 3(c) the transparent graphene sheets are randomly folded due to their defective structure formed upon electrochemical reduction. Fig. 3(d) shows the HRTEM of ErGO which consists of 8–9 layers of graphene sheets. The interlayer distances between the sheets are about 0.38 nm which corresponds to the (002) plane of ErGO.

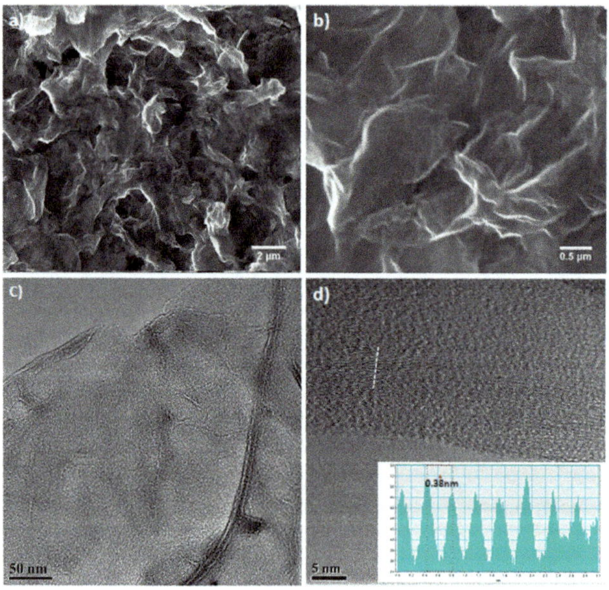

Fig. 3 (a and b) SEM images of ErGO on a GCE. (c) TEM image of ErGO. (d) HRTEM image of ErGO and the inset shows the distances between successive layers at the folded edge of ErGO.

Effect of GO loading on the ORR performance

To gain an insight into the role of the GO catalyst loading during the ORR electrochemical process we investigated its electrocatalytic performance as a function of mass loading. We carried out linear sweep voltammetry (LSV) measurements on a rotating disk electrode (RDE) with GO loadings varying from 5 μg to 40 μg on a GCE in an O_2 saturated 0.1 M KOH electrolyte solution. All five loadings showed a substantial reduction process in the presence of oxygen. As shown in Fig. 4 the onset potentials of the ORR did not show any substantial changes with mass loading, however, a systematic decrease of the current density was observed at increased mass loadings (30 and 40 μg), which is consistent with the insulating nature of GO. The typical two-step pathway was observed for the GO electrode with onset potentials at around −0.28 V and −0.65 V, indicating a successive two-electron reaction pathway.

The ORR performance

The reduction of oxygen in alkaline solutions can proceed by two major pathways: a direct four electron pathway and a two electron 'peroxide' pathway according to eqn (2) and (3).[47,48]

$$O_2 + 2H_2O + 4e^- \rightarrow 4OH^- \quad (2)$$

$$O_2 + H_2O + 2e^- \rightarrow HO_2^- + OH^- \quad (3)$$

The peroxide (HO_2^-) can be further reduced according to the further $2e^-$ reduction in eqn (4)

$$HO_2^- + H_2O + 2e^- \rightarrow 3OH^- \quad (4)$$

or through the disproportionation reaction in eqn (5)

$$HO_2^- \rightarrow \frac{1}{2}O_2 + OH^- \quad (5)$$

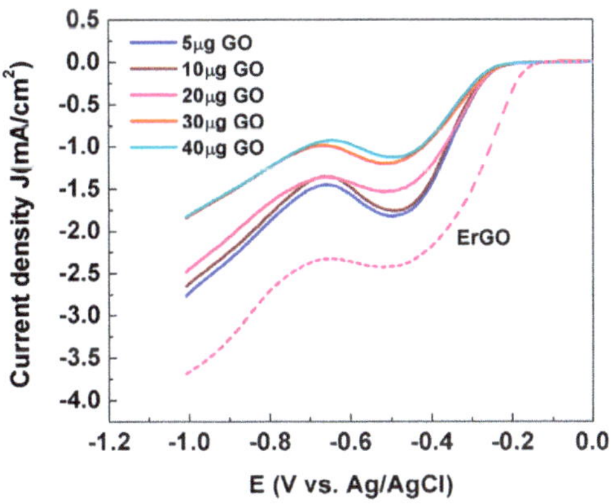

Fig. 4 Linear-sweep voltammetry curves of GO with various mass loadings and ErGO (20 μg) in an O_2 saturated 0.1 M KOH electrolyte at a scan rate of 10 mV s^{-1} and a rotation rate of 1600 rpm.

It is worth noting that summation of the two sequential two electron reductions (involving peroxide) (eqn 3 and 4) leads to the respective four electron reduction.

The electrocatalytic activities of the GO/GCE, ErGO/GCE, and NrGO/GCE electrodes were evaluated by CV and RDE voltammetry. A catalyst with a higher electrocatalytic activity towards the ORR will demonstrate an earlier onset potential and a higher peak current. Accordingly, its RDE voltammograms should also show an earlier current drop with a positively shifted onset potential and a higher steady-state current.

The cyclic voltammograms (CVs) of GO, ErGO and NrGO in an aqueous O_2 saturated 0.1 M KOH solution are shown in Fig. 5(a). As can be seen, the onset potential of ORR for the GO electrode subjected to 22 reducing CV cycles is at −0.18 V (*versus* AgCl) with the cathodic reduction peak at around −0.40 V (*versus* Ag/AgCl). After electrochemical reduction of GO both the onset potential and the ORR reduction peak potential shifted positively to −0.11 V and −0.26 V respectively, accompanied by a concomitant increase in the peak current density (Fig. 5(a)). These results clearly demonstrate a significant enhancement in the ORR electrocatalytic activity for the ErGO with respect to the graphene oxide electrode. Compared with GO, the ErGO shows a dramatically increased capacitance similar to NrGO, this is mainly related to the increased conductivity of the

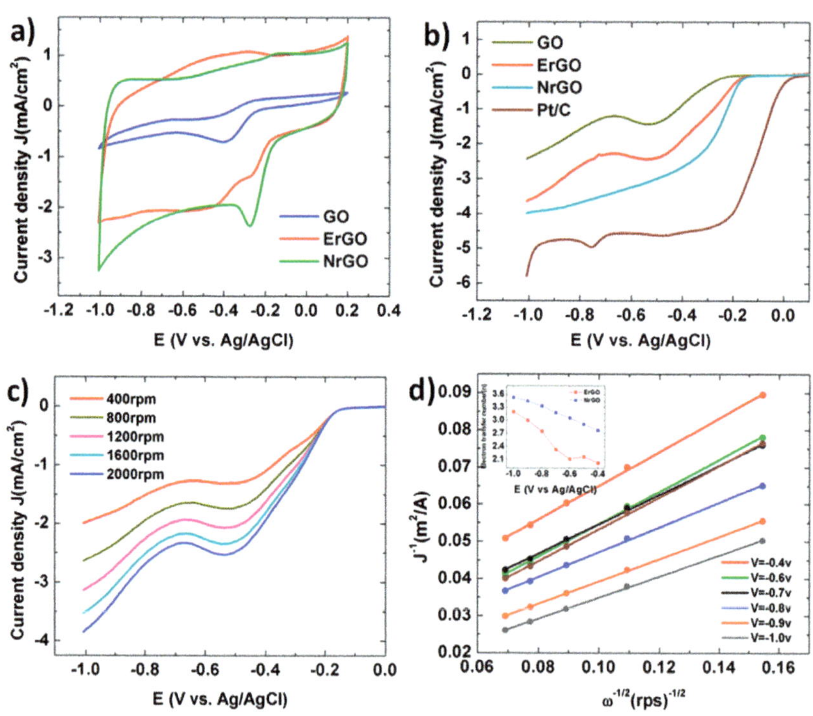

Fig. 5 (a) CV curves of oxygen reduction on the GO, ErGO, and NrGO electrodes in an O_2 saturated 0.1 M KOH solution at a scan rate of 100 mV s^{-1}. (b) RDE curves for oxygen reduction on the GO, ErGO and Pt/C electrodes in an O_2 saturated 0.1 M KOH solution at a scan rate of 10 mV s^{-1}. (c) Rotating disk voltammograms of ErGO in O_2 saturated 0.1 M KOH at different rotation rates. (d) The corresponding Koutecky–Levich plots (J^{-1} *vs.* $\omega^{-0.5}$) at different potentials and the inset shows the *n* numbers of ErGO and NrGO.

material. The change in shape of the CV curves demonstrates the increase of the non-Faradaic currents induced by the increase in effective surface area of the electrically conductive graphene that is generated by the reduction of GO.

To further investigate the ORR performance, we carried out linear sweep voltammetry (LSV) measurements on a rotating disk electrode. As shown in Fig. 5(b), the ORR of the GO electrode commenced at around −0.18 V (onset potential), whereas the ORR onset potential of the ErGO electrode subjected to 22 reducing CV cycles significantly shifted positively to −0.11 V with the limiting diffusion current at −1 V being about 1.7 times higher than that of the GO electrode. For the GO and ErGO, a clear reduction pre-wave was observed at low overpotentials, followed by a second reduction wave starting at around −0.7 V, which is indicative of a prevalent $2e^-$ reduction pathway.[56,57] The absence of the peak in the LSV curve of NrGO at −0.7 V (Fig. 5(b)) suggests that NrGO has a more efficient oxygen reduction due to the presence of nitrogen related electrocatalytic active sites for oxygen reduction.

Fig. S4† presents the as measured LSV curves of 20 μg of GO coated on a GCE before and after electrochemical reduction. In the same graph we present the LSV curve of ErGO after subtracting the background signal. The LSV background signal was obtained from a N_2 saturated 0.1 M KOH solution. The significantly more positive onset potential of ErGO (−0.18 V for GO *vs.* −0.11 V for ErGO) and the higher current clearly demonstrate that the improved electrocatalytic activity after electrochemical reduction is not due to the background signal or due to a difference in the mass loading.

A rotating disk electrode (RDE) was used to reveal the ORR kinetics of ErGO in an O_2 saturated 0.1 M KOH solution. As shown in Fig. 5(c) as the rotation speed increases the oxygen reduction current also increases. The inset of Fig. 5(d) shows the electron transfer numbers (n) for ErGO and NrGO. In the case of ErGO the electron transfer number (n) was calculated to be 2–3.3 in the voltage range of −0.4 V to −1.0 V from the slopes of the Koutecky–Levich plots. At low overpotentials (<−0.7 V) the value of n is close to two and the reduction of O_2 produces HO_2^-. At more negative potentials the n value gradually increases, which indicates that the peroxide formed reduces further to OH^- in this potential range. In the case of the NrGO, the electron transfer numbers (n) were substantially larger ranging from 2.7 to 3.5 in the voltage range of −0.4 V to −1.0 V, which suggests that the NrGO is more electro active when compared to ErGO. The absence of the peak in the LSV curve of NrGO at −0.7 V (Fig. 5(b)) suggests that NrGO has a larger number of electrocatalytic active sites for oxygen reduction. Although both catalysts operate under combined $4e^-$ and $2e^-$ ORR processes, ErGO follows a more predominant $2e^-$ pathway (Fig. 5(b)).

Fig. 6 shows that reduction of GO for more than 22 cycles didn't show any substantial improvement in the ORR performance.

Various surface functional groups, such as hydroxyl, carboxylic, and carbonyl are present on the surface of graphene. A long standing question is which type of functional group is responsible for the electrocatalysis of oxygen reduction.[49] Quinone functional groups have been postulated as catalytic sites for $2e^-$ ORR on carbon materials, which means that the reduction of oxygen stops at the peroxide stage. However for ErGO the presence of additional oxygen groups may catalyse the further reduction of hydrogen peroxide (HO_2^-) at more negative potentials as exemplified by the increased n numbers. A similar increment in the n numbers

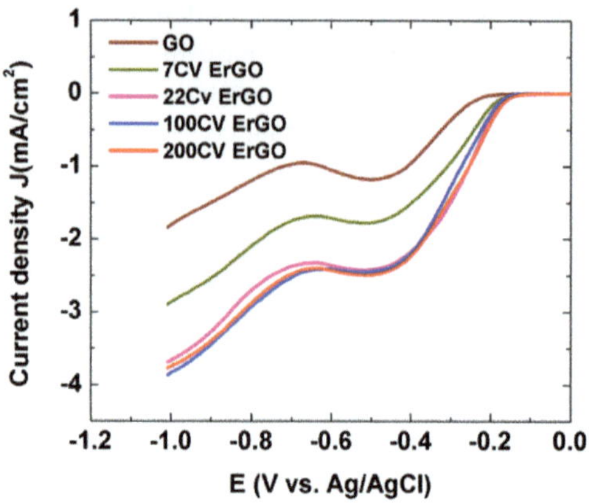

Fig. 6 Rotating disk voltammograms in O_2 saturated 0.1 M KOH at a scan rate of 10 mV s^{-1} and 1600 rpm of GO subjected to different reduction cycles.

was observed between a polished and an oxidised GC electrode where the reduction current of the anodised GC significantly exceeded the limiting current of the $2e^-$ process.[50] It should be noted that the presence of cationic species such as Na^+ accrued from the reduction process may have an influence on the electron transfer process and play a role in the ORR.

Electrochemically active surface area

We estimated the electrochemical active surface area (EASA) of the GO and ErGO electrodes using the following eqn (6).[51,52]

$$\mathrm{EASA} = \frac{C_{\mathrm{dl}}}{C_{\mathrm{s}}} \tag{6}$$

where C_{dl} is the electrochemical double-layer capacitance of the catalytic surface, and C_{s} (specific capacitance) is the double layer capacitance of an atomically smooth planar surface of the material per unit area under same electrolyte conditions. To measure C_{dl}, a potential range in which no apparent Faradaic processes occur was chosen. The C_{dl} was determined by measuring the non-Faradaic capacitive current associated with double-layer charging from the scan-rate dependence of the cyclic voltammograms (CVs) as shown in Fig. 7. The double layer charging current (i_{c}) is equal to the product of scan rate (ν) and double layer capacitance (C_{dl}) as given by eqn (7).[51–54]

$$i_{\mathrm{c}} = \nu C_{\mathrm{dl}} \tag{7}$$

Thus a plot of i_{c} *versus* ν gives a straight line with a slope equal to the C_{dl} as shown in Fig. 7(b). The estimated C_{dl} values of GO and ErGO are 0.023 mF and 0.35 mF. Specific capacitances for carbon electrode materials have been measured in alkaline and acidic solutions and typical reported values range between $C_{\mathrm{s}} = 0.005–0.055$ mF cm^{-2} in alkaline and acidic solutions.[55] We used a specific capacitance of $C_{\mathrm{s}} = 0.020$ mF cm^{-2} for both GO and ErGO as representative reported values. The C_{dl} and EASA values of each material are shown in Table S1.†

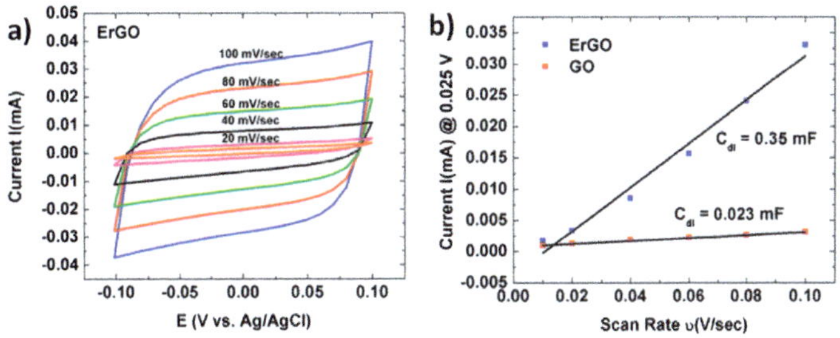

Fig. 7 (a) Cyclic voltammograms of ErGO in the potential range of −0.1 V to +0.1 V *vs.* Ag/AgCl (non-Faradaic region) at scan rates of 100 mV s^{-1}, 80 mV s^{-1}, 60 mV s^{-1}, 40 mV s^{-1}, 20 mV s^{-1} and 10 mV s^{-1} in a N_2 saturated 0.1 M KOH solution. (b) The cathodic charging current measured at 0.025 V *vs.* Ag/AgCl as a function of scan rate for ErGO and GO.

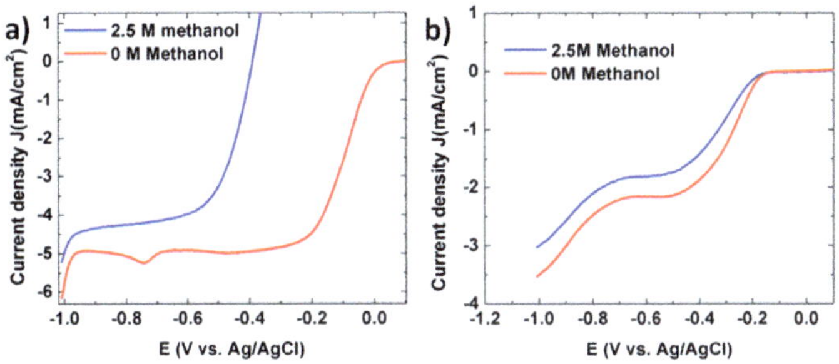

Fig. 8 (a) Polarization curves for ORR on commercial Pt/C catalyst in an O_2-saturated 0.1 M KOH electrolyte containing 2.5 M methanol at a rotation rate of 1600 rpm and scan rate of 10 mV s^{-1}. (b) Polarization curves for ORR on ErGO electrode in an O_2 saturated 0.1 M KOH electrolyte containing 2.5 M methanol at a rotation rate of 1600 rpm and scan rate of 10 mV s^{-1}.

The estimated EASA values of GO and ErGO are 1.05 cm^2 and 15.9 cm^2. The EASA value of ErGO is 15 times higher than that of GO.

Methanol poisoning

To examine the possible crossover effect in the presence of other fuel molecules (*e.g.* methanol) linear sweep voltammetry responses for the ORR at the ErGO and Pt/C electrodes were obtained. As shown in Fig. 8(a), for the commercial Pt/C catalyst, a sharp increase in the onset ORR potential and a net positive current were observed in the presence of 2.5 M methanol. These changes indicate that the methanol-oxidation reaction (MOR), rather than the ORR, was the preferred reaction on the Pt/C electrode. In contrast, the corresponding onset potential and overall performance of the ErGO electrode remained almost unchanged in the presence of MeOH (Fig. 8(b)). The results unambiguously show that the ErGO has a much better methanol tolerance than the commercial Pt/C catalyst.

Impedance studies

EIS was used to probe the interfacial processes of the ORR electrode reactions. Fig. 9 shows the EIS spectra in the form of Nyquist plots for the GO and ErGO

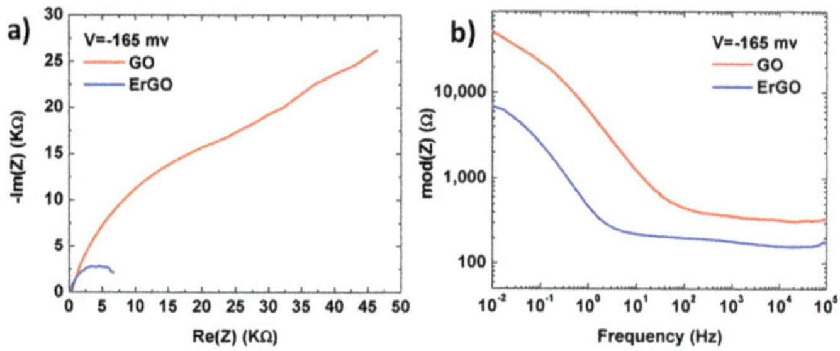

Fig. 9 (a) Nyquist plots and (b) Bode plots of GO and ErGO obtained at a bias of −0.165 V over the frequency range of 100 kHz to 10 mHz.

electrodes at a bias of −0.165 V. The ErGO shows a small semicircle at the intermediate and low frequency region whereas the GO shows a tremendously large arc, suggesting a very slow or negligible reaction rate of ORR and an extremely large charge transfer resistance compared to the ErGO. It is clear that the impedance values measured from the GO electrode are higher than those of the ErGO illustrating the resistive and conductive character of GO and ErGO respectively.

Conclusions

We have shown that nitrogen-free electrochemically reduced graphene oxide (ErGO) displays a significantly enhanced catalytic activity towards the oxygen reduction reaction in alkaline solutions compared to the initially unreduced GO electrode. However its ORR activity is inferior to that of nitrogenated graphene oxide (NrGO) prepared by hydrothermal treatment. The electrochemical partial reduction of GO was confirmed by X-ray diffraction and X-ray photoelectron spectroscopy. Electrochemical impedance spectroscopy (EIS) verified the enhanced electron transfer ability of the electrochemically reduced graphene oxide (ErGO) compared to GO. Kinetic studies revealed that although both the ErGO and NrGO catalysts operate under combined $4e^-$ and $2e^-$ ORR processes, the ErGO follows a more predominant $2e^-$ pathway. This ORR pathway in ErGO has been linked to the presence of quinone functional groups, which favour the $2e^-$ ORR process.

Acknowledgements

This work was financially supported by the VCRS studentship from the University of Ulster. XPS was performed at the National EPSRC XPS User's Service (NEXUS) at Newcastle University, an EPSRC Mid-Range Facility.

References

1 F. Cheng and J. Chen, *Chem. Soc. Rev.*, 2012, **41**, 2172.
2 L. Qu, Y. Liu, J. Baek and L. Dai, *ACS Nano*, 2010, **4**, 1321.
3 H. Wang, T. Maiyalagan and X. Wang, *ACS Catal.*, 2012, **2**, 781.

4 Z. Yang, Z. Yao, G. Li, G. Fang, H. Nie, Z. Liu, X. Zhou, X. Chen and S. Huang, *ACS Nano*, 2012, **6**, 205.
5 Z. Liu, F. Peng, H. Wang, H. Yu, W. Zheng and J. Yang, *Angew. Chem., Int. Ed.*, 2011, **50**, 3257.
6 C. Zhu and S. Dong, *Nanoscale*, 2013, **5**, 1753.
7 Z. Sheng, H. Gao, W. Bao, F. Wang and X. Xia, *J. Mater. Chem.*, 2012, **22**, 390.
8 N. Daems, X. Sheng, I. F. J. Vankelecom and P. P. Pescarmona, *J. Mater. Chem. A*, 2014, **2**, 4085.
9 X. Kong, C. Chen and Q. Chen, *Chem. Soc. Rev.*, 2014, **43**, 2841.
10 J. Park, Y. J. Jang, Y. J. Kim, M. Song, S. Yoon, D. H. Kim and S. Kim, *Phys. Chem. Chem. Phys.*, 2014, **16**, 103.
11 R. Li, Z. Wei, X. Gou and W. Xu, *RSC Adv.*, 2013, **3**, 9978.
12 Z. Yao, H. Nie, Z. Yang, X. Zhou, Z. Liu and S. Huang, *Chem. Commun.*, 2012, **48**, 1027.
13 S. Wang, L. Zhang, Z. Xia, A. Roy, D. W. Chang, J. Baek and L. Dai, *Angew. Chem., Int. Ed.*, 2012, **51**, 4209.
14 J. Liang, Y. Jiao, M. Jaroniec and S. Z. Qiao, *Angew. Chem., Int. Ed.*, 2012, **51**, 11496.
15 Y. Zheng, Y. Jiao, L. Ge, M. Jaroniec and S. Z. Qiao, *Angew. Chem., Int. Ed.*, 2013, **52**, 3110.
16 L. Zhang, J. Niu, L. Dai and Z. Xia, *Langmuir*, 2012, **28**, 7542.
17 L. Zhang and Z. Xia, *J. Phys. Chem. C*, 2011, **115**, 11170.
18 A. Ganguly, S. Sharma, P. Papakonstantinou and J. Hamilton, *J. Phys. Chem. C*, 2011, **115**, 17009.
19 D. Li, M. B. Muller, S. Gilje, R. B. Kaner and G. G. Wallace, *Nat. Nanotechnol.*, 2008, **3**, 101.
20 H. Shin, K. K. Kim, A. Benayad, S. Yoon, H. K. Park, I. Jung, M. H. Jin, H. Jeong, J. M. Kim, J. Choi and Y. H. Lee, *Adv. Funct. Mater.*, 2009, **19**, 1987.
21 A. Ambrosi, C. K. Chua, A. Bonanni and M. Pumera, *Chem. Mater.*, 2012, **24**, 2292.
22 D. Long, W. Li, L. Ling, J. Miyawaki, I. Mochida and S. Yoon, *Langmuir*, 2010, **26**, 16096.
23 Y. Su, Y. Zhang, X. Zhuang, S. Li, D. Wu, F. Zhang and X. Feng, *Carbon*, 2013, **62**, 296.
24 A. Ambrosi and M. Pumera, *Chem.–Eur. J.*, 2013, **19**, 4748.
25 X. Peng, X. Liu, D. Diamond and K. T. Lau, *Carbon*, 2011, **49**, 3488.
26 A. J. Haque, H. Park, D. Sung, S. Jon, S. Choi and K. Kim, *Anal. Chem.*, 2012, **84**, 1871.
27 Y. Cao, M. Zhu, P. Li, R. Zhang, X. Li, Q. Gong, K. Wang, M. Zhong, D. Wu, F. Lin and H. Zhu, *Phys. Chem. Chem. Phys.*, 2013, **15**, 19550.
28 X. Xu, D. Huang, K. Cao, M. Wang, S. M. Zakeeruddin and M. Grätzel, *Sci. Rep.*, 2013, **3**, 1489.
29 D. Huang, B. Zhang, Y. Zhang, F. Zhan, X. Xu, Y. Shen and M. Wang, *J. Mater. Chem. A*, 2013, **1**, 1415.
30 J. Yang and S. Gunasekaran, *Carbon*, 2013, **51**, 36.
31 X. Liu, X. Qi, Z. Zhang, L. Ren, G. Hao, Y. Liu, Y. Wang, K. Huang, X. Wei, J. Li, Z. Huang and J. Zhong, *RSC Adv.*, 2014, **4**, 13673.
32 M. A. Raj and S. A. John, *J. Phys. Chem. C*, 2013, **117**, 4326.

33 Y. Shao, J. Wang, M. Engelhard, C. Wang and Y. Lin, *J. Mater. Chem.*, 2010, **20**, 743.

34 L. Yang, D. Liu, J. Huang and T. You, *Sens. Actuators, B*, 2014, **193**, 166.

35 K. Sheng, Y. Sun, C. Li, W. Yuan and G. Shi, *Sci. Rep.*, 2012, **2**, 247.

36 X. Peng, X. Liu, D. Diamond and K. T. Lau, *Carbon*, 2011, **49**, 3488.

37 Y. Cao, M. Zhu, P. Li, R. Zhang, X. Li, Q. Gong, K. Wang, M. Zhong, D. Wu, F. Lin and H. Zhu, *Phys. Chem. Chem. Phys.*, 2013, **15**, 19550.

38 M. A. Raj and S. A. John, *J. Phys. Chem. C*, 2013, **117**, 4326.

39 V. Mani, B. Devadas and S. Chen, *Biosens. Bioelectron.*, 2013, **41**, 309.

40 J. Ping, Y. Wang, Y. Ying and J. Wu, *Anal. Chem.*, 2012, **84**, 3473.

41 C. Zhu, S. Guo, Y. Fang, L. Han, E. Wang and S. Dong, *Nano Res.*, 2011, **4**, 648.

42 P. Papakanstantinou, S. Surbhi, S. K. Bikkarolla, A. Ganguly, D. James, *WIPO PATENTS*, WO 2012/114108 A1, 2012.

43 S. Stankovich, D. A. Dikin, R. D. Piner, K. A. Kohlhaas, A. Kleinhammes, Y. Jia, Y. Wu, S. T. Nguyen and R. S. Ruoff, *Carbon*, 2007, **45**, 1558.

44 C. Chen, Q. Zhang, M. Yang, C. Huang, Y. Yang and M. Wang, *Carbon*, 2012, **50**, 3572.

45 H. Zhang, T. Kuila, N. H. Kim, D. S. Yu and J. H. Lee, *Carbon*, 2014, **69**, 66.

46 K. N. Kudin, B. Ozbas, H. C. Schniepp, R. K. Prud'homme, I. A. Aksay and R. Car, *Nano Lett.*, 2008, **8**, 36.

47 M. R. Tarasevich, A. Sadkowski and E. Yeager, in *Comprehensive Treatise of Electrochemistry*, ed. B. E. Conway, J. O'M. Bockris, E. Yeager, S. U. M. Khan and R. E. White, Plenum Press, New York, 1983, vol. 7, p. 301

48 E. Yeager, *Electrochim. Acta*, 1984, **29**, 1527.

49 B. Sljukic, C. E. Banks and R. G. Compton, *J. Iran. Chem. Soc.*, 2005, **2**, 1.

50 K. Vaik, D. J. Schiffrin and K. Tammeveski, *Electrochem. Commun.*, 2004, **6**, 1.

51 S. Trasatti and O. A. Petrii, *Pure Appl. Chem.*, 1991, **63**, 711.

52 R. Boggio, A. Carugati and S. Trasatti, *J. Appl. Electrochem.*, 1987, **17**, 828.

53 C. C. L. McCrory, S. Jung, J. C. Peters and T. F. Jaramillo, *J. Am. Chem. Soc.*, 2013, **135**, 16977.

54 J. D. Benck, Z. Chen, L. Y. Kuritzky, A. J. Forman and T. F. Jaramillo, *ACS Catal.*, 2012, **2**, 1916.

55 (*a*) T. A. Centeno and F. Stoeckli, *J. Power Sources*, 2006, **154**, 314; (*b*) D. Qu, *J. Power Sources*, 2002, **109**, 403; (*c*) R. Signorelli, D. C. Ku, J. G. Kassakian and J. E. Schindall, *Proc. IEEE*, 2009, **97**, 1837; (*d*) M. D. Stoller, C. W. Magnuson, Y. Zhu, S. Murali, J. W. Suk, R. Piner and R. S. Ruoff, *Energy Environ. Sci.*, 2011, **4**, 4685.

56 C. Song and J. Zhang, *Electrocatalytic Oxygen Reduction Reaction*, Springer London, 2008, p. 89.

57 I. Kruusenberg, J. Leis, M. Arulepp and K. Tammeveski, *J. Solid State Electrochem.*, 2010, **14**, 1269.

Faraday Discussions

DISCUSSIONS

Applications, composites, and devices: general discussion

Santosh Kumar Bikkarolla, Mark Baxendale, Chris Ewels, Toshiaki Enoki, Katsumi Kaneko, Nazario Martín, Petrus Santa-Cruz, Rebecca Edwards, Zeba Khanam, David Zitoun, Pulickel Ajayan, Varsha Khare, Alexander Zöpfl, Thurid Gspann, Hitoshi Ogihara, Milo Shaffer, Karl Coleman, Mary Chan-Park, Pagona Papakonstantinou, Sehmus Ozden, Oana Andreea Bârsan, Alan Windle and Ian Kinloch

DOI: 10.1039/C4FD90046D

Karl Coleman opened the discussion of the paper by Mary Chan-Park: Is the source of the SWCNTs important? I ask this in the context that the ratio of SC to metallic will be different from different suppliers and methods of production.

Mary Chan-Park responded: We have not tried other suppliers of nanotubes but in the literature, gel electrophoresis sorting has been applied to HiPco® and CoMoCAT® tubes. For the radical preferential attack, the diameter effect is important.

Karl Coleman commented: During the electrophoresis does the source of the NTs effect the movement/separation on the gel.

Mary Chan-Park answered: The different types can mostly work, and the small diameter tubes are easier to separate than the large diameter tubes in general.

Thurid Gspann said: There appears to be a length issue when dealing with carbon nanotubes. On the one hand for most applications, mechanical as well as electrical, CNT ends represent the ultimate defect. In other words CNTs should be as long as possible. On the other hand, purification, solubilisation and separation processes either shorten CNTs or require them to be of a length usually less than 5 micron. What is the estimated maximum length of CNTs to be separated by gel electrophoresis? And do you see any possibilities of increasing this length to say 100 micron range?

Mary Chan-Park responded: I think the length of tubes separated by gel electrophoresis is around 0.5 micron but the measured lengths vary from 300 to 800 nm (see Fig. 5 in my manuscript).

Mark Baxendale remarked: Are the channels of your FETs made from individual semiconductor nanotubes, networks comprising only semiconductor nanotubes, or networks comprising both semiconductors and a small population of metallic nanotubes?

Mary Chan-Park answered: The channels are networks of mainly semiconducting tubes but with small amounts of metallic tubes.

Mark Baxendale asked: Is it possible to say whether the FET response is that due to the intrinsic properties of the constituent nanotubes and not simply a manifestation of the inter-nanotube junctions in the network?

Mary Chan-Park responded: It should be due to both the intrinsic properties and also affected by intertube junctions. If the tubes are not sorted, they cannot switch. Our network devices show switching of 3 to 4 orders. But the intertube resistance also affects the behaviour.

Mark Baxendale asked: From your work, is possible to comment the expected response of an FET with an individual semiconductor nanotube channel?

Mary Chan-Park answered: If we do the single tube devices, the percentage of switchable devices would correlate to the percent purity of the semiconducting tubes as determined by uv-vis spectroscopy.

Karl Coleman asked: Does the conducting performance depend on the density of the tubes?

Mary Chan-Park responded: Yes, but if there are too many tubes, the on/off ratio will be sacrificed. We typically keep the on : off ratio at the 10^{3-4} for comparison and that somewhat dictates the density of the tubes.

Milo Shaffer said: A range of mechanisms have been proposed to account for the selectivity of chemical reactions for different types of nanotube, including the availability of electrons at the Fermi level and the influence of bond strain.[1] In your system, which do you think are the most important, and what are the implications for the competition between selection by electronic character and by diameter (or any other parameter)?

Do you titrate in the radical reagent, and if so, do you see an evolution of the selectivity with stoichiometry?

1. S. Hodge *et al.*, *Chem. Soc. Rev.*, 2012, **41**, 4409–4429.

Mary Chan-Park responded: There are two main mechanisms – diameter dependence and chirality due to confinement of wave vector in the quasi-1D cylinders.

There are both effects but the arc discharge tubes are in a narrow diameter range of around 1.4nm to 1.6nm. In this range, the Fermi level electron transfer to the radicals predominate in the reactions.

It is hard to titrate the radical because they cannot be gradually controlled easily.

Rebecca Edwards remarked: I notice that there are a relatively large number of steps after the gel electrophoresis; were these steps important to the resulting properties?

Mary Chan-Park answered: Yes, the steps are mainly related to the removal of gel from the nanotube. In fact, solution spectroscopy data are relatively easier to get. To prove that devices can work, the tubes need to be sorted *and* they need to be cleaned. These steps are for cleaning and are very important.

Nazario Martín asked: Can you comment on the similarities and differences between the functionalisation depending on the source of the radicals? Have you tried phenyl radicals instead of naphthyl?

Mary Chan-Park replied: We tried the phenyl azo compounds before instead of the naphthyl azo compounds. They cannot differentiate the electronic structures of nanotubes.

Nazario Martín asked: Did you produce your radical *in situ*? They are typically quite unstable. Why did you not follow well established methodologies from diazonium salts (Tour reaction) for radical functionalisation of CNTs, graphene and fullerenes *etc.*?

Mary Chan-Park answered: The radicals are produced *in situ* by the probe sonication of azo naphthalene compounds. We have tried diazonium salts before but they are hard to control, and cannot work well for separation of large diameter nanotubes.

Karl Coleman opened the discussion of the paper by Oana Andreea Bârsan: Was the response you observed an ohmic response when you assembled your network?

Oana Andreea Bârsan responded: Our SWCNT networks showed a dominant semi-conductive behaviour with a non-ohmic response.

Karl Coleman said: When recording the electrical data from the device, did the behaviour change with addition of the polymer – did you see a different profile before and after addition of polymer?

Oana Andreea Bârsan answered: The resistance of the SWCNT films was recorded before, during, and after polymer impregnation. While blank measurements showed an increase in resistance of 3.1 times (before cooling) due to removal of dopants upon heating, all of the films showed an average increase in resistance of 4.3 times (before cooling) after the polymer impregnation process.

Mark Baxendale asked: It might be better to do impedance spectroscopy – an AC probe will give you more information given the non-ohmic response; for example, it is possible to extract the junction and bundle resistances, and junction capacitance.

Oana Andreea Bârsan responded: This is a very good idea. Thank you.

Sehmus Ozden asked: Are your CNTs randomly oriented or covalently interconnected?

Oana Andreea Bârsan replied: The SWCNTs we use are not functionalized so there are no covalent bonds between them.

Mary Chan-Park said: Did you anneal your sample before adding the polymer? At what temperature and for how long?

Oana Andreea Bârsan responded: The SWCNT films were subjected to a 400 °C treatment for two hours under a nitrogen atmosphere as part of the film preparation process.

Katsumi Kaneko asked: As scanning of AFM should give the SWCNT network structure, did you observe the AFM images before and after the scanning?

The information should give important nanolevel information on the stability of the network.

Oana Andreea Bârsan answered: Yes, we scanned the same area on a ~30 nm thick SWCNT film up to 5 consecutive times and we didn't observe any changes in the topography images.

Petrus Santa-Cruz said: The inhomogeneity shown in the peeled-off composite films (shown in Fig. 5 of your paper) seems to be greater than in SWCT films (Fig. 1d and 1f in the paper), that looks very homogeneous. What is the cause of this segregation? How does this homogeneity compare with samples prepared by traditional mixing of the SWCNTs with the uncured polymer mixture, prior to film deposition?

Oana Andreea Bârsan answered: The apparent inhomogeneity in Fig. 5 is due to the polymer that has crept in at the bottom of the SWCNTs' network during the polymer impregnation process, forming droplet areas of pure polymer. Around these areas, ropes of SWCNTs are visible sticking out of the composite film that most likely got pulled while peeling the composite off the substrate. The homogeneity at the bottom side of this composite film is not representative of the homogeneity of the SWCNTs' network inside the cured polymer matrix.

We have not prepared samples by the traditional mixing technique using exactly the same components; the high viscosity of the polymer mixture requires the addition of a solvent in the mixture and SWCNTs are extremely difficult to disperse in any type of solvent. Carbon nanotube-epoxy composites have been prepared using this technique, mostly with multi-walled CNTs, and these composites present CNTs agglomerates tens to hundreds of micrometres large (see references 10, 21, 22, 23 in the paper).

Karl Coleman asked: Do you think the change of resistance with film thickness is real?

Oana Andreea Bârsan answered: I believe you are referring to Fig. 7c, where the polymer impregnation effect (final/initial resistance ratios) is shown as a function of SWCNT film thickness. Considering that initially, a SWCNT film 15 nm thin has a resistance 30 times higher than a 135 nm SWCNT film (Fig. 7b), it is possible that the small variations in resistance ratios after polymer impregnation seen in Fig. 7c are simply due to experimental errors. This is because each sample is unique and once the polymer film is applied on the surface of a SWCNT film and the sample is heated, it is very difficult to control the spreading of the polymer on the SWCNT film surface during the polymer impregnation process.

Pulickel Ajayan enquired: Is the discrepancy you see in the resistance of your deposited nanotube films due to varying contact resistance? Have you done four probe electrical measurements of films of different thicknesses to resolve the issue?

Oana Andreea Bârsan replied: Yes, we have measured the resistance of SWCNT films of different thicknesses using both four point probe measurements and two Au electrodes with a multimeter. Even if the techniques are completely different, the results are similar as you can see in Fig. 7 b of the paper.

Zeba Khanam asked: Why choose SWCNT instead of MWCNTs? Is there any specific reason for that?

Oana Andreea Bârsan replied: We chose SWCNTs because they have a simple structure with no inner layers and no charge transport between the inner tubes. They also have a very narrow diameter distribution (1-2 nm) and a very large aspect ratio (usually above 1000). These characteristics made them promising for our goal of preparing a uniform conductive CNT network.

Zeba Khanam commented: Its not the case that we always need a surfactant or solvent to disperse the CNTs in epoxy. The supercritical carbon dioxide treatment at high pressure enhances the dispersion even without using any dispersion agent.[1]

1. M. G. H. Zaidi, Modification in mechanical, thermal and electrical characteristics of epoxy through dispersion of multi-walled carbon nanotubes in supercritical carbon dioxide, *Carbon Lett.*, 2013, **14**(4), 218–227.

Karl Coleman remarked: Is your film transparent?

Oana Andreea Bârsan replied: The films' transmittance depends on their thickness (see Fig. 7a showing absorbance as a function of film thickness). The typical ~30 nm thin SWCNT films have a transmittance of ~85%.

Chris Ewels opened the discussion of the paper by Ian Kinloch: In the context of short-fibre theory and associated stress transfer, what do you think is the importance of rippling and folding in graphene in these composites?

Ian Kinloch replied: It is known in fibre composites that fibre waviness does lead to a decrease in uniaxial properties, because the waviness effectively introduces a range of fibre alignment into the system. We would expect this behaviour to be also present in graphene composites. It is also plausible that if the graphene folds such that its overall diameter is beneath the critical diameter that the flake will not provide reinforcement.

Pulickel Ajayan asked: For the multi-layered structure of graphene, is the changing Raman peak really a result of the shear between layers or due to some form of defects? Have you done the same experiment using multi-layer graphene grown by CVD? Are the results similar?

Ian Kinloch responded: We believe that the change of the Raman peak is due to the shear of the layers and this shear produces reversible stacking defects into the structure.[1]

1. L. Gong *et al.*, *ACS Nano*, 2013, **7**(8), 7287–7294.

Petrus Santa-Cruz addressed a question to both Ian Kinloch and Pulickel Ajayan: How do you control the size of the graphene flake during electrochemical exfoliation?

Pulickel Ajayan answered: The size of the exfoliated flakes depends more on the starting material (crystallite size) compared to the electrochemical conditions, in my opinion. This is different from chemical exfoliation where the flakes can be continuously reduced in size as a function of treatment conditions and time.

Ian Kinloch replied: The diameter of the graphene flakes is approximately the same as the diameter of the starting graphite particle size. The thickness of the flakes is controlled by the diameter of the starting graphite and the number of times the exfoliation process is run. We find that 5 μm diameter graphite particles exfoliate to few layer graphene after 2–3 exfoliation cycles, whereas 20 μm diameter graphite particles require 5 exfoliation cycles. We currently cannot exfoliate graphene with a diameter greater than 20 μm. Please see reference 1 for more details.

1. A. M. Abdelkader *et al.*, *ACS Appl. Mater. Interfaces*, 2014, **6**(3), 1632–1639.

Karl Coleman commented: If you have a mixture of flake sizes do you see that the larger flakes dominate the mechanical properties?

Ian Kinloch answered: To the first approximation, any flakes above the critical diameter will fully reinforce the composite whereas any flakes below the critical diameter will not reinforce the composite. In reality, this cut-off would be broader, with the degree of reinforcement dropping off as the diameter of the flake decreases beneath the critical diameter.

Pagona Papakonstantinou asked: Did your small size flakes have more oxygen than the larger flakes?

Ian Kinloch responded: We believe that the oxygen content is the same in both sized flakes and is of the order of ~5 at%.

Pagona Papakonstantinou commented: It was mentioned that the amount of graphene that could be loaded is limited – would it be possible to explain or expand on this?

Ian Kinloch answered: The maximum loading of graphene, as with all reinforced composites, will be limited by a number of factors:

1) Having sufficient matrix to coat and bond the reinforcment into a coherent structure.

2) The increase of matrix viscosity with the addition of graphene making the system unprocessable.

3) Not being able to properly disperse the graphene.

Pulickel Ajayan asked: From the reinforcement point of view as a filler in composites, how do you compare a nanotube and a graphene nanoribbon of same dimensions; which do you think would provide better reinforcement?

Ian Kinloch responded: The answer to this question is not immediately obvious as there are several competing effects. The graphene nanoribbon will have a higher intrinsic modulus than the nanotube because the modulus depends on the cross-sectional area of the reinforcement and the nanotube has to account for a hole down its middle. There should also be better stress transfer for the nanoribbon since every atom its in contact with two polymer interfaces, while for the nanotube the atoms just have one side in contact with the polymer. However, the nanotubes will have a higher bending stiffness and thus persistence length than the nanoribbons due to the nanotubes' cylindrical shape and thus the nanotubes will have less waviness in the composite.

Pulickel Ajayan commented: Regarding reinforcement effects of the tubes *versus* ribbons: I suppose what is different is not just the effect of a central hole in nanotubes but also the exposed edges present in graphene but not in nanotubes. The interfacial chemistry between the edges and the matrix should be quite different.

Karl Coleman commented on the works of Ian Kinloch and Oana Andreea Bârsan :Which technique for producing the composite as presented in your two papers (pre-making the network or not) do you think will win out?

Oana Andreea Bârsan replied: It is difficult to compare the two techniques without a target application in mind. Both techniques are viable but for different purposes. If we consider applications such as electrostatic shielding, electromagnetic interference shielding *etc.*, the typical mixing approach might win out because it requires the use of small amounts of CNTs. Mechanical properties of the composite materials can also be improved using small amounts of CNTs, which would lead to a cheaper material. On the other hand, if applications such as transistors or flexible, transparent conductors are the target, pre-making the conductive network might prove to be more efficient.

Karl Coleman opened the discussion of the paper by David Zitoun: You make your own graphene oxide – what are the issues with commercial sources? Is it manganese contamination?

David Zitoun responded: You're right, chemical contamination is the main concern (especially Mn).

Karl Coleman commented: Your device seems to have more consistent cycling performance when using rGO – the capacity of the GO sample is dipping below the rGO. Is the GO turning into rGO?

David Zitoun answered: The hypothesis of an electrochemical reduction of GO to rGO makes sense at the working potential (between 0 and 1 V *vs* Li^+/Li). Only Raman spectroscopy after cycling would give a clear answer.

Karl Coleman remarked: Should we read anything into the fact that the capacity curves for GO and rGO crossover?

David Zitoun answered: Not really, it means that some of the capacity of Si cannot be accessed at the beginning in the rGO (too many mechanical constraints) while GO is a much more flexible matrix. However, it is too flexible when it comes to long-term stability.

Pulickel Ajayan said: You have used a half cell. So why is the performance of the Si–GO anode better than Si–rGO?

David Zitoun responded: The active material performs as the electrode *versus* metallic Li (half-cell) in excess. The limiting factor is still the Si which needs to be well encapsulated (Si–rGO), otherwise, some of the Si "delaminate".

Pulickel Ajayan asked: Is the weight the total weight of the anode reported Si and GO or just Si alone?

David Zitoun responded: We've reported the values normalized to Si weight; GO does not show any Li intercalation in our experiments.

Santosh Kumar Bikkarolla communicated: You have mentioned glassy carbon should be used as the counter electrode instead of platinum wire. I am wondering about the conductivity of a glassy carbon electrode due to the amorphous nature. Instead of glassy carbon electrode it will more appropriate to use graphite rod as counter electrode which is more conducting due to crystalline nature as done by

Liang *et al.*.[1] Can you provide any reference where a glassy carbon has been used as the counter electrode?

1. Y. Liang *et. al.*, *Nat. Mater.*, 2011, **10**, 780–786.

David Zitoun communicated in reply: Glassy carbon is used routinely in electrochemistry. The company Pine is for instance selling rotating disc electrodes and counter electrodes. The use of GC as counter electrode is a directive from the Argonne labs in Los Alamos. However, it is fair to mention that many articles still use platinum as counter electrode.

Rebecca Edwards opened the discussion of the paper by Alexander Zöpfl: It states in your paper that you get reproducible electrical properties from your spin-coated rGO films, does that include reproducibility across different batches of rGO?

Alexander Zöpfl responded: The reproducible electrical resistance is given in a range of 100–250 kΩ. This range covers the resistance of different batches of reduced graphene oxide (rGO). The deviation reflects only partially properties of the material itself, it is more a result of the spin coating process. The preparation of rGO *via* Hummers' method followed by a reduction shows only little variance from batch to batch. In our opinion, one of the most critical influences is the starting graphite material. Here, we observed *e.g.* great differences between powder and flake graphite (which was used in this work) in flake size, dispersibility, and also in the Raman spectra. For our different batches from the same starting material, we could not observe changes in Raman characteristics or the final sensing behaviour which are tolerable for any practical application. To achieve highly reproducible electrical resistance was not the focus of our work. Here, we concentrated more on the optimisation of only some spin coating parameters. The substrate surface (properties, structure), the spinning solution (concentration of rGO, spreading), and the spin coating process itself (speed, acceleration, time) have been investigated. In an industrial process one has also to consider parameters like *e.g.* the adjustment of the substrate chips in the spin coater or the reproducible deposition of the spinning solution at identical positions, which can influence the final sensor properties.

David Zitoun commented: Do you have a way to measure the time response of your sensors? What is the role of the reduced graphene oxide in the sensor?

Alexander Zöpfl replied: The reduced graphene oxide (rGO) is used as asensor material in a chemiresistive setup. The material undergoes a very fast and high sensitive change in its electrical conductance upon gas adsorption already at low temperature (25–85 °C). This makes it superior to conventional metal oxide sensor materials which need to be operated at higher temperatures. Reduced graphene oxide shows more p-type semiconducting behavior and therefore mainly positive charge carriers (holes) are involved. The electron transfer from rGO to adsorbed gas molecules leads to more positive charge carriers and therefore the conductance of rGO is improved, as observed *e.g.* for NO_2.

The thickness of the sensing layer is crucial for fast and reversible detection of gases. Therefore, a 2D material which practically consists only of a surface without any bulk phase is clearly favourable to get a fast and reversible response. Our sensor obviously consists of multiple layers which is clearly reflected in the sensors response (see *e.g.* Fig. SI8b in the paper) comprising of a fast and huge change in signal in the beginning and a smaller and slower change afterwards, which is a consequence of the diffusion of the analyte into the sensor layer.

David Zitoun asked: Can you synchronize your gas sensing/mixing lines with the potentiostat to measure the time response of the sensor?

Alexander Zöpfl replied: The gas mixing device and the sensing sourcemeter could be synchronized to measure the response time of the sensor. However, it is also important to know that the gas is completely mixed before it reaches the sensor. The two main parameters in this context are the size of the measurement cell and the flow rate. Many publications lack such practical information. Considering a real application, we chose a decent flow rate of 100 sccm, a relatively small gas chamber (see Fig. SI1b in the paper) and an additional mixing tube right after the mass flow controller (see Fig. SI1a in the paper). Yet, we still could observe an impact of the flow rates on the measured signal change and therefore only time responses below 1 min have been reported. Much shorter response times in the region of seconds have been obtained when using the sensor outside the gas chamber under ambient conditions. Due to environmental effects on the sensor setup these measurements are difficult to compare and were not mentioned in the manuscript.

David Zitoun asked: There is another mechanism of sensing with Pd based on volumetric expansion of Palladium hydride – you get the opposite response? What is your mechanism?

Alexander Zöpfl responded: The signal originates from catalytic effects of the metal nanoparticles on the graphene. Similar observations have been made for single walled carbon nanotubes and nanoribbons which were decorated with Pd.[1,2] It is supposed that hydrogen atoms dissociate within Pd and lower its work function, allowing electrons to jump easily from Pd to the graphene structure. The electron transfer results in an increase of resistance and respectively a decrease in conductance because of the p-type semiconducting behaviour of reduced graphene oxide.

1. S. Mubeen, T. Zhang, B. Yoo, M. A. Deshusses, and N. V. Myung, *J. Phys. Chem. C*, 2007, **111**(17), 6321–6327.
2. J. L. Johnson, A. Behnam, S. J. Pearton, and A. Ural, *Adv. Mater.*, 2010, **22**, 4877–4880.

Karl Coleman commented: You are analysing single component gases – what happens if you have a mixture?

Alexander Zöpfl answered: In fact, we are already dealing with a gas mixture of oxygen and nitrogen of the synthetic air carrier gas and the analyte gas. Humidity also was not excluded for the low temperature measurements. An array of different modified sensors is proposed to distinguish between different single component gases. The selectivity of each sensor for each gas was tested, and from the results it is expected that an addition of further analyte gases to the mixture results in a different signal pattern for all of the sensors. Data analysis by the model of principal component analysis should allow the detection of a single analyte in a complex gas mixture. This could not be tested within this study due to the limitation in the gas mixing device which only consists of three mass flow controllers.

David Zitoun opened the discussion of the paper by Santosh Kumar Bikkarolla: 99% of the literature is working on cells with acidic electrolytes – why are you using alkaline? How long do you test your materials for?

Santosh Kumar Bikkarolla responded: There is a paper Koper[1] which states: "Several reasons have emerged as to why one should prefer an alkaline electrolyte to an acidic one. Many materials are more stable in alkali."

1. M. T. M. Koper, *Nat. Chem.*, 2013, **5**, 255–256.

Rebecca Edwards remarked: Given that you are using graphene oxide (and partially reduced GO) in alkaline solutions do you think there could be an effect on the catalysis during the reaction brought about by the removal of oxidative debris from the graphene oxide surface (as described by Thomas *et al.*[1])?

1. H. R. Thomas *et al.*, *Chem. Mater.*, 2013, **25**, 3580–3588.

Santosh Kumar Bikkarolla answered: In the work presented in Thomas *et al.*,[1] as well as that in Rourke *et al.*,[2] oxidation debris were successfully separated from GO by using base-washing protocols. These protocols involved either (i) treatment in weak basic solutions for long periods (more than 3 h) or (ii) smaller periods assisted by heat treatment or (iii) treatment in strong basic solutions.

In our experiments the ORR performance of graphene oxide or electrochemically reduced graphene oxide was carried out on electrodes produced by drop casting ink on glassy carbon with nafion as a binder at a weak basic solution of 0.1 M KOH solution at room temperature. Removal of oxidation debris from the surface of the electrode cannot take place under our operation conditions, however very mild deoxygenation of the electrode surface may take place only when experiments are carried out over the duration of many hours.

1. H. R. Thomas *et al.*, *Chem. Mater.*, 2013, **25**, 3580–3588.
2. J. P. Rourke *et al.*, *Angew. Chemie Int. Ed.*, 20131, **50**, 3173–3177.

David Zitoun commented: But the issue isn't the ORR in an alkaline cell, as shown in the review article.[1]

1. M. T. M. Koper, *Nat. Chem.*, 2013, **5**, 255–256.

Santosh Kumar Bikkarolla replied: We have to disagree on this point. The efficient ORR from inexpensive catalysts is a serious bottleneck in the commercialization of current fuel cells working in either acidic or alkaline environments.

This problem is well documented in the literature and an enormous amount of research effort has been dedicated to the replacement of Pt catalysts (see references[1,2,3]). The paper by Koper[4] does not elaborate on the ORR of inexpensive catalysts.

1. Y. Liang *et al.*, Co_3O_4 nanocrystals on graphene as a synergistic catalyst for oxygen reduction reaction, *Nat. Mater.*, 2011, **10**, 780–786.
2. J. Suntivich *et al.*, Design principles for oxygen-reduction activity on perovskite oxide catalysts for fuel cells and metal-air batteries, *Nat. Chem.*, 2011, **3**, 546–550.
3. F. Cheng *et al.*, Rapid room-temperature synthesis of nanocrystalline spinels as oxygen reduction and evolution electrocatalysts, *Nat. Chem.*, 2011, **3**, 79–84.
4. M. T. M. Koper, Hydrogen electrocatalyis: A basic solution, *Nat. Chem.*, 2013, **5**, 255–256.

Santosh Kumar Bikkarolla asked David Zitoun: You have mentioned that silver is a non-expensive electrocatalyst which can allow fuel cells to be commercialized. Would it be possible for you to provide some references which mention this statement? In addition can give some references which study the ORR performance of silver in fuel cell conditions in acidic media?

David Zitoun answered: There are some references describe the ORR in alkaline medium. One of the most up-to-date reviews is by Neburchilov *et al.*[1]

1. V. Neburchilov, H. Wang, J. J. Martín, W. Qu, *J. Power Sources*, 2010, **195**, 1271.

Hitoshi Ogihara commented: The properties of rGO and GO electrocatalysts such as types of surface functional groups, electrical conductivity, and electrochemically active surface area are different. What is the most important factor effecting the performance for ORR activity of the electrocatalysts?

Pagona Papakonstantinou answered: Graphene oxide, GO (>30 at% oxygen) is an insulator. The electrocatalyst needs to be conductive to allow efficient electron transfer for ORR. For this reason GO has low ORR activity. So conductivity is a very important factor.

One should note that in many cases, the reduction of GO is usually accompanied by nitrogen incorporation even in small amounts inherited by commonly used reduction agents (eg N_2H_4, NH_3–OH). Overall the reduction process reduces the work function of the material. Addition/combination of heteroatoms (oxygen is always present) helps to reduce further the work function and enhances the efficiency of ORR from 2e towards 4e.

Zeba Khanam communicated: I want to know the meaning of term background signal in LSV? Kindly elaborate more with respect to ErGO.

Pagona Papakonstantinou communicated in reply: The background response is measured by taking LSVs in an Ar or N_2 saturated solution. By subtracting this background signal from the one obtained in an O_2 saturated solution we get the true ORR response of the material under investigation.

Zeba Khanam communicated: I could not understand the results of the impedance studies. The GO shows a large arc and large charge transfer resistance with higher impedance than ErGO. Why it is so? Kindly explain.

Pagona Papakonstantinou communicated in reply: The impedance spectra mirror the ORR activity because a small bias has been applied close to the onset potential . A high ORR activity is reflected by a small semicircle which indicates a small charge transfer resistance as it allows fast shuttling of electrons during ORR. You may see this also in Fig. 9b of the paper which provides the modulus of impedance ($\sqrt{(Z_{real}^2 + Z_{imag}^2)}$) as a function of frequency. In the low frequency regime it is obvious that the modulus of Z is smaller for the electrochemically reduced graphene oxide. Basics of impedance spectra can be found on the application notes of various companies specializing on EIS.

Petrus Santa-Cruz presented some slides of his groups' recent work: At the LandPhoton Group (Chemistry Department of UFPE-Brazil), we have developed a photonically-marked CNT with rare earth coordination complexes, in which the CNT act as an antenna in this system. We named the luminescent CNT as LCND, from Light Conversion Nanostructured Device.

Previously we have designed and produced photonic molecular devices by the supramolecular association of chemical species in a supramolecular way,[1] leading to Light Conversion Molecular Devices (LCMD) as similar to those described by Lehn.[2]

For the new use of CNT as antenna in the LCND I am presenting at this meeting, we employed green chemistry methodology to functionalize in two steps, first by a click reaction assisted by ultraviolet radiation, and then, by complexation assisted by microwave irradiation. This process was presented in our poster at this meeting entitled "Photonic carbon meta-nanotubes as light conversion nanostructured devices (LCNDs)".

Here we chose the europium for the active part of the complex, and present a MWCNT–Eu^{3+} LCND with a better quantum efficiency than the non-hybrid complex, as shown in our poster. In addition, the final material presents excellent dispersion in water, broadening the applications for these photonically-marked CNTs, to track then in nature or biological systems, as shown in my slides.

The highly efficient photonic CNT–Eu hybrid system is produced in two steps: in the first one a previous photonic functionalization occurs with 4-Azido benzoic acid under UV-B radiation, and in a second step, the europium complex is growth under microwave irradiation. LCMD molecular devices are based on the "antenna effect" of ligands. We propose that in these our nanostructured LCND devices, the covalently functionalized carbon nanotubes may act as a nanostructured antenna, since we have experienced an enhanced luminescence quantum yield.

1. G. F. de Sá, O. L. Malta, C. Donega, A. M. Simas, R. L. Longo, P. A. Santa-Cruz, E. F. Silva, Spectroscopic Properties and Design of Highly Luminescent Lanthanide Coordination Complexes, *Coord. Chem. Rev.*, 2000, **196**, 165–195.
2. N. Sabbatini, M. Guardigli, J.-M. Lehn, Luminescent Lanthanide Complexes as Photochemical Supramolecular Devices, *Coord. Chem. Rev.* 1993, **123**, 201–228.

Varsha Khare opened a general discussion of the topics raised at the meeting: In the discussion at this meeting, the use of carbon materials in controlling environmental pollution was hardly mentioned. As scientists, we have responsibilities to society *e.g.* to provide safe and clean water free from toxicity.

Zeba Khanam communicated: As a research scholar in the Dept. of Environmental Science working with CNTs, I would be interested in the comments of the delegates on the applications of carbon nanomaterials in environmental science, in particular with reference to the papers presented at this meeting.

Pulickel Ajayan commented: We are nearly 30 years into the emergence of carbon nanomaterials, starting with the discovery of fullerenes. It seems that the field of carbon nanoscience has had a great run with a new material emerging as the savior when another wanes in popularity; such as nanotubes replacing fullerenes and graphene stealing the limelight from nanotubes. Can we predict what might be the "next kid on the block"?

Alan Windle replied: For the materials scientists and technologists, the challenge to turn nanocarbons into useful materials where they positively impact industry and thus society is really just getting underway. They are all going to be rather busy with what we have got. But what are the successors to buckyballs, CNTs and graphene ? In one sense we are probably rather running out of dimensions to do much new with carbon. One way of answering this question is to say what fascinates me in the carbon allotrope area. I think I would go unstable and start taking a deeper interest in linear acetylenic carbon (carbyne to its friends),[1] which is a carbon-only chain with alternating triple and single bonds. It is of course highly unstable and predicted to exothermically cross link, perhaps explosively, which would add to the fun. Staying with instability, there are also the diamondoids (*e.g.* reference 2), small cage structures larger than adamantane but with an upper limit at what one might call a nano diamond crystal. They are stabilised by non-carbon additions and thus not really allotropes, but with carbon positions which tend to map onto the diamond lattice.

However, will there be another interesting, distinct carbon allotrope? Probably, but I cannot guess what, although there is an important pointer from the history of the three types of nanocarbon we currently focus on. Whatever the new form will turn out to be, it will be likely to have left calling cards for us to recognise. C_{60}, probably the one nano allotrope which did really come from left field, had been leaving its fingerprint for decades before 1985 as an anomalously large peak at 720 amu in mass spectra on equipment which could handle such heavy molecules. Carbon nanotubes had been seen and recognised, as sub micron fibrils (the nano word had yet to be applied) almost since electron microscopes were invented, certainly since 1952. ICI ran reformers at one of their plants in North England. Regularly, they had to close the plant down, in order to scrape a strange fibrous carbon deposit from the inside of the nickel steel reaction tubes. They buried kg of the black stuff, so perhaps future industrial archeologists will discover a carbon nanotube deposit on the site, even mine it. The key papers which introduced the science of soot to the scientific *cognoscenti* were from Iijima, who in the words of Rick Smalley, definitely discovered nanotubes best, if not first. As for graphene, the molecule of graphite – if one can accept 'molecule' when its molecular weight is undefinable – its isolation using 'sellotape' was presaged by Tony Kelly (arguably the father of composite science) who used exactly that technique to prepare graphite samples thin enough so they could be observed in the TEMs of the 1960s. Perhaps he did get down to a single layer, but his interest was in interlayer dislocations. Of course, the recent Nobel prizes for graphene were rewards for putting everything together, and carrying our brilliant electrical measurements on graphene, and relating the results, which will be of great commercial value, to the physics derived from the Dirac, *K* points. These three nanocarbon structures, the pinnacles of progress in carbon nanoscience, have happened because people had fun, which somehow encouraged lateral thinking – the true mother of invention.

1. R. H. Baughman, *Science*, 2006, **312**(5776), 1009.
2. M. N. R. Ashfield, P. W. May, C. A. Pego, and N. M. Everitt, *Chem. Soc. Rev.* 1994, **23**, 21.

Sehmus Ozden remarked: Graphene nanoribbons and graphene quantum dots can be synthesized using solution chemistry. I wonder if it might be possible to synthesize carbon nanotubes *via* solution chemistry?

Toshiaki Enoki asked: Most of carbon nanomaterials, which have been investigated, are mostly sp^2-based carbon nanomaterials. In this meeting, we discussed sp^2-based carbon nanomaterials, but sp^3 carbon nanomaterials should be added to the carbon nanomaterials in the future. Actually, when the size of diamond becomes nanodimensional, the structural instability is expected to give interesting phenomena. The issues surrounding this should be investigated.

Faraday Discussions

PAPER

Concluding remarks: summary and some pointers for the future

Alan Windle*

Received 24th October 2014, Accepted 24th October 2014
DOI: 10.1039/c4fd90040e

Introduction

Faraday Discussions often have a significance beyond the presentation of the latest incremental advances in knowledge, for they can reset priorities in a field, and sometimes signal the denouement of various models which had become sacred cows to some. Whether we will look back to the markers put down during this meeting, and say that it was a watershed for carbon nanoscience, only time will tell.

The title which has brought us here "New Advances in Carbon Nanomaterials" is significant, not so much the "New Advances" bit – it is perhaps difficult to argue for an advance which is otherwise than new, although one can all too often find the converse! – rather, it is the choice of the word "Nanomaterials" which set the scene for much of the science we have shared. The word "materials" has the implication of utility, one might argue that a material is a substance which is useful as a solid, so the use of this word does mean that the science being pursued should have a practical objective in view. The 'applications angle' has underpinned so many of the papers presented at this meeting, keeping it "on message" throughout.

To me, one of the delights was the way in which the meeting ranged effortlessly across the different carbon nanoforms, from functionalised C_{60} to functionalized graphene, from single wall carbon nanotubes for display applications to structural composites based on graphene. It should not be surprising that there is huge commonality in the science.

In the wake of UK's Nobel Prizes for graphene, we have seen locally a polarization in the rush for 'graphene gold' to fund new research and to build new laboratories. The view that graphene is something very new and that, by implication, C_{60} and nanotubes are yesterday's science, has been heard. Such views were perhaps necessary to secure government attention and funding, but there was nothing of that nature in this meeting. Our background quest was for unifying principles and relevant techniques, and the discussion format helped to achieve science at its best. We also ranged across chemistry, physics and

Department of Materials Science, Charles Babbage Road, Cambridge, CB30FS, UK. E-mail: ahw1@cam.ac.uk

materials science with the occasional nod towards biology, testing our agility of understanding across traditional disciplines, but challenged to make the effort by Professor Ajayan's masterful introduction which launched the meeting (DOI: 10.1039/c4fd90039a).

Making nanotubes by CVD

Perhaps not surprisingly the first papers scheduled focused on the synthesis of carbon nanostructures, Gspann *et al.* (DOI: 10.1039/c4fd00066h) described the latest advances in floating catalyst synthesis of carbon nanotube materials, both fibres and non-woven mats, spun directly from the reaction zone,[1] while Professor Baxendale (DOI: 10.1039/c4fd00071d) explained how he had used a CVD process to create nano-scale iron particles covered with an aura of carbon nanotubes, which showed ferromagnetic properties. Similar particles had also been seen in the paper by Gspann *et al.*, although in that work the emphasis was on their elimination from the fibre product as undesirable catalyst residues. It was noted however by both sets of authors that these larger, nanotube covered, iron particles tended to form close to the reactor walls. The role of sulphur in floating catalyst synthesis was introduced at this stage, and the theme rattled through most of the meeting. The exact role of sulphur in carbon nanotube floating catalyst synthesis is not fully understood. It seems that the sulphur could partially coat the growing iron catalyst particles to ensure release of the growing graphene coating as a nanotube, a sort of surfactant role[2] with echoes of catalyst chemistry where a partial sulphur covering can achieve optimum catalyst activation.[3] Additionally, sulphur is known as a 'promoter' for hydrocarbon reaction, and it is possible that sulphur enhances the edge growth of graphitic layers by a mechanism yet to be molecularly elucidated. A significant development of the CVD process was reported in a poster by Anthony *et al.* It has eventually proved possible, building on earlier work,[4] to grow CNTs from the surface of carbon fires without compromising the strength of the latter. This advance links to one of the key hopes for carbon nanotubes; the improvement of electrical conductivity in carbon fibre composites for aircraft to enhance their ability to sustain lightning strikes. It is hoped that carbon fibres with hairy coatings of nanotubes could provide sufficient inter-fibre electrical percolation to give the required conductivity.

Low temperature chemical synthesis

Another route to carbon nanostructures, which is as yet unexploited, is through the traditional 'wet' methods of low temperature chemical synthesis. There are a growing number of reports in the literature *e.g.* ref. 5, albeit rather isolated, of classic synthetic routes to making nanotubes, often through the clever construction of a hemispherical end (half a fullerene) whose active edge is then grown into a nanotube, possibly by reversion to CVD methods. If one views a carbon nanotube, at least a single wall one, as an ultimately rigid polymer molecule, then one might see classic synthesis as a sensible method to develop, with the prospect of scalability to industrial levels. We have had a glimpse or two of this in this meeting. In one paper (Peña Alvarez *et al.*, DOI: 10.1039/c4fd00103f) aromatic rings were put around fullerenes, with the resultant structure being tracked by Raman. These ring molecules seem tantalizingly close to a key building

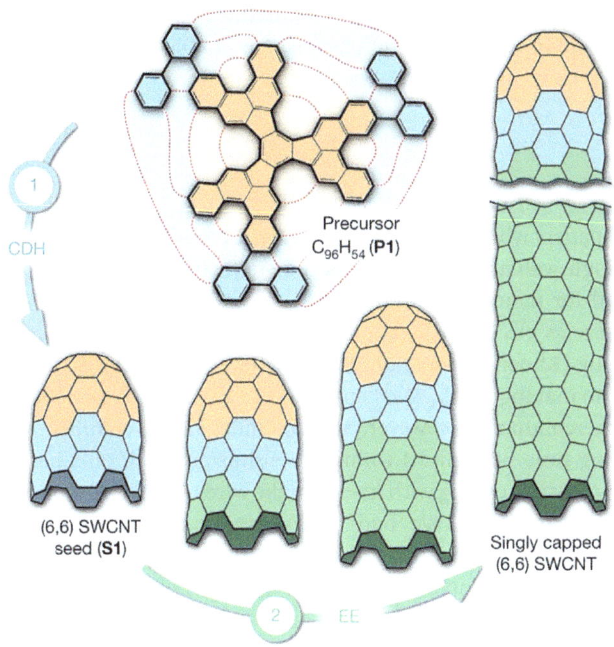

Fig. 1 Schematic of a potential synthesis route.[8]

block for nanotubes, and have been applied in this respect;[6] in another paper (Vo *et al.*, DOI: 10.1039/c4fd00131a), chevron-like graphene nanoribbons had been solution synthesized[7] and their electrical properties measured with a view to their application in electronic devices. All I can say is "watch this space"; there may be a disruptive technology in the wings. If enough synthetic polymer chemists can be excited by nanotubes, not to mention the funding bodies, then the somewhat isolated reports in the literature might join up and we would have a revolution, which could leave our beloved CVD approach looking rather old-hat. The very recent Nature paper by Sanchez-Valencia *et al.*[8] demonstrates the synthesis of a flat version of the end cap of a nanotube on a platinum surface followed by an anneal, which causes the planar molecule to jump into three dimensions to make the nanotube seed. The seed can then extended by more conventional CVD. They call it a two-step bottom-up synthesis. The beauty of such an approach is that the tubes are all monochiral, indeed all (6,6) armchair, and presumably future variation of the initial synthesis protocols will enable one to dial-in any desired chirality. On the other hand the need for a flat surface catalyst may ask questions as to the process's eventual scalability. As this paper had hardly appeared at the time of the meeting, a schematic from it appears as Fig. 1.

It is important however, that we do not lose sight of the fact that collections of mixed single nanotubes can be separated into chiralities, or at least into metallic and semiconducting fractions. The method described in this meeting (Mesgari *et al.*, DOI: 10.1039/c4fd00092g) involved the selected attack of metallic nanotubes followed by separation of the semiconducting fraction by electrophoresis. Issues which remain with all these separation methods are the maximum length of nanotubes that can be successfully separated, which may not be very high, and also the problem of scalability. Micrograms may enable the science, but hardly an industry.

Doping

We enjoyed papers which examined what might be called 'proximity effects'. The doping of single wall nanotubes by naphthalene, especially the di-nitro derivative, was achieved for nanotube bundles, with the point being made that CNTs are seen to behave as electron acceptors, but only because of a background level of oxygen doping (Khoerunnisa *et al.*, DOI: 10.1039/c4fd00119b). There was also a clever route to making Cs_3C_{60} (McLennan *et al.*, DOI: 10.1039/c4fd00085d), a mid range T_c superconductor. However, superconductivity in co-fullerenes has been around a long time now, and I think my enthusiasm will only be re-awakened when I see workers wrestling with how to make wires out of this type of compound and how to terminate it. We had demonstrations of the skills of synthetic chemistry in studies of functionalization and grafting. The action of dilute hydrochloric acid in generating hydroxyl functionalized pits, a study which demonstrated the power of XPS in expert hands (Buono *et al.*, DOI: 10.1039/c4fd00061g) was one, while taking this further, the exploitation of key functionalized sites as sites for grafting other chemical moieties onto nanotubes (Hu *et al.*, DOI: 10.1039/c4fd00116h), another.

Surfaces and edges

The discussion was never destined to get far away from surfaces, and so it proved, and also, as a sub-set of surfaces – edges. Surface treatment of graphene with both oxygen and nitrogen plasmas, to etch away the layer from the substrate, depends not only on the heat treatment but also on whether the plasma is based on oxygen or nitrogen, *i.e.* whether it is a sputter or etching process, the nitrogen sputtering being the more aggressive (Hui *et al.*, DOI: 10.1039/c4fd00118d). It seems that nitrogen can stabilize regions of high curvature in encouraging the formation of pentagons or pentagon pairs[9,10] which, coupled with nitrogen's ability to stabilize edges, may help us understand observations that carbon nanostructures synthesized in the presence of nitrogen seem to take on an unusually wide variety of forms (Ewels *et al.*, DOI: 10.1039/c4fd00111g), a sort of 'Cambrian explosion' of nanostructures. Not surprisingly the edges of nanoribbons are very important too, and we were treated to an exact and beautifully illustrated description of the influence of hydrogen-terminated edges in graphene (Fujii *et al.*, DOI: 10.1039/c4fd00073k). The level of understanding achieved so far is encouraging, and we should confidently expect continued progress on the detail in the not-to-distant future. The role of sulphur, oxygen and nitrogen were all discussed at various points in the meeting in relation to their influence on nanocarbon forms (see published discussion contributions, DOIs: 10.1039/c4fd90042a, 10.1039/c4fd90044h, 10.1039/c4fd90045f and 10.1039/c4fd90046d), with various views being promoted regarding similarities and differences in their various modes of action.

Fullerenes

Fullerenes may not yet have hit the big time as materials, but they have certainly joined the synthetic chemist's armoury as another sort of molecular group to be covalently added into other molecules over which they have traditional command as molecular engineers. This discussion brought to the fore several papers on the modification of fullerenes (Kratzer *et al.*, DOI: 10.1039/c4fd00069b; Yamakoshi *et al.*, DOI: 10.1039/c4fd00076e) including the issue of chirality in these structures (Girón *et al.*, DOI: 10.1039/c4fd00065j). I am not sure where C_{60} as a molecule is going in terms of future materials. It is perhaps reminiscent of the pure mathematicians who sell their subject as 'applicable maths' if not actually applied. Dangling fullerenes on other molecules is certainly applicable chemistry – long may it continue.

Theory

We have had two particularly intensive theory papers. There was the beautiful Moiré fringe work, not new in conception, but beautifully applied to graphene layers (Cosma *et al.*, DOI: 10.1039/c4fd00146j), and a study of the stability of endohedral fullerenes to establish stability criteria (Rodríguez-Fortea and Poblet, DOI: 10.1039/c4fd00096j). However, molecular modeling has been an underground theme, emerging on occasions in a number of papers. In the last decade computational modeling has come of age with its reliability established, or at least its limitations understood and accepted by the community at large rather than being used to belittle the whole approach. Where processes occur on the nano-scale – and the nucleation and growth of nanotubes from 1 nm metal particles is a nice example – then modern computational power enables the process to be simulated in chemical detail, using highly realistic inter atomic force fields, with model atom counts being small enough to permit simulation over useful time scales. Experimental *in situ* analysis is difficult, especially for high temperature processes, and I expect that within the next ten years many of these experimental studies will be *modelling led*.

Applications

If a material is only really a material when it is proved useful, it is comforting that we have had our attention focused on a diverse range of applications for carbon nanostructures: ink-jet printing (Arapov *et al.*, DOI: 10.1039/c4fd00067f), electrical conductivity (Bârsan *et al.*, DOI: 10.1039/c4fd00087k), mechanical properties (Vallés *et al.*, DOI: 10.1039/c4fd00112e), batteries (Miroshnikov *et al.*, DOI: 10.1039/c4fd00089g), and sensors (Zöpfl *et al.*, DOI: 10.1039/c4fd00086b; Bikkarolla *et al.*, DOI: 10.1039/c4fd00088a). In the area of fibres made from carbon nanotubes, a recent book[11] has sections dealing with the application of nanotubes to electromagnetics, motors, robots, biomedical devices, composites and antennae. A typical criticism leveled at the 'nano-revolution' is disguised as the questions: when is it really going to deliver?, when will society actually feel its benefit? It is likely that several of the applications mentioned will make it into our domestic lives, but development of new technology often takes decades – by which time all the scientific hype has moved elsewhere. However, a very recent announcement regarding the application of carbon nanotubes means that there is probably the first 'killer application' for nano carbon, for nanotubes in this instance. The report concerns a Chinese company which makes touch screens and who are major suppliers for Apple. They use nanotubes grown on substrates as arrays and push these flat, in an automated process, to achieve a layer which is conductive yet sufficiently transparent to replace ITO technology. The driver for its introduction is explained as intrinsic cheapness. Perhaps this nanotube technology will, in turn, be overturned by emergent graphene approaches; we shall see. Two paragraphs of the article from the business journal "DIGITIMES"[12] (July 2014) are reproduced in Fig. 2 below.

It appears that soon, all of us who handle Apple devices at least, will be stroking screens containing carbon nanotubes as the active component. An output of 12 million CNT-based touch screens per month is a very clear pointer to the future impact of nanotechnology. Carbon nanoscience is now delivering in a big way, so perhaps the critics might fall silent.

Finally, the question on many peoples' lips is, will – and if so when – graphene supplant carbon nanotubes? To address this knotty one, it is perhaps helpful to divide applications between devices and bulk materials. The carbon atom may have the ability to bond in different ways to give a rich range of properties greatly exploitable in devices. However, some of the carbon bonds are especially rigid and strong, and the atom is amongst the lightest. It is not obvious that electronics will actually be exploiting the strength of the covalent bonds nor the very low density of carbon materials. The actual mass of nanotubes in a touch screen is vanishingly small, and thus the lightness of the atom is not being exploited. Lightness comes into its own with bulk materials and is preeminent in applications such as aerospace. If one is synthesising materials in the form of thin films on substrates, one has a process where the scale-up to tonnage quantities is daunting. A key for the development of graphene as a reinforcement for structural composites, which was the driver behind one of our application papers (Vallés *et al.*, DOI: 10.1039/

CNTouch receives nanotube touch panel orders from major smartphone vendors in China

Ninelu Tu, Taipei; Joseph Tsai, DIGITIMES [Wednesday 16 July 2014]

CNTouch, a subsidiary of Foxconn Electronics (Hon Hai Precision Industry), has received nanotube-based touch panel orders for entry-level and mid-range smartphones from China-based major vendors. In addition to production lines in Tianjin, China, the company has established new production lines in Guiyang, China, according to sources from the upstream supply chain.

*****

Currently, CNTouch's Tianjin production lines are able to produce **three million 4-inch panels per month** and its production lines in Guiyang are currently able to deliver about **three million 5-inch panels per month**. After its Guiyang plants are completed, the site will be able to output **six million 5-inch panels per month**, the sources noted

Fig. 2 Extract from "DIGITIMES" report.

c4fd00112e), will be its ready availability, underlying the importance of developing routes to the exfoliation of graphite mined from the earth. There has been recent exciting progress in this area,[13] and it may be there that graphene finds an important exploitation route. As for electronics, it seems that the huge new effort in the UK to drive graphene research is really opening up a more general field of two-dimensional materials, based on layered solids, or van der Waals solids as they are sometimes called. It may be that the layers employed in devices of the future are boron nitride, or even a sulphide, exploiting the science following the graphene revolution.

And now

Faraday Discussion 173 appeared very successful, although the real test will be the degree to which it has led us, as participants, to recalibrate our research enthusiasms for the future. One will watch citations levels of the 24 papers in the years ahead with great interest.

Acknowledgements

I am grateful to the RSC for keeping alive the Faraday Discussion meeting format over so many decades and for the organizing committee for making Discussion 173 such an excellent example of the genre. I also thank Dr Thurid Gspann for helpful comments on the manuscript.

References

1 Y. L. Li, *et al.*, *Direct spinning of carbon nanotube fibres from CVD synthesis*, *Science*, 2004, **788**, 221–226.
2 G. G. Tibbetts, *et al.*, *Carbon*, 1994, **32**, 569.
3 M. S. Kim, N. M. Rodriguez and R. T. K. Baker, *J. Catal.*, 1993, **143**, 449.
4 H. Qian, E. S. Greenhalgh, M. S. P. Shaffer and A. Bismark, *J. Mater. Chem.*, 2010, **20**(23), 4751–4762.
5 A. Mueller, K. Y. Amsharov and M. Jansen, *Fullerenes, Nanotubes, Carbon Nanostruct.*, 2012, **20**, 401–404.
6 H. Omachi, *et al.*, *Nat. Chem.*, 2013, **5**, 572–576.
7 T. H. Vo, *et al.*, *Nat. Commun.*, 2014, **5**, 3189.
8 J. R. Sanchez-Valencia, *et al.*, *Nature*, 2014, **512**, 61–64, DOI: 10.1038/nature13607.
9 N. Hellgren, *et al.*, *Phys. Rev. B: Condens. Matter*, 1999, **59**, 5162.
10 C. P. Ewels, *Nano Lett.*, 2006, **6**, 890–895.
11 *Nanotube Superfiber Materials*, ed. M. J. Schulk, V. N. Shanov and Z. Yin, Elsevier, Amsterdam, 2014.
12 N. T. Taipei and J. Tsai, *DIGITIMES*, Wednesday 16th July 2014.
13 K. R. Paton, *et al.*, *Scalable production of large quantities of defect-free fewlayer graphene by shear exfoliation, Nat. Mater.*, 2014, **13**, 624–630, DOI: 10.1038/nmat3944.

Poster titles

Addend distribution on fullerene cages in radical addition products of IPR open-shell fullerenes C_{74} (D_{3h}) and C_{76} (T_d), **Ayrat R. Khamatgalimov and Valeri I. Kovalenko**, *A.E. Arbuzov Institute of Organic and Physical Chemistry, RAS, Russia*

Regioselectivity in radical addition products of higher fullerene C_{90}, **Ayrat R. Khamatgalimov, Regina Tuktamysheva and Valeri I. Kovalenko**, *A.E. Arbuzov Institute of Organic and Physical Chemistry, RAS, Russia*

Carbon nanotubes supporting CeO_2 as a superior catalyst for the selective catalytic reduction of NO with NH_3 at medium temperature, **Xiongbo Chen, Chaoping Cen, Ping Fang, Zhongbiao Wu and Haiqiang Wang**, *South China Institute of Environmental Sciences, the Ministry of Environmental Protection of PRC, China*

The platform effect of graphene oxides in ionic liquids with copper nanocomposites on CO_2 transport, **Il Seok Chae and Yong Soo Kang**, *Department of Energy Engineering, Hanyang University, Seoul, South Korea*

Nanostructured bulk graphene/lanthanum strontium titanium oxide nanocomposites with high thermoelectric performance and broad thermal window, **Yue Lin, Colin Norman, Deepanshu Srivastava, Feridoon Azough, Li Wang, Robert Freer and Ian Kinloch**, *The School of Materials, The University of Manchester, United Kingdom*

Radical substructures rule instability of IPR isomers of higher fullerene molecules, **Valeri Kovalenko, Ayrat R. Khamatgalimov, Marat R. Faskhutdinov and Regina R. Matkazina**, *Arbuzov Institute of organic and physical chemistry Russian Academy of Sciences, Kazan, Russia*

The electronic structure of amorphous and graphitic carbon nanoparticles, **Johannes Margraf, Volker Strauss, Dirk M. Guldi and Timothy Clark**, *Computer Centre and Institute of Physical Chemistry, University of Erlangen, Germany*

Continuous synthesis of carbon nanotubes on carbon fibre, **David B. Anthony, Hui Qian, Emile Greenhalgh, Alexander Bismarck and Milo S. P. Shaffer**, *Imperial College London, United Kingdom*

Functionalised CNT–lipid nanostructures as drug delivery carriers, **Nicholas P. Gaunt, Matthew J. Baker, Gary Bond, Yogita Patil-Sen and Chandrashekhar V. Kulkarni**, *Centre for Materials Science, School of Forensics and Investigative Sciences, UCLan, United Kingdom*

Regioselective aromatic halogenation in carbon nanotube, **Scott A. Miners, Graham A. Rance and Andrei N. Khlobystov**, *School of Chemistry, University of Nottingham, United Kingdom*

Pressure modulation of cyclo[*n*]paraphenylenes' supramolecular complexes, **Miriam Peña-Alvarez, Paula Mayorga, Juan T. Lopez-Navarrete, Valentín G. Baonza, Mercedes Taravillo and Juan Casado**, *Universidad Complutense Madrid, Spain*

Application of carbon nanomaterials to the binder of fibre preform in carbon fibre reinforced composites, **Dong Gi Seong, Jong Rok Ha, Wonoh Lee, Moon Kwang Um and Byung Sum Kim**, *Korea Institute of Materials Science, South Korea*

Energy storage using carbon nanotube/steel electrodes, **Jagdeep S. Sagu and Upul Wijayantha**, *Loughborough University, United Kingdom*

Photophysics of threaded polyyne chains, **Levon D. Movsisyan, Gregory M. Greetham, Michael Towrie, Anthony W. Parker, Amber L. Thompson and Harry L. Anderson**, *Department of Chemistry, University of Oxford, United Kingdom*

Synthesis of magnetic particles on graphene oxide and their near-field electromagnetic wave absorption properties, **Byung Mun Jung, Sang-Bok Lee, Jin-Woo Yi, Seungchan Cho and Sang-Kwan Lee**, *Composite Research Centre, Korea Institute of Materials Science, South Korea*

Chitosan gel based fluorescent carbon dots as a protective coat over alginate beads for effective low pH drug delivery, **Neelam Gogoi and Devasish Chowdhury**, *Physical Sciences Division, Institute of Advanced Study in Science and Technology (IASST), India*

Construction of NBMOs and localization of HOMOs in finite-length zigzag nanotubes, **Noriyuki Mizoguchi**, *Meiji Pharmaceutical University, Japan*

Mesooporous graphene/IL hybrid material coated sand: an efficient mode for heavy metal detoxification, **Varsha Khare, Shivraj Raj Poudel, Dae-Seob Song and Sung-Hoon Ahn**, *Mechanical and Aerospace Engineering, Seoul National University, South Korea*

Determination of the gas evolution potential in graphene based symmetric supercapacitor operating in the alkaline aqueous electrolyte, **Krzysztof Wasiński, Mariusz Walkowiak, Paulina Półrolniczak and Grzegorz Lota**, *Institute of Non-Ferrous Metals Division in Poznan Central Laboratory of Batteries and Cells, Poland*

A cost-effective method for characterisation of functionalised carbon nanomaterials, **Kris Seunarine and Chris J. Spacie**, *Haydale Graphene Industries PLC, United Kingdom*

Scalable synthesis of nitrogen doped graphene nanosheets for oxygen reduction reaction in fuel cells, **John Benson and Pagona Papakonstantinou**, *University of Ulster, United Kingdom*

Surfactant-stabilised pristine graphene for use as a gene delivery vehicle, **Catriona McCallion, Nick Kay, Alain Pluen and Alexander Golovanov**, *University of Manchester, United Kingdom*

Fabrication of non enzymatic glucose biosensor using copper oxide nanorods decorated functionalised graphene hybrid nanocomposites, **Pranati Nayak, P. N. Santhosh and S. Ramaprabhu**, *AENL, India*

Polyamide 6/graphene composites: effects of *in situ* polymerisation on the structure and properties of graphene oxide and reduced graphene oxide, **Aidan O'Neill and Dorian Dixon**, *University of Ulster, United Kingdom*

Reversible electrochemical lithium insertion in SnO_2/reduced graphene oxide composites, **Paulina Półrolniczak, Krzysztof Wasiński, Mariusz Walkowiak and Agnieszka Martyła**, *Institute of Non-Ferrous Metals Division in Poznań Central Laboratory of Batteries and Cells, Poland*

Elemental sulphur trapped by expanded graphite and graphene applied as electrode material in rechargeable Li–S battery, **Paulina Półrolniczak, Krzysztof Wasiński and Mariusz Walkowiak**, *Institute of Non-Ferrous Metals Division in Poznań Central Laboratory of Batteries and Cells, Poland*

Hydrothermal carbonized chitosan for oxygen reduction reaction in alkaline media, **Maria Rybarczyk, Milena Jablonska and Marek Lieder**, *Gdansk University of Technology, Poland*

Photonic carbon meta-nanotubes as light conversion nanostructured devices (LCNDs), **Elaine Vaz, Janaína dos Anjos, Audrey Andrade, Yadira Vega-Cant, Fernando Rodríguez-Macías and Petrus Santa-Cruz**, *Federal University of Pernambuco - UFPE, Brazil,Brazil*

Wide-area strain sensors based upon graphene–polymer composite coatings probed by Raman spectroscopy, **Arun Prakash Aranga Raju, Amanda Lewis, Brian Derby, Robert Young, Ian Kinloch, Recep Zan and Kostya Novoselov**, *School of Materials, University of Manchester, United Kingdom*

Potential of single layer graphene oxide for the degradation of emerging organic contaminants in wastewater, **Viet Linh Nguyen, Rosa Busquets, Raymond L.D. Whitby and Andrew B Cundy**, *Kingston University, United Kingdom*

Anticorrosive studies of carbon nanocomposite coating over aluminium in NaCl solution, **Zeba Khanam, M G H Zaidi and Vir Singh**, *GB Pant University of Ag and Tech, India*

Studies on graphene oxide in poly(methyl methacrylate) and polystyrene, **Stephen Boothroyd, David Johnson, Mike Weir, Richard Thompson, N. Clarke and Karl Coleman**, *Durham University, United Kingdom*

Theoretical investigation on carbon nucleations on nickel carbides at initial stages of single-walled carbon nanotube formations, **Zhimin Yang, Qiang Wang, Xiaoye Shan, Hongjun Zhu and Shuo-Wang Yang**, *Institute Of High Performance Computing, Singapore*

Gold atoms on hydroxylated graphite surfaces: a theoretical study, **Carlo Buono, Philip R Davies, Neil Robinson and David J Willock**, *Cardiff University, United Kingdom*

The enhanced electrochemical property of Li_2MnSiO_4 nanofiber by carbon coating, **Byeong Chan Jang, Bo Ra Kim and Cheong Kim**, *Korea National University of Transportation, South Korea*

Electrophoretic deposition of TiO_2/graphene electrodes for flexible dye sensitized solar cell, **Mridula Baro, Cherianath Vijayan and Sundara Ramaprabhu**, *IIT Madras, India*

Chemical free unzipping carbon nanotube at high impact, **Sehmus Ozden, Pedro A. S. Autreto, Chandra Sekhar Tiwary, Suman Khatiwada, Leonard Machado, Douglas S. Galvao, Robert Vajtai, Enrique V. Barrera and Pulickel M. Ajayan**, *Department of Material Science and NanoEngineering, Rice University, United States*

3D carbon-based macroporous architectures, **Sehmus Ozden, Narayanan T. Narayanan, Chandra Sekhar Tiwary, Amelia H. C. Hart, Robert Vajtai and Pulickel M. Ajayan**, *Department of Material Science and NanoEngineering, Rice University, United States*

Dispersion of single-walled carbon nanotubes with peptides, **Bnar J. Ahmed, Hannah A, Bolt, Karl S. Coleman and Stephen S. Cobb**, *Department of Chemistry, Durham Universit, UK*

The Skinner Prize for the best poster was jointly awarded to Mr Johannes Margraf of the University of Erlangen, Germany, for his poster on the electronic structure of amorphous and graphitic carbon nanoparticles and to Mr Arun Prakash Aranga Raju of the University of Manchester, UK, for his poster on wide-area strain sensors based on graphene–polymer composite coatings probed by Raman spectroscopy.

List of participants

Miss Bnar Ahmed, *Durham University, UK*
Professor Pulickel Ajayan, *Rice University, United States*
Mr Fahad Alanazi, *University of Manchester, UK*
Mr Salman Alanazi, *University of Manchester, UK*
Dr David Anthony, *Imperial College London, UK*
Dr Tariq Aqeel, *Public Authority for Applied Education and Training (PAAET Kuwait), Kuwait*
Mr Arun Prakash Aranga Raju, *University of Manchester, UK*
Mr Kirill Arapov, *Eindhoven University of Technology, The Netherlands*
Ms Mridula Baro, *IIT Madras, India*
Dr Oana Andreea Bârsan, *Eindhoven University of Technology, The Netherlands*
Professor Giancarlo Battaglin, *DSMN-Ca' Foscari University of Venezia, Italy*
Dr Mark Baxendale, *Queen Mary, University of London, UK*
Mr John Benson, *University of Ulster, UK*
Mr Santosh Kumar Bikkarolla, *School of Engineering, Engineering Research Institute, University of Ulster, UK*
Dr Stephen Boothroyd, *Durham University, UK*
Mr Carlo Buono, *Cardiff University, UK*
Dr Rosa Busquets, *School of Pharmacy and Chemistry, Kingston University, UK*
Professor Juan Casado, *University of Malaga, Spain*
Professor Chaoping Cen, *South China Institute of Environmental Sciences, the Ministry of Environmental Protection of PRC, China*
Dr Il Seok Chae, *Hanyang University, South Korea*
Professor Mary Chan-Park, *Nanyang Technological University, Singapore*
Dr Xiongbo Chen, *South China Institute of Environmental Sciences, the Ministry of Environmental Protection of PRC, China*
Miss Ming Chung, *Johnson Matthey Plc, UK*
Professor Karl Coleman, *Durham University, UK*
Dr Pedro Costa, *KAUST, Saudi Arabia*
Mr Hamish Crawford, *Royal Society of Chemistry, UK*
Dr Philip Davies, *Cardiff University, UK*
Mr Ben Dobson, *Durham University, UK*
Miss Rebecca Edwards, *Applied Graphene Materials, UK*
Professor Toshiaki Enoki, *Tokyo Institute of Technology, Japan*
Dr Chris Ewels, *Institute of Materials (IMN), Nantes, France*
Professor Vladimir Falko, *Lancaster University, UK*
Dr Ping Fang, *South China Institute of Environmental Sciences, the Ministry of Environmental Protection of PRC, China*
Mr Nicholas Gaunt, *UCLan, UK*
Miss Neelam Gogoi, *Institute of Advanced Study in Science and Technology (IASST), India*
Dr Thurid Gspann, *University of Cambridge, UK*
Dr Noreen Hanly, *University of Limerick, UK*

Professor Malcolm Heggie, *University of Surrey, UK*
Professor Dr Andreas Hirsch, *University of Erlangen-Nüremberg, Germany*
Ms Sheng Hu, *Imperial College London, UK*
Miss Lok Shu Hui, *McMaster University, Canada*
Mr Byeong Chan Jang, *Korea National University of Transportation, Korea*
Dr David Johnson, *Durham University, UK*
Dr Byung Mun Jung, *Korea Institute of Materials Science, Korea*
Professor Dr Katsumi Kaneko, *Shinshu University, Japan*
Dr Ayrat Khamatgalimov, *A.E. Arbuzov Institute of Organic and Physical Chemistry of Kazan Scientific Center of Russian Academy of Sciences, Russia*
Ms Zeba Khanam, *GB Pant University of Ag and Tech, India*
Professor Varsha Khare, *Seoul National University, South Korea*
Professor Ian Kinloch, *University of Manchester, UK*
Professor Dr Valeri Kovalenko, *Arbuzov Institute of Organic and Physical Chemistry, Russian Academy of Sciences, Russia*
Mrs Renata Kovalenko, *Arbuzov Institute of organic and physical chemistry Russian Academy of Sciences, Russia*
Professor Dame Irena Kratochvílová, *Institute of Physics ASCR, Czech Republic*
Dr Yue Lin, *The University of Manchester, UK*
Mr Johannes Margraf, *University Erlangen, Germany*
Professor Dr Nazario Martin, *University Complutense of Madrid, Spain*
Ms Catriona McCallion, *University of Manchester, UK*
Dr Ronan McHale, *Thomas Swan & Co. Ltd, UK*
Mr Scott Miners, *University of Nottingham, UK*
Professor Dr Noriyuki Mizoguchi, *Meiji Pharmaceutical University, Japan*
Dr Lyndsey Mooring, *BAE Systems, UK*
Mr Robert Mora, *University of Durham, UK*
Mr Levon Movsisyan, *University of Oxford, UK*
Professor Erich Müller, *Imperial College London, UK*
Mrs Pranati Nayak, *IIT Madras, India*
Dr Huynh Thien Ngo, *CYS-MANA, National Institute for Materials Science (NIMS), Japan*
Professor Dr Thanh TK Nguyen, *UCL Healthcare Biomagnetic and Nanomaterials Labs, UK*
Mr Aidan O'Neill, *University of Ulster, UK*
Dr Hitoshi Ogihara, *Tokyo Institute of Technology, Japan*
Dr Sehmus Ozden, *Material Science and NanoEngineering, Rice University, United States*
Professor Pagona Papakonstantinou, *University of Ulster, UK*
Mr Rami Pashameah, *Hull University, UK*
Miss Miriam Peña-Alvarez, *Universidad Complutense Madrid, Spain*
Dr Paulina Półrolniczak, *Institute of Non-Ferrous Metals Division in Poznań Central Laboratory of Batteries and Cells, Poland*
Professor Kosmas Prassides, *Durham University, UK*
Miss Chloe Pritchard, *Royal Society of Chemistry, UK*
Dr Victor Putz, *US Air Force, United States*
Dr Antonio Rodriguez-Fortea, *Rovira i Virgili University, Spain*
Professor Matthew Rosseinsky, *University of Liverpool, UK*
Ms Maria Rybarczyk, *Gdansk University of Technology, Poland*
Mr Jagdeep Sagu, *Loughborough University, UK*
Professor Dr Petrus Santa-Cruz, *Federal University of Pernambuco -UFPE, Brazil*
Dr Dong Gi Seong, *Korea Institute of Materials Science, Korea*
Mr Kris Seunarine, *Haydale Limited, UK*

Professor Milo Shaffer, *Imperial College London, UK*
Dr David Sheppard, *Horiba UK Limited, UK*
Ms Claire Springett, *Royal Society of Chemistry, UK*
Dr Thomas Trevethan, *University of Surrey, UK*
Dr Ayse Turak, *McMaster University, Canada*
Mr Michael Tynan, *Durham University, UK*
Dr Krzysztof Wasiński, *Institute of Non-Ferrous Metals Division in Poznan Central Laboratory of Batteries and Cells, Poland*
Professor Alan Windle, *Cambridge University, UK*
Dr Rachel Wood, *Royal Society of Chemistry, UK*
Dr Yoko Yamakoshi, *Laboratorium für Organische Chemie, ETH-Zürich, Switzerland*
Dr Shuo-Wang Yang, *Institute Of High Performance Computing, Singapore*
Dr David Zitoun, *Bar Ilan University, Israel*
Mr Alexander Zöpfl, *University of Regensburg, Germany*